Ecology of Saprotrophic Basidiomycetes

British Mycological Society Symposia Series

Series Editor:

Pieter van West
University of Aberdeen
Aberdeen Oomycete Group
Institute of Medical Sciences
Aberdeen, UK

Cover image: photograph courtesy of T.D. Rotheray depicting interaction between *Hypholoma fasciculare* (left) and *Resinicium bicolor* growing from wood blocks across the surface of non sterile soil. The interaction is being grazed by collembola (*Folsomia candida*), see chapter 11 for further details.

ECOLOGY OF SAPROTROPHIC BASIDIOMYCETES

Editors

Lynne Boddy
Cardiff School of Bioscience
Cardiff University
Cardiff, UK

Juliet C. Frankland
Centre for Ecology and Hydrology
Lancastes, UK

Pieter van West
University of Aberdeen
Aberdeen Oomycete Group
Institute of Medical Sciences
Aberdeen, UK

ELSEVIER

Amsterdam – Boston – Heidelberg – London – New York – Oxford
Paris – San Diego – San Francisco – Singapore – Sydney – Tokyo
Academic Press is an imprint of Elsevier

ACADEMIC PRESS

Academic Press is an imprint of Elsevier
84 Theobald's Road, London WC1X 8RR, UK
Radarweg 29, PO Box 211, 1000 AE Amsterdam, The Netherlands
Linacre House, Jordan Hill, Oxford OX2 8DP, UK
30 Corporate Drive, Suite 400, Burlington, MA 01803, USA
525 B Street, Suite 1900, San Diego, CA 92101-4495, USA

First edition 2008

ISBN: 978-0-12-374185-1
ISSN: 0275-0287

For information on all Academic Press publications visit our website at books.elsevier.com

Printed and bound in the United Kingdom
08 09 10 11 12 10 9 8 7 6 5 4 3 2 1

Preface

In 2007 it is 25 years since the publication of *Decomposer Basidiomycetes: Their Biology and Ecology* (Frankland *et al.*, 1982). This was a seminal book but, while most of what was written remains true, our breadth and depth of understanding of many areas has extended dramatically—particularly the mycelium, though inevitably some areas have lagged behind, for example enzymology in the natural environment. New vistas have opened up with the advent of powerful computing, modelling and molecular approaches. Undoubtedly, understanding the ecology of basidiomycetes is tantamount to understanding the role of fungi in natural ecosystems since they are major agents of decomposition and nutrient cycling, and yet this important branch of mycology does not seem to have attracted the attention that it deserves. This volume should help to redress the balance.

The book is not a collection of articles that hang loosely together under an umbrella heading, but rather a set of tightly connected chapters selected to cover the area of saprotrophic basidiomycete ecology. It essentially falls into three main sections covering: (1) the basidiomycete life-style; (2) interactions with each other and with other fungi and with other organisms; and (3) structure and function of basidiomycete communities.

The crucial role of the mycelium in the ecology of fungi, only touched on in Frankland *et al.* (1982) but highlighted in Jennings and Rayner (1984), is now well established. The first four chapters therefore set the scene by discussing the major aspects of basidiomycete mycelial structure and function, and the effect of the abiotic environment on function. Fricker *et al.* (Chapter 1) demonstrate that mycelial network architecture is of major significance in the acquisition of and distribution of nutrients, in interplay between different regions of the mycelium and in survival when parts of the network are destroyed. They also consider the costs and benefits of different architectures to large mycelial networks, and how network construction may be analysed using modern graph theory approaches. The enzymes produced by saprotrophic basidiomycetes to obtain energy and mineral nutrients for growth are then reviewed (Chapter 2), Petr Baldrian pointing out that the main gaps in our current knowledge are in the ecology of enzyme production under natural conditions. When energy and nutrients have been obtained they must be routed to the site of need, hence translocation mechanisms are reviewed by Watkinson *et al.* in Chapter 3, with particular emphasis on environmentally essential nitrogen. Mycelial network architecture, enzymology and nutrient uptake and translocation are all influenced by the abiotic environment, and Naresh Magan (Chapter 4) considers the effect of temperature, water availability and their interactions on growth.

Past overemphasis on fruit bodies in fungal ecology does not mean that their role is insignificant, and the importance of sexual reproduction is cogently argued in Chapter 5. Here Moore *et al.* not only review physiological factors favouring production and development of fruit bodies, but also summarize principles of fungal developmental biology. They also reveal dramatic changes in the seasonal pattern of fruiting in the U.K., correlated with climate change, which must surely reflect changes in mycelial activity. Stenlid then addresses issues in population biology (Chapter 6), notably aspects of fungal individuality, the size and dynamics of individual mycelia and how their integrity is controlled through somatic incompatibility, gene flow and long distance dispersal.

The second section covers interactions of basidiomycete mycelia with other saprotrophs, ectomycorrhizas and root pathogens (Chapter 7), bacteria (Chapter 8) and invertebrates (Chapter 9). Woodward and Boddy emphasise the role of interactions in ecosystem functioning, particularly nutrient cycling and release, effects on decomposition rates and the potential of saprotrophs as biological control agents. They note that interactions can be mediated: (1) at a distance; (2) following contact at the hyphal level; and (3) following contact at the mycelial level, and that antagonism at a distance and at the mycelial level are effected by volatile and diffusible chemicals including enzymes, toxins and other anti-fungal metabolites. de Boer and van der Wal introduced the, until recently, little studied area of bacterial–fungal interactions, showing that bacteria can affect functioning of saprotrophic basidiomycetes both negatively and positively. Then Boddy and Jones review the wide field of interactions with invertebrates. Following the extensive reviews in Anderson *et al.* (1984) and Wilding *et al.* (1989), considerable research has been performed in many such areas, not least effects of grazing on both groups of organisms. They point out that attention must now turn to effects on ecosystem processes and consequences of global environmental change.

The third section opens with a consideration of litter layer basidiomycetes (Chapters 10 and 11). Lindahl and Boberg emphasise the use of modern molecular approaches to reveal community structure during different stages of community development, and the role of these fungi in carbon and nitrogen cycling in boreal forests. This is complemented by the chapter by Lodge *et al.* on the greatly neglected tropical rain forest. Here mat-forming basidiomycetes play a crucial role in litter stability, prevention of erosion as well as decomposition and nutrient cycling.

Basidiomycete community structure development and function has been studied most intensively in wood, as reflected by the five chapters covering this topic in Frankland *et al.* (1982) and the major treatise by Rayner and Boddy (1988). This is probably not only because of the commercial value and ecological importance of wood, and the 100,000 or more species involved in wood decomposition, but also because the three-dimensional structure of communities can be relatively easily revealed. The plethora of studies on temperate angiosperm wood have revealed much about the ecological strategies adopted by wood-decay Basidiomycota, habitat factors influencing community development, and community development pathways during early to middle stages of decomposition in the standing tree and on the forest floor, but much remains to be discovered about late stages of decay (Boddy and Heilmann-Clausen, Chapter 12). Boreal wood decay communities are, by contrast with temperate forests, less diverse and more similar on a global scale, with about 2,500 species in Fennoscandia (Stenlid *et al.*, Chapter 13). Classic inventory approaches, complemented by pure culture studies of mycelia and recently by molecular detection methods, show parallels with temperate forests in terms of fungal community structure and development. Attention is now being widened to consideration of the distribution of wood-decay fungi at the landscape and global scale (Heilmann-Clausen and Boddy, Chapter 14). It is becoming increasingly clear, from molecular and mating studies, that many species, previously thought to have a wide distribution, actually circumscribe several biological taxa, each with a much more restricted distribution. Thus continental drift, glaciations and other long-term geological and geographical factors have more impact on the current distribution patterns of fungi than hitherto realized.

The role of basidiomycetes in grasslands has been a neglected area of study. Griffith and Roderick point out its significance in Chapter 15, where they review the four functional groupings—litter decomposers, dung fungi, terricolous species and root endophytes—that encompass the several hundred basidiomycete saprotrophs which are preferentially found in grassland. Basidiomycetes in aquatic ecosystems are also often forgotten. Gareth Jones (the only author who also contributed to Frankland *et al.*, 1982) and Rattaket Choeyhklin provide an up-to-date review of marine and freshwater basidiomycetes which, although few in number compared to their terrestrial counterparts, colonize a wide range of substrata, particularly those that are woody. Nearly every sampling study has found a genus new to science, with unique adaptations to their habitat.

The final chapter, by Heilmann-Claussen and Vesterholt, focuses on the urgent need for fungal conservation, highlighting different selection approaches, and explaining the difficulties in using criteria designed for red-listing other organisms.

Note on terminology. While editing, we have not aimed at total uniformity between chapters in all terms. Most notably 'Basidiomycota' and 'basidiomycetes', 'saprotrophs' and 'saprobes' have been used interchangeably. The term 'habitat' has been used to define 'the place' where a fungus lives, therefore including aspects of the physico-chemical environment. 'Substratum', albeit only occasionally used, alludes to the medium in which the fungus is growing within a habitat, while 'substrate' has been reserved for a specific chemical molecule utilized by the organism. 'Resource' is a general term meaning 'food source', and so has been used interchangeably with substratum and substrate, depending on context.

Acknowledgement. Finally, we would like to thank Mrs. Julie Harris for unstinting secretarial support and Jacob Heilmann-Clausen for help with taxonomy.

Lynne Boddy, Juliet C. Frankland and Pieter van West

Anderson, J.M., Rayner, A.D.M. and Walton, D.W. (eds.). (1984). *Invertebrate–Microbe Interactions.* Cambridge University Press, Cambridge.
Frankland, J.C., Hedger, J.N. and Swift, M.J. (eds.). (1982). *Decomposer Basidiomycetes: Their Biology and Ecology,* 355 pp. Cambridge University Press, Cambridge.
Jennings, D.H. and Rayner, A.D.M. (eds.). (1984). *Ecology and Physiology of the Fungal Mycelium*. Cambridge University Press, Cambridge.
Rayner, A.D.M. and Boddy, L. (1988). *Fungal Decomposition of Wood: Its Biology and Ecology.* Wiley, Chichester.
Wilding, N., Collins, M.M., Hammond, P.M. and Webber, J.F. (eds.). (1989). *Insect–Fungus Interactions*. Academic Press, London.

Contents

List of Contributors

Petr Baldrian, Laboratory of Biochemistry of Wood-Rotting Fungi, Institute of Microbiology of the ASCR, v. v. i. Videnska 1083, 14220 Praha 4, Czech Republic

Dan Bebber, Department of Plant Sciences, University of Oxford, South Parks Road, Oxford OX1 3RB, England, UK

Johanna Boberg, Department of Forest Mycology and Pathology, Swedish University of Agricultural Sciences, Box 7026, SE-750 07 Uppsala, Sweden

Lynne Boddy, Cardiff School of Biosciences, Cardiff University, Biomedical Building, Museum Avenue, Cardiff CF10 3VS, Wales, UK

Karla Claudio –Campos, c/o Dr. Robert Ross, Department of Biology, University of Puerto Rico, 205 Antonio R. Barceló, Puerto Rico 00736, USA

Rattaket Choeyklin, Biological Science Program, Faculty of Science, Burapha University, 169 Long-Hard Bangsaen Road, Tambon Saensook, Amphur Muang, Chonburi, 20131 Thailand

Anders Dahlberg, Swedish Species Information Centre, Swedish University of Agricultural Sciences, Box 7007, S-750 07 Uppsala, Sweden

Peter Darrah, Department of Plant Sciences, University of Oxford, South Parks Road, Oxford OX1 3RB, England, UK

Wietse de Boer, Netherlands Institute of Ecology (NIOO-KNAW), Centre for Terrestrial Ecology, Department of Terrestrial Microbial Ecology, P.O. Box 40, 6666 ZG Heteren, The Netherlands

Juliet C. Frankland, Centre for Ecology and Hydrology, Lancaster LA1 4AP, England, UK

Mark D. Fricker, Department of Plant Sciences, University of Oxford, South Parks Road, Oxford OX1 3RB, England, UK

Alan C. Gange, School of Biological Sciences, Royal Holloway, University of London, Egham, Surrey TW20 0EX, UK

Edward G. Gange, Belvedere, Southampton Road, Whaddon, Salisbury, Wilts SP5 3DZ, UK

Gareth S. Griffith, Institute of Biological Sciences, University of Wales Aberystwyth, Aberystwyth SY23 3DA, Wales, UK

Jacob Heilmann-Clausen, HabitatVision, Skælskørvej 22, 4180 Sorø, Denmark

E.B. Gareth Jones, BIOTEC Central Research Unit, National Centre for Genetic Engineering and Biotechnology, NASDA, 113 Paholyothin Road, Khlongl, Khlong Luang, Pathum Thani 12120, Thailand

T. Hefin Jones, Cardiff School of Biosciences, Cardiff University, Biomedical Building, Museum Avenue, Cardiff CF10 3US, Wales, UK

Kerstin Kühnert, Nikolausstr. 2, 97082 Würzburg, Germany

Rachel Leisso, Montana State University, Plant Sciences and Plant Pathology, 119 Plant Sciences Building, Bozeman, MT 5971-3150, USA

Björn Lindahl, Department of Forest Mycology and Pathology, Swedish University of Agricultural Sciences, Box 7026, SE-750 07 Uppsala, Sweden

D. Jean Lodge, Research Botanist, USDA-Forest Service, Center for Forest Mycology Research, PO Box 1377, Luquillo, Puerto Rico 00773–1377, USA

Jordan Macy, Department of Natural Resources, 219 James Hall, University of New Hampshire, Durham, NH 03824, USA

Naresh Magan, Applied Mycology Group, Cranfield Health, Cranfield University, Silsoe, Bedford MK45 4DT, UK

William H. McDowell, Department of Natural Resources, 219 James Hall, University of New Hampshire, Durham, NH 03824, USA

David Moore, Faculty of Life Sciences, 1.800 Stopford Building, The University of Manchester, Manchester M13 9PT, UK

Reijo Penttilä, Finnish Environment Institute, PO Box 140, FIN-00251 Helsinki, Finland

Kevin Roderick, Institute of Biological Sciences, University of Wales Aberystwyth, Aberystwyth SY23 3DA, Wales, UK

Jan Stenlid, Department of Forest Mycology and Pathology, Swedish University of Agricultural Sciences, Box 7026, SE-750 07 Uppsala, Sweden

Monika Tlalka, Department of Plant Sciences, University of Oxford, South Parks Road, Oxford OX1 3RB, England, UK

Annemieke van der Wal, Laboratory for Ecological Risk Assessment, National Institute for Public Health and the Environment, Antonie van Leeuwenhoeklaan 9, PO Box 1, 3720 BA Bilthoven, The Netherlands

Pieter van West, University of Aberdeen, Aberdeen Oomycete Group, College of Life Sciences and Medicine, Institute of Medical Sciences, Foresterhill, Aberdeen AB25 2ZD, Scotland, UK

Jan Vesterholt, Botanical Museum, University of Copenhagen, DK-1123 Copenhagen K, Denmark

Sarah Katherine Ward, Department of Animal Health and Biomedical Sciences, University of Wisconsin-Madison, 1656 Linden Drive, Madison, WI 53706, USA

Sarah Watkinson, Department of Plant Sciences, University of Oxford, South Parks Road, Oxford OX1 3RB, England, UK

Steve Woodward, University of Aberdeen, School of Biological Sciences, Department of Plant and Soil Sciences, Cruickshank Building, St. Machar Drive, Aberdeen AB24 3UU, Scotland, UK

Section 1:
Basidiomycete Life-Style

CHAPTER 1

Mycelial Networks: Structure and Dynamics

Mark D. Fricker, Dan Bebber and **Lynne Boddy**

Contents

Abstract

To survive saprotrophic fungi must be able to capture organic resources discontinuously dispersed in space and time. Some basidiomycetes can only achieve this by production of sexual and asexual spores or sclerotia — categorized as 'resource-unit-restricted', whereas 'non-resource-unit-restricted' basidiomycetes can also spread between organic resources as mycelium. Mycelial distribution and foraging within organic resources and among relatively homogeneously and heterogeneously distributed resources is reviewed. 'Non-resource-unit-restricted' Basidiomycota have evolved different patterns of mycelial spread appropriate to discovery of resources of different sizes and distributions. They show remarkable patterns of re-allocation of biomass and mineral nutrients on discovery and colonization of new resources. Network architecture is a significant factor in the acquisition and distribution of nutrients, and in survival when parts of the network are destroyed. The costs and benefits of different architectures to large mycelial networks are considered.

British Mycological Society Symposia Series
Published by Elsevier Ltd.

1. INTRODUCTION

For most Basidiomycota in terrestrial ecosystems the predominant body form is the mycelium, comprising an interconnecting series of apically extending tubes —hyphae. Hyphae provide a large surface:volume, ideal for secreting enzymes for extracellular digestion of resources (Chapter 2), and for subsequent uptake of small molecules. Mineral nutrients, carbon and energy sources are presumed to be taken up largely at hyphal tips, be they embedded within an organic resource or foraging externally for new resources, and translocated from these sources to sites of demand (sinks; Chapter 3). Nutrient acquisition and other aspects of physiology are affected by the local environment (Chapter 2), and mycelia exhibit remarkable physiological and morphological plasticity. Moreover, since mycelial activity in one region can be supported by supply of water and nutritional resources from elsewhere, growth can sometimes occur in inhospitable places and adverse conditions. The interconnectedness of mycelia is of crucial significance to the organization and ecological roles of fungi (Rayner *et al.*, 1995).

In terrestrial ecosystems, the organic resources on which saprotrophic Basidiomycota depend are usually discrete, varying in size from small to large plant fragments, e.g. bud scales, leaves and large woody components. These resources are distributed heterogeneously in both space and time. For example, the relatively homogeneous carpet of forest floor leaf litter comprises spatially discrete leaves, input largely over a 6–8 week period in autumn by broadleaf deciduous trees, or more evenly during the year by many conifers. Branches are patchily distributed on the forest floor, falling throughout the year, though often with larger inputs following high winds. To survive saprotrophic fungi must be able to capture these discontinuously dispersed resources. Some Basidiomycota can only achieve this by production of sexual and asexual spores or sclerotia, and have been categorized as 'resource-unit-restricted', whereas 'non-resource-unit-restricted' Basidiomycota can also spread between organic resources as mycelium. Spores, although allowing rapid spread, sometimes over long distances, contain only relatively small food reserves from which to produce a mycelium for invasion of the organic resource upon which it has landed. Sclerotia often provide larger resources and also allow survival in time. Growth as mycelium, in contrast, allows the fungus to draw upon a much larger supply of nutrients.

This chapter considers mycelia growing within organic resources, and the ways in which they search and colonize them when discontinuous. It also examines the significance of network architecture, and the costs and benefits of large mycelial networks.

2. MYCELIA WITHIN ORGANIC SUBSTRATA

There is little information on mycelia within organic resources. Exceptions are maps of the extent of mycelia, inferred from interaction zone lines (see Chapters 7 and 11), and location of hyphae in relation to type of rot (Rayner and Boddy,

1988). The size of mycelia ranges from a few millimetres to many metres, in the case of longitudinally extensive (30 m or more) primary colonizers of attached branches and standing trunks (Boddy, 2001). The three-dimensional shape of the mycelial boundary is largely governed by the anatomy of the resource and by surrounding antagonistic fungi. For example, in wood, decay columns tend to be larger longitudinally than in other directions, reflecting difficulty of radial and tangential spreads. The diamond shaped cankers on sycamore (*Acer pseudoplatanus*) caused by *Gibberella Zeae* (Ascomycota) result from spread between nutrient rich ray cells (Bevercombe and Rayner, 1980). Crucially lacking, however, is knowledge of the interconnectedness of different parts of the mycelium, and even the amount of mycelial biomass at different locations within organic resources. That there is spatial heterogeneity of mycelial distribution within wood decay columns is suggested by the common observation that when wood is incubated in a humid environment mycelium often grows out rapidly and profusely from the edges, and more slowly and less densely from more central regions.

3. MYCELIA FORAGING BETWEEN RELATIVELY HOMOGENEOUSLY DISTRIBUTED RESOURCES

Fungi have evolved a variety of foraging and behavioural responses to encounters with new resources. Fungi that utilize individual, relatively homogeneous resources, e.g. a leaf litter layer, effectively colonize as if individual components are simply parts of a larger resource. Mycelia form large patches with no particular pattern, e.g. *Collybia* spp. and *Marasmius* spp., or form fairy rings, e.g. *Clitocybe nebularis* (Dowson *et al.*, 1989). Nothing is known of the network architecture of mycelial patches, but fairy rings of *C. nebularis* extend through the leaf litter layer as an ever increasing annulus of mycelium ~30–40 cm wide (Dowson *et al.*, 1989; Figure 1a–d). The band is differentiated into three distinct zones: (1) the leading edge comprises mycelial cords (linear organs of predominantly parallel hyphae) spreading across the leaf litter layer and up to 6 cm into soil beneath; (2) a central region of dense mycelium which ramifies throughout, and presumably causes, intensely bleached leaf litter but does not extend into the mineral soil; (3) mycelium at the trailing edge which becomes progressively fragmented before completely disappearing. (Fruit bodies are produced from the middle of zone 2.) This outwardly extending annulus does not form as a result of lack of nutrients in central areas, as these are replenished every autumn, nor are toxic metabolites likely to be the cause, since when part of the annulus was transplanted into this region it grew well (Dowson *et al.*, 1989). Rather, these mycelia exhibit highly polarized growth, such that when a turf containing all zones of the annulus was relocated elsewhere, growth continued in the original direction of travel with limited lateral growth (Dowson *et al.*, 1989). Young mycelia of *C. nebularis* form patches, but what triggers annulus formation is unknown. Presumably ring formation is related to size and might be expected to start when a patch is over 80 cm diameter (i.e. double the width of the mycelial band).

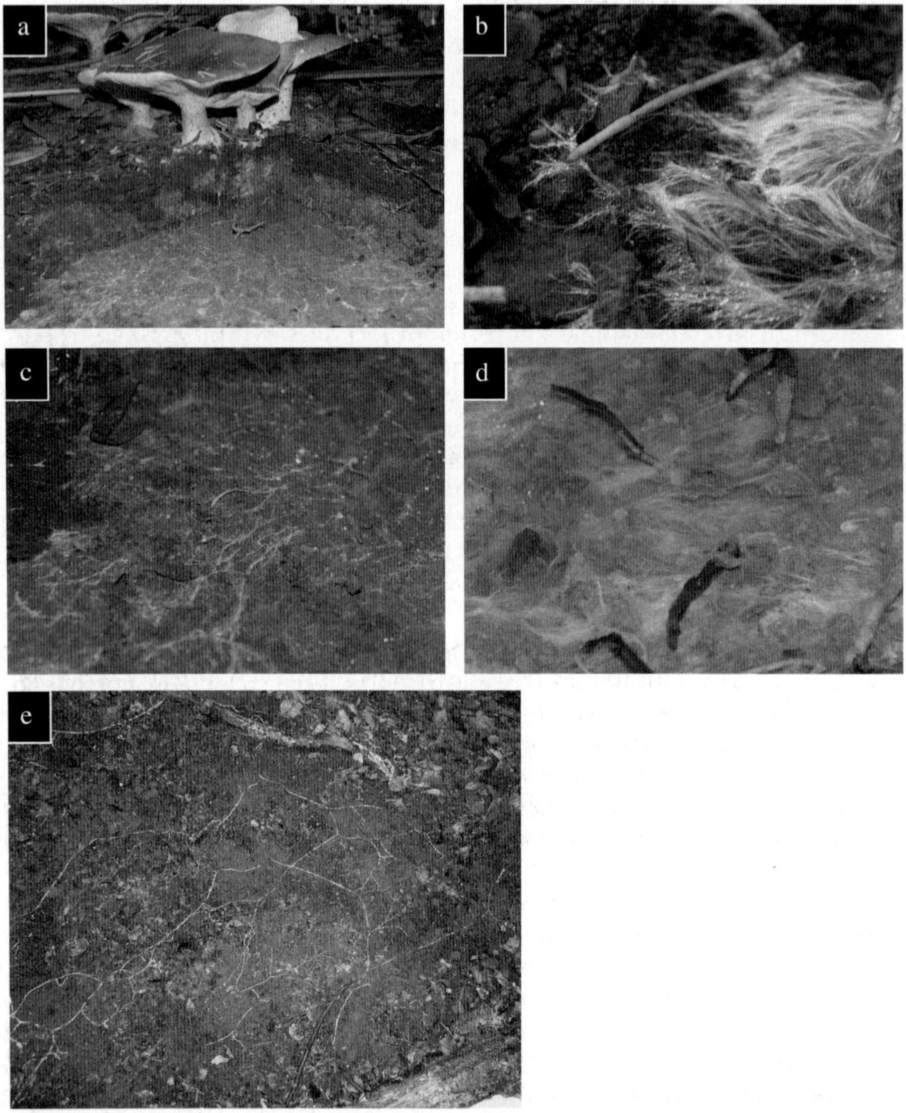

Figure 1 (a–d) Mycelium of a *Clitocybe nebularis* fairy ring which had developed under a paving slab in a garden. (a) Location of fruit bodies in relation to mycelium. Note aggregation into cords, but still with diffuse mycelium, towards the leading edge (left). (b) Mycelium aggregating into fine cords at leading edge. (c) Thicker cords amidst dense fine mycelium. (d) Very dense, fine mycelium in central zone of annulus. (e) Mycelial network of *Megacollybia platyphylla* in a mixed deciduous woodland, revealed by removal of surface litter. Digital images (a)–(d) courtesy of David Moore.

4. MYCELIA FORAGING BETWEEN RESOURCES DISTRIBUTED HETEROGENEOUSLY IN SPACE AND TIME

Fungi that utilize spatially discrete resources, with centimetre- or even metre-scale separations, have developed a variety of foraging strategies. They commonly form linear mycelial aggregates termed rhizomorphs, e.g. *Marasmius androsaceus* and *Armillaria* spp., or cords, e.g. *Hypholoma fasciculare* and *Phanerochaete velutina* (e.g. Boddy, 1984, 1993, 1999; Hedger, 1990; Cairney, 1992, 2005; Rayner *et al.*, 1995; Boddy and Jones, 2006). Rhizomorphs are linear organs, with a thick melanized rind, the whole organ extending from the tip (Rayner *et al.*, 1985). Mycelial cords are also insulated from the environment with a thick rind, but they develop from a mycelial margin of diffuse hyphae, each of which extends apically. They can all draw on water, nutrients and energy held within other parts of the mycelium to sustain growth outside the organic resource(s) to which they are connected. In addition, although mycelial cords are insulated from the environment, they are able to absorb water and soluble nutrients via individual hyphae at the mycelial margin or that sometimes develop elsewhere forming patches, and they may colonize small litter components *en route* to large organic resources (Boddy, 1999; Watkinson *et al.*, 2006).

Fungi producing extra-resource mycelium risk loss of a large amount of biomass, as a result of invertebrate grazing, antagonistic microorganisms or death due to an unfavourable microenvironment. This can be minimized by a variety of different strategies. These include: (1) active growth and search for new resources; (2) a 'sit and wait' strategy, in which a mycelial network awaits arrival of resources, e.g. by branch fall, and then active colonization, often responding elsewhere in the system; and (3) most commonly, a combination of both. With all these strategies the mycelial networks are continuously remodelled in response to environmental cues, which can be abiotic (e.g. nutrient sources, microclimate or destructive events) and biotic (e.g. interaction with other fungi or grazing by invertebrates). Remodelling occurs through a complex combination of growth, branching, hyphal fusion and regression of different mycelial regions. Throughout the network, not only does morphology alter but also a complex set of physiological processes associated with uptake, storage and redistribution of nutrients change (Bebber *et al.*, 2006; Watkinson *et al.*, 2006). Both morphological and physiological changes are highly coordinated so that responses to local environmental changes can propagate through the mycelial network.

4.1 Search and Response Behaviour

Fungi have evolved a wide variety of patterns of mycelial outgrowth from resources into soil and litter (Figure 2; Boddy, 1999; Boddy and Jones, 2006). These have been quantified in terms of radial extension rate, hyphal coverage, and surface and mass fractal dimension (D_S and D_M, respectively) (Boddy, 1999; Boddy *et al.*, 1999; Boddy and Donnelly, 2007). These range between mycelia characterized by diffuse, slowly extending search fronts, with a high D_M (close to the maximum of 2 in two dimensions), e.g. *H. fasciculare* (Figure 2b) and *Stropharia*

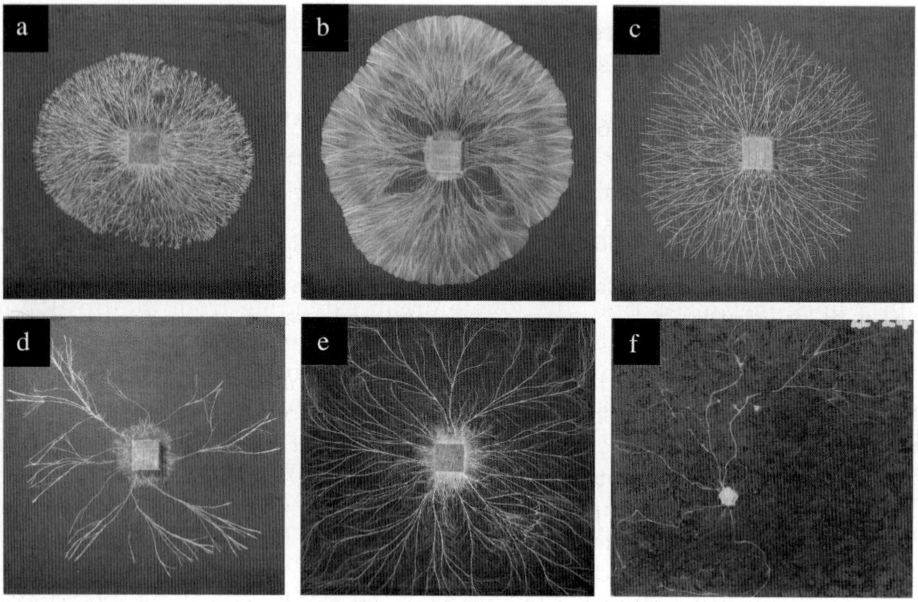

Figure 2 Patterns of mycelial outgrowth of four cord-forming basidiomycota across compacted soil in 24 × 24 cm trays from *x* cm (a–e) and *y* cm (f) Beech (*Fagus sylvatica*) wood inocula. (a) *Coprinus picaceus*, (b) *Hypholoma fasciculare*, (c) *Phallus impudicus*, (d) *Resinicium bicolor* and (e and f) *Phanerochaete velutina*. Digital images (a)–(d) courtesy Alaa Alawi, and digital images (e) and (f) from photographs taken by Rory Bolton.

spp., and open systems characterized by well-defined, rapidly extending cords throughout the system, with a lower D_M (between 1 and ~1.8), e.g. *Agrocybe praecox*, *Coprinus picaceus*, *Phallus impudicus*, *P. velutina* and *Resinicium bicolor*. The former can be considered to be short-range foragers that are likely to be successful in discovering and exploiting abundant, relatively homogeneously distributed resources as they search areas intensively (Figure 2b), and the latter long-range foragers that would be less successful at capitalizing on relatively homogeneously supplied nutrients, but would successfully discover large, more sparsely distributed resources. Mycelial systems tend to become more open with time as they become larger (D_M decreases; Donnelly *et al.*, 1995; Boddy *et al.*, 1999; Figure 2a, c and d; patterns are modified by the quantity and quality of the resource from which the mycelium is extending (Bolton and Boddy, 1993; Donnelly and Boddy, 1997a; Boddy *et al.*, 1999; Zakaria and Boddy, 2002; Figure 2e and f), soil structure and nutrient status (Donnelly and Boddy, 1998; Boddy *et al.*, 1999; Zakaria and Boddy, 2002), microclimate (Donnelly and Boddy, 1997b; Owen, 1997; Wells *et al.*, 2001), interaction with mycelia of other species (Donnelly and Boddy, 2001) and invertebrate grazing (Kampichler *et al.*, 2004; Harold *et al.*, 2005; Bretherton *et al.*, 2006; Tordoff *et al.*, 2006; Wood *et al.*, 2006; Chapter 9).

When new resources are encountered the mycelium responds with dramatic changes in morphology (network architecture) and often with considerable reallocation of biomass. When the new resources are substantially larger than those from which the mycelium emanated, mycelium connecting the new resource with the original resource usually aggregates to form thick cords, while radial extension slows or ceases, and non-resource-connected mycelium regresses (Dowson *et al.*, 1986, 1988; Bolton *et al.*, 1991; Boddy, 1993, 1999; Bolton, 1993; Donnelly and Boddy, 1997a; Figure 3a–c). Subsequently mycelium grows out from the newly colonized resource, and foraging continues, though the amount of time before foraging continues depends on the sizes of the original and new resource (Bolton, 1993; Boddy and Jones, 2006). With short-range foragers (e.g. *H. fasciculare*), there are similar, although less dramatic, changes to system architecture even when newly encountered resources are similar in size to the original resource.

Not only does the mycelium respond by changes to network architecture but also with physiological responses: there is highly coordinated uptake, storage and redistribution of nutrients throughout the network (Watkinson *et al.*, 2006; Chapter 3). Mineral nutrients (e.g. nitrogen and phosphorous) can be transported from wood resources to support growth at the mycelial margin, and nutrients scavenged as mycelia extend through soil can be translocated away from sites of uptake to sites of demand or storage, and commonly accumulate in wood resources connected within the mycelial system (Wells and Boddy, 1990; Wells *et al.*, 1990, 1998, 1999; Cairney, 1992 Hughes and Boddy, 1994; Olsson and Gray, 1998). Rates of translocation can be rapid (sometimes $>25\,\text{cm}\,\text{h}^{-1}$), the largest fluxes being through cords interconnecting resources (e.g. Wells and Boddy, 1990). Many factors, including the overall nutritional status of the mycelial network, and the distribution and quantity of colonized and newly encountered organic resources, affect the balance between, and the main sites of, uptake, storage and demand for carbon and mineral nutrients (Abdalla and Boddy, 1996; Hughes and Boddy, 1996; Wells *et al.*, 1998, 1999; Boddy and Jones, 2006).

4.2 Persistent Mycelial Networks: 'Sit and Wait' Strategy

Saprotrophic cord- and rhizomorph-forming Basidiomycota produce extensive long-lived mycelial networks on the forest floor, eventually covering several square metres to many hectares (Thompson and Rayner, 1982; Thompson and Boddy, 1988; Smith *et al.*, 1992; Ferguson *et al.*, 2003; Cairney, 2005; Figure 1e). The largest recorded to date is a genet of *Armillaria ostoyae* spanning 965 ha, with a maximum separation of 3,810 m and estimated as 1,900–8,650 years old (Ferguson *et al.*, 2003). The true extent and degree of connectivity within a genet is not known, however, since parts of mycelia can be separated from each other during development, and can also rejoin if parts of the same genet meet again. Similar systems are also found in the canopy of tropical forests where they effectively form a net (Hedger, 1990). Whether on the forest floor or in the canopy, these large, persistent networks allow capture of resources arriving by litter fall or root death at any time.

Although persistent, established systems are dynamic both as continued extension at growing fronts (Thompson and Rayner, 1983) and as renewed mycelial growth from mature cords. Arrival of new resources can result in reallocation of

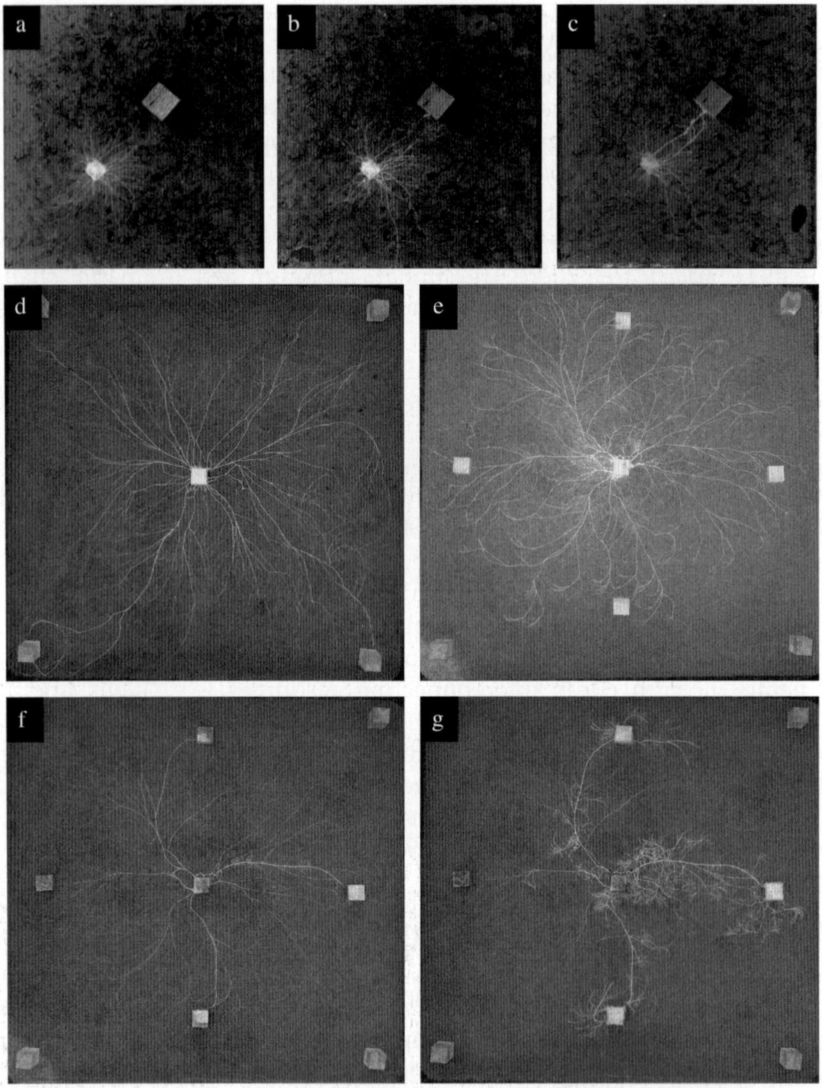

Figure 3 Reallocation of mycelial biomass of *Phanerochaete velutina* following colonization of new wood resources. (a–c) Extending from a 0.5 cm^3 beech (*Fagus sylvatica*) wood inoculum to an 8 cm^3 wood resource, in 24 × 24 cm trays of non-sterile soil, after, respectively, 11, 15 and 20 days. Note regression of much of the mycelium not connected to a new resource, and thickening of connected cords (c). (d–g) Growth in 57 × 57 cm soil trays, with four new wood resources (located half way along each microcosm side) added after 36 days in (e)–(g). (d) Control with no additional wood resources. Images were captured 78–85 days after adding the central wood inoculum. Note thickening of cords connecting inoculum with new resources (f and g), and thinning of other areas compared with 78 days control having no additional resources (e). Outgrowth from the newly colonized lower resource is evident from 78 days (Perspex blocks in the corners of trays were for support of other replicates in stacks). Proliferation of mycelium occurred along cords linking the central wood inoculum with new resources between 78 days (f) and 85 days (g). Digital images (a)–(c) from photographs taken by rory bolton. Digital images (d)–(g) Courtesy of Jon Wood.

biomass, with thickening of cords connecting resources, and regression of non-connective fine mycelium (Wood *et al.*, 2006; Figure 3d–g). Moreover, sometimes renewed growth occurs elsewhere as ephemeral patches of highly branched fine hyphae or along cords interconnecting new and original resources (Wells *et al.*, 1997; Wood *et al.*, 2006; Figure 3g). The patches have been shown, using ^{32}P orthophosphate, to be sites of nutrient uptake (Wells *et al.*, 1977), and presumably developed to satisfy the increased demand for nutrients to produce mycelial biomass and enzymes during early stages of colonization and decomposition. Carbon and mineral nutrients are continually rerouted to sites of need in mycelial systems interconnecting a variety of resources in different states of decay (Wells *et al.*, 1998).

5. ANALYSIS OF NETWORK ARCHITECTURE AND FUNCTION

Within the mycelial networks of saprotrophic Basidiomycota there is considerable scope for communication, since hyphae maintain continuity with their immediate 'ancestors' and if contact is made with neighbouring regions can become connected via *de novo* formation of cross-links (anastomoses). This results both radially and tangentially in systems with many connected loops (Figure 4).

The mycelium has evolved differently in different species resulting in a range of network architectures, adapted differently for differing balances of exploration, transport efficiency and resilience to damage. Highly interconnected mycelia are costly to construct but offer alternate transport routes and thus resilience to damage. Sparse networks with fewer interconnections can extend further for a

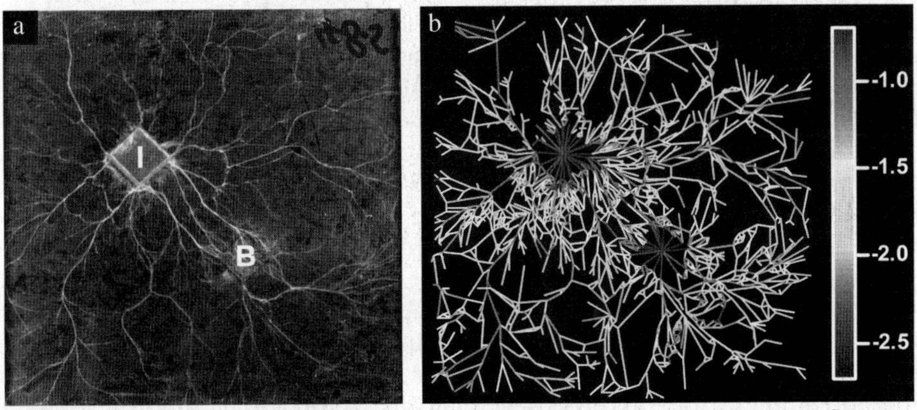

Figure 4 (a) Mycelium of *Phanerochaete velutina* after 25 days growth from a 4 cm^3 beech wood inoculum (I) across a tray (24 × 24 cm) of compressed non-sterile woodland soil. The fungus has met and colonized a second wood block (B). Digital image courtesy of rory bolton. (b) The digitized network of the same mycelium coloured by $\log_{10}$ of link cross-sectional area. The number of nodes $V = 1{,}738$, links $E = 2{,}617$ and the number of separate parts $G = 1$. The number of closed loops (cycles) in the illustrated system is $E - V + G = 880$, and the fraction of all possible cycles present $\alpha = 0.25$.

given construction cost, but risk the loss of pathways should one part of the network become damaged. Networks can vary not only in their connectedness but also in the strength of their connections. Thick cords confer greater transport capacity and resistance to breakage, but are more costly to produce. While these concepts have been implicit in discussions of fungal foraging strategies and fractal descriptions of mycelia, the architecture of mycelial networks has been little explored until recently (Bebber *et al.*, 2006; Bebber *et al.*, 2006, in press; Fricker and Bebber, in press; Fricker *et al.*, 2007).

5.1 Quantifying Network Characteristics

D_M is a useful metric for comparing space filling by mycelia (Boddy and Donnelly, in press), but it only expresses a small fraction of the complex architecture of mycelial systems. Tools for analysing networks are, however, emerging from graph theory and statistical mechanics (Albert and Barabási, 2001; Strogatz, 2001; Dorogovtsev and Mendes, 2002; Newman, 2003; Amaral and Ottino, 2004), that are applicable to mycelial networks (Bebber *et al.*, 2006, in press; Fricker and Bebber, in press; Fricker *et al.*, 2007), and have already proved valuable for understanding the properties of many physical systems that can be described as sets of connected entities, including biological networks such as protein–protein interactions and food webs (Bork *et al.*, 2004; Dunne *et al.*, 2004).

A network is simply a set of nodes, or vertices (V), connected by a set of links (E), or edges. Weights, that define properties such as resistance to breakage or transport capacity, can be associated with either nodes or links, or both. The nodes of a fungal mycelium are the tips, branch points and fusions of hyphae or cords, while the links are the hyphae or cords themselves. Various weights can be assigned to the nodes and links. For example, the mass of a cord can be approximated by its volume, the length multiplied by the cross-sectional area. Similarly, assuming that cords are composed of bundles of hyphae rather than being hollow tubes, the resistance to flow could be a function of the length divided by the cross-sectional area, i.e. long thin tubes have a greater resistance to flow.

The properties of nodes are often defined by the links to which they are connected. In the case of networks without link weights (the majority of examples in the literature lack this information), the number of links per node (termed degree k) is often used to describe something about the connectedness of the network. This measure is unlikely to be of interest in describing mycelial networks, as the majority of nodes will be of degree $k = 3$ (the branches and fusions) or 1 (the cord tips). Instead, the sum of link weights per node, known as the node strength, is likely to be of greater interest. For example, calculating the node strength for link cross-sectional area could indicate which nodes are likely to be important in transport.

The number of nodes, links and separate parts G, known as disconnected components or subgraphs, $G = 1$ (for an unbroken network), form the network that can be used to calculate the number of closed loops (cycles) in the network via the simple relation $E - V + G$ (Figure 4). This cyclomatic number is extremely important, for it indicates the number of alternate pathways among points in the network that determine both the resilience to damage and the capability of parallel flow. The cyclomatic number is typically normalized to the maximum number possible for a network of a given size to give the meshedness or alpha coefficient, allowing networks with differing numbers of nodes to be compared.

5.2 Modelling Transport

The fungal mycelium is essentially a transport network for nutrients, water and metabolites (Chapter 3). Modelling of transport in the mycelium has been attempted using various approaches, including partial differential equations and autonomous agents (Edelstein and Segel, 1983; Deutsch *et al.*, 1993). Since these methods ignore the network structure of the mycelium, greater insights may be obtained by taking an explicitly network-based approach into the analysis of transport. One way to achieve this is to calculate shortest path distances from each node to every other. If the effective physiological distance, or transport resistance, from one end of a cord to the other is modelled as the cord length divided by its cross-sectional area, the shortest path from one node to another will be the route with the smallest sum of these distances. The shortest path is therefore effectively the path of least resistance. Analysis of shortest paths of *P. velutina*, growing from wood blocks over soil, shows that the shortest paths from the wood blocks to other nodes of the fungus are smaller than they would be in a network with identical topology (i.e. number and location of links and nodes) but with uniform cord transport capacity (Bebber *et al.*, submitted). The fungus has therefore allocated resources to cords in a way that increases its transport efficiency. The only nodes for which the fungus is less efficient than in the randomized system are those at the periphery of the mycelium, where very fine hyphae are located (Figure 4). Here the fungus has optimized mycelial distribution for searching for new resources rather than optimizing for transport.

The routes taken by shortest paths can reveal other aspects of network transport. For example, the importance of a node can be estimated by its betweenness centrality, which is the proportion of shortest paths that pass through that node (Freeman, 1977). The proportion of paths that pass through the node with the greatest betweenness centrality is the central point dominance. In fungal networks, resources such as wood blocks usually have the greatest betweenness centrality.

The shortest path is usually only one of the several routes that could be taken from one node to another. Transport through a real network will often make use of these alternate routes, in the same way that electricity will flow through each of a set of resistors in parallel. Use of shortest paths to characterize transport necessarily ignores the importance of these parallel pathways through the network. Methods for solving current flow (e.g. Wu, 2004) through networks of electrical resistors can in principle be used to model flow through mycelial networks, for example, by applying a voltage to the inoculum and grounding the hyphal tips. This may provide more realistic models of flow than simply using the shortest path.

5.3 Modelling Resilience

In nature, fungal mycelia are threatened by damage from physical disturbance and targeted attack by grazers such as Collembola (Chapter 9). Network architecture plays an important role in resilience to damage, through both route redundancy and the probability of link breakage. Assuming a spatially random mode of attack, the probability of a link being attacked is proportional to its length. If, when attacked, the probability of link breakage is inversely proportional to its cross-sectional area, then the joint probability of a link being attacked and broken is proportional to length divided by cross-sectional area. The effect of attack on transport can be followed by examining the global efficiency, the sum of the reciprocals of all shortest paths (Latora and Marchiori, 2001), as the network disintegrates. Paths that are no longer traversable due to the formation of multiple disconnected components are infinitely long, and thus contribute zero to the global efficiency. Global efficiency therefore declines with increasing proportions of broken links. Another way of characterizing resilience is through the reachability, or availability, of a network (Ross and Harary, 1959). Reachability is the proportion of shortest paths that still exist (i.e. are not infinitely long). Reachability is one for a network that has not been fragmented (i.e. consists of one subgraph), since all nodes are mutually available. Reachability does not depend on path length and is therefore independent of the link breakage probability function, whereas the efficiency will be greater if thick cords are less likely to break than thin ones.

Another way to measure resilience is to measure the proportion of the original network that remains connected to the wood block as increasing numbers of links are broken. In nature, disconnection from a food supply is likely to result in death of the disconnected part. When networks of *P. velutina* were tested against model networks with uniform link weights, more of the fungal network remained attached to the inoculum when a given proportion of links were broken (Bebber *et al.*, in press). This demonstrated that the allocation of resources to cords not only increases transport efficiency, but also increases the resilience of the network to this kind of random attack. Inspection of network models that have been attacked in this way suggests that the secret to this increased resilience is the maintenance of a connected core structure as peripheral cords are broken. This pattern is intriguingly similar to that obtained in real networks after attack by certain species of Collembola (Chapter 9). Other mycophagous species may attack networks in other ways, depending, for example, on the size of their mouthparts.

5.4 Changes in Network Architecture over Time

As already mentioned, in peripheral regions cords are thin and at growing fronts hyphae are not aggregated, and therefore have high resistance to transport and long path lengths to the inoculum. As the network develops, some links become strengthened, such that the effective path lengths become dramatically shortened over time, while other links are removed, leading to an overall decrease in the material cost density over time (Bebber *et al.*, submitted). The expectation would be that such strengthening would be accompanied by an increase in the overall construction cost of the network. However, thinning and removal of extraneous cords actually results in a decrease in the volume of material per unit area covered by the network (Bebber *et al.*, submitted). The mechanism by which certain cords are selected for reinforcement while others are broken down remains unknown. A possible conceptual model is one of Darwinian evolution, in which multiple cords are produced but only the 'fittest', in some sense, survive and produce further growth.

5.5 Future Research Direction

One of the most important avenues for further research will be the comparison of network structures and dynamics among the many different cord-forming fungal species. Like any organism, a fungus must partition limited resources among competing requirements. For example, a very dense, highly connected network might have high transport capacity and resilience to damage or attack because of multiple transport pathways. However, it would incur a large material cost of construction per unit area of explored space, and would cover new ground slowly. Conversely, a sparse system could extend further for the same material cost, but would have fewer alternate routes and therefore lower resilience to disconnection. Variation in these tradeoffs among species could reveal important axes of niche differentiation in fungi.

Further, fungi provide one of the few real network systems that can be experimentally manipulated, and that can actually rebuild themselves following damage. Analyses of cord-forming Basidiomycota mycelial systems are therefore likely to reveal a range of evolutionary solutions to network design that may inform the development of other types of transport and infrastructure network, e.g. road, rail and telephone networks.

ACKNOWLEDGEMENTS

We thank N.E.R.C. for funding much of the research reported here. The research on fungal foraging patterns has been conducted over many years and with many collaborators to whom we extend our thanks: Saleh Abdalla, Alaa Alawi, Rory Bolton, Sam Bretherton, Clare Culshaw, Peter Darrah, Damian Donnelly, Chris Dowson, Simon Harold, Melanie Harris, Carolyn Hughes, Juliet Hynes, Hefin Jones, Christian Kampichler, Jonathan Leake, Louise Owen, Alan Rayner, Johann Rolschewski, Timothy Rotheray, Phillip Springham, Joanna Thomas, Wendy Thompson, Monica Tlalka, George Tordoff, Sarah Watkinson, John Wells, Jonathan Wood and Jamil Abd Zakaria.

REFERENCES

Abdalla, S.H.M. and Boddy, L. (1996). Effect of soil and litter type on outgrowth patterns of mycelial systems of *Phanerochaete velutina*. *FEMS Microbiology Ecology*, **20**, 195–204.

Albert, R. and Barabási, A.-L. (2001). Statistical mechanics of complex networks. *Reviews of Modern Physics*, **74**, 47–97.

Amaral, L.A.N. and Ottino, J.M. (2004). Complex networks: Augmenting the framework for the study of complex systems. *European Physical Journal B*, **38**, 147–162.

Bebber, D.P., Hynes, J., Darrah, P.R., Boddy, L. and Fricker, M.D. (in press). Biological solutions to transport network design. *Proceedings of the Royal Society B*.

Bebber, D.P., Tlalka, M., Hynes, J., Darrah, P.R., Ashford, A., Watkinson, S.C., Boddy, L. and Fricker, M.D. (2006). Imaging complex nutrient dynamics in mycelial networks. In: *Fungi and the Environment* (G.M. Gadd, S.C. Watkinson and P. Dyer, eds.), pp. 1–21. Cambridge University Press, Cambridge, UK.

Bevercombe, G.P. and Rayner, A.D.M. (1980). Diamond bark disease of sycamore in Britain. *New Phytologist*, **86**, 379–392.

Boddy, L. (1984). The micro-environment of basidiomycete mycelia in temperate deciduous woodlands. In: *Ecology and Physiology of the Fungal Mycelium* (D.H. Jennings and A.D.M. Rayner, eds.), pp. 261–289. Cambridge University Press, Cambridge, UK.

Boddy, L. (1993). Saprotrophic cord-forming fungi: Warfare strategies and other ecological aspects. *Mycological Research*, **97**, 641–655.

Boddy, L. (1999). Saprotrophic cord-forming fungi: Meeting the challenge of heterogeneous environments. *Mycologia*, **91**, 13–32.

Boddy, L. (2001). Fungal community ecology and wood decomposition processes: From standing tree to complete decay of coarse woody debris. In: *Ecology of Woody Debris in Boreal Forests* (B.G. Jonsson and N. Kruys, eds.). *Ecological Bulletins*, **49**, 43–56.

Boddy, L. and Donnelly, D.P. (in press). Fractal geometry and microorganisms in the environment. In: *Biophysical Chemistry of Fractal Structures and Process in Environmental Systems* (N. Senesi and K. Wilkinson, eds.). Wiley, Chichester, UK.

Boddy, L. and Jones, T.H. (2006). Mycelial responses in heterogeneous environments: Parallels with macroorganisms. In: *Fungi in the Environment* (G.M. Gadd, S.C. Watkinson and P. Dyer, eds.), pp. 112–140. Cambridge University Press. Cambridge, UK.

Boddy, L., Wells, J.M., Culshaw, C. and Donnelly, D.P. (1999). Fractal analysis in studies of mycelium in soil. *Geoderma*, **88**, 301–328.

Bolton, R.G. (1993). Resource acquisition by migratory mycelial cord systems of *Phanerochaete velutina* and *Hypholoma fasciculare*, PhD Thesis, University of Wales, Cardiff.

Bolton, R.G. and Boddy, L. (1993). Characterisation of the spatial aspects of foraging mycelial cord systems using fractal geometry. *Mycological Research*, **97**, 762–768.

Bolton, R.G., Morris, C.W. and Boddy, L. (1991). Non-destructive quantification of growth and regression of mycelial cords using image analysis. *Binary*, **3**, 127–132.

Bork, P., Jensen, L.J., Von Mering, C., Ramani, A.K., Lee, I. and Marcotte, E.M. (2004). Protein interaction networks from yeast to human. *Current Opinion in Structural Biology*, **14**, 292–299.

Bretherton, S., Tordoff, G.M., Jones, T.H. and Boddy, L. (2006). Compensatory growth of *Phanerochaete velutina* mycelial systems grazed by *Folsomia candida* (Collembola). *FEMS Microbiology Ecology*, **58**, 33–44.

Cairney, J.W.G. (1992). Translocation of solutes in ectomycorrhizal and saprotrophic rhizomorphs. *Mycological Research*, **96**, 135–141.

Cairney, J.W.G. (2005). Basidiomycete mycelia in forest soils: Dimensions, dynamics and roles in nutrient distribution. *Mycological Research*, **109**, 7–20.

Deutsch, A., Dress, A. and Rensing, L. (1993). Formation of morphological differentiation patterns in the ascomycete *Neurospora crassa*. *Mechanisms of Development*, **44**, 17–31.

Donnelly, D.P. and Boddy, L. (1997a). Resource acquisition by the mycelial-cord-former *Stropharia caerulea*: Effect of resource quantity and quality. *FEMS Microbiology Ecology*, **23**, 195–205.

Donnelly, D.P. and Boddy, L. (1997b). Development of mycelial systems of *Stropharia caerulea* and *Phanerochaete velutina* on soil: Effect of temperature and water potential. *Mycological Research*, **101**, 705–713.

Donnelly, D.P. and Boddy, L. (1998). Developmental and morphological responses of mycelial systems of *Stropharia caerulea* and *Phanerochaete velutina* to soil nutrient enrichment. *New Phytologist*, **138**, 519–531.

Donnelly, D.P. and Boddy, L. (2001). Mycelial dynamics during interactions between *Stropharia caerulea* and other cord-forming, saprotrophic basidiomycetes. *New Phytologist*, **151**, 691–704.

Donnelly, D.P., Wilkins, M.F. and Boddy, L. (1995). An integrated image analysis approach for determining biomass, radial extent and box-count fractal dimension of macroscopic mycelial systems. *Binary*, **7**, 19–28.

Dorogovtsev, S.N. and Mendes, J.F.F. (2002). Evolution of networks. *Advances in Physics*, **51**, 1079–1187.

Dowson, C.G., Rayner, A.D.M. and Boddy, L. (1986). Outgrowth patterns of mycelial cord-forming basidiomycetes from and between woody resource units in soil. *Journal of General Microbiology*, **132**, 203–211.

Dowson, C.G., Rayner, A.D.M. and Boddy, L. (1988). Foraging patterns of *Phallus impudicus*, *Phanerochaete laevis* and *Steccherinum fimbriatum* between discontinuous resource units in soil. *FEMS Microbiology Ecology*, **53**, 291–298.

Dowson, C.G., Rayner, A.D.M. and Boddy, L. (1989). Spatial dynamics and interactions of the woodland fairy ring fungus, *Clitocybe nebularis*. *New Phytologist*, **111**, 699–705.

Dunne, J.A., Williams, R.J. and Martinez, N.D. (2004). Food-web structure and network theory: The role of connectance and size. *Proceedings of the National Academy of Sciences of the United States of America*, **99**, 12917–12922.

Edelstein, L. and Segel, L.A. (1983). Growth and metabolism of mycelial fungi. *Journal of Theoretical Biology*, **104**, 187–210.

Ferguson, B.A., Dreisbach, T.A., Parks, C.G., Filipo, G.M. and Schmitt, C.L. (2003). Coarse-scale population structure of pathogenic *Armillaria* species in a mixed-conifer forest in the Blue Mountains of northeast Oregon. *Canadian Journal of Forest Research*, **33**, 612–623.

Freeman, L.C. (1977). A set of measures of centrality based on betweenness. *Sociometry*, **40**, 35–41.

Fricker, M.D. and Bebber, D.P. Network organization of mycelial fungi. *The Mycota*, Springer Verlag, Berlin, in press.

Fricker, M.D., Bebber, D., Darrah, P.R., Tlalka, M., Watkinson, S.C., Boddy, L., Yiasoumis, L., Cartwright, H.M., Meškauskas, A., Moore, D., Smith, M.D., Nakagaki, T., Lee, C.F. and Johnson, N. (2007). Inspiration from microbes: From patterns to networks. In: *Complex Systems and Inter-Disciplinary Science* (B.W. Arthur, R. Axtell, S. Bornholdt, J. Efstathiou, P.M. Hui, P. Maini, M. Nowak, R. Sole, D. Stauffer, F. Reed-Tsochas and K. Turner, eds.). World Scientific Publishing Co., London, in press.

Harold, S., Tordoff, G.M., Jones, T.H. and Boddy, L. (2005). Mycelial responses of *Hypholoma fasciculare* to collembola grazing: Effect of inoculum age, nutrient status and resource quality. *Mycological Research*, **109**, 927–935.

Hedger, J. (1990). Fungi in the tropical forest canopy. *The Mycologist*, **4**, 200–202.

Hughes, C.L. and Boddy, L. (1994). Translocation of ^{32}P between wood resources recently colonised by mycelial cord systems of *Phanerochaete velutina*. *FEMS Microbiology Ecology*, **14**, 201–212.

Hughes, C.L. and Boddy, L. (1996). Sequential encounter of wood resources by mycelial cords of *Phanerochaete velutina*: Effect on growth patterns and phosphorus allocation. *New Phytologist*, **133**, 713–726.

Kampichler, C., Rolschewski, J., Donnelly, D.P. and Boddy, L. (2004). Collembolan grazing affects the growth strategy of the cord-forming fungus *Hypholoma fasciculare*. *Soil Biology and Biochemistry*, **36**, 591–599.

Latora, V. and Marchiori, M. (2001). Efficient behaviour of small-world networks. *Physical Review Letters*, **87**, 198701-1–198701-4.

Newman, M.E.J. (2003). The structure and function of complex networks. *SIAM Review*, **45**, 167–256.

Olsson, S. and Gray, S.N. (1998). Patterns and dynamics of ^{32}P phosphate and ^{14}C labelled AIB translocation in intact basidiomycete mycelia. *FEMS Microbiology Ecology*, **26**, 109–120.

Owen, S.L. (1997). Comparative development of the mycelial cord-forming fungi *Coprinus picaceus* and *Phanerochaete velutina*, with particular emphasis on pH and nutrient reallocation, PhD Thesis, University of Wales, Cardiff.

Rayner, A.D.M. and Boddy, L. (1988). *Decomposition of Wood: Its Biology and Ecology*. Wiley, Chichester, UK.

Rayner, A.D.M., Griffith, G.S. and Ainsworth, A.M. (1995). Mycelia interconnectedness. In: *The Growing Fungus* (N.A.R. Gow and G.M. Gadd, eds.), pp. 21–40. Chapman and Hall, London.

Rayner, A.D.M., Powell, K.A., Thompson, W. and Jennings, D.H. (1985). Morphogenesis of vegetative organs. In: *Developmental Biology of Higher Fungi* (D. Moore, L.A. Casselton, D.A. Wood and J.C. Frankland, eds.), pp. 249–279. Cambridge University Press, Cambridge, UK.

Ross, I.C. and Harary, F. (1959). A description of strengthening and weakening members of a group. *Sociometry*, **22**, 139–147.

Smith, M.L., Bruhn, J.N. and Anderson, J.B. (1992). The fungus *Armillaria bulbosa* is among the largest and oldest living organisms. *Nature*, **356**, 428–431.

Strogatz, S.H. (2001). Exploring complex networks. *Nature*, **410**, 268–276.

Thompson, W. and Boddy, L. (1988). Decomposition of suppressed oak trees in even-aged plantations. II. Colonisation of tree roots by cord and rhizomorph producing basidiomycetes. *New Phytologist*, **93**, 277–291.

Thompson, W. and Rayner, A.D.M. (1982). Spatial structure of a population of *Tricholomopsis platyphylla* in a woodland site. *New Phytologist*, **92**, 103–114.

Thompson, W. and Rayner, A.D.M. (1983). Extent development and functioning of mycelial cord systems in soil. *Transactions of the British Mycological Society*, **81**, 333–345.

Tordoff, G.M., Jones, T.H. and Boddy, L. (2006). Grazing by *Folsomia candida* (Collembola) affects the mycelial morphology of the cord-forming basidiomycetes *Hypholoma fasciculare*, *Phanerochaete velutina* and *Resinicium bicolor* differently during early outgrowth onto soil. *Mycological Research*, **110**, 335–345.

Watkinson, S.C., Bebber, D., Darrah, P., Fricker, M., Tlalka, M. and Boddy, L. (2006). The role of wood decay fungi in the carbon and nitrogen dynamics of the forest floor. In: *Fungi in Biogeochemical Cycles* (G.M. Gadd, ed.), pp. 151–181. Cambridge University Press, Cambridge, UK.

Wells, J.M. and Boddy, L. (1990). Wood decay, and phosphorus and fungal biomass allocation, in mycelial cord systems. *New Phytologist*, **116**, 285–295.

Wells, J.M., Donnelly, D.P. and Boddy, L. (1997). Patch formation and developmental polarity in mycelial cord systems of *Phanerochaete velutina* on nutrient-depleted soil. *New Phytologist*, **136**, 653–665.

Wells, J.M., Harris, M.J. and Boddy, L. (1998). Temporary phosphorus partitioning in mycelial systems of the cord-forming basidiomycete *Phanerochaete velutina*. *New Phytologist*, **140**, 283–293.

Wells, J.M., Harris, M.J. and Boddy, L. (1999). Dynamics of mycelial growth and phosphorus partitioning in developing *Phanerochaete velutina*: Dependence on carbon availability. *New Phytologist*, **142**, 325–334.

Wells, J.M., Hughes, C. and Boddy, L. (1990). The fate of soil-derived phosphorus in mycelial cord systems of *Phanerochaete velutina* and *Phallus impudicus*. *New Phytologist*, **114**, 595–606.

Wells, J.M., Thomas, J. and Boddy, L. (2001). Soil water potential shifts: Developmental responses and dependence on phosphorus translocation by the cord-forming basidiomycete *Phanerochaete velutina*. *Mycological Research*, **105**, 859–867.

Wood, J., Tordoff, G.M., Jones, T.H. and Boddy, L. (2006). Reorganization of mycelial networks of *Phanerochaete velutina* in response to new woody resources and collembola grazing. *Mycological Research*, **110**, 985–993.

Wu, F.Y. (2004). Theory of resistor networks: The two-point resistance. *Journal of Physics A*, **37**, 6653–6673.

Zakaria, A.J. and Boddy, L. (2002). Mycelial foraging by *Resinicium bicolor*: Interactive effects of resource quantity, quality and soil composition. *FEMS Microbiology Ecology*, **40**, 135–142.

CHAPTER 2

Enzymes of Saprotrophic Basidiomycetes

Petr Baldrian

Contents

Abstract

Decomposer fungi utilize dead organic matter that is mainly composed of cell wall polysaccharides and other biopolymers. These include cell wall polymers of plant origin (cellulose, hemicelluloses, lignin, pectin), cell wall polysaccharides of fungi (chitin) and nutrient reserve polysaccharide (starch) as well as proteins. Utilization of these polymers necessitates production of extracellular enzymes; the polysaccharide-based biopolymers are usually degraded by hydrolytic enzymes causing endo and/or exocleavage. Lyases and specific oxidases are also produced. Wood-rotting cellulolytic fungi have evolved complex systems of nonenzymatic cellulose cleavage based on the production of reactive oxygen species, but the detailed functioning and relative importance of this decomposition mechanism is still unclear. Lignin decomposition is catalyzed by a set of oxidases and peroxidases with auxiliary enzymes providing hydroxyl radicals, but it also includes the provision of enzyme cosubstrates such as organic acids or aryl alcohols. The composition of ligninolytic systems is thus very complex and species-specific. Compared to decomposition of wood, far less is known about

British Mycological Society Symposia Series
Published by Elsevier Ltd.

basidiomycete species decomposing litter. Some litter-decomposing fungi are apparently physiologically related to ligninolytic wood-rotters but the composition and regulation of their ligninolytic systems is not so well characterized, and little is known of their enzymology in the natural soil environment. However, it seems clear that litter decomposers are able to degrade lignin as well as cellulose and hemicelluloses and probably also chitin and starch. Their ligninolytic system also plays an important role in the transformation of humic substances including humus formation and mineralization. The main gaps in our current knowledge are in the ecology of enzyme production under natural conditions and in estimating the role of decomposer basidiomycetes in complex biological processes in soils.

1. INTRODUCTION

Extracellular enzymes that degrade biopolymers are one of the defining features of saprobic basidiomycetes, and are a selective advantage in the environments of these species. They utilize dead organic matter that is mainly composed of cell wall polysaccharides and other biopolymers, including cell wall polymers of plant origin (cellulose, hemicelluloses, lignin, pectin), cell wall polysaccharides of fungi (chitin) and nutrient reserve polysaccharide (starch) as well as proteins. Decomposition of some biopolymers, cellulose, starch and hemicelluloses provides carbon and energy for growth, while others—most notably lignin—have to be degraded only to get access to other substrates. Since the decomposition of proteins (that serve as an easily available source of carbon and nitrogen) is widespread among many organisms, the landmarks of the saprotrophic lifestyle are the extracellular enzymes that degrade nonprotein biopolymers.

Extracellular enzyme production has only been studied in detail in two ecological groups of saprotrophic basidiomycetes, wood-rotters and soil-inhabiters. This chapter will focus on the enzymes produced by these basidiomycetes and on their involvement, roles and occurrence in the environment.

2. DECOMPOSITION OF BIOPOLYMERS

2.1 Decomposition of Cellulose

2.1.1 Enzyme-Catalyzed Decomposition

Cellulose is the main polymeric component of the plant cell wall and is the most abundant polysaccharide on earth. The chemical composition is simple: it consists of D-glucose residues linked by β-1,4-glycosidic bonds to form linear polymeric chains of over 10,000 glucose residues. Cellulose contains both highly crystalline regions where individual chains are linked to each other, and less-ordered amorphous regions (Hon, 1994). Although chemically simple, the extensive intermolecular bonding pattern of cellulose generates a crystalline structure that can result in a very complex morphology.

For the efficient decomposition of cellulose, an enzymatic system must include endo and exotype hydrolases and β-glucosidases, the complementary activities acting synergistically. Endo-1,4-β-glucanases (EG, **EC 3.2.1.4**) belong to endotype enzymes that hydrolyse cellulose molecules preferentially in the amorphous parts of the fibril. Endoglucanase activity probably exists in all wood-degrading fungi including brown-rot fungi (Highley, 1988). Cellobiohydrolases (CBH, **EC 3.2.1.91**) are exotype enzymes that attack cellulose fibres from both reducing and nonreducing ends. No CBH activity has been detected in most brown-rot fungi, which makes their ability to degrade pure crystalline cellulose questionable. Exceptions can be found in a few of the *Coniophoraceae* (Nilsson and Ginns, 1979; Schmidhalter and Canevascini, 1993a, 1993b). The product of CBH action, cellobiose, is hydrolyzed by 1,4-β-glucosidases (**EC 3.2.1.21**) to two glucose units. Cellobiose and glucose can be taken up and assimilated by the hyphae. The cellobiose is probably hydrolyzed to glucose by cell wall bound or intracellular β-glucosidases. All cellulolytic enzymes share the same chemical specificity for β-1,4-glycosidic bonds, which they cleave by a general acid-catalyzed hydrolysis. Most cellulolytic hydrolases are proteins of 30–50 kDa molecular mass with acidic pH optima between 2.5 and 4.5.

Oxidative enzymes participating in cellulose hydrolysis were first detected in *Phanerochaete chrysosporium* in the 1970s (Westermark and Eriksson, 1975). Biochemically the enzyme is currently described as cellobiose dehydrogenase (CDH, **EC 1.1.3.25**). CDH will oxidize cellobiose, lactose and mannobiose that all have β-1,4-bonds, whereas monosaccharides and α-1,4-glycosidic bond containing maltose are not oxidized (Henriksson *et al.*, 2000). As electron acceptors CDH reduces quinones, phenoxy radicals, cytochrome *c*, complexed Fe^{3+}, manganese and molecular oxygen, which leads to the production of hydrogen peroxide; production of the latter having been demonstrated in *P. chrysosporium* CDH (Kremer and Wood, 1992). These properties enable this enzyme to combine cellulose and lignin decomposition. CDHs have been characterized from several white-rot fungi, e.g. *Heterobasidion annosum, P. chrysosporium, Pycnoporus cinnabarinus, Schizophyllum commune,* and *Trametes versicolor* (Henriksson *et al.*, 2000), and are also known from the brown-rot fungus *Coniophora puteana* (Schmidhalter and Canevascini, 1992) and several ascomycetes.

Cellulolytic enzyme systems in white-rot basidiomycetes have received less attention than in asco- and deuteromycetes, but *P. chrysosporium* or its anamorph, *Sporotrichum pulverulentum* (Eriksson and Hamp, 1978) and *Ischnoderma resinosum* (Sutherland, 1986) have been studied. The work with *P. chrysosporium* has shown that in this fungus the cellulase systems are analogous to systems well known in the ascomycete *Trichoderma reesei* (Lynd *et al.*, 2002). *P. chrysosporium* produces a cellulase system with seven CBH, of which CBHI-4 is the major cellobiohydrolase (van den Wymelenberg *et al.*, 1993). Until now, only a 28-kDa endoglucanase (EG28) lacking cellulose-binding motif (CBM) has been isolated from *P. chrysosporium* and synergism between the EG28 and the CBH was demonstrated. Birch *et al.* (1995) reported differential splicing in the CBM-encoding region of the *cbh1.2* gene, depending on whether microcrystalline or amorphous cellulose was used as the substrate. They proposed that differential splicing of the *cbhI*-like genes could yield cellobiohydrolase and endoglucanase activity. Apart from CBH and endoglucanase activities, *P. chrysosporium* also produces β-glucosidase and CDH (Lynd *et al.*, 2002).

Amongst brown-rot fungi, cellulolysis by *Gloeophyllum trabeum* is best understood. It has a hydroquinone-driven system for the production of extracellular reactive oxygen species, a β-glucosidase, a xylanase, and three endoglucanases, one of which is a processive enzyme that was reported to degrade crystalline cellulose (Cohen *et al.*, 2005). The relative role of the enzymatic and nonenzymatic decomposition of cellulose in this species is unclear.

2.1.2 Nonenzymatic Decomposition of Cellulose

Recently, several systems of nonenzymatic decomposition of cellulose by brown-rot fungi have been proposed that generate free reactive radicals in the modifications of the Fenton reaction—the decomposition of hydrogen peroxide in the presence of ferrous ions (Goodell, 2003). The two systems with the best experimental support are based on the studies with *G. trabeum* and *C. puteana*.

G. trabeum exhibits the rapid ability to degrade an aliphatic polyether via extracellular one-electron oxidation (Kerem *et al.*, 1998), it produces simple aromatic compounds, 4,5-dimethoxycatechol and 2,5-dimethoxyhydroquinone, these compounds may serve as ferric chelators, oxygen-reducing agents and redox-cycling compounds (Kerem *et al.*, 1999). It also produces the necessary NADH-dependent quinone reductases (Jensen *et al.*, 2002). The above phenolic compounds participate in the endocleavage of cellulose by the quinone cycling mechanism generating reactive hydroxyl radicals.

Certain brown-rot fungi accumulate oxalic acid causing a noticeable decrease of pH probably because of the presence of oxalate decomposing enzymes (Shimada *et al.*, 1997). It has been suggested that these brown-rot fungi may use oxalic acid as a proton donor for enzymatic and nonenzymatic hydrolysis of polysaccharides and as a chelator for a Fe^{2+}–H_2O_2 system generating hydroxyl radicals (Shimada *et al.*, 1997). Hyde and Wood (1997), who studied *C. puteana*, suggested that Fe^{3+} is reduced by CDH within fungal cells, and that the Fe^{2+} diffuses at some distance from the hyphae, where a Fe^{2+}–oxalate complex is formed and a Fenton reaction-based hydroxyl radical formation occurs.

Another brown-rot fungus, *Piptoporus betulinus*, produces three endoglucanases and two β-glucosidases (Valášková and Baldrian, 2006a). Nonenzymatic decomposition of cellulose has not been *P. betulinus*, and it is questionable how this species decomposes crystalline cellulose since the enzymatic system is unable to attack it (Valášková and Baldrian, 2006a).

2.2 Hemicellulose Degrading Enzymes

Hemicellulose is a low molecular weight linear or branched polymer usually containing several different sugar units and substituted side chains. Xylans, consisting of D-xylose units, and glucomannans, consisting of D-glucose and D-mannose units, are the main hemicelluloses of angiosperm and conifer trees, respectively, while other lignocellulosic materials may additionally contain considerable amounts of arabinogalactans and galactans (Dekker, 1985). Branched polymers contain neutral and/or acidic side groups that render hemicelluloses noncrystalline or poorly crystalline. Hemicelluloses thus usually form a matrix together with pectins and proteins in primary plant cell walls and with lignin in secondary cell walls.

Enzymatic decomposition of hemicelluloses requires a complex set of different enzymes reflecting structural variability. Hemicellulose hydrolysis proceeds through the action of endotype enzymes that liberate shorter fragments of substituted oligosaccharides, which are further degraded by side-group cleaving enzymes and exotype enzymes. Cleavage results in liberation of monomeric sugars and acetic acid. Similarly to cellulose hydrolysis, the hydrolases act synergistically to convert hemicellulose polymer into soluble units.

Xylanases have been widely studied due to their biotechnological importance (Subramaniyan and Prema, 2002). Endo-1,4-β-xylanases (**EC 3.2.1.8**) catalyse the random hydrolysis of β-1,4-glycosidic bonds in xylans. A range of different endoxylanases with different specificities have been found in both white-rot and brown-rot fungi. They show the highest activity against polymeric xylan, and the rate of the hydrolysis decreases with decreasing chain length (Coughlan and Hazlewood, 1993). 1,4-β-xylosidase (**EC 3.2.1.37**), which is needed for the release of xylose monomers, has also been reported from several white-rot and brown-rot basidiomycetes. The effective native xylan decomposition seems to involve another three enzyme types: 1,4-β-glucuronidases (**EC, 3.2.1.131**), β-arabinofuranosidases (**EC 3.2.1.55**) and acetyl xylan esterases (**EC 3.2.1.72**). They all differ in specificity with respect to the neighbouring substituents and chain length. For example, in conifers, where the xylan has arabinose as a substituent, xylan decomposition requires a β-arabinofuranosidase but not the esterase. β-Glucuronidases have been characterized from *P. chrysosporium* (Castanares *et al.*, 1995) and *S. commune* (Johnson *et al.*, 1989), and β-arabinofuranosidase from *P. chrysosporium* (Coughlan and Hazlewood, 1993).

Complete hydrolysis of glucomannans also requires a wide set of enzymes. Endo-1,4-β-mannanases (**EC 3.2.1.78**) hydrolyse randomly the 1,4-β-mannopyranosyl linkages of gluco- and galactoglucomannans, releasing oligomeric fragments. *Acetyl(gluco)mannan esterase* removes the acetyl groups and 1,4-β-galactosidase (**EC 3.2.1.22**) removes galactose. 1,4-β-mannosidase (**EC 3.2.1.25**) and β-glucosidase cleave the β-1,4 linkages between oligomeric fragments. However, few studies have been conducted on these enzymes from wood-rotting basidiomycetes, although they obviously effectively remove mannan from the cell walls of wood (Eriksson *et al.*, 1992). Apparently, many species produce a wide set of hemicellulose-degrading enzymes. For example, endomannanase, endoxylanase, β-mannosidase and β-xylosidase were produced by the white-rot fungi *Pleurotus ostreatus*, *T. versicolor* and the brown-rotter *P. betulinus* while the latter also produced β-galactosidase activity (Valášková and Baldrian, 2006b). Hemicellulase production does not appear to be as strictly controlled by substrate induction as cellulase production (Valášková and Baldrian, 2006a, 2006b).

Hemicelluloses are bound to lignin by three types of covalent linkages (Špániková and Biely, 2006). The first involves *p*-coumaric or ferulic acid, linked by ether bonds to lignin, and esterically to hemicellulose sugars. This linkage could be cleaved by feruloyl esterase (**EC 3.1.1.73**) that is typical of filamentous fungi and has been demonstrated in the ligninolytic basidiomycetous yeast *Aureobasidium pullulans* (Rumbold *et al.*, 2003) but not yet in other ligninolytic species. The second is represented by ether linkages between OH-groups of saccharides and lignin alcohols. The third are ester linkages between 4-*O*-methyl-D-glucuronic acid or D-glucuronic acid residues of glucuronoxylans and hydroxyl groups of lignin alcohols, which can potentially be cleaved by *glucuronoyl esterase* reported from *S. commune* (Špániková and Biely, 2006). The physiological role of the recently found *Pleurotus sapidus carboxylesterase* has not yet been assigned (Zorn *et al.*, 2005).

The nonenzymatic systems described above for the cleavage of cellulose, work equally well for hemicellulose decomposition. Although the above definitions of enzymes with respect to the reactions they catalyze seem to be clear, in reality the substrate specificity often overlaps. Since some endoglucanases, recently denominated "processive endoglucanases" (Gilad *et al.*, 2003) show oligosaccharide-releasing activity, they can be classified as both endoglucanase and cellobiohydrolase. Also the cellulolytic and hemicellulolytic enzymatic systems cannot be separated completely since the substrates are chemically analogous and individual enzyme molecules frequently exhibit activities with more than one substrate (Copa-Patino and Broda, 1994; Cohen *et al.*, 2005). In the case of *P. betulinus* purified "β-glucosidase" also exhibited β-galactosidase, β-mannosidase and β-xylosidase activities and a weak cellobiohydrolase activity (Valášková and Baldrian, 2006a). The classification according to the major activity can thus be misleading with respect to the actual physiological role, and even the purified enzymes described so far probably exhibit additional activities that have not been searched for.

2.3 Decomposition of Lignin

Lignin is a branched polymer of substituted phenylpropane units joined by carbon–carbon and ether linkages. The monolignol precursors *p*-coumaryl, coniferyl and sinapyl alcohol form *p*-hydroxyphenyl-, guaiacyl-, and syringyl type units in lignin. The major linkage in lignin, the arylglycerol-β-aryl ether substructure, comprises about half of the total interunit linkages. Lignin of gymnosperms contains mainly guaiacyl type units with some *p*-coumaryl units, whereas angiosperm lignin consists of both guaiacyl and syringyl type lignin with few *p*-coumaryl residues (Sjöström, 1993).

The ligninolytic systems consist of oxidases, peroxidases and hydrogen peroxide producing enzymes. Ligninolytic oxidase—laccase—oxidizes its substrates using molecular oxygen, while the peroxidases need the supply of extracellular hydrogen peroxide which is formed by the oxidation of different extracellular metabolites.

Laccases (**EC 1.10.3.2**) have been known for many years in plants, fungi, and insects and have recently been found in bacteria. Although they exhibit low redox potentials they can oxidize a wide variety of substrates and were thus proposed to play a variety of roles, including synthesis of pigments, fruit-body morphogenesis and detoxification (Thurston, 1994; Mayer and Staples, 2002; Baldrian, 2006). Their role in lignin decomposition has recently been reviewed (Leonowicz *et al.*, 2001). Laccases are typically proteins of 50–70 kDa with acidic pH optima (3.0–5.5 for 2,6-dimethoxyphenol and 4.0–6.0 for guaiacol). They are produced by white-rot and litter-decomposing basidiomycetes and by a range of ascomycetes (Baldrian, 2006). Although there are some records of laccases in mycorrhizal basidiomycetes, and in the brown-rotter *C. puteana* (Lee *et al.*, 2004), it seems that their significance for these functional groups is limited (Baldrian, 2006). Recently it was proposed that several compounds produced during lignin transformation can act as redox mediators and thus make it possible to mediate the oxidation of compounds with high redox potential (Camarero *et al.*, 2005; Baldrian, 2006), which is important for the understanding of lignin decomposition.

Lignin peroxidase and manganese peroxidase were discovered in the mid-1980s in *P. chrysosporium* and described as true ligninases because of their high redox potential (Gold *et al.*, 2000; Martinez, 2002). Lignin peroxidase degrades nonphenolic lignin units (up to 90% of the polymer), whereas Mn-peroxidase generates Mn^{3+}, which acts as a diffusible oxidizer on phenolic or nonphenolic lignin units (Jensen *et al.*, 1996; Hofrichter, 2002).

Lignin peroxidases (LiP, **EC 1.11.1.14**) have molecular masses of approximately 40 kDa and very low pH optima, approximately 2.5–3.0. Fungi commonly produce several enzyme isoforms or isoenzymes of LiP but the significance of such multiplicity is not known (Hatakka, 2001; Hildén *et al.*, 2006). LiP preferentially catalyzes the cleavage of the $C\alpha$–$C\beta$ bond in the propyl side chain of lignin (Kirk and Farrell, 1987). Veratryl alcohol, an aromatic compound produced by some white-rot fungi during secondary metabolism, is necessary for LiP catalysis. It acts as a cation radical redox mediator of remote substrates, protects LiP from inactivation by H_2O_2 and completes the catalytic cycle of the enzyme (Hatakka, 2001). LiP is produced by far fewer white-rot fungi than Mn-peroxidase and laccase, e.g. *P. chrysosporium*, *Bjerkandera* spp., *Trametes* spp. and *Phlebia* spp.

Mn-peroxidases (MnP, **EC 1.11.1.13**) are somewhat larger heme proteins than LiPs with molecular masses of 47–60 kDa, glycosylated, and have usually acidic pH optima. Although MnP is able to oxidize phenolic substrates, it most frequently oxidizes Mn^{2+} to Mn^{3+} that is stabilized by organic acids such as oxalate, malate, lactate or malonate. The chelated Mn^{3+} is diffusible and can oxidize a wide range of substrates including phenols, nonphenolic aromatic compounds, carboxylic acids, thiols and unsaturated aliphatic compounds (e.g. fatty acids). The initial oxidation can be followed by a sequence of radical-based or oxidative reactions leading to lignin decomposition and mineralization (Hatakka, 2001; Hofrichter, 2002). The prerequisite of MnP action is a massive production of organic acids by fungi that can be as high as 45 mM malate, 3.5 mM fumarate and 10 mM oxalate (Hofrichter *et al.*, 1999). Production of MnP is widespread among white-rot basidiomycetes and it is usually produced in several isoforms (Hofrichter, 2002).

Versatile peroxidase (VP, **EC 1.11.1.7**) has been described in *Pleurotus* spp. and other basidiomycetes as a third type of ligninolytic peroxidase that combines the catalytic properties of LiP, MnP and plant/microbial peroxidases oxidizing phenolic compounds (Martínez *et al.*, 1996; Heinfling *et al.*, 1998; Ruiz-Duenas *et al.*, 1999). The role of *heme-thiolate haloperoxidases*, newly detected in a range of litter-decomposing basidiomycetes in lignin decomposition, is still questionable. The enzyme is able to catalyze several reactions including oxidations and halogenations of aromatic compounds, but exhibits an unusually high pH optimum around 7 (Hofrichter and Ullrich, 2006).

Other extracellular enzymes involved in wood lignin decomposition are oxidases generating H_2O_2, and mycelium-associated dehydrogenases that reduce lignin-derived compounds. The former include the extracellular aryl-alcohol oxidase (AAO, **EC 1.1.3.7**) described in *Bjerkandera adusta* (Muheim *et al.*, 1990) and other fungi, and *P. chrysosporium glyoxal oxidase* (Kersten, 1990). Pyranose oxidases (**EC 1.1.3.10**) produce hydrogen peroxide in the hyphal periplasmic space of several white-rot fungi, e.g. *P. chrysosporium*, *T. versicolor* or *Phlebiopsis gigantea*, during oxidation of glucose or xylose, the major sugars derived from wood (Giffhorn, 2000). Fungal aryl-alcohol dehydrogenases (AAD, **EC 1.1.1.91**) and *quinone reductases (QR)* are also involved in lignin decomposition (Gutiérrez *et al.*, 1994; Guillén *et al.*, 1997).

Decomposition of lignin by saprotrophic basidiomycetes is a complex process including several enzymatic and nonenzymatic reactions (Figure 1). Laccases or ligninolytic peroxidases produced by white-rot fungi oxidize the lignin polymer, thereby generating aromatic radicals (**1**) (Eriksson *et al.*, 1992). These are involved in different nonenzymatic reactions that include demethoxylation (**2**), C4-ether breakdown (**3**), aromatic ring cleavage (**4**) and $C\alpha$–$C\beta$ breakdown (**5**) (Hatakka, 2001; Martínez *et al.*, 2005). The aromatic aldehydes released from $C\alpha$–$C\beta$ breakdown of lignin, or synthesized *de novo* by fungi are the substrate for AAO that generate H_2O_2 in cyclic redox reactions (**6, 7**). AAD are also involved (Guillén *et al.*, 1994; Gutiérrez *et al.*, 1994). Phenoxy radicals from C4-ether breakdown (**3**) can be reduced by oxidases to phenolic compounds (**8**), as reported for AAO (Marzullo *et al.*, 1995) or repolymerize back with the lignin polymer (**9**). The phenolic compounds can be reoxidized by laccases or peroxidases (**10**). Phenoxy radicals can also be subjected to $C\alpha$–$C\beta$ breakdown (**11**), yielding *p*-quinones. Quinones from the reactions (**7**) and (**11**) contribute to oxygen activation in redox cycling reactions involving laccases, peroxidases and QR (**12, 13**) (Guillén *et al.*, 1997). This results in reduction of Fe^{2+} present in wood (**14**), either by superoxide cation radical or directly by the semiquinone radicals, and its reoxidation with concomitant reduction of H_2O_2 to hydroxyl free radicals (OH) (**15**) (Guillén *et al.*, 2000; Hammel *et al.*, 2002). The hydroxyl radical is a strong oxidizer that can attack lignin (**16**) and probably participates in the initial phases of wood decay when the cell wall is intact and pore size prevents the penetration of ligninolytic enzymes (Flournoy *et al.*, 1991, 1993; Evans *et al.*, 1994). In the final steps, simple products from lignin decomposition enter hyphae and are incorporated into intracellular catabolic routes (Martínez *et al.*, 2005).

The above mechanism, however, refers only to white-rot fungi capable of production of ligninolytic enzymes. Brown-rot fungi can only affect lignin by the formation of hydroxyl radicals (see Section 2.1.2.) that can remove methoxyl

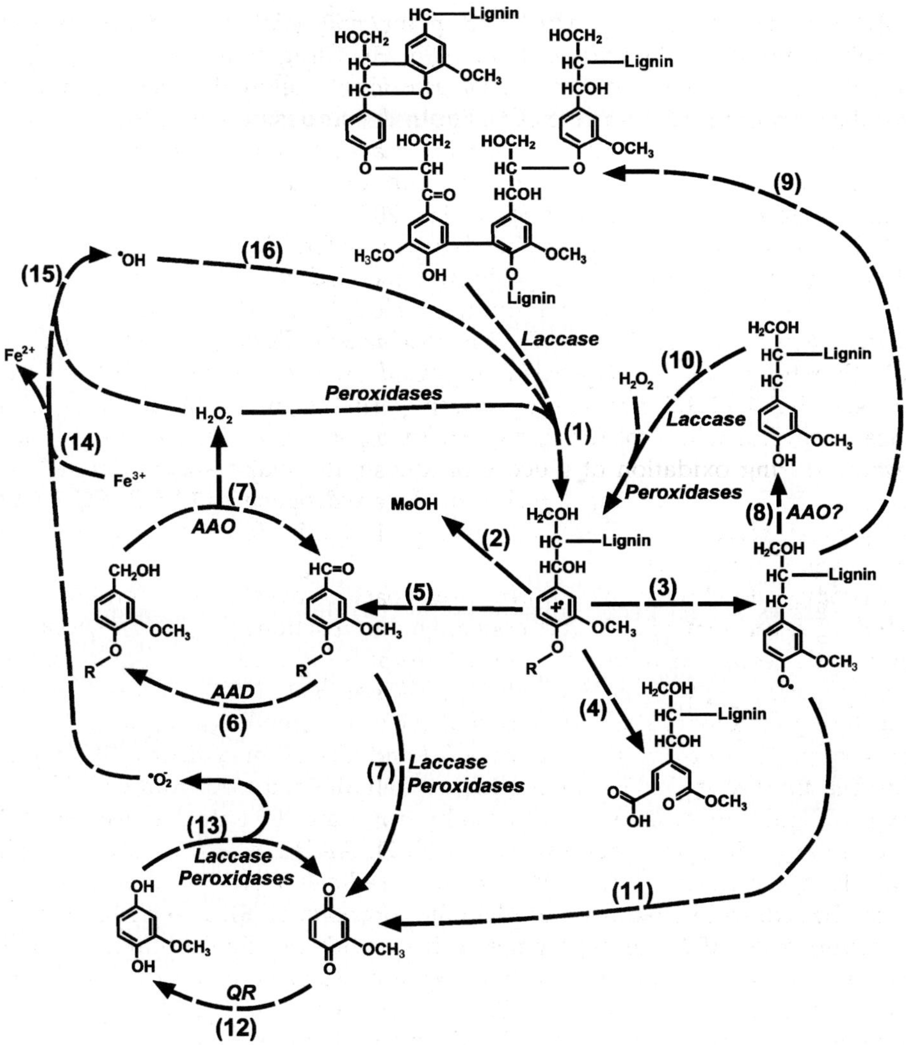

Figure 1 Schematic representation of processes involved in the decomposition of lignin by white-rot basidiomycetes (see text for explanations).
Source: Based on Hatakka (2001) and Martínez *et al.* (2005).

groups from lignin and produce methanol, and thus they leave a residue that consists mainly of modified lignin. Demethoxylation of methoxyl groups and aromatic hydroxylation makes the modified lignin more reactive (Hatakka, 2001).

White-rot fungi can be classified into four groups based on the extracellular lignolytic enzymes produced (Hatakka, 1994): (1) LiPs, MnPs and laccase; (2) MnPs and laccase; (3) LiPs and laccase; (4) no peroxidases, only laccases. Recently, it was confirmed that *P. chrysosporium* does not produce true laccase, but a functionally related enzyme, perhaps belonging to a very specific fifth group (Baldrian, 2006).

2.4 Enzymes Degrading Pectic Compounds, Starch and Chitin

2.4.1 Pectic Compounds

Pectic substances are complex polysaccharides being major components of middle lamellae, in the form of calcium pectate and magnesium pectate. Pectic substances account for 0.5–4.0% of the fresh weight of plants. Their relative molecular masses range from 25 to 360 kDa. Pectic substances mainly consist of galacturonans and rhamnogalacturonans — in which the C-6 carbon of galactate is oxidized to a carboxyl group, the arabinans and the arabinogalactans. Based on the increasing methoxylation of galacturonate groups, pectic substances can be divided into pectic acids, pectinic acids and pectin (Jayani *et al.*, 2005). The pectinolytic enzymes may all be divided into three groups: (1) protopectinases degrade insoluble protopectin, resulting in highly polymerized soluble pectin; (2) esterases catalyze the de-esterification of pectin by the removal of methoxy esters; and (3) depolymerases catalyze the hydrolytic cleavage of the α-1,4-glycosidic bonds in the D-galacturonic acid moieties of the pectic substances. Depolymerases act on pectic substances by two different mechanisms: hydrolysis, in which they catalyze the hydrolytic cleavage with the introduction of water across the oxygen bridge, and trans-elimination lysis, in which they break the glycosidic bond by a trans-elimination reaction (Conner, 2001).

Pectinolytic enzymes are produced by phytopathogenic basidiomycetes that have to penetrate pectin-rich plant tissues (Jayani *et al.*, 2005) and are also present in wood-rotting saprotrophs. Wood can contain up to 4% pectin and its decomposition enables penetration of bordered and pinoid pits and accelerates colonization. Endopolygalacturonase (**EC 3.2.1.15**) found in several brown-rot and white-rot basidiomycetes has a pH optimum between 3.0 and 4.5 (Green and Clausen, 1999; Xavier-Santos *et al.*, 2004). The purified enzymes from *Postia placenta*, *P. chrysosporium* and *Chondrostereum purpureum* have molecular masses of 34–42 kDa and are produced as multiple isoenzymes (Miyairi *et al.*, 1985; Shanley *et al.*, 1993; Clausen and Green, 1996). In addition to endopolygalacturonase, pectate lyase (**EC 4.2.2.2**) was also detected in *C. purpureum* (Miyairi *et al.*, 2002) and pectin lyase (**EC 4.2.2.10**) but not pectate lyase in *Trametes. trogii* (Levin and Forchiassin, 1998). Although only a few species have been tested, pectinolytic enzymes seem to be widespread among wood-rotting saprotrophs.

2.4.2 Starch

Starch serves as a storage polysaccharide contained in wood, roots and leaves, mainly in parenchyma cells (Willför *et al.*, 2005). The main enzyme responsible for its decomposition is amylase, 1,4-α-glucosidase (**EC 3.2.1.3**) which has been identified in several wood-rotting basidiomycetes including *Ceriporiopsis subvermispora* (Sethuraman *et al.*, 1998), *P. chrysosporium* (Dey *et al.*, 1991) and

S. commune (Shimazaki *et al.*, 1984). It is also produced by soil-inhabiting saprotrophs, e.g. *Coprinus* sp. (Stephens *et al.*, 1991). Production can be induced, e.g. by the interaction of saprotrophic fungi and invertebrates (Dyer *et al.*, 1992), but the ecological significance of the enzyme is not yet fully clarified.

2.4.3 Chitin

Chitin, an unbranched homopolymer of 1,4-β-linked *N*-acetyl-D-glucosamine, is widely distributed in nature, present mainly as a cell wall component in fungi (including basidiomycetes) and in the exoskeletons of insects and other invertebrates. It is believed to be the second most abundant polysaccharide on earth, next to cellulose (Duo-Chan, 2006). The main difference from all the previously mentioned biopolymers is the presence of 7% nitrogen in the molecule. The chitin content of fungal mycelium harvested from wood has been estimated to be around 2%. Assuming an *N* content of 0.25–3% in the mycelium of wood-degrading basidiomycetes, 5–50% of the total *N* in wood-degrading mycelium would be found in chitin (Lindahl and Finlay, 2006). Chitin in fungal mycelium is therefore an important nitrogen source.

The decomposition of chitin is initiated by endochitinases (**EC 3.2.1.14**) that hydrolyse the bonds between *N*-acetylglucosamine residues at random locations within the chitin macromolecule, disrupting the structural integrity of chitin and producing oligosaccharides of varying length. β-*N*-acetylhexosaminidases (**EC 3.2.1.52**) further degrade the oligosaccharides, releasing monosaccharides from the nonreducing ends. A third type of enzyme—*chitobiosidase*, cleaves disaccharides of *N*-acetylglucosamine from the ends of chitin chains (Lorito, 1998; Lindahl and Finlay, 2006).

Compared to phytopathogenic fungi, the occurrence of chitinase in saprotrophic basidiomycetes has not yet attracted much attention. Endochitinase and β-*N*-acetylhexosaminidase have been purified from the white-rotter *P. cinnabarinus*. The enzymes had molecular masses of 60 and 36 kDa and acidic pH optima at 2.5–4.5 (Ohtakara, 1988). Production of extracellular chitinolytic enzymes has also been documented for the root-infecting basidiomycetes *H. annosum* and *Armillaria ostoyae* and several ectomycorrhizal (ECM) fungi (Hodge *et al.*, 1995). All reports about the presence of chitinolytic enzymes however do not necessarily mean that the fungus uses them for saprotrophic purposes. In some cases the enzymes can be merely involved in the reconfiguration of the chitin-containing cell wall. Recently, Lindahl and Finlay (2006) studied the production of chitinolytic enzymes by three secondary wood colonizers *Coniophora arida, Hypholoma capnoides* and *Resinicium bicolor*. All of the three tested fungi produced endochitinases, chitobiosidases and *N*-acetylhexosaminidases during colonization of wood. *N*-acetylhexosaminidase activity, and in some cases also chitobiosidase and endochitinase activities, were higher during secondary overgrowth of another fungus than during primary colonization of noncolonized wood. The results suggest that wood-degrading fungi degrade their own cell walls as well as the hyphae of earlier colonizers.

3. SPECIFIC ASPECTS OF ENZYMOLOGY OF BASIDIOMYCETES FROM DIFFERENT HABITATS

3.1 Decaying Wood

Most of the enzymology of saprotrophic basidiomycetes in natural resources has been studied in wood-rotting species, especially those causing white rot (Hatakka, 2001; Martínez *et al.*, 2005). Two main types of rot caused by basidiomycetes have been differentiated based on morphological appearance of the wood—white rot and brown rot, though they can also cause other types under certain circumstances (Table 1). Those species causing white rot and brown rot are taxonomically closely related, both often being found in the same genus. Those causing brown rot are most commonly found on conifers, and represent only 7% of wood-rotting basidiomycetes (Gilbertson, 1980). With white rot all wood components are utilized, resulting in a bleached appearance as wood decays, though different components can be removed at different rates (Table 1). With brown rot lignin is slightly modified, allowing utilization of polysaccharides, resulting in a brown material consisting of oxidized lignin, which represents a potential source of aromatic compounds for the stable humic material fraction in forest soils (Hatakka, 2001). The white-rot basidiomycetes and the brown-rotter *C. puteana* decreased both the relative content of lignin and the ratio of syringyl/guaiacyl units compared to undecayed wood of *Eucalyptus* sp. (del Río *et al.*, 2001).

The ability to decompose wood and the types of wood decay caused relates not only to the enzymes and other reactive species produced by the fungi, but also the location to which they are delivered. It is generally accepted that the mobility of enzymes in decaying lignocellulosic material is limited to cell lumina, the largest anatomical pores, and to pores formed during the decay process. During both white rot and brown rot of wood by *P. chrysosporium* and *Postia placenta* respectively, pores of less than 50 Å were inaccessible to molecules with a molecular weight above 12 kDa, i.e. for enzymes (Flournoy *et al.*, 1991, 1993). The initiators of both cellulose and lignin breakdown are thus suggested to be small molecular mass compounds that can readily diffuse from fungal hyphae and penetrate into the wood cell walls (Hatakka, 2001; Hammel *et al.*, 2002; Reading *et al.*, 2003). Due to (1) limited diffusibility of enzymes, (2) the reactivity of small molecular mass compounds generated by fungi (Hammel *et al.*, 2002), (3) the small likelihood of concerted action of peroxidases and hydrogen peroxide generating enzymes over larger distances and (4) the fact that a considerable part of the lignocellulose-degrading apparatus is cell-wall associated (Valášková and Baldrian, 2006b) most of the wood-decay reactions apparently occur near fungal hyphae. Thus the penetration of a resource is a necessary prerequisite for the capture and utilization of the substrate (Boddy, 2000). With white rot, enzymic decomposition results in formation of erosion grooves around hyphae, which eventually coalesce causing gradual thinning of the cell wall from the lumen outwards, i.e. from the S3 layer to the S1 layer (Rayner and Boddy, 1988). With brown rot, carbohydrates are removed from the S2 layer, even though hyphae are in the lumen on the S3 layer.

Table 1 Comparison of structural and chemical features caused by saprotrophic basidiomycetes during different types of wood decay

	White rot simultaneous rot	White rot selective delignification	Brown rot	Soft rot
Properties of decayed wood	Bleached appearance, lighter in colour than sound wood, moist, soft, spongy, strength loss after advanced decay; brittle fracture	Bleached appearance, lighter in colour than sound wood, moist, soft, spongy, strength loss after advanced decay; fibrous appearance	Brown, dry, crumbly, powdery, brittle consistency, breaks up into cubes, drastic loss of strength at initial stage of decay; very uniform ontogeny of wood decay	Soft consistency in wet environments; brown and crumbly in dry environments; generally uniform ontogeny of wood decay
Host and wood type	Hardwood, rarely softwood	Hardwood and softwood	Softwoods; seldom hardwoods; forest ecosystems and timber	Generally hardwoods (softwoods very slightly degraded); forest ecosystems, waterlogged woods
Cell wall constituents degraded	Cellulose, lignin and hemicellulose	Initial attack selective for hemicelluloses and lignin, later also cellulose	Cellulose and hemicelluloses; lignin slightly modified; in some cases, extended decomposition of hardwood (including middle lamella)	Cellulose and hemicelluloses, lignin slightly altered
Anatomical features	Cell wall attacked progressively from	Lignin decomposition in	Decomposition at a great distance from	Cell wall attack in the proximity of

Table 1 (*Continued*)

	White rot simultaneous rot	White rot selective delignification	Brown rot	Soft rot
	lumen; erosion furrows associated with hyphae	middle lamella and secondary wall; middle lamella dissolved by diffusible agents (not in contact with hyphae), radial cavities in cell wall	hyphae (diffusion mechanism); entire cell wall attacked rapidly with cracks and clefts	hyphae starts from cell lumen; longitudinal cylindrical cavities in secondary wall or secondary wall erosions from cell lumen
Examples of causal agents	Basidiomycetes (e.g. *Trametes versicolor*, *Irpex lacteus*, *Phanerochaete chrysosporium*, *Heterobasidion annosum*) and some xylariaceous ascomycetes	Basidiomycetes (e.g. *Ganoderma australe*, *Phlebia tremellosa*, *Ceriporiopsis subvermispora*, *Pleurotus* spp., *Phellinus pini*)	Basidiomycetes exclusively (e.g. *Gloeophyllum* spp., *Laetiporus sulphureus*, *Piptoporus betulinus*, *Postia placenta*, *Serpula lacrymans*, *Coniophora puteana*)	Mainly ascomycetes; some white-rot (*Inonotus hispidus*) and brown-rot (*Rigidoporus crocatus*) basidiomycetes cause facultative soft-rot decay

Source: Based on Eriksson *et al*. (1992), Zabel and Morell (1992) and Schwarze *et al*. (2000).

All models suggested that brown rot decay involves generation of hydroxyl radicals, which are small enough to enter the wood cell wall (Hammel *et al.*, 2002; Goodell, 2003). However, none of the proposed models has been fully verified experimentally. Many of the proposed low molecular mass compounds have been isolated from cultures of both brown-rot and white-rot fungi which makes it difficult to understand their specific role in brown rot decay. These compounds can be phenolates or other types of iron-chelating compounds (siderophores), oxalate and simple aromatic compounds (Hatakka, 2001; Goodell, 2003).

3.2 Litter and Soil

Compared to wood, far less is known about the occurrence, properties and roles of enzymes in forest litter and soil. Soil and litter is a heterogeneous environment which may hamper detection and estimation of enzyme activities. Another problem is the difficulty of linking enzyme activities in soil/litter to specific species producing them. There have been a few studies of ligninolytic activities and enzymes of litter-decomposing basidiomycetes. Basidiomycetes in the genera *Stropharia* and *Agrocybe* mineralized ^{14}C-(ring)-labelled synthetic lignin at about half the rate of wood-inhabiting white-rot fungi (Steffen *et al.*, 2000). The main ligninolytic enzymes in litter-decomposing fungi such as *Agaricus bisporus* (Bonnen *et al.*, 1994), *Collybia dryophila* (Steffen *et al.*, 2002a) and *Stropharia rugosoannulata* (Steffen *et al.*, 2002b) are laccase and MnP. Many other litter-decomposing species produce laccase but the production of ligninolytic peroxidases has not been reported till date (Baldrian, 2006).

Unlike the activity of other extracellular enzymes, ligninolytic enzymes can be linked to the presence of fungi, although not only to basidiomycetes. Relatively high activities of laccase—the dominant ligninolytic enzyme—were detected in angiosperm and coniferous forest litter and soils, compared to agricultural or meadow soils (Rosenbrock *et al.*, 1995; Criquet *et al.*, 2000; Ghosh *et al.*, 2003). Several laccase enzymes (most of them from basidiomycetes) were usually responsible for the measured activity in temperate forest soil (Luis *et al.*, 2004). Laccase activity reflects the course of decomposition of organic substances and thus varies in time. It increased during leaf litter decomposition in Mediterranean broadleaved litter (Fioretto *et al.*, 2000) and the pattern of detected isoenzymes varied during the succession (Di Nardo *et al.*, 2004). Laccase actually also reflects the presence of mycelia. In evergreen oak litter, laccase activity reflected the annual dynamics of fungal biomass that is probably driven by microclimate, while Mn-peroxidase occurred only in autumn (Criquet *et al.*, 2000). Significantly increased activity of oxidases and peroxidases occurs in the soil under fairy rings of saprobic basidiomycetes, e.g. *Marasmius oreades* or *Agaricus arvensis*, compared to soil devoid of visible mycelia (Gramss, 1997; Gramss *et al.*, 2005). Laccase activity and the diversity of laccase gene sequences decreases with depth in soil profiles, associated with decrease in fungal biomass (Luis *et al.*, 2004). Laccase activity has a high small-scale variability in soil, and the distribution of mRNA transcripts does not always reflect the laccase gene pool (Luis *et al.*, 2005a, 2005b).

In addition to ligninolysis, ligninolytic enzymes in soils are also involved in the transformation of soil humic compounds: humic acids (HA), fulvic acids and humin (Kästner and Hofrichter, 2001). Extracellular peroxidase activities correlate with HA decomposition (Kästner and Hofrichter, 2001). MnP may have a more important role than LiP in the decomposition process (Dehorter and Blondeau, 1992), being able to depolymerize and mineralize HA (Steffen *et al.*, 2000, 2002a; Hofrichter, 2002). The interaction of laccases with humic substances probably leads both to depolymerization of humic substances and their synthesis from monomeric precursors, and the balance of these two processes can be influenced by the nature of the humic compounds (Zavarzina *et al.*, 2004). Fakoussa and Frost (1999) observed the decolourization and decrease of molecular weight of HA, accompanied by the formation of fulvic acids during the growth of *T. versicolor* cultures producing mainly laccase, while HA synthesis occurred *in vitro* with laccase (Katase and Bollag, 1991). Adsorption of laccases to soil humic substances or inorganic soil constituents changes their temperature activity profiles (Criquet *et al.*, 2000) and inhibits their activity (Claus and Filip, 1990; Zavarzina *et al.*, 2004; Baldrian, 2006).

There are unfortunately only a few studies of enzyme production by litter-decomposing basidiomycetes in leaf litter, though production of, e.g. cellulases, is often attributed to them (Criquet *et al.*, 2002). Laccase, Mn-peroxidase, mannanase and xylanase are produced by *Mycena galopus* in *Picea sitchensis* litter (Ghosh *et al.*, 2003). This species is particularly vigorous in degrading lignin and cellulose, leading to the formation of a "white-rot" litter (Frankland, 1998). The litter-decomposer *Lepista nuda* produced laccase, endoglucanase, β-glucosidase and β-xylosidase in *Fagus sylvantica* buried leaves in soil (Colpaert and van Laere, 1996).

3.3 Biopolymer-Degrading Enzymes in Ectomycorrhizal Fungi

ECM fungi have some saprotrophic abilities. Several attempts have been made to detect ligninolytic enzymes including laccases in ECM fungi (Cairney and Burke, 1998; Burke and Cairney, 2002). Gene fragments with high similarity to laccase from wood-rotting fungi have been found in species in several genera including *Amanita, Cortinarius, Hebeloma, Lactarius, Paxillus, Piloderma, Russula, Tylospora* and *Xerocomus* (Chen *et al.*, 2003; Luis *et al.*, 2004). Laccases have been purified from *Cantharellus cibarius, Lactarius piperatus, Russula delica* and *Thelephora terrestris* (Baldrian, 2006). However, syringaldazine oxidation was rare (Burke and Cairney, 2002) and the laccase activities are much less than for white-rotters and litter decomposers (Baldrian, 2006). Activity of another ligninolytic enzyme, Mn-peroxidase, has to date been confirmed only in *Tylospora fibrillosa*, a species also containing a putative sequence of laccase and possibly also lignin peroxidase (Chambers *et al.*, 1999; Chen *et al.*, 2001, 2003).

T. terrestris and *Suillus bovinus* colonizing beech litter produced significant activities of β-glucosidase and β-xylosidase but not endoglucanase; the saprotroph *L. nuda*, however, always produced more (Colpaert and van Laere, 1996). Cellulolytic and xylanolytic activity also occurs, e.g. with *Pisolithus tinctorius* (Cao and Crawford, 1993; Colpaert and van Laere, 1996), cellobiohydrolase and glucuronidase has been detected in mycorrhizal root tips (Courty *et al.*, 2005). Chitinases are also produced by some mycorrhizal basidiomycetes but less than by saprotophic or parasitic root-colonizers (Hodge *et al.*, 1995). Extracellular enzyme systems provide an additional means of nutrient acquisition by mycorrhizal species but probably are of far less importance to decomposition processes than are those of saprotrophs.

4. CONCLUSIONS AND FUTURE PERSPECTIVES IN THE ENZYMOLOGY OF DECOMPOSER BASIDIOMYCETES

In the past the potential use of ligninolytic enzymes in biotechnology led to the accumulation of greater knowledge of many basic and applied aspects of the enzymology of lignin decomposition (Hatakka, 2001; Martínez *et al.*, 2005). Nonetheless some areas are only just beginning to be studied, e.g. role of esterases cleaving the bonds between lignin and hemicelluloses. Other aspects of the enzymology of saprotrophic basidiomycetes have attracted only scattered attention without even a recent review of cellulose decomposition by these species.

Many questions on both the physiology and ecology of enzyme production remain to be answered. For brown-rot fungi we do not know how most of them degrade crystalline cellulose. What is the relative importance of the nonenzymatic versus enzymatic depolymerization of polysaccharides? What nonenzymatic systems are used by different species? Are there as yet undiscovered systems? Even less is known of the enzymology of litter-decomposing fungi. Several species seem to have a physiology similar to white-rot fungi, but what is the difference in metabolism in nutrient-rich soil patches compared to soil devoid of resources? Can we expect some nonligninolytic decomposers in soils equivalent to "brown-rotters"? What is the exact difference in enzyme production between saprotrophic and mycorrhizal basidiomycetes inhabiting the same habitat?

Many unresolved questions remain concerning the ecology of enzyme production and substrate transformation in natural conditions, in wood and even more in soil and litter. Owing to the complexity of the microbial communities in nature it is not yet possible to make a clear link between expressed genes (i.e. from individual species) and enzyme activities or to relate individual catalyzed reactions to complex environmental processes and to nutrient flow. The solution of these problems and the struggle for understanding of how the complex processes of substrate transformation are regulated in nature represent perhaps one of the greatest challenges for microbial ecologists. This may be aided by the developing abilities to link genes to individual species.

ACKNOWLEDGEMENTS

This work was supported by the Grant Agency of the Czech Academy of Sciences (B600200516), by the Czech Science Foundation (526/05/0168), by the Ministry of Education, Youth and Sports of the Czech Republic (LC06066) and by the Institute of Microbiology of the ASCR, v.v.i. in the frames of its institutional research concept.

REFERENCES

Baldrian, P. (2006). Fungal laccases: Occurrence and properties. *FEMS Microbiology Reviews*, **30**, 215–242.

Birch, P.R.J., Sims, P.F.G. and Broda, P. (1995). Substrate-dependent differential splicing of introns in the regions encoding the cellulose binding domains of two exocellobiohydrolase I-like genes in *Phanerochaete chrysosporium*. *Applied and Environmental Microbiology*, **61**, 3741–3744.

Boddy, L. (2000). Interspecific combative interactions between wood-decaying basidiomycetes. *FEMS Microbiology Ecology*, **31**, 185–194.

Bonnen, A.M., Anton, L.H. and Orth, A.B. (1994). Lignin degrading enzymes of the commercial button mushroom, *Agaricus bisporus*. *Applied and Environmental Microbiology*, **60**, 960–965.

Burke, R.M. and Cairney, J.W.G. (2002). Laccases and other polyphenol oxidases in ecto- and ericoid mycorrhizal fungi. *Mycorrhiza*, **12**, 105–116.

Cairney, J.W.G. and Burke, R.M. (1998). Do ecto- and ericoid mycorrhizal fungi produce peroxidase activity? *Mycorrhiza*, **8**, 61–65.

Camarero, S., Ibarra, D., Martinez, M.J. and Martinez, A.T. (2005). Lignin-derived compounds as efficient laccase mediators for decolorization of different types of recalcitrant dyes. *Applied and Environmental Microbiology*, **71**, 1775–1784.

Cao, W. and Crawford, D.L. (1993). Carbon nutrition and hydrolytic and cellulolytic activities in the ectomycorrhizal fungus *Pisolithus tinctorius*. *Canadian Journal of Microbiology*, **39**, 529–535.

Castanares, A., Hay, A.J., Gordon, A.H., McCrae, S.I. and Wood, T.M. (1995). D-xylan-degrading enzyme system from the fungus *Phanerochaete chrysosporium*: Isolation and partial characterisation of an alpha-(4-*O*-methyl)-D-glucuronidase. *Journal of Biotechnology*, **43**, 183–194.

Chambers, S.M., Burke, R.M., Brooks, P.R. and Cairney, J.W.G. (1999). Molecular and biochemical evidence for manganese-dependent peroxidase activity in *Tylospora fibrillosa*. *Mycological Research*, **103**, 1098–1102.

Chen, D.M., Bastias, B.A., Taylor, A.F.S. and Cairney, J.W.G. (2003). Identification of laccase-like genes in ectomycorrhizal basidiomycetes and transcriptional regulation by nitrogen in *Piloderma byssinum*. *New Phytologist*, **157**, 547–554.

Chen, D.M., Taylor, A.F.S., Burke, R.M. and Cairney, J.W.G. (2001). Identification of genes for lignin peroxidases and manganese peroxidases in ectomycorrhizal fungi. *New Phytologist*, **152**, 151–158.

Claus, H. and Filip, Z. (1990). Effects of clays and other solids on the activity of phenoloxidases produced by some fungi and actinomycetes. *Soil Biology and Biochemistry*, **22**, 483–488.

Clausen, C.A. and Green, F. (1996). Characterization of polygalacturonase from the brown-rot fungus *Postia placenta*. *Applied Microbiology and Biotechnology*, **45**, 750–754.

Cohen, R., Suzuki, M.R. and Hammel, K.E. (2005). Processive endoglucanase active in crystalline cellulose hydrolysis by the brown rot basidiomycete *Gloeophyllum trabeum*. *Applied and Environmental Microbiology*, **71**, 2412–2417.

Colpaert, J.V. and van Laere, A. (1996). A comparison of the extracellular enzyme activities of two ectomycorrhizal and a leaf-saprotrophic basidiomycete colonizing beech leaf litter. *New Phytologist*, **134**, 133–141.

Conner, R.C. (2001). Pectinolytic and cellulolytic enzymes in the microbial modification of plant tissues. *Journal of Applied Bacteriology*, **84**, 147–160.

Copa-Patino, J.L. and Broda, P. (1994). *Phanerochaete chrysosporium* β-D-glucosidase/β-D-xylosidase with specificity for (1→3)-β-D-glucan linkages. *Carbohydrate Research*, **253**, 265–275.

Coughlan, M.P. and Hazlewood, G.P. (1993). β-1,4-D-xylan-degrading enzyme systems: Biochemistry, molecular biology and applications. *Biotechnology and Applied Biochemistry*, **17**, 259–289.

Courty, P.E., Pritsch, K., Schloter, M., Hartmann, A. and Garbaye J. (2005). Activity profiling of ectomycorrhiza communities in two forest soils using multiple enzymatic tests. *New Phytologist*, **167**, 309–319.

Criquet, S., Farnet, A.M., Tagger, S. and Le Petit, J. (2000) Annual variations of phenoloxidase activities in an evergreen oak litter: Influence of certain biotic and abiotic factors. *Soil Biology and Biochemistry,* **32**, 1505–1513.

Criquet, S., Tagger, S., Vogt, G. and Le Petit, J. (2002). Endoglucanase and beta-glycosidase activities in an evergreen oak litter: Annual variation and regulating factors. *Soil Biology and Biochemistry,* **34**, 1111–1120.

Dehorter, B. and Blondeau, R. (1992). Extracellular enzyme activities during humic acid degradation by the white rot fungi *Phanerochaete chrysosporium* and *Trametes versicolor. FEMS Microbiology Letters,* **94**, 209–215.

Dekker, R.F.H. (1985). Biodegradation of hemicelluloses. In: *Biosynthesis and Biodegradation of Wood Components* (T. Higuchi, ed.), pp. 505–533. Academic Press, Orlando.

del Río, J.C., Gutiérrez, A., Martínez, M.J. and Martínez, A.T. (2001). Py-GC/MS study of *Eucalyptus globulus* wood treated with different fungi. *Journal of Analytical and Applied Pyrolysis,* **58/59**, 441–452.

Dey, S., Maiti, T.K., Saha, N., Banerjee, R. and Bhattacharyya, B.C. (1991). Extracellular protease and amylase activities in ligninase-producing liquid cultures of *Phanerochaete chrysosporium. Process Biochemistry,* **26**, 325–329.

Di Nardo, C., Cinquegrana, A., Papa, S., Fuggi, A. and Fioretto, A. (2004). Laccase and peroxidase isoenzymes during leaf litter decomposition of *Quercus ilex* in a Mediterranean ecosystem. *Soil Biology and Biochemistry,* **36**, 1539–1544.

Duo-Chan, L. (2006). Review of fungal chitinases. *Mycopathologia,* **161**, 345–360.

Dyer, H.C., Boddy, L. and Preston-Meek, C.M. (1992). Effect of the nematode *Panagrellus redivivus* on growth and enzyme production by *Phanerochaete velutina* and *Stereum hirsutum. Mycological Research,* **96**, 1019–1028.

Eriksson K.E.L., Blanchette, R.A. and Ander, P. (1992). *Microbial and Enzymatic Degradation of Wood Components.* Springer, Berlin.

Eriksson, K.E. and Hamp, S.G. (1978). Regulation of endo-1,4-β-glucanase production in *Sporotrichum pulverulentum. European Journal of Biochemistry,* **90**, 183–190.

Evans, C.S., Dutton, M.V., Guillén, F. and Veness, R.G. (1994). Enzymes and small molecular mass agents involved with lignocellulose degradation. *FEMS Microbiology Reviews,* **13**, 235–240.

Fakoussa, R.M. and Frost, P.J. (1999). In vivo-decolorization of coal-derived humic acids by laccase-excreting fungus *Trametes versicolor. Applied Microbiology and Biotechnology,* **52**, 60–65.

Fioretto, A. Papa, S., Curcio, E., Sorrentino, G. and Fuggi, A. (2000). Enzyme dynamics on decomposing leaf litter of *Cistus incanus* and *Myrtus communis* in a Mediterranean ecosystem. *Soil Biology and Biochemistry,* **32**, 1847–1855.

Flournoy, D.S., Kirk, T.K. and Highley, T.L. (1991). Wood decay by brown-rot fungi: Changes in pore structure and cell-wall volume. *Holzforschung,* **45**, 383–388.

Fluornoy, D. S., Paul, J.A, Kirk, T.K. and Highley, T.L. (1993). Changes in the size and volume of pores in sweetgum wood during simultaneous rot by *Phanerochaete chrysosporium* Burds. *Holzforschung,* **47**, 297–301.

Frankland, J.C. (1998). Fungal succession: Unravelling the unpredictable. *Mycological Research,* **102**, 1–15.

Ghosh, A., Frankland, J.C., Thurston, C.F. and Robinson, C.H. (2003). Enzyme production by *Mycena galopus* mycelium in artificial media and in *Picea sitchensis* F-1 horizon needle litter. *Mycological Research,* **107**, 996–1008.

Giffhorn, F. (2000). Fungal pyranose oxidases: Occurrence, properties and biotechnical applications in carbohydrate chemistry. *Applied Microbiology and Biotechnology,* **54**, 727–740.

Gilad, R., Rabinovich, L., Yaron, S., Bayer, E.A., Lamed, R., Gilbert, H.J. and Shoham, Y. (2003). Cell, a noncellulosomal family 9 enzyme from *Clostridium thermocellum,* is a processive endoglucanase that degrades crystalline cellulose. *Journal of Bacteriology,* **185**, 391–398.

Gilbertson, R.L. (1980). Wood rotting fungi of North America, *Mycologia,* **72**, 1–49.

Gold, M.H., Youngs, H.L. and Gelpke, M.D.S. (2000). Manganese peroxidase. *Metal Ions in Biological Systems,* **37**, 559–586.

Goodell, B. (2003). Brown-rot fungal degradation of wood: Our evolving view. *ACS Symposium Series,* **845**, 97–118.

Gramss, G. (1997). Activity of oxidative enzymes in fungal mycelia from grassland and forest soils. *Journal of Basic Microbiology*, **37**, 407–423.
Gramss, G., Voigt, K.D. and Bergmann, H. (2005). Factors influencing water solubility and plant availability of mineral compounds in the tripartite fairy rings of *Marasmius oreades* (BOLT.: FR.) FR. *Journal of Basic Microbiology*, **45**, 41–54.
Green, F. and Clausen, C.A. (1999). Production of polygalacturonase and increase of longitudinal gas permeability in southern pine by brown-rot and white-rot fungi. *Holzforschung*, **53**, 563–568.
Guillén, F., Gómez-Toribio, V., Martíinez, M.J. and Martínez, A.T. (2000). Production of hydroxyl radical by the synergistic action of fungal laccase and aryl alcohol oxidase. *Archives of Biochemistry and Biophysics*, **383**, 142–147.
Guillén, F., Martínez, A.T, Martínez, M.J. and Evans, C.S. (1994). Hydrogen-peroxide-producing system of *Pleurotus eryngii* involving the extracellular enzyme aryl-alcohol oxidase. *Applied Microbiology and Biotechnology*, **41**, 465–470.
Guillén, F., Martínez, M.J., Munoz, C. and Martínez, A.T. (1997). Quinone redox cycling in the ligninolytic fungus *Pleurotus eryngii* leading to extracellular production of superoxide anion radical. *Archives of Biochemistry and Biophysics*, **339**, 190–199.
Gutiérrez, A., Caramelo, L., Prieto, A., Martínez, M.J. and Martínez, A.T. (1994). Anisaldehyde production and aryl-alcohol oxidase and dehydrogenase activities in ligninolytic fungi of the genus *Pleurotus*. *Applied and Environmental Microbiology*, **60**, 1783–1788.
Hammel, K.E., Kapich, A.N., Jensen, K.A. and Ryan, Z.C. (2002). Reactive oxygen species as agents of wood decay by fungi. *Enzyme and Microbial Technology*, **30**, 445–453.
Hatakka, A. (1994). Lignin-modifying enzymes from selected white-rot fungi: Production and role in lignin degradation. *FEMS Microbiology Reviews*, **13**, 125–135.
Hatakka, A. (2001). Biodegradation of lignin. In: *Lignin, Humic Substances and Coal* (M. Hofrichter and A. Steinbüchel, eds.), Vol. 1, pp. 129–179. Wiley-VCH, Weinheim.
Heinfling, A., Ruiz-Duenas, F.J., Martínez, M.J., Bergbauer, M., Szewzyk, U. and Martínez, A.T. (1998). A study on reducing substrates of manganese-oxidizing peroxidases from *Pleurotus eryngii* and *Bjerkandera adusta*. *FEBS Letters*, **428**, 141–146.
Henriksson, G., Johansson, G. and Pettersson, G. (2000). A critical review of cellobiose dehydrogenases. *Journal of Biotechnology*, **78**, 93–113.
Highley, T.L. (1988). Cellulolytic activity of brown-rot and white-rot fungi on solid media. *Holzforschung*, **42**, 211–216.
Hildén K.S., Mäkelä, M.R., Hakala, T.K., Hatakka, A. and Lundell, T. (2006). Expression on wood, molecular cloning and characterization of three lignin peroxidase (LiP) encoding genes of the white rot fungus *Phlebia radiata*. *Current Genetics*, **49**, 97–105.
Hodge, A., Alexander, I.J. and Gooday, G.W. (1995). Chitinolytic enzymes of pathogenic and ectomycorrhizal fungi. *Mycological Research*, **99**, 935–941.
Hofrichter, M. (2002). Review: Lignin conversion by manganese peroxidase (MnP). *Enzyme and Microbial Technology*, **30**, 454–466.
Hofrichter, M., Vares, T., Kalsi, M., Galkin, S., Scheibner, K., Fritsche, W. and Hatakka, A. (1999). Production of manganese peroxidase and organic acids and mineralization of C-14-labelled lignin (C-14-DHP) during solid-state fermentation of wheat straw with the white rot fungus *Nematoloma frowardii*. *Applied and Environmental Microbiology*, **65**, 1864–1870.
Hofrichter, M. and Ullrich, R. (2006). Heme-thiolate haloperoxidases: Versatile biocatalysts with biotechnological and environmental significance. *Applied Microbiology and Biotechnology*, **71**, 276–288.
Hon, D.N.-S. (1994). Cellulose: A random walk along its historical path. *Cellulose*, **1**, 1–25.
Hyde, S.M. and Wood, P.M. (1997). A mechanism for production of hydroxyl radicals by the brown-rot fungus *Coniophora puteana*: Fe(III) reduction by cellobiose dehydrogenase and Fe(II) oxidation at a distance from the hyphae. *Microbiology*, **143**, 259–266.
Jayani, R.S., Saxena, S. and Gupta, R. (2005). Microbial pectinolytic enzymes: A review. *Process Biochemistry*, **40**, 2931–2944.
Jensen, K.A., Bao, W.L., Kawai, S., Srebotnik, E. and Hammel, K.E. (1996). Manganese-dependent cleavage of nonphenolic lignin structures by *Ceriporiopsis subvermispora* in the absence of lignin peroxidase. *Applied and Environmental Microbiology*, **62**, 3679–3686.

Jensen, K.A, Ryan, Z.C., van den Wymelenberg, A., Cullen, D. and Hammel, K.E. (2002). An NADH: quinone oxidoreductase active during boidegradation by the brown rot basidiomycete *Gloeophyllum trabeum*. *Applied and Environmental Microbiology*, **68**, 2699–2703.

Johnson, D.G., Silva, M.C., Mackenzie, C.R., Schneider, H. and Fontana, J.D. (1989). Microbial degradation of hemicellulosics materials. *Applied Biochemistry and Biotechnology*, **20/21**, 245–258.

Kästner, M. and Hofrichter, M. (2001). Biodegradation of lignin. In: *Lignin, Humic Substances and Coal* (M. Hofrichter and A. Steinbüchel, eds.), Vol. 1, pp. 349–378. Wiley-VCH, Weinheim.

Katase, T. and Bollag, J.M. (1991). Transformation of trans-4-hydroxycinnamic acid by a laccase of the fungus *Trametes versicolor*: Its significance in humification. *Soil Science*, **151**, 291–296.

Kerem, Z., Bao, W. and Hammel, K.E. (1998). Rapid polyether cleavage via extracellular one-electron oxidation by a brown-rot basidiomycete. *Proceedings of the National Academy of Sciences of the United States of America*, **95**, 10373–10377.

Kerem, Z., Jensen, K.A., Hammel, K.E. (1999). Biodegradative mechanism of the brown rot basidiomycete *Gloeophyllum trabeum*: Evidence for an extracellular hydroquinone-driven fenton reaction. *FEBS Letters*, **446**, 49–54.

Kersten, P.J. (1990). Glyoxal oxidase of *Phanerochaete chrysosporium*: Its characterization and activation by lignin peroxidase. *Proceedings of the National Academy of Sciences of the United States of America*, **87**, 2936–2940.

Kirk, T.K. and Farrell, R.L. (1987). Enzymatic "combustion": The microbial degradation of lignin. *Annual Review of Microbiology*, **41**, 465–505.

Kremer, S.M. and Wood, P.M. (1992). Evidence that cellobiose oxidase from *Phanerochaete chrysosporium* is primarily an Fe(III) reductase: Kinetic comparison with neutrophil NADPH oxidase and yeast flavocytochrome-B2. *European Journal of Biochemistry*, **205**, 133–138.

Lee, K.H., Wi, S.G., Singh, A.P. and Kim, Y.S. (2004). Micromorphological characteristics of decayed wood and laccase produced by the brown-rot fungus *Coniophora puteana*. *Journal of Wood Science*, **50**, 281–284.

Leonowicz, A., Cho, N.S., Luterek, J., Wilkolazka, A., Wojtas-Wasilewska, M., Matuszewska, A., Hofrichter, M., Wesenberg, D. and Rogalski, J. (2001). Fungal laccase: Properties and activity on lignin. *Journal of Basic Microbiology*, **41**, 185–227.

Levin, L. and Forchiassin, F. (1998). Culture conditions for the production of pectinolytic enzymes by the white-rot fungus *Trametes trogii* on a laboratory scale. *Acta Biotechnologica*, **18**, 157–166.

Lindahl, B.D. and Finlay, R.D. (2006). Activities of chitinolytic enzymes during primary and secondary colonization of wood by basidiomycetous fungi. *New Phytologist*, **169**, 389–397.

Lorito M. (1998). Chitinolytic enzymes and their genes. In: *Trichoderma* and *Gliocladium* (G.E. Harman and C.P. Kubicek, eds.), Vol. 2, pp. 73–99. Taylor & Francis, London.

Luis, P., Kellner, H., Martin, F. and Buscot, F. (2005a). A molecular method to evaluate basidiomycete laccase gene expression in forest soils, *Geoderma*, **128**, 18–27.

Luis, P., Kellner, H., Zimdars, B., Langer, U., Martin, F. and Buscot, F. (2005b). Patchiness and spatial distribution of laccase genes of ectomycorrhizal, saprotrophic, and unknown basidiomycetes in the upper horizons of a mixed forest cambisol, *Microbial Ecology*, **50**, 570–579.

Luis, P., Walther, G., Kellner, H., Martin, F. and Buscot, F. (2004). Diversity of laccase genes from basidiomycetes in a forest soil. *Soil Biology and Biochemistry*, **36**, 1025–1036.

Lynd, L.R., Weimer, P.J., van Zyl, W.H. and Pretorius, I.S. (2002). Microbial cellulose utilization: Fundamentals and biotechnology. *Microbiology and Molecular Biology Reviews*, **66**, 506–577.

Martinez, A.T. (2002). Molecular biology and structure-function of lignin-degrading heme peroxidases. *Enzyme and Microbial Technology*, **30**, 425–444.

Martínez, M.J., Ruiz-Dueñas, F.J., Guillén, F. and Martínez, A.T. (1996). Purification and catalytic properties of two manganese-peroxidase isoenzymes from *Pleurotus eryngii*. *European Journal of Biochemistry*, **237**, 424–432.

Martínez, A.T., Speranza, M., Ruiz-Duenas, F.J., Ferreira, P., Guillén, F., Martínez, M.J., Gutiérrez, A. and del Rio C.J. (2005). Biodegradation of lignocellulosics: Microbial, chemical, and enzymatic aspects of the fungal attack of lignin. *International Microbiology*, **8**, 195–204.

Marzullo, L., Cannio, R., Giardina, P., Santini, M.T. and Sannia, G. (1995). Veratryl alcohol oxidase from *Pleurotus ostreatus* participates in lignin biodegradation and prevents polymerization of laccase-oxidized substrates. *Journal of Biological Chemistry*, **270**, 3823–3827.

Mayer, A.M. and Staples, R.C. (2002). Laccase: New functions for an old enzyme. *Phytochemistry,* **60**, 551–565.
Miyairi, K., Nishida, K., Takarae, M., Shikanai, Y. and Okuno, T. (2002). Purification and characterization of pectate lyase I and II from fungus *Stereum purpureum. Journal of Applied Glycosciences,* **49**, 99–106.
Miyairi, K., Okuno, T. and Sawai, K. (1985). Purification and properties of endopolygalacturonase I from *Stereum purpureum*, a factor inducing silver-leaf symptoms on apple trees. *Agricultural and Biological Chemistry,* **49**, 111–118.
Muheim, A., Waldner, R., Leisola, M.S.A. and Fiechter, A. (1990). An extracellular aryl alcohol oxidase from the white-rot fungus *Bjerkandera adusta. Enzyme and Microbial Technology,* **12**, 204–209.
Nilsson, T. and Ginns, J. (1979). Cellulolytic activity and the taxonomic position of selected brown-rot fungi. *Mycologia*, **71**, 170–177.
Ohtakara, A. (1988). Chitinase and beta-*N*-acetylhexosaminidase from *Pycnoporus cinnabarinus. Methods in Enzymology,* **161**, 462–470.
Rayner, A.D.M. and Boddy, L. (1988). Fungal communities in the decay of wood. *Advances in Microbial Ecology,* **10**, 115–166
Reading, N.S., Welch, K.D. and Aust, S.D. (2003). Free radical reactions of wood-degrading fungi. *ACS Symposium Series*, **845**, 16–31.
Rosenbrock, P., Buscot, F. and Munch, J.C. (1995). Fungal succession and changes in the fungal degradation potential during the initial stage of litter decomposition in a black alder Forest [*Alnus glutinosa* (L) Gaertn]. *European Journal of Soil Biology,* **31**, 1–11.
Ruiz-Duenas, F. J., Martínez, M.J. and Martínez, A.T. (1999). Molecular characterization of a novel peroxidase isolated from the ligninolytic fungus *Pleurotus eryngii. Molecular Microbiology,* **31**, 223–236.
Rumbold, K., Biely, P., Mastihubová, M., Gudelj, M., Gübitz, G., Robra, K.H. and Prior, B.A. (2003). Purification and properties of a feruloyl esterase involved in lignocellulose degradation by *Aureobasidium pullulans. Applied and Environmental Microbiology,* **69**, 5622–5626.
Schmidhalter, D.R. and Canevascini, G. (1992). Characterization of the cellulolytic enzyme system from the brown-rot fungus *Coniophora puteana. Applied Microbiology and Biotechnology,* **37**, 431–436.
Schmidhalter, D.R. and Canevascini, G. (1993a). Isolation and characterization of the cellobiose dehydrogenase from the brown-rot fungus *Coniophora puteana* (Schum ex Fr.) Karst. *Archives of Biochemistry and Biophysics*, **300**, 559–563.
Schmidhalter, D.R. and Canevascini, G. (1993b). Purification and characterization of two exocellobiohydrolases from the brown-rot fungus *Coniophora puteana* (Schum ex Fr.) Karst. *Archives of Biochemistry and Biophysics*, **300**, 551–558.
Schwarze, F.W.M.R., Engels, J. and Mattheck, C. (2000). *Fungal Strategies of Decay in Trees.* Springer, Berlin.
Sethuraman, A., Akin, D.E. and Eriksson, K.E.L. (1998). Plant-cell-wall-degrading enzymes produced by the white-rot fungus *Ceriporiopsis subvermispora. Biotechnology and Applied Biochemistry,* **27**, 37–47.
Shanley, N.A., Vandenbroek, L.A.M., Voragen, A.G.J. and Coughlan, M.P. (1993). Isolation and characterization of an endopolygalacturonase from *Phanerochaete chrysosporium. Journal of Biotechnology,* **28**, 179–197.
Shimada, M., Akamtsu, Y., Tokimatsu, T., Mii, K. and Hattori, T. (1997). Possible biochemical roles of oxalic acid as a low molecular weight compound involved in brown-rot and white-rot wood decays. *Journal of Biotechnology,* **53**, 103–113.
Shimazaki, T., Hara, S. and Sato, M. (1984). Production, purification and some properties of extracellular amylase of *Schizophyllum commune. Journal of Fermentation Technology,* **62**, 165–170.
Sjöström, E. (1993). *Wood Chemistry. Fundamentals and Applications*. Academic Press, San Diego, CA.
Steffen, K.T., Hatakka, A. and Hofrichter, M. (2002a). Degradation of humic acids by the litter-decomposing basidiomycete *Collybia dryophila. Applied and Environmental Microbiology,* **68**, 3442–3448.

Steffen, K.T., Hofrichter, M. and Hatakka, A. (2000). Mineralisation of C-14-labelled synthetic lignin and ligninolytic enzyme activities of litter-decomposing basidiomycetous fungi. *Applied Microbiology and Biotechnology*, **54**, 819–825.

Steffen, K.T., Hofrichter, M. and Hatakka, A. (2002b). Purification and characterization of manganese peroxidases from the litter-decomposing basidiomycetes *Agrocybe praecox* and *Stropharia coronilla*. *Enzyme and Microbial Technology*, **30**, 550–555.

Stephens, S.K., Elliott, T.J. and Wood, D.A. (1991). Extracellular enzyme mutants of *Coprinus bilanatus*. *Enzyme and Microbial Technology*, **13**, 976–981.

Subramaniyan, S. and Prema, P. (2002). Biotechnology of microbial xylanases: Enzymology, molecular biology, and application. *Critical Reviews in Biotechnology*, **22**, 33–64.

Sutherland, J.B. (1986). Regulation of cellulolytic activity in the white-rot fungus *Ischnoderma resinosum*. *Mycologia*, **78**, 52–55.

Špániková, S. and Biely, P. (2006). Glucuronoyl esterase: Novel carbohydrate esterase produced by *Schizophyllum commune*. *FEBS Letters*, **580**, 4597–4601.

Thurston, C.F. (1994). The structure and function of fungal laccases. *Microbiology*, **140**, 19–26.

Valášková, V. and Baldrian, P. (2006a). Degradation of cellulose and hemicelluloses by the brown rot fungus *Piptoporus betulinus*: Production of extracellular enzymes and characterization of the major cellulases. *Microbiology*, **152**, 3613–3622.

Valášková, V. and Baldrian, P. (2006b). Estimation of bound and free fractions of lignocellulose-degrading enzymes of wood-rotting fungi *Pleurotus ostreatus*, *Trametes versicolor* and *Piptoporus betulinus*. *Research in Microbiology*, **157**, 119–124.

van den Wymelenberg, A., Covert, S. and Cullen, D. (1993). Identification of the gene encoding the major cellobiohydrolase of the white rot fungus *Phanerochaete chrysosporium*. *Applied and Environmental Microbiology*, **59**, 3492–3494.

Westermark, U. and Eriksson, K.E. (1975). Purification and properties of cellobiose:quinone oxidoreductase from *Sporotrichum pulverulentum*. *Acta Chemica Scandinavica*, **B29**, 419–424.

Willför, S., Sundberg, A., Hemming, J. and Holmbom, B. (2005). Polysaccharides in some industrially important softwood species. *Wood Science and Technology*, **39**, 245–257.

Xavier-Santos, S., Carvalho, C.C., Bonfa, M., Silva, R., Capelari, M. and Gomes, E. (2004). Screening for pectinolytic activity of wood-rotting basidiomycetes and characterization of the enzymes. *Folia Microbiologica*, **49**, 46–52.

Zabel, R. and Morell J. (1992). *Wood Microbiology: Decay and its Prevention*. Academic Press, London.

Zavarzina, A.G., Leontievsky, A.A., Golovleva, L.A. and Trofimov, S.Y. (2004) Biotransformation of soil humic acids by blue laccase of *Panus tigrinus* 8/18: An *in vitro* study. *Soil Biology and Biochemistry*, **36**, 359–369.

Zorn, H., Bouws, H., Takenberg, M., Nimtz, M., Getzlaff, R., Breithaupt, D.E. and Berger, R.G. (2005). An extracellular carboxylesterase from the basidiomycete *Pleurotus sapidus* hydrolyses xanthophyll esters. *Biological Chemistry*, **386**, 435–440.

CHAPTER 3

Mycelial Networks: Nutrient Uptake, Translocation and Role in Ecosystems

Monika Tlalka, Dan Bebber, Peter R. Darrah and **Sarah C. Watkinson**

Contents

Abstract

Sequestration and release of carbon in the decomposer subsystem of the forest floor are key ecosystem functions of saprotrophic basidiomycetes. Both are the result of fungal metabolic processes commonly regulated by nitrogen availability. Saprotrophic basidiomycetes are the primary wood decomposer organisms in N-limited boreal and temperate forests. To

British Mycological Society Symposia Series
Published by Elsevier Ltd.

predict the ecosystem effects of atmospheric nitrogen deposition in forests, we need better understanding of the fungal adaptive responses that link carbon conversions to nitrogen dynamics. Some Basidiomycota clades have evolved the capacity to develop mass flow nutrient channels—cords—in response to nutrient context. Rapid bidirectional nutrient transport in cords enables these fungi to operate extensive and persistent resource supply networks, and to exploit the spatiotemporally uncoupled carbon and nitrogen resources of the upper soil horizons of the forest floor. Both the initiation of cord development and the velocity, direction and magnitude of amino acid flows within the corded network are regulated in response to the amounts and geometry of its carbon and nitrogen supply. Predictive models of fungal metabolic, physiological and developmental responses to environmental nitrogen, at cell and organism scale, can be realistically parameterized with data from experimentally manipulated saprotrophic mycelia in microcosms and ecosystems. In future, the whole-genome sequence of the basidiomycete cord-forming wood decay fungus *Serpula lacrymans* will provide a model for -omics technologies to dissect the extracellular and intracellular nutrient responses that underlie the functions of basidiomycete networks in ecosystems.

1. INTRODUCTION

1.1 The Last 25 Years

The British Mycological Symposium, volume *Decomposer Basidiomycetes* (Frankland *et al.*, 1982), identified lines of enquiry into basidiomycete biology and ecology that proved immensely fruitful in the following 25 years. Three chapters in that volume are still particularly relevant to current questions in basidiomycete nutrient uptake and translocation, and its importance in ecosystems. Quantitative investigation (Frankland, 1982) established the abundance of basidiomycete mycelium and its significance in wood decomposition in ecosystems. A conceptual basis was established with Swift's (1982) model of the decomposer subsystem of woodland ecosystems, in which basidiomycetes are primary in the detrital food web, utilizing lignocellulose in carbon/energy capture, and collecting mineral nutrients. The fundamental importance of population genetics was established as a determining factor in the physical structuring of basidiomycete mycelial individuals in populations, affecting their ecosystem function (Rayner and Todd, 1982).

1.2 Translocation through Mycelial Cords Underlies Key Ecosystem Functions of Basidiomycetes

Since 1982, we have come to recognize the importance of basidiomycete mycelia, particularly species that form conducting cords or rhizomorphs, as conduits for nutrients through soil (Cairney, 2005). Although fungal translocation had been well described physiologically, a breakthrough in appreciating its ecological significance was the publication (Simard *et al.*, 1997) of results of a field experiment that demonstrated net carbon flow between ectomycorrhizal tree seedlings in the field. This led the editors of *Nature* to coin the phrase 'wood-wide web' for the concept of an underground fungal resource network that could be tapped by plants, and could have important effects on aboveground productivity and diversity. The path of carbon and nitrogen through cords of ectomycorrhizal mycelium has been mapped using imaging techniques including real-time radioisotope digital imaging (Leake *et al.*, 2004). The notion of basidiomycete mycelium as an underground infrastructure for ecosystems, implied by the metaphor of the 'wood-wide web', is further supported by the finding that achlorophyllous plants, formerly believed to be parasitic on plants, are mycoheterotrophic (Leake, 2005). They 'cheat' the fungal wood-wide web, drawing off carbon/energy instead of supplying it by photosynthesis. Some mycoheterotrophs rely on specific mycelium for seedling establishment as well as for sustained growth (Bidartondo and Bruns, 2001). Thus, the basidiomycete networks of woodland contribute to plant diversity as well as productivity.

Fungal networks conduct mineral nutrients, including nitrogen, phosphorus, potassium, calcium and magnesium (Jentschke *et al.*, 2001), as well as carbon compounds—'the energy currency of the ecosystem' (Read, 1997). Cords channel bidirectional flows (Lindahl *et al.*, 2001), though the mechanism and pathways remain obscure. They extend the resource-gathering range of individual fungi, enabling individual mycelia to extend over areas of metres, inferred in woodland from the pattern of clonal distribution (Rayner and Todd, 1982; Burnett, 2003), and directly visible in cord-formers that grow over the surface of masonry in buildings (Ramsbottom, 1953; Money, 2007). Translocation of both carbon and mineral nutrients, including amino acids, has been shown in both ectomycorrhizal (He *et al.*, 2003) and saprotrophic cords (Tlalka *et al.*, 2007; Watkinson *et al.*, 2006; Bebber *et al.*, 2007). By mobilizing resources and information gathered over the whole network, and deploying accumulated resources at a localized site of biosynthesis such as sites of colonization, attack or sporophore construction, cords confer 'network-enabled capability' (Smith, 2006) on the mycelium. Because of these adaptive movements of their resources through mycelial networks, fungi contribute significantly to carbon sequestration by importing carbon into the soil from plants (Leake *et al.*, 2004) and from decomposing lignocellulosic litter (Frey *et al.*, 2003).

1.3 The Translocation Mechanism is likely to be Similar in Saprotrophic and Ectomycorrhizal Basidiomycetes

The functional overlap between symbiotic and saprotrophic networks has only recently been recognized (Read and Perez-Moreno, 2003). Species with both strategies for carbon capture from plants—utilizing dead remains and living roots—co-exist in forest organic horizons and litter layers (Tedersoo *et al.*, 2003). Experiments with isotope markers in microcosms show that nutrient exchange can occur between mycelium of decomposers and symbionts (e.g. Lindahl *et al.*, 1999).

Although species that exploit plant litter for minerals are in a minority of ectomycorrhiza, they are among the most abundant fungi in nitrogen-limited forests (Read and Perez-Moreno, 2003). Interestingly, it is in these podsolized boreal forest soils that the most marked spatial separation of saprotrophic and ectomycorrhizal basidiomycete species and their associated nutrient cycling functions has been reported (Lindahl *et al.*, 2007).

It is now appropriate to consider the physiology and underlying cellular machinery of both ectomycorrhizal and saprotrophic basidiomycete translocation together, because we know from molecular phylogeny that the difference between symbiotic and saprotrophic biology does not correspond with a fundamental taxonomic or phylogenetic separation in basidiomycetes. Ectomycorrhizal symbiosis is an evolutionarily unstable relationship that has been both gained and lost (Hibbett *et al.*, 2000). Recently published basidiomycete phylogenies (Hibbett *et al.*, 2000; Binder and Hibbett, 2002; Hibbett and Binder, 2002; Moncalvo *et al.*, 2002) indicate that the saprotrophs *Coniophora* and *Serpula* are in the Boletales clade with the ectomycorrhizal *Suillus* and *Paxillus*, while another pair of sister taxa is the ectomycorrhizal *Tomentella* and saprotrophic *Phanerochaete*. *Tomentella*, which produces resupinate sporophores on dead wood, has been shown only recently by molecular analysis of root tips, to be an ectomycorrhizal genus (Kõljalg *et al.*, 2000). Some mycorrhizal basidiomycetes that decompose litter also have the ability to obtain mineral nutrients saprotrophically. For example, *Paxillus involutus* can obtain nitrogen and phosphorus from tree leaf litter and return it, via mycelial cords connected to the ectomycorrhizal mantle, to roots of tree seedlings (Perez-Moreno and Read, 2000).

Translocation mechanisms in ectomycorrhizal basidiomycetes can be interpreted from experiments with similar networks in the more experimentally tractable saprotrophic basidiomycetes. Microcosm studies of the physiology and mycelial topology of foraging cord formers show similar context-cued cord development and network organization centred on rich unit carbon sources (roots or pieces of dead wood). These form the hubs from which mycelium grows out to sweep the surrounding area for further resources. Mineral nutrients are acquired by diffuse assimilatory mycelium which is both supplied by and feeds back into the mass flow cord system (Olsson *et al.*, 2002; Boddy and Jones, 2007).

1.4 Translocation might Function as a Homeostatic Mechanism in Fungi Adapted to Utilize Spatially and Temporally Separate Carbon and Nitrogen Resources

In boreal and temperate forests carbon and nitrogen sources may be spatially separate, and limiting nitrogen must be gathered where and when it becomes available, from soil, canopy through-fall, lea litter or other detritus. Biogeographical data (Read, 1991) show that plants with basidiomycete ectomycorrhizas predominate in nitrogen-limited forest biomes, and that ectomycorrhizal symbionts with cord-forming mycelium (Agerer, 2001) capable of acquiring nitrogen in organic form are common in such ecosystems (Read and Perez-Moreno, 2003). Basidiomycete nitrogen-scavenging networks adapt host plants to growing in slow-mineralizing acid soils, where almost all available nitrogen is held in the biota or in very slowly turning over tannic compounds in humus (Richter and Markewitz, 2001; Dighton, 2003; Bardgett, 2005). Wood decomposing basidiomycetes of such habitats are mainly brown rot fungi. These generate humus by a decay process that leaves a slow-turnover lignin-rich residue, with which nitrogen becomes bonded. Up to 30% of carbon sequestered in boreal forest is in this form (Ryvarden and Gilbertson, 1993). Nitrification hardly occurs (Davidson *et al.*, 1992; Stark and Hart, 1997), and when any available nitrogen is released it is immediately absorbed by competing plants and microbes (Kaye and Hart, 1997).

The rapid uptake and directed translocation of amino acid, observed by real-time radioisotope imaging in corded mycelia of *Phanerochaete velutina* (Tlalka *et al.*, 2002; Watkinson *et al.*, 2006; Bebber *et al.*, 2007), may be adaptive to competitive nitrogen capture by the fungal network (Ettema and Wardle, 2002). It confers spatial advantage for enabling the fungus to remove captured nitrogen to wood resources where competition may be less. Similar preferential phosphorus allocation to carbon-rich resources has been shown with isotope labelling in microcosms (Wells *et al.*, 1999).

A conceptual model of nitrogen dynamics in such environments (Lindahl *et al.*, 2002) places fungal mycelium as a central controlling network for all the major mineral nutrient fluxes between soil, ectomycorrhizal and saprotrophic mycelial networks. The conducting activity of such mycelial networks is critical to exploiting their heterogeneous environment, in which, in the authors' words, 'carbon and nitrogen resources are spatially uncoupled'. The implication is that fungi in these habitats are adapted to gather carbon and energy from one part of their mycelial networks and nitrogen from another. By translocation within the network, a mycelium is enabled to reconcile these two essential resources for biosynthesis. Fungal nitrogen translocation can have a rate-determining effect on ecosystem carbon flux. Nitrogen import through mycelium to N-poor lignocellulose carbon resources results in faster decomposition rates (Beare *et al.*, 1992; Frey *et al.*, 2000). This occurs as a side-effect of the fungal adaptation which enables the individual mycelium to import nitrogen scavenged from soil into newly acquired carbon resource in the form of plant litter at the soil surface. In other words, fungi are adapted by natural selection, not to perform ecosystem services, but to maximize their own fitness. We do not know how translocation is regulated for metabolic homeostasis, but it evidently requires a coordinated network-wide response to differences in internal levels of different nutrients. A requirement to sense and respond to spatial differences in internal carbon and nitrogen levels offers an explanation for the widespread morphogenetic sensitivity of fungi to intracellular C:N ratio (Watkinson, 1999).

Cords developing secondarily in an established mycelial network, as occurs in *Serpula lacrymans* and *Coniophora puteana*, form by linear aggregation and differentiation of hyphae around a central low resistance channel, with surrounding tissue of which the anatomy and function is still obscure, embedded in accretions of extracellular matrix material (Jennings and Watkinson, 1982). Cords are elicited under conditions of N limitation or when hyphae themselves become the main source of nutrients. On uniformly nutrient-rich media, and when assimilating or invading cellulosic resources, mycelium takes the form of separate hyphae. Under nitrogen limitation in initially defined media, and in mycelium connecting two separate cellulose resources, cords differentiate. Nutrient perfusion prevents the cord development. Physiological evidence points to intracellular amino acid as a cord-inducing signal (Watkinson, 1999). Nitrate as sole nitrogen source induces early autolysis and cord formation, and suppresses secondary metabolism.

Metabolic fungal adaptations to nitrogen limitation include amino acid uptake transporters with a range of substrate preferences and affinities for scavenging uptake from dilute solutions or nitrogen-rich resource. Global metabolic regulation of transcription of genes is also involved in nitrogen assimilation or dissimilation, to adapt to local or transient nitrogen starvation or repletion (Caddick, 2004).

2. INCORPORATING FUNGAL PROCESSES IN PREDICTIVE MODELS OF FOREST FLOOR CARBON AND NITROGEN DYNAMICS

2.1 Fungal Networks Acting as Nitrogen Reservoirs and Distribution Systems might Control the Rates of Forest Floor Carbon Fluxes

Basidiomycetes are emerging as potentially of central importance in ecosystem, and even global, carbon cycling. As the main decomposers (Boddy and Watkinson, 1995) and mutualistic plant symbionts (Leake *et al.*, 2004) in N-limited soils, where the largest proportion of terrestrial carbon is sequestered (Post *et al.*, 1982), they may act as gatekeepers of carbon fluxes. Their monopoly of available nitrogen would put them in control of inputs of limiting nitrogen to plants, and to the decomposer subsystem. It is thus appropriate to consider how fungal processes might be incorporated into integrated models of terrestrial carbon cycling (Falkowski *et al.*, 2000). Fungal mycelium has the physiological potential to act as both an expandable reservoir and a distribution system for elevated nitrogen inputs (Watkinson *et al.*, 2006), and may account for the responsive nitrogen absorption of the forest floor (Currie, 1999). We need better understanding of the coordinate regulation of carbon and nitrogen metabolism in fungi, to predict the effects of anthropogenic nitrogen deposition on soil carbon content (Bardgett, 2005).

Theoretical models of responses to environmental change are being developed to cover the range of scales, from micrometres to metres, that basidiomycete ectomycorrhizal fungi occupy. A critique of some models that have been developed suggests that, at organism scale, functional equilibrium and stoichiometric models are relatively easy to parameterize with accessible realistic data (Collins Johnson *et al.*, 2006). Such models might thus have good predictive value to inform management.

Below we describe some approaches that we are using to collect ecologically relevant data on the responses of saprotrophic cord-forming wood decay basidiomycetes to their carbon and nitrogen resources, ranging from cell to ecosystem.

2.2 Experimental Approaches using Saprotrophic Cord forming Wood Decay Basidiomycetes to Investigate Effects of Carbon and Nitrogen Distribution on Cord Development and Translocation

2.2.1 Soil Nitrogen and Network Topology in Microcosms

Network topology has been found by mathematical network analysis to be adapted to exploit distant resources, and to survive interactions with other organisms that may delete parts of the network (Fricker *et al.*, in press; Chapter 1). Network topology is species-specific in cord-forming wood decay fungi, reflecting short- and long-range foraging strategy, in other words, preference for small or large carbon resources (Boddy, 1999; Boddy and Jones, 2007). In the long-range foraging species *P. velutina*, individuals established in large dead logs extend corded mycelium into the surrounding soil, with diffuse fans of assimilating hyphae at their extremities (Boddy and Jones, 2007).

While it is the carbon resource that acts as the hub for extension growth, the extension rate and morphology of outgrowing mycelium are modified by the nitrogen environment. The ratio of carbon to nitrogen in the substratum determines the extent of cording of the mycelium, with nitrogen limitation acting as a cue for cord development in several species of saprotrophic cord-forming wood decay fungi. The morphogenetic effect of nitrogen content and C:N ratio have been investigated in axenic defined nutrient agar medium in *P. velutina*, *C. puteana* (common in European woodland) and *S. lacrymans*, an inhabitant of calcareous pine forest, from which the aggressive dry rot fungus of buildings has arisen (Kauserud *et al.*, 2004). Interestingly, the switch to cord development occurred at similar carbon:nitrogen ratio in all three species. Factorial experiments demonstrated that carbon and nitrogen contents interacted and higher nitrogen was limiting at higher carbon. The effect on morphogenesis was different from that on biomass. Cord development was quantitatively related to nitrogen limitation, while biomass increased with both nitrogen and carbon.

Using mycelial C:N ratio to cue the development of nutrient conduits might be an adaptation to foraging under nitrogen limitation. The organism must prioritize carbon capture, as it cannot grow at all without an energy source. However, once carbon is sufficient, nitrogen is necessary for protein synthesis, for further growth and enzyme synthesis to exploit the wood carbon base. Mycelial topology is expected to be adapted to maximize the success of nitrogen scavenging in the nitrogen-limited environment. The developmental options (Boddy and Jones, 2007) are to toggle between diffuse, assimilatory mycelial growth and the development of hydrophobic (Olsson *et al.*, 2002) mass flow channels resistant to attack, which extend the scavenging range by supplying carbon to distant foraging fronts to capture nitrogen over a wider area.

Subverting the link between amino acid uptake into the cell and developmental response can alter foraging network topology and prevent the organism from capturing fresh carbon resources (Dobson *et al.*, 1993). This paramorphogenetic effect of α-aminoisobutyric acid (AIB, a non-metabolizable amino acid analogue) is exploited in its use to control the spread of dry rot from infected to uninfected timber elements in buildings. Mycelium charged with a high concentration of AIB extends only very slowly, and remains as a localized symmetrical cushion of hyphae at the original carbon resource base. This morphology is typical of juvenile mycelium on a rich nutrient resource (Tlalka *et al.*, 2007).

To investigate the effects of soil nitrogen pollution on woodland floor mycelial networks, we grew *P. velutina*, *C. puteana* and *S. lacrymans* from wood blocks, previously colonized for 2 or 7 months, over moist sand, with and without ammonium nitrate. Mycelia grown from the more decayed (less C-rich) blocks into nitrogen-free sand extended in mainly corded form, with diffuse assimilating mycelium limited to fans at the cord tips. Nitrogen (10 mg ammonium nitrate in 100 ml water added to 500 g sand) induced predominantly diffuse mycelium with few cords, initiated later. With nitrogen, the mycelium covered the plate rapidly and completely, compared with nitrogen-free sand where half the plate remained uncovered after 4 weeks (Figure 1). The concentration of nitrogen was

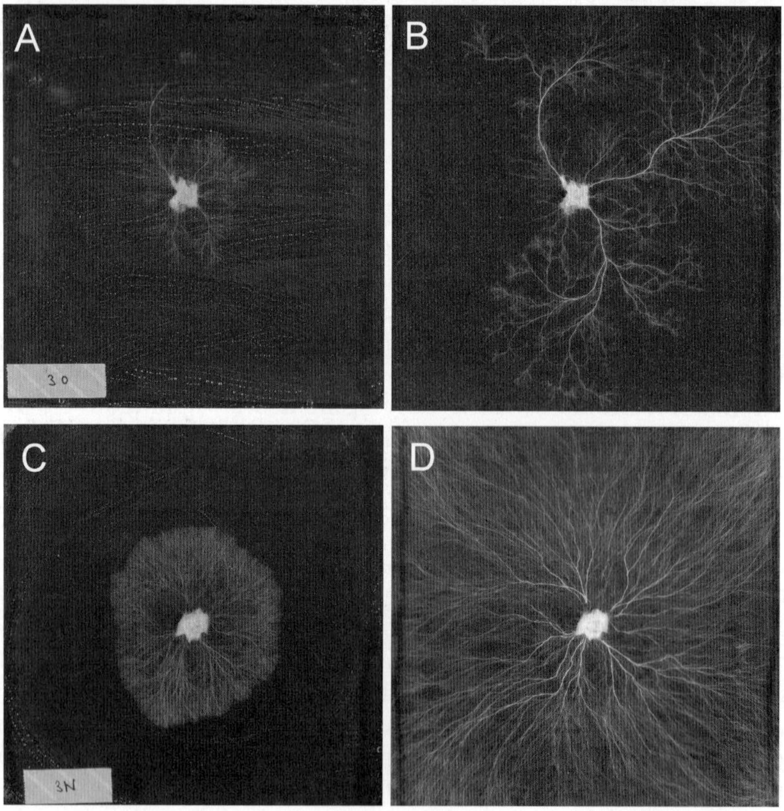

Figure 1 Balance between corded and diffuse mycelium growing from colonized wood blocks over sand. The 24 cm square dishes contained 500 g sand, to which was added either 100 ml deionized water or 10 mg of ammonium nitrate in 100 ml deionized water. Mycelia were grown from beech wood blocks that had been colonized for 7 months, and were photographed at 2 and 4 weeks. (A) water, 2 weeks; (B) water, 4 weeks; (C) ammonium nitrate, 2 weeks; (D) ammonium nitrate, 4 weeks.

approximately 10-fold that recorded at the NERC ECN site at Wytham Wood, Oxfordshire. However, the results indicated a potential effect of eutrophication of woodland soils on saprotrophic fungal networks, which requires further investigation.

2.2.2 Nitrogen Limitation and Cord Initiation in Uniform Defined Media

Defined axenic culture is remote from ecological conditions, but it can parameterize a functional model to predict trigger points for N-induced switches in mycelial network topology. Using baseline data from defined media, carbon and nitrogen nutrient status of mycelium, and its likely effects on development in the natural habitat might be predicted. Figure 2 shows the relationship between nitrogen concentration in defined uniform nutrient agar media and biomass at 1 week of *C. puteana* mycelium, which was chosen for this experiment because it grows in culture with a relatively symmetrical geometry, compared with the highly variable morphology of *P. velutina*. Morphology was recorded as corded or uncorded. The curve of biomass against log nitrogen concentration was biphasic, suggesting a switch from high affinity, scavenging cell membrane transporters to a lower affinity system in the more nitrogen-rich environment. Interestingly, the discontinuity in the curve coincided with a clear switch, constant across three replicates, between corded and diffuse hyphae.

2.2.3 Photon Counting Scintillation Imaging of Changes in the Direction of Amino Acid Flow in Mycelium, Induced by Carbon Resource Capture

Nitrogen translocation into local carbon resources has been observed in the field (Frey *et al.*, 2000) and at organism scale in microcosms (Watkinson *et al.*, 1981). In

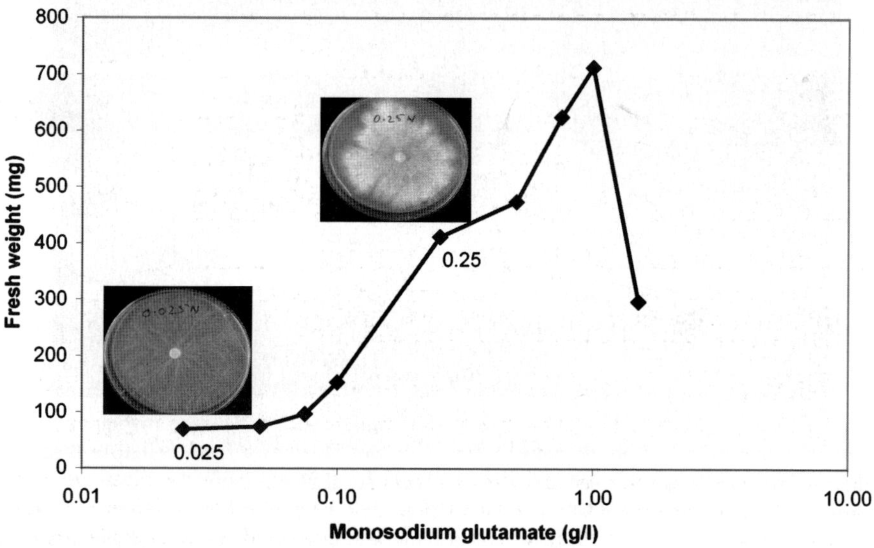

Figure 2 Biomass increase with nitrogen supply, accompanied by suppression of cord development in *Coniophora puteana* at nitrogen content over 0.25 mg l^{-1}. Basal medium, g l^{-1}: sucrose, 20; KH_2PO_4, 1; $MgSO_4 \cdot 7H_2O$, 0.5; $FeSO_4 \cdot 7H_2O$, 0.01.

both cases, nitrogen import through mycelium enhanced the rate of cellulose decay. We do not know the time scale or sequence of cellular events leading from colonization of a localized insoluble cellulose resource to the induction of homeostatic amino acid translocation. By using photon counting scintillation imaging (PCSI) (Tlalka *et al.*, 2002, 2007; Bebber *et al.*, 2007; Fricker *et al.*, 2007), we can observe the effect of a fresh cellulose resource on the distribution of the free translocated amino acid pool in a mycelium, in real time. The microcosm used for the PCSI image (Figure 3) was designed to mimic the advancing edge of a mycelium in the forest floor, meeting and colonizing a fresh fragment of plant litter. A cellulose filter paper disc was placed near the advancing margin of a *P. velutina* mycelium in which the free amino acid pool had been labelled by prior

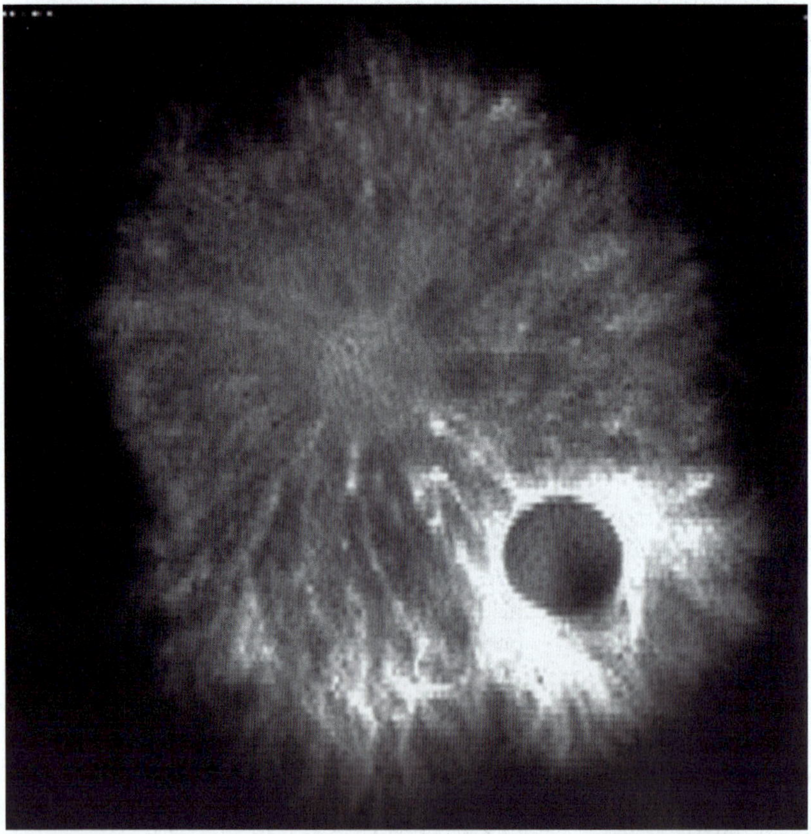

Figure 3 Photon counting scintillation image of a *Phanerochaete velutina* mycelium, to show distribution of translocated amino acid following colonization of a fresh cellulose resource. mycelium was grown over a scintillation screen from an agar inoculum disc to which ^{14}C-labelled α-aminoisobutyric (AIB) acid at subtoxic concentration (10 μl, 0.9 mmol) had been added. AIB is taken up and translocated but not metabolized, so remains unchanged and acts as a marker for movement of the free amino acid pool through the mycelium. The radiolabel, which was initially evenly distributed throughout the system, was translocated preferentially towards the new carbon resource.

addition of ^{14}C-AIB to the fungus. Within a few hours, the amino acid distribution throughout the colony changed, becoming asymmetrical as translocation was directed towards the added fresh resource.

The mechanism by which the organism perceives the local carbon resource is not known, but appears to be related to a nutrient signal because glass fibre discs, with similar hydrophilic physical characteristics to cellulose but no carbon content, did not induce this effect.

2.2.4 Capacity of Mycelium to Respond to Carbon and Nitrogen Resource Asymmetries by Translocation

The hypothesis that translocation operates to equilibrate separately acquired carbon and nitrogen for metabolic homeostasis was investigated by a series of experiments with the tractably symmetrical fungi *C. puteana* and *S. lacrymans*, both in Boletales, Coniophorales. Limiting nitrogen levels were established by testing each species over ranges of defined media. To mimic the effect of a network encountering a localized carbon or nutrient source and thus experiencing a patch of nutrient disequilibrium, cultures were grown on split plates with separate N and C sources.

When *C. puteana* mycelium, pre-grown on a permeable cellulose membrane over nitrogen-free agar medium, was transferred to a split plate with uniform high carbon as sucrose on one side but only nitrogen on the other, there was a striking differentiation of behaviour on each side of the plate. Mycelial extension, accompanied by cord development, accelerated on the N-limited side, apparently supplied with nitrogen by the cords that developed across the carbon-only medium. Extension ceased on the N-rich side, and cord development did not occur. However, biomass increased threefold, and metabolism appeared to alter, the mycelium releasing a dark brown pigment into the medium, presumably as a result of the onset of secondary metabolism.

In a short-term experiment, the pattern of equilibration of translocated amino acid between N-poor and N-rich sides of split plates was measured by scintillation counting of radiolabelled AIB added centrally to the inoculum. At 6 h there was much greater, but highly variable, allocation to the N-poor side, accompanied by a slight increase in biomass on that side. However, later, the distribution of AIB became much more uniform, and there was no consistent preferential allocation after 12 h. This could reflect an initial preferential allocation to carbon-rich cells, followed by futile circulation of the non-metabolized amino acid which remained in the mycelium and was not unloaded for biosynthesis.

2.2.5 Amino Acid Translocation between Compatible Individuals following Fusion

The discovery by Bebber (unpublished data) that fusion between two mycelial individuals is quickly followed by rapid amino acid flow from one to another suggests that equilibration of nitrogen throughout fused networks might be an important adaptive feature of foraging networks. This would enable formerly separate individuals to take advantage of others' foraging successes. Glass and Kaneko (2003) have drawn attention to the homeostatic and resource-sharing potential of fusion between individual networks, including the evolutionary significance of the limits placed by the vegetative incompatibility loci of fungi on the frequency of fusions between the individuals of a population.

In this context it is interesting that the cord-forming wood decay species *S. lacrymans* has been found to have unusually few compatibility types (Kauserud *et al.*, 2006), so that chance-met individuals are likely to be able to fuse and share resources. Fungal vegetative compatibility genes that limit vegetative fusions between individuals in the field, by determining the number of vegetative compatibility types in the population of a species (Burnett, 2003), are believed to have evolved in response to selection against transmission of cytoplasmic viral infection when mycelia fuse (Glass and Kaneko, 2003). However, under resource heterogeneity, an individual that can link with a neighbour to extend joint network capacity acquiring distant limiting resources might be at a selective advantage. This is particularly likely in woodland where long-lived networks may frequently be temporarily fragmented under climatic stress, such as summer drought. Rapid reconnection to restore network capability might put the species at a competitive advantage.

Compatible and incompatible mycelia were grown together to compare the extent of amino acid translocation in fused and unfused networks (Figure 4). The incompatible dikaryons S7 and S16 of *S. lacrymans*, characterized as belonging to the A and D vegetative compatibility types, respectively (Kauserud *et al.*, 2006), were grown in sand microcosms from paired colonized wood blocks. The inhibitory translocated amino acid AIB was locally applied to the margin of one individual when half-grown. The effect of AIB at very high concentration was to slow extension growth and induce a dense, regular margin without cords. In fused mycelia grown from paired blocks both colonized by S7, the effect of locally added amino acid took effect evenly throughout the system, while the incompatible mycelia grown from paired blocks colonized by S7 and S16 grew alongside each other but did not fuse, and inhibition was limited to the network containing AIB, which was subsequently engulfed by the uninhibited partner network (Figure 4).

2.2.6 Resource Sharing by Fusion may Enhance Biomass Production on Asymmetric Carbon and Nitrogen Resources

A short-term experiment suggested that homeostatic translocation may occur, and result in greater fitness, when compatible individuals with complementary nutrient requirements fuse. Compatible and incompatible pairings of S7 and S16 were set up on split plates. Most of the available carbon was on one side of the plate, and all the nitrogen on the other. Biomass on each side, and on control uniform media with either or both carbon and nitrogen, was recorded as a measure of fitness. Cords formed to bridge the gap between the sides of the plate in the compatible pairings only. After 1 week, connected mycelium on the carbon-only side weighed significantly more than unconnected mycelium (Figure 5). It would be interesting to investigate this effect further with realistic time scales and resources.

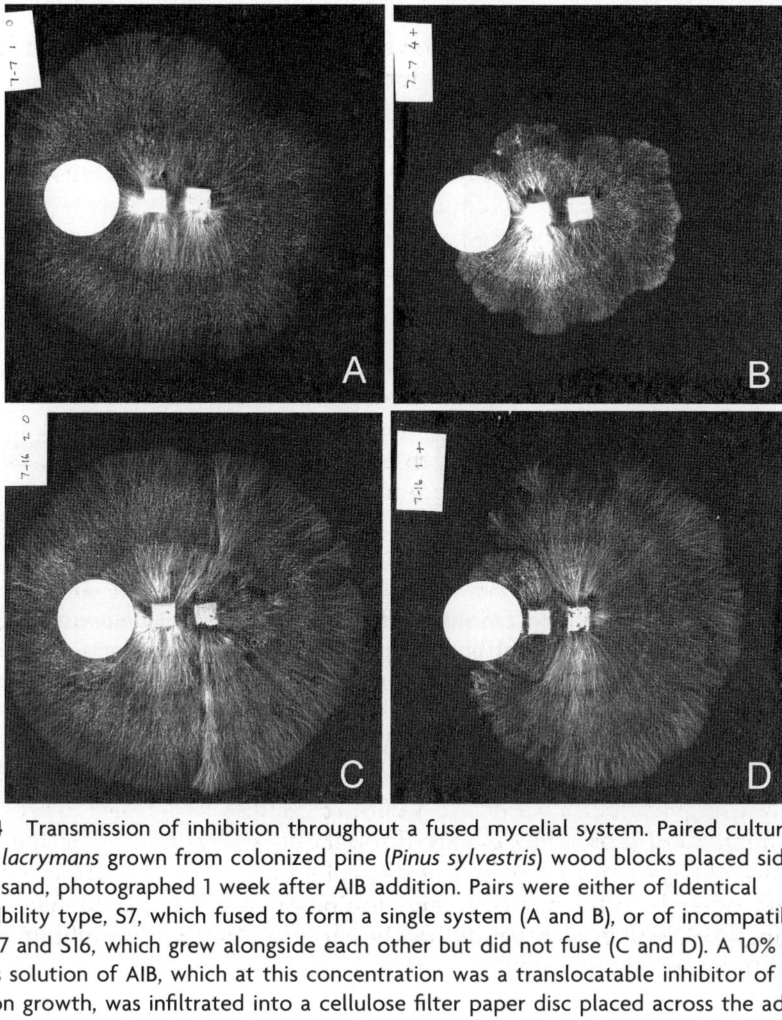

Figure 4 Transmission of inhibition throughout a fused mycelial system. Paired cultures of *Serpula lacrymans* grown from colonized pine (*Pinus sylvestris*) wood blocks placed side by side on sand, photographed 1 week after AIB addition. Pairs were either of Identical compatibility type, S7, which fused to form a single system (A and B), or of incompatible types, S7 and S16, which grew alongside each other but did not fuse (C and D). A 10% (w/v) aqueous solution of AIB, which at this concentration was a translocatable inhibitor of extension growth, was infiltrated into a cellulose filter paper disc placed across the advancing margin of one mycelium in 2 and 4; water alone was added in 1 and 3 incompatible mycelia from wood blocks colonized by S7 and S16 formed a barrage at the point of contact and grew beside each other at similar rates. The inhibition of extension caused by the presence of AIB in the translocated free amino acid pool was transmitted between compatible, but not incompatible, pairs of mycelia, indicating amino acid sharing on fusion between compatible individuals.

3. FUTURE PROSPECTS

3.1 '-omics' Technologies, from Cell to Field

Future experiments on *S. lacrymans*, making use of the whole-genome sequence made available in 2008, would enable adaptive developmental and physiological

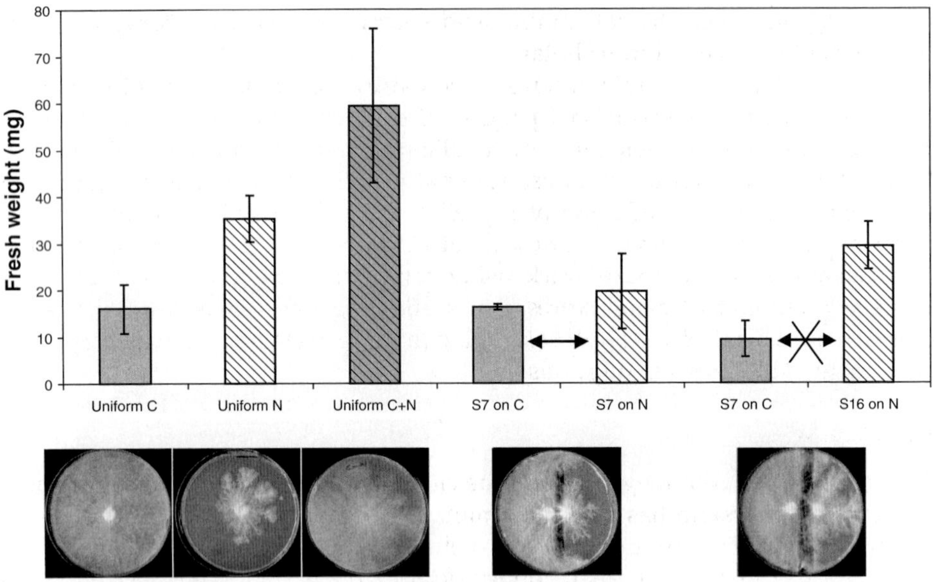

Figure 5 Evidence that compatible individuals, each of which had access to either a carbon or a nitrogen source, but not both, can share nutrients and enhance joint biomass by fusion, while incompatible individuals cannot. Split plates containing, on one side only nitrogen, as $1\,g\,l^{-1}$ monosodium glutamate, on the other only carbon, as $40\,g\,l^{-1}$ sucrose, were inoculated with paired mycelia of *Serpula lacrymans*, using either compatible (S7 with S7) or incompatible (S7 with S16) pairs. Plates were inoculated with small inoculum discs placed next to, and on either side of, an air gap. Biomass is in mg fresh wt, means of three replicates, 1 week after inoculation. cords formed to link compatible mycelia while incompatible ones formed a barrage.

responses described here to be investigated at metabolic, cell and molecular levels. An intriguing possibility is that fungi optimized for developmental response to nutrients will prove to be enlightening models for dissecting the cellular links between carbon and nitrogen nutrient sensing and cell development. Recently huge advances have been made in our understanding of nitrogen and carbon nutrient sensing and developmental pathways in yeasts (Cooper, 2004). In *Saccharomyces cerevisiae* (Roosen *et al.*, 2005) the protein kinase Sch9 is a glucose and nitrogen-responsive regulator involved in nutrient signalling, that affects cell size and longevity. It controls the transcription of clusters of genes encoding proteins with functions in proteolysis, stress response and filamentous and invasive growth forms. Together with the transcription factor TOR, it is believed to compose a nutritional integrator system controlling a morphological switch that induces invasive pseudohyphal growth under N-limitation. Amino acids are also sensed by Ssy1, a non-functional homologue of a membrane transporter, believed to undergo conformational change on amino acid binding. Evidence suggests that intracellular amino acids inhibit the sensing of extracellular amino acids, of interest if this system is conserved in cord-forming wood decay fungi in which the sole amino acid source for mycelium foraging over a non-nutrient surface is intracellular.

In ecosystems and microcosms, microarray technology would make it possible to access transcriptional profiles of different foraging phases, and link the expression of gene clusters to metabolic responses. Kasuga *et al.* (2005) have demonstrated the validity of transcriptional profiling under realistic conditions as an ecological tool. Prediction of mycelial network behaviour would be facilitated by rapidly assessing, *in situ*, the nitrogen status of mycelium in its forest floor habitat. In preliminary work using sand/wood microcosms, spatial and temporal mapping of gene expression is allowing investigation of differential expressions in parts of a *P. velutina* foraging mycelium acting as sources and sinks for nitrogen (Tlalka *et al.*, unpublished).

3.2 Imaging

The power of live cell imaging would be enhanced by reporter constructs, once a transformation system has been determined. A promising genetic system is the expression of the intracellular serine proteinase (Watkinson *et al.*, 2001) which has already been cloned. Patterns of pulsation may have a role in network coordination (Tlalka *et al.*, 2003, 2007; Fricker *et al.*, 2007). The expression of genes for specific metabolic steps, combined with analysis of pulse pattern, may help to illuminate the processes that underlie signal reception, propagation and response. The mechanism of pulsation is not understood. Pulsation was most marked in translocating cords (Bebber *et al.*, 2007), and did not necessarily accompany rhythmic hyphal growth. *S. lacrymans* was not among the species with pronounced pulsation in transport, despite the rhythmic growth evident in Figure 4, where two successive waves of denser growth appeared to have occurred during the experiment.

3.3 Modelling

Modelling approaches, such as those described by Fricker *et al.* (Chapter 1) and Davidson (2006), are being developed to describe and predict mycelial behaviour under realistic parameters. These cover a range of scales. At cell scale, individual hyphae have been shown, by theoretical models parameterized from confocal imaging (using fluorescence recovery after photobleaching, 'FRAP', techniques), to be capable of supplying tip growth by vacuolar amino acid transport from cytoplasm several tens of micrometres behind the tip (Darrah *et al.*, 2006), and by diffusion facilitated by movements of the dynamic pleiomorphic vacuolar system (Cole *et al.*, 1998). Fungal vacuoles store N-rich amino acids including arginine (Klionsky *et al.*, 1990), and there is evidence from arbuscular mycorrhizal fungi that spatially differentiated metabolism in hyphae underlies vacuolar amino acid transport between soil and root (Govindarajulu *et al.*, 2005). Organism level network development is also being analysed with agent-based models that contribute the power of modern computing to explore and predict developmental responses (Meskauskas *et al.*, 2004; Bebber *et al.*, 2007; Chapter 1) over a wider parameter space than would ever be accessible by experimental approaches.

Mathematical analysis of networks in microcosms is revealing the way on which the conductivity conferred by cords is arranged within the network to maximise different adaptive attributes, including robustness to attack and the acquisition of distant resources (Chapter 1).

3.4 Ecosystem Function

The effects of elevated nitrogen on the decay of woody litter have usually been measured by adding nitrogen compounds to soil, and the results are typically variable (Knorr *et al.*, 2005), presumably because of the chemical and physical complexities of the soil–fungus–wood path, and differential effects on component processes involved in net wood decay. To analyse nitrogen effects on wood decay in the field, it would be more informative to measure the effect of the nitrogen content of decomposer mycelium on wood mass loss. *P. velutina* is a suitable organism for this because it can be established in woodland soils from pre-colonized wood blocks placed on the woodland floor (Boddy, personal communication). In this way, large numbers of even-aged mycelia from identical carbon resources can be set up for statistically valid assessment of nitrogen partitioning between mycelium, underlying soil and colonized wood. Using this approach, tests are in progress for a predicted correlation between mycelial intracellular nitrogen and wood decay.

ACKNOWLEDGEMENTS

Most of the work described here was funded by NERC Grant NER/A/S/2002/00882 and BBSRC Grant 43/P19284. We thank Lynne Boddy for detailed protocols for soil-based microcosms, and the *P. velutina* isolate used in these experiments. Strains S7 and S16 of *S. lacrymans* were provided by Olaf Schmidt, University of Hamburg. Dan Eastwood (University of Warwick) works with Monika Tlalka on gene expression in *P. velutina*. We are grateful for photography by John Baker, and technical assistance from Caroline O'Brien. Our thanks to Mark Fricker for helpful discussion.

REFERENCES

Agerer, R. (2001). Exploration types of ectomycorrhizae. *Mycorrhiza*, **11**, 107–114.

Bardgett, R.D. (2005). *The Biology of Soil: A Community and Ecosystem Approach*. Oxford University Press, Oxford.

Beare, M.H., Parmelee, R.W., Hendrix, P.F., Cheng, W., Coleman, D. and Crossley, D.A. (1992). Microbial and faunal interactions and effects on litter nitrogen and decomposition in ecosystems. *Ecological Monographs*, **62**, 569–591.

Bebber, D.P., Tlalka, M., Hynes, J., Darrah, P.R., Ashford, A.E., Watkinson, S.C., Boddy, L. and Fricker, M.D. (2007). Imaging complex nutrient dynamics in mycelial networks. In: *Fungi in the Environment* (G.M. Gadd, P. Dyer and S.C. Watkinson, eds.), pp. 3–21. Cambridge University Press, Cambridge.

Bidartondo, M.I. and Bruns, T.D. (2001). Extreme specificity in epiparasitic Monotropoideae (Ericaceae): Widespread phylogenetic and geographical structure. *Molecular Ecology*, **10**, 2285–2295.

Binder, M. and Hibbet, D.S. (2002). Higher-level phylogenetic relationships of homobasidiomycetes (mushroom-forming fungi) inferred from four rDNA regions. *Molecular Phylogeny and Evolution*, **22**, 76–90.

Boddy, L. (1999). Saprotrophic cord-forming fungi: Meeting the challenge of heterogeneous environments. *Mycologia*, **91**, 13–32.

Boddy, L. and Jones, H.T. (2007). Mycelial responses in heterogeneous environments: Parallels with macroorganisms. In: *Fungi in the Environment* (G.M. Gadd, S.C. Watkinson and P. Dyer, eds.), pp. 112–140. Cambridge University Press, Cambridge.

Boddy, L. and Watkinson, S.C. (1995). Wood decomposition, higher fungi, and their role in nutrient redistribution. *Canadian Journal of Botany*, **73**(Suppl. 1), S1377–S1383.

Burnett, J. (2003). *Fungal Populations and Species*. Oxford University Press, Oxford.

Caddick, M.X. (2004). Nitrogen regulation in mycelial fungi. In: *The Mycota: Biochemistry and Molecular Biology III* (R. Bramble and G.A Marzluf, eds.), 2nd edn., pp. 349–368. Springer, Berlin.

Cairney, J.W.G. (2005). Basidiomycete mycelia in forest soils: Dimensions, dynamics and roles in nutrient distribution. *Mycological Research*, **109**, 7–20.

Cole, L., Orlovich, D.A. and Ashford, A.E. (1998). Structure, function, and motility of vacuoles in filamentous fungi. *Fungal Genetics and Biology*, **24**, 86–100.

Collins Johnson, N., Hoeksema, J.D., Bever, J.D., Bala Chaudhury, V., Gehring, C., Klironomos, J., Koide, R., Miller, M.R., Moore, J., Moutoglis, P., Schwartz, M., Simards, S., Swenson, W., Umbanhowar, J., Wilson, G. and Zabinski, C. (2006). From Lilliput to Brobdingnag: Extending models of mycorrhizal function across scales. *Bioscience*, **56**, 889–900.

Cooper, T.G. (2004). Integrated regulation of the nitrogen–carbon interface. In: *Topics in Current Genetics 7, Nutrient-Induced Responses in Eukaryotic Cells* (J. Winderickx and P.M. Taylor, eds.), pp. 225–257. Springer, Berlin.

Currie, W.S. (1999). The responsive C and N biogeochemistry of the temperate forest floor. *Trends in Ecology and Evolution*, **14**, 316–320.

Darrah, P.R., Tlalka, M., Ashford, A.E., Watkinson, S.C. and Fricker, M.D. (2006). The vacuole system is a significant intracellular pathway for longitudinal solute transport in basidiomycete fungi. *Eukaryotic Cell*, **5**, 1111–1125.

Davidson, F.A. (2006). Mathematical modeling of the form and function of fungal mycelia. In: *Fungi in the Environment* (G.M. Gadd, S.C. Watkinson and P. Dyer, eds.), pp. 57–73. Cambridge University Press, Cambridge.

Davidson, E.A., Hart, S.C. and Firestone, K. (1992). Internal cycling of nitrate in soils of a mature coniferous forest. *Ecology*, **73**, 1148–1156.

Dighton, J. (2003). *The Role of Fungi in Ecosystem Processes*. Marcel Dekker, New York, NY.

Dobson, J., Power, J.M., Singh, J. and Watkinson, S.C. (1993). The effectiveness of 2-aminoisobutyric acid as a translocatable fungistatic agent for the remedial treatment of dry rot caused by *Serpula lacrymans* in buildings. *International Biodeterioration and Biodegradation*, **31**, 129–141.

Ettema, C.H. and Wardle, D. (2002). Spatial soil ecology. *Trends in Ecology and Evolution*, **17**, 177–183.

Falkowski, P., Scholes, R.J., Boyle, E., Canadell, J., Canfield, D., Elser, J., Guber, N., Hibbard, K., Hogberg, P., Liner, S., Mackenzie, F.T., Moore, B., Pederson, T., Rosenthal, Y., Seitzinger, S., Smetacek, V. and Steffen, W. (2000). The global carbon cycle: A test of our knowledge of earth as a system. *Science*, **290**, 291–296.

Frankland, J.C. (1982). Biomass and nutrient cycling by decomposer basidiomycetes. In: *Decomposer Basidiomycetes: Their Biology and Ecology* (J.C. Frankland, J.N. Hedger and M.J. Swift, eds.), pp. 241–261. Cambridge University Press, Cambridge.

Frankland, J.C., Hedger, J.N. and Swift, M.J., eds. (1982). *Decomposer Basidiomycetes: Their Biology and Ecology.* Cambridge University Press, Cambridge.

Frey, S.D., Elliott, E.T., Paustian, K. and Peterson, G.A. (2000). Fungal translocation as a mechanism for soil nitrogen inputs to surface residue decomposition in a no-tillage agroecosystem. *Soil Biology and Biochemistry*, **32**, 689–698.

Frey, S.D., Six, J. and Elliott, E.T. (2003). Reciprocal transfer of carbon and nitrogen by decomposer fungi at the soil–litter interface. *Soil Biology and Biochemistry*, **35**, 71001–1004.

Fricker, M.D., Bebber, D., Tlalka, M., Watkinson, S.C., Boddy, L., Yiasoumis, L., Cartwright, H.M., Meskaukas, A., Moore, D., Smith, D.M.D., Nakagaki, T., Lee, C.F. and Johnson, N. Inspiration from microbes: From patterns to networks. In: *Complex Systems and Inter-Disciplinary Science*

(B.W. Arthur, R. Axtell, S. Bornholdt, J. Efstathiou, P.M. Hui, P. Maini, M. Nowak, R. Sole, D. Stauffer, F. Reed-Tsochas and K. Turner, eds.). World Scientific Publishing Company, London, in press.

Fricker, M., Tlalka, M., Bebber, D., Watkinson, S.C. and Darrah, P.R. (2007). Fourier-based spatial mapping of oscillatory phenomena in fungi. Technical Advance. *Fungal Genetics and Biology*, in press.

Glass, N.L. and Kaneko, I. (2003). Fatal attraction: Nonself recognition and heterokaryon incompatibility in filamentous fungi. *Eukaryotic Cell*, **2**, 1–8.

Govindarajulu, M., Pfeffer, P.E., Jin, H., Abubaker, J., Douds, D.D., Allen, J.W., Bücking, H., Lammers, P.J. and Shachar-Hill, Y. (2005). Nitrogen transfer in the arbuscular mycorrhizal symbiosis. *Nature*, **435**, 819–823.

He, X.-H., Critchley, C. and Bledsoe, C. (2003). Nitrogen transfer within and between plants through common mycorrhizal networks. *Critical Reviews in Plant Sciences*, **22**, 531–567.

Hibbett, D.S. and Binder, M. (2002). Evolution of complex fruiting-body morphologies in homobasidiomycetes. *Proceedings of the Royal Society B*, **269**, 1963–1969.

Hibbett, D.S., Gilbert, L.-Z. and Donoghue, M.J. (2000). Evolutionary instability of ectomycorrhizal symbioses in basidiomycetes. *Nature*, **407**, 506–508.

Jennings, L. and Watkinson, S.C. (1982). The structure and development of mycelial strands in *Serpula lacrymans*. *Transactions of the British Mycological Society*, **78**, 465–474.

Jentschke, G., Brandes, B., Kuhn, A.J., Schroder, W.H. and Godbold, D.L. (2001). Interdependence of phosphorus, nitrogen, potassium and magnesium translocation by the ectomycorrhizal fungus *Paxillus involutus*. *New Phytologist*, **149**, 327–337.

Kasuga, T., Townsend, J.P., Tian, C., Gilbert, L.B., Mannhaupt, G., Taylor, J.W. and Glass, N.L. (2005). Long-oligomer microarray profiling in *Neurospora crassa* reveals the transcriptional program underlying biochemical and physiological events of conidial germination. *Nucleic Acids Research*, **33**, 6469–6485.

Kauserud, H., Högberg, N., Knudsen, H., Elborne, S.A. and Schumacher T. (2004). Molecular phylogenetics suggests a North American link between the anthropogenic dry rot fungus *Serpula lacrymans* and its wild relative *S. himantioides*. *Molecular Ecology*, **13**, 3137–3146.

Kauserud, H., Saetre, G.-P., Schmidt, O., Decock, C. and Schumacher, T. (2006). Genetics of self/nonself recognition in *Serpula lacrymans*. *Fungal Genetics and Biology*, **43**, 503–510.

Kaye, J.P. and Hart, S.C. (1997). Competition for nitrogen between plants and soil microorganisms. *Trends in Ecology and Evolution*, **12**, 139–143.

Klionsky, D.J., Herman, P.K. and Emr, S.D. (1990). The fungal vacuole: Composition, function and biogenesis. *Microbiological Reviews*, **54**, 226–292.

Knorr, M., Frey, S.D. and Curtis, P.S. (2005). Nitrogen additions and litter decomposition, a meta-analysis. *Ecology*, **86**, 3252–3257.

Kõljalg, U., Dahlberg, A., Taylor, A.F.S., Larsson, E., Hallenberg, N., Stenlid, J.K., Larsson, K.-H., Fransson, P.M., Kårén, O. and Jonsson, L. (2000). Diversity and abundance of resupinate thelephoroid fungi as ectomycorrhizal symbionts in Swedish boreal forests. *Molecular Ecology*, **9**, 1985–1996.

Leake, J. (2005). Plants parasitic on fungi: Unearthing the fungi in mycoheterotrophs and debunking the 'saprophytic' plant myth. *Mycologist*, **19**, 113–122.

Leake, J., Johnson, D., Donnelly, D. and Boddy, L. (2004). Networks of power and influence: The role of mycorrhizal mycelium in controlling plant communities and agro-ecosystem functioning. *Canadian Journal of Botany*, **82**, 1016–1045.

Lindahl, B., Finlay, R. and Olsson, S. (2001). Simultaneous, bidirectional translocation of ^{32}P and ^{33}P between wood blocks connected by mycelial cords of *Hypholoma fasciculare*. *New Phytologist*, **150**, 189–194.

Lindahl, B.D., Ihrmark, K., Boberg, J., Trumbore, S.E., Högberg, P., Stenlid, J. and Finlay, R.D. (2007). Spatial separation of litter decomposition and mycorrhizal nitrogen uptake in a boreal forest. *New Phytologist*, in press.

Lindahl, B., Stenlid, J., Olsson, S. and Finlay, R. (1999). Translocation of ^{32}P between interacting mycelia of a wood decomposing fungus and ectomycorrhizal fungi in microcosm systems. *New Phytologist*, **144**, 183–193.

Lindahl, B., Taylor A.F.S. and Finlay, R.D. (2002). Defining nutritional constraints on carbon cycling in boreal forests — Towards a less 'phytocentric' perspective. *Plant and Soil*, **242**, 123–135.

Meskauskas, A., Fricker, M.D. and Moore, D. (2004). Simulating colonial growth of fungi with the neighbour-sensing model of hyphal growth. *Mycological Research*, **108**, 1241–1256.

Moncalvo, J.-M., Lutzoni, F.M., Rehner, S.A., Johnson, J. and Vilgalys, R. (2002). One hundred and seventeen clades of euagarics. *Molecular Phylogeny and Evolution*, **23**, 357–400.

Money, N. (2007). Plagues upon houses and cars: The unnatural history of *Meruliporia incrassata, Serpula lacrymans* and *Sphaerobolus stellatus*. In: *Fungi in the Environment* (G.M. Gadd, S.C. Watkinson and P. Dyer, eds.), pp. 288–308. Cambridge University Press, Cambridge.

Olsson, P.A., Jakobsen, I. and Wallander, H. (2002). Foraging and resource allocation strategies of mycorrhizal fungi in a patch environment. In: *Mycorrhizal Ecology* (M.G.A. van der Heijden and I. Sanders, eds.), pp. 93–115. Springer, Berlin. *Ecological Studies* **157**.

Perez-Moreno, J. and Read, D.J. (2000). Mobilization and transfer of nutrients from litter to tree seedlings via the vegetative mycelium of ectomycorrhizal plants. *New Phytologist*, **145**, 301–309.

Post, W.M., Emanuel, W.R., Zinke, P.J. and Stangenberger, A.G. (1982). Soil carbon pools and world life zones. *Nature*, **298**, 15–156.

Ramsbottom, J. (1953). *Mushrooms and Toadstools: A Study of the Activities of Fungi*, pp. 233–246. Collins, London.

Rayner, A.D.M. and Todd, N.K. (1982). Population structure in wood-decomposing basidiomycetes. In: *Decomposer Basidiomycetes: Their Biology and Ecology* (J.C. Frankland, J.N. Hedger and M.J. Swift, eds.), pp. 109–128. Cambridge University Press, Cambridge.

Read, D.J. (1991). Mycorrhizas in ecosystems. *Experientia*, **47**, 376–391.

Read, D.J. (1997). The ties that bind. *Nature*, **388**, 517–518.

Read, D.J. and Perez-Moreno, J. (2003). Mycorrhizas and nutrient cycling in ecosystems—A journey towards relevance? *New Phytologist*, **157**, 475–492.

Richter, D.D. and Markewitz, D. (2001). *Understanding Soil Change*. Cambridge University Press, Cambridge.

Roosen, J., Engelen, K., Marchal, K., Mathys, J., Griffioen, G., Cameroni, E., Thevelein, J.M., De Virgilio, C., De Moor, B. and Winderickx, J. (2005). PKA and Sch9 control a molecular switch important for the proper adaptation to nutrient availability. *Molecular Microbiology*, **55**, 862–880.

Ryvarden, L. and Gilbertson, R.L. (1993). *European Polypores, Part 1*. Fungiflora, Oslo.

Simard, S.W., Perry, D.A., Jones, M.D., Myrold, D.D., Durall, D.M. and Molina, R. (1997). Net transfer of carbon between ectomycorrhizal tree species in the field. *Nature*, **388**, 579–582.

Smith, E.A. (2006). Complexity: Implications for network-enabled effects-based operations. *Pointer: Journal of the Singapore Armed Forces*, **32**, 13–25.

Stark, J.M. and Hart, S.C. (1997). High rates of nitrification and nitrate turnover in undisturbed coniferous forests. *Nature*, **385**, 61–64.

Swift, M.J. (1982). Basidiomycetes as components of forest ecosystems. In: *Decomposer Basidiomycetes: Their Biology and Ecology* (J.C. Frankland, J.N. Hedger and M.J. Swift, eds.). Cambridge University Press, Cambridge.

Tedersoo, L., Kõljalg, U., Hallenberg, N. and Larsson, K.H. (2003). Fine scale distribution of ectomycorrhizal fungi and roots across substrate layers including coarse woody debris in a mixed forest. *New Phytologist*, **159**, 153–165.

Tlalka, M., Darrah, P.R., Hensman, D., Watkinson, S.C. and Fricker, M.D. (2003). Noncircadian oscillations in amino acid transport have complementary profiles in assimilatory and foraging hyphae of *Phanerochaete velutina*. *New Phytologist*, **158**, 325–335.

Tlalka, M., Watkinson, S.C., Darrah, P.R., Bebber, D. and Fricker, M.D. (2007). Emergence of self-organised oscillatory domains in fungal mycelia. *Fungal Genetics and Biology*, in press.

Tlalka, M., Watkinson, S.C., Darrah, P.R. and Watkinson, S.C. (2002). Continuous imaging of amino-acid translocation in intact mycelia of *Phanerochaete velutina* reveals rapid, pulsatile fluxes. *New Phytologist*, **153**, 173–184.

Watkinson, S.C. (1999). Metabolism and hyphal differentiation in large basidiomycete colonies. In: *The Fungal Colony* (N.A.R. Gow, D. Robson and G.M. Gadd, eds.), pp.127–157. Cambridge University Press, Cambridge.

Watkinson, S.C., Bebber, D., Darrah, P.R., Fricker, M.D., Tlalka, M. and Boddy, L. (2006). The role of wood decay fungi in the carbon and nitrogen dynamics of the forest floor. In: *Fungi in Biogeochemical Cycles* (G.M. Gadd, ed.), pp. 151–181. Cambridge University Press, Cambridge.
Watkinson, S.C., Burton, K.S and Wood, D.A. (2001). Characteristics of intracellular peptidase and proteinase activities from the mycelium of a cord-forming wood decay fungus, *Serpula lacrymans. Mycological Research*, **105**, 698–704.
Watkinson, S.C., Davison, E.M. and Bramah, J. (1981). The effect of nitrogen availability on growth and cellulolysis by *Serpula lacrymans. New Phytologist*, **89**, 295–305.
Wells, J.M., Harris, M.J. and Boddy, L. (1999). Dynamics of mycelial growth and phosphorus partitioning in developing mycelial cord systems of *Phanerochaete velutina*: Dependence on carbon availability. *New Phytologist*, **142**, 325–334.

CHAPTER 4

Ecophysiology: Impact of Environment on Growth, Synthesis of Compatible Solutes and Enzyme Production

Naresh Magan

Contents

Abstract

Basidiomycetes are important components of the decomposer microbial community involved in nutrient cycling and have been exploited successfully in the production of cultivated mushrooms. Environmental factors such as water availability, temperature and pH and their interactions have significant impacts on basidiomycete colonisation and fruiting potential. This chapter examines the effect that temperature, water availability and their interactions have on growth of different temperate and tropical basidiomycetes. Temperature ranges vary widely between temperate and tropical species and profiles for growth are presented. Basidiomycetes are generally more sensitive to matric than solute stress. This has been shown for *Trametes* species, as well as with *Agaricus bisporus* and *Pleurotus* species. The mechanisms of water-stress tolerance have received much attention, especially for cultivated species, where methods for optimising fruiting has been of interest. Sugar alcohols are involved in water-stress tolerance and the role of these in providing a gradient of water movement within mycelium of a range of cultivated and decay species. Basiodiomycetes produce significant amounts of extracellular enzymes to enable them to play such an important role in decomposition processes. Production of cellulases, ligninases and laccases occur over a wide range of temperature and water potentials.

British Mycological Society Symposia Series
Published by Elsevier Ltd.

At below the wilting point of plants laccases, cellulases and lignin peroxidases are produced in soil. The ability to produce copious amounts of these enzymes and their stability has resulted in their biotechnological exploitation.

1. INTRODUCTION

Decomposition of lignocellulosic resources in boreal, temperate and tropical climates is determined by the types of decay fungi present, and the prevailing environmental conditions (Cartwright and Findlay, 1958; Boddy and Rayner, 1988). Basidiomycota are frequently the dominant decomposer organisms and are thus critical in nutrient cycling in ecosystems. These fungi can also cause significant economic damage as pathogens and decomposers of buildings, other human artefacts and food stuff, by virtue of their ability to utilise cellulose, hemicellulose and lignin. Their mycelial habit and the production of aggregated mycelial structures—cords and rhizomorphs, allow them to forage for and colonise heterogeneously distributed bulky resources (Chapter 1). Saprotrophic Basidiomycota are exposed during these processes to a range of fluctuating environmental conditions, especially of water availability, temperature, pH and gaseous regime. These factors individually, and interacting, have an impact on the ability to grow, colonise different lignocellulosic matrices, produce the necessary enzymes for decomposition and to be competitive. This chapter considers the impact of abiotic factors, especially temperature and water availability, on the activity of saprotrophic Basidiomycota, their mechanisms of tolerance and ability to produce the necessary enzymes which enable them to be such important components of decomposition processes and nutrient recycling. Examples are chosen from temperate and tropical species in natural ecosystems, as well as cultivated species such as *A. bisporus* and *Pleurotus* spp.

2. EFFECTS OF TEMPERATURE AND WATER AVAILABILITY ON GROWTH

2.1 Temperature and Growth

Temperature is a major determinant of ecological niche (Hudson, 1986; Magan, 1997). It exerts an influence on fungi largely via its effects on enzyme-catalysed reactions. The overall response of a fungus to different temperatures represents the combined effect of numerous different chemical reactions, each of which exhibits its own characteristic relationship to temperature (Rayner and Boddy, 1988). The growth responses of fungi to temperature are variable, with many wood-rotting Basidiomycota are largely mesothermic, having minimum, optimum and maximum temperatures for growth in the region of 5, 25 and 40°C respectively (Cartwright and Findlay, 1958). Some wood-inhabiting fungi have a tolerance to unusually high temperatures. Thermophiles with an optimum between 40 and 50°C have been found in piled wood chips (Hudson, 1986). However, the optimum temperature for growth of most wood-decay fungi from temperate regions is between 25 and 30°C. For example, none of the 11 wood-rotting basidiomycetes studied by Boddy (1983) grew at 40°C and only five grew at 35°C. Many tropical species, especially in the genus *Trametes*, grow optimally in the range 30–40°C, compared to 20–30°C for temperate species (Mswaka and Magan, 1999). This highlights that significant differences exist between tropical and temperate basidiomycetes. Temperature relationships of four North American *Ganoderma* species correlated with geographic distribution: *G. colossum* collected from sub-tropical regions of Florida had a high-optimum temperature range of 35–40°C (Adaskaveg and Gilbertson, 1989).

It is possible that fungi from the tropics may exhibit higher temperature optima for growth and tolerance than their temperate counterparts, but this has seldom been examined. The influence of high temperatures on fungi decaying *Pinus contorta* slash was investigated by Loman (1965). Two fungi, *Phlebia phlebiodes* and *Lenzites sepiaria*, which have high-temperature optima and a wide temperature tolerance, were dominant in the upper 5 cm of exposed slash. *Stereum sanguinolentum* and *Coniophora puteana*, which had lower temperature optima and narrower temperature tolerances, were mainly active at greater depths in slash pieces. This demonstrated that temperature markedly affects the distribution of fungi within a habitat.

Detailed studies have been carried out on a range of tropical species, especially *Trametes* and related species from Zimbabwe, on 3% malt extract agar (MEA) and on wood substrates (Mswaka and Magan, 1998, 1999). These showed that maximum growth rates were largely similar for isolates of the same species, regardless of their origins (Fig. 1). Generally, the *Trametes* species that were collected from hotter regions of Zimbabwe had broad growth temperature ranges (Mswaka and Magan, 1999). For example, *T. cervina* and *T. cingulata* grew on 3% MEA at >50°C indicating thermotolerance. *T. cervina* showed the highest growth rate (approximately 18 mm/day) whilst *T. menziesii* and *Lenzites elegans* grew very slowly under the same conditions (approximately 6 and 8 mm/day respectively). There appeared to be three groups based on their optimum and maximum temperatures for growth. Species such as *T. menziesii*, *T. versicolor*, *Phaeotrametes decipiens* and *Irpex stereoides* fell into the low-temperature group with an optimum between 25 and 30°C and no growth above 37°C. The intermediate group of *T. pocas*, *T. modesta* and *T. elegans* had optima of 30–37°C with no growth at 45°C, whereas *T. cervina*, *T. cingulata* and *T. socotrana* were considered high-temperature fungi with growth optima in the range 37–40°C and growth ceasing at 55°C. Interestingly, the maximum growth rate of *T. versicolor* (ca.10 mm/day) was 2.5 mm/day faster than that recorded for temperate isolates of this same species (Hennington, 1968; Boddy, 1983).

Thermophilism in fungi relates to those fungi which grow only within the range 20-50°C or higher, with the maximum temperature at which a fungus (*Humicola lanuginosa*) is known to grow at $\sim$60°C (Lacey and Magan, 1991). Although the *Trametes* species had high-temperature optima, they cannot be strictly classified as thermophiles because of their ability to grow at <20°C.

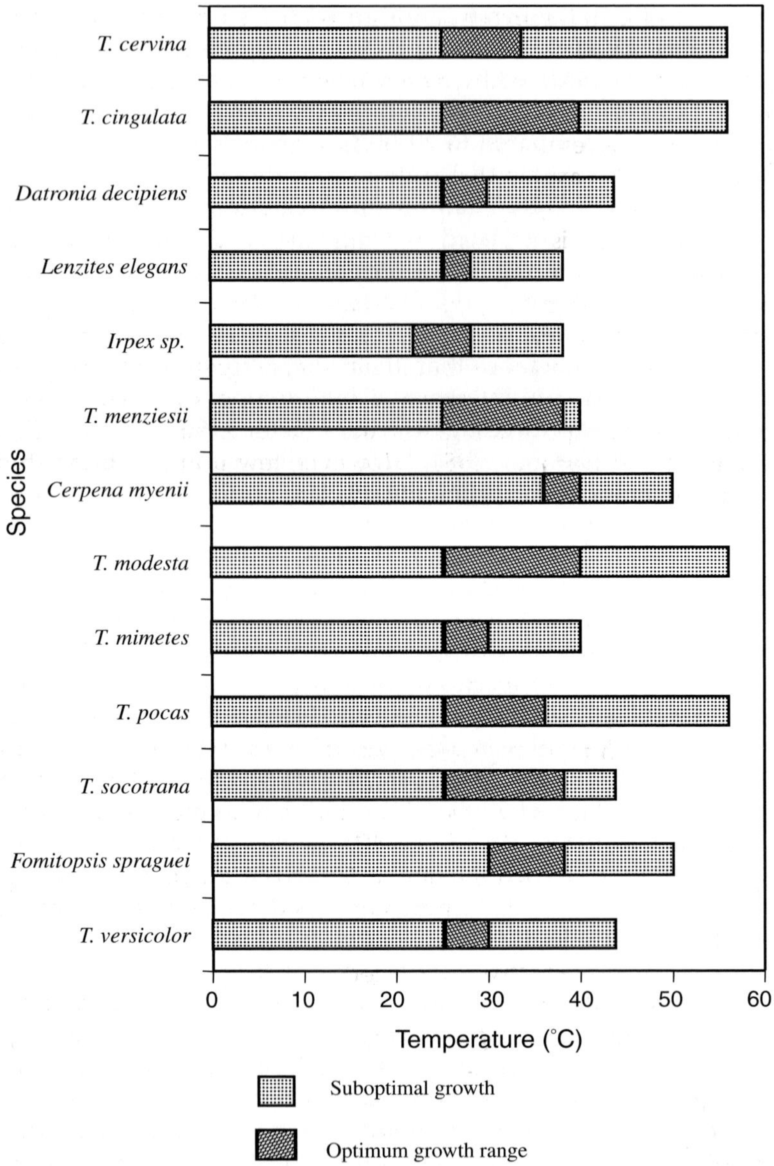

Figure 1 The temperature range and optimal range for growth of different tropical basidiomycetes.
Source: Adapted from Mswaka and Magan (1999).

2.2 Water Relations

Water is essential as a solvent for metabolic processes, for transport of metabolites, enzymes and organelles and, via turgor pressure, a vital skeletal ingredient and the driving force behind skeletal growth. Within solid organic resources, the availability of water to the decomposer organism is affected by two main forces, matric potential—exerted by the substratum, and osmotic potential due to dissolved salts (Boddy, 1983; Magan, 1997). Matric potential is a result of forces associated with the interfaces between water and the solid matrix, and osmotic potential is a result of the presence of solutes within the water. An advantage of using water potential as a measure of water availability is that the osmotic, matric and turgor components of the total water availability can be quantified. The water potential of a resource such as wood is an important and highly variable characteristic, often determining the growth of the organisms which decay it (Luard, 1982).

Wood-inhabiting fungi vary considerably in their abilities to grow at low-water content. Boddy (1983) studied the water relations of 12 temperate species of wood-inhabiting basidiomycetes and found that all grew very slowly or not at all on malt agar at −4.4 MPa water potential. Others, working mainly on temperate basidiomycetes, have reported similar results (Tresner and Hayes, 1971; Clarke *et al.*, 1980; Eamus and Jennings, 1986). However, few if any of these studies have compared the tolerance/sensitivity of temperate or tropical basiodiomycetes to matric and solute stress.

In general, most studies examining the effect of ionic solutes (e.g. NaCl, KCl), non-ionic solutes (e.g. glycerol) and matric water stress (e.g. PEG 8000) on basidiomycetes have shown a decrease in growth rate with an increase in imposition of water stress (Fig. 2). Growth of most basidiomycetes, including *A. bisporus*, *Pleurotus* and *Trametes* species, are more sensitive to a reduction in matric than solute potential (Magan *et al.*, 1995; Beecher *et al.*, 2000; Lee *et al.*, 2000).

Growth of most basiodiomycetes ceased completely or was extremely slow between −4 and −7 MPa osmotic potential. For example, growth of *Rhizoctonia solani* ceased at −5 MPa (Dubé *et al.*, 1971) and that of the dry rot species *Serpula lacrymans* ceased between −3 and −6 MPa on agar (Clarke *et al.*, 1980). Tresner and Hayes (1971) found that 94 out of 107 basidiomycetes ceased growth at −3 MPa while growth of 12 temperate wood-decay basidiomycetes ceased between −4.4 and −7.1 MPa when KCl was used as a solute (Boddy, 1983).

Temperature has interactive effects with water potential on basidiomycete growth (Fig. 3; Mswaka and Magan, 1999); there was a significant decrease in matric-stress tolerance at marginal temperatures for growth. However, sometimes growth rates may not be a good indicator of ecological competence. *I. stereoides* grew significantly slower than other species, but had a wider optimum matric and solute potential growth range and lower minima than many other basidiomycetes examined (Mswaka and Magan, 1999). Although the natural habitat for this fungus is moist evergreen forest, it is found exclusively on attached dead branches and stems of its angiosperm host, which are usually desiccated because of their exposed position. Thus, tolerance of low-water stress is an ecologically important adaptation by the fungus to its niche. Again, an ascomycete, *Xylaria hypoxylon*, has been shown to have the ability to maintain low-water potentials in the wood that it occupies (Boddy, 1986), thereby preventing the more active decomposers such as *T. versicolor* from replacing it. It is possible that *I. stereoides*, itself a slow growing fungus, may be using the same

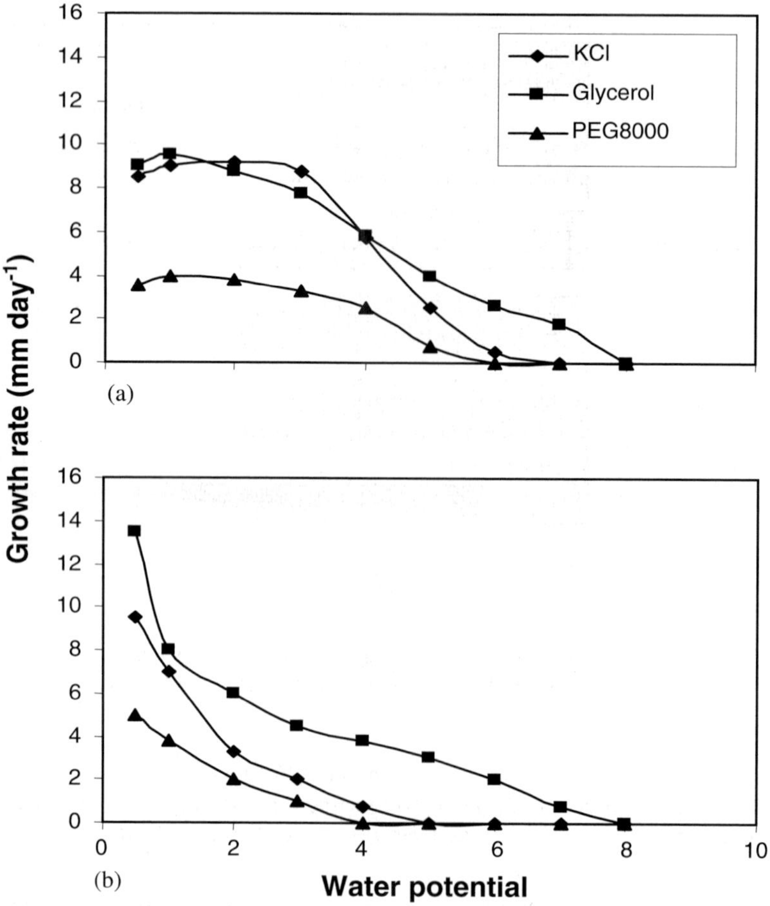

Figure 2 Comparison of effect of (a) solute potential (modified ionically with KCl or non-ionically with glycerol) and (b) matric potential (modified with PEG 8000) on relative growth rates of (a) *T. socotrana* and (b) *Lenzites elegans*.

ecological strategy as that used by *X. hypoxylon*. A small reduction in solute/matric potential resulted in a stimulation of the growth of *I. stereoides* indicating that this fungus is inhibited by very high-water availability in a resource.

Species isolated from hot and dry regions might be expected to have a better tolerance of extreme fluctuations in water stress. Many tropical species such as *T. cingulata* and *T. modesta* react to low-water stress by the formation of chlamydospores, an adaptation which favours their survival under desiccating conditions. These chlamydospores probably germinate at the onset of the wet season, allowing the fungus to continue to decompose its resource. Similarly, *Hyphodontia paradoxa*, which is unable to grow at low-water potentials but also produces chlamydospores (Boddy, 1983), was still viable after 81 days at greater than −100 MPa, but non-viable after 343 days (Miller and Meyer, 1934). However, the limiting water content for wood decay is reported to lie below 30%

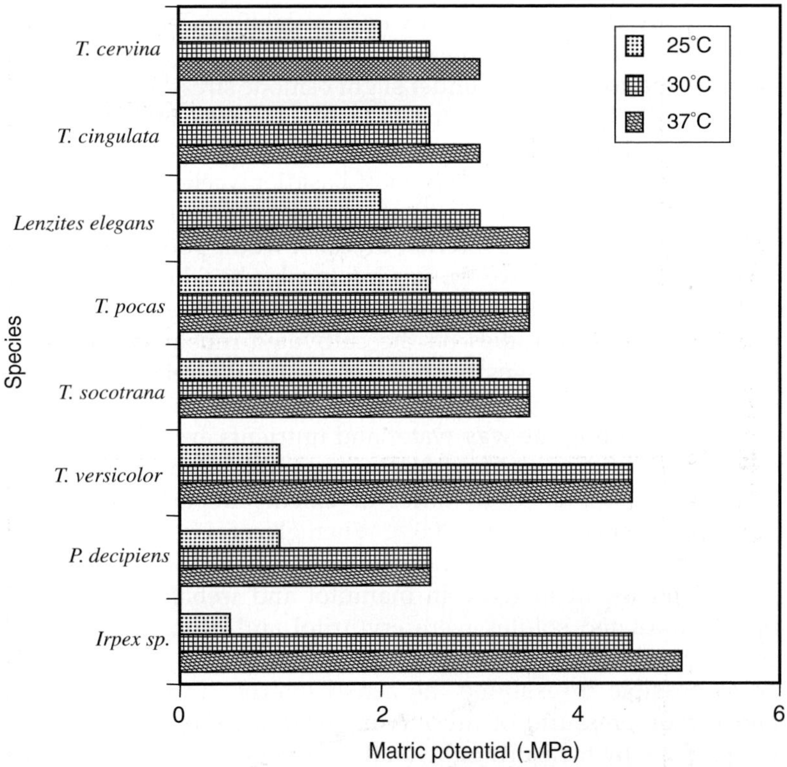

Figure 3 Effect of temperature on the matric potential growth range of some tropical wood-decay fungi on 3% MEA. *Source*: adapted from Mswaka and Magan (1999).

(percentage oven-dry weight) in undecayed wood, which corresponds to water potentials in the region −4 MPa (Boddy and Rayner, 1988). At matric potentials below this, water is only present in the transient micropores which are too small to allow hyphae or enzyme molecules to enter and effectively decompose the wood (Griffin, 1981). Such information is critical for understanding the ecological role of saprotrophic species in colonisation and decay of wood.

2.3 Compatible Solutes, Tolerance of Water Stress and Translocation of Water

For fungi to grow under solute or matric stress, compatible solutes are needed to enable enzyme systems to function, and basidiomycetes are no exception in this respect (Magan, 1997). The key compatible solutes are the high-molecular weight sugar alcohols (polyols) mannitol and arabitol, and the low-molecular weight erythritol and glycerol. Glycerol and erythritol are outstanding in this respect and their accumulation is a major determinant of the water relations of xerotolerant and xerophilic micro-organisms (Brown, 1978; Magan, 1997). For desiccation tolerance, trehalose synthesis is particularly important. Potassium ions are easily accumulated by fungal cells and can serve as a compatible cytoplasmic osmoticum of low toxicity (Harris, 1981). However, at high concentrations, ionic solutes can become toxic (Luard, 1983). This was shown by studies in which stimulation of growth of some species occurred under slight osmotic stress (e.g. −0.5 to −1.5 MPa; Magan *et al.*, 1995). This may be due to active accumulation of external KCl in the hyphae of the fungus (Brownell and Schneider, 1983), although less is known about K^+ ion accumulation in mycelium of basidiomycetes than other fungi.

At equivalent concentrations, mannitol and arabitol are less effective as compatible solutes in reducing internal mycelial water potential when compared to erythritol and glycerol. However, as saprotrophic basidiomycetes are relatively sensitive the major polyols which are accumulated are indeed arabitol and mannitol. Many detailed studies on the cultivated mushroom *A. bisporus* have examined the effect of changes in solute stress on the relative accumulation of polyols in the mycelium, and subsequently in fruit bodies. This is because of the interest in understanding the way water and nutrients are translocated over long distances into fruit bodies and to optimise the production of the latter. Compatible solutes are accumulated to different extents depending on the external osmoticum (Table 1; Beecher *et al.*, 2000). When glycerol is present in the medium it can readily enter the mycelium and improve the tolerance of water stress. This was accompanied by an increase in mannitol and trehalose. In ionic-modified media the accumulated solutes were erythritol and glucose. No trehalose was found in mycelia in this treatment.

It is also possible to examine the actual internal water potential (osmotic, matric and turgor pressure) of mycelia in relation to tolerance to water stress. This has been done by measuring the total water potential of the mycelial sample using a thermocouple psychrometer, then freezing the sample in liquid nitrogen, followed by thawing of the sample to release the ions which could then be measured (Beecher *et al.*, 2000). By subtraction, the turgor potential can be determined. For *A. bisporus*, with both ionic water stress and matric stress, there

Table 1 Mean sugar and polyol concentrations (mM) in mycelia of *A. bisporus* colonies grown on media modified to different water potentials with glycerol and KCl

Treatments	Malt extract agar (−0.5 MPa)	Glycerol-modified (−2.5 MPa)	KCl-modified (−2.5 MPa)
Sugars			
Trehalose	7.1	13.9	0
Glucose	1.1	0.5	3.1
Polyols			
Arabitol	0.4	1.2	0.3
Mannitol	6.0	35.0	5.6
Erythritol	1.1	0.5	3.1
Glycerol	4.3	218.0	3.8

Note: Analyses carried out using HPLC with a RI detector.
Source: From Magan *et al.* (1995).

was an increase in turgor pressure and a significant decrease in osmotic and total water potential with increasing water-stress treatment (Fig. 4). This confirmed that matric stress induced the highest mean turgor at 25°C. Thus, as greater water stress is imposed the internal water potential is changed by the synthesis of polyols, enabling the fungus to continue to function.

Interestingly, when polyol accumulation in fruit bodies of *A. bisporus* was examined, it appeared that in all the different tissues mannitol was by far the

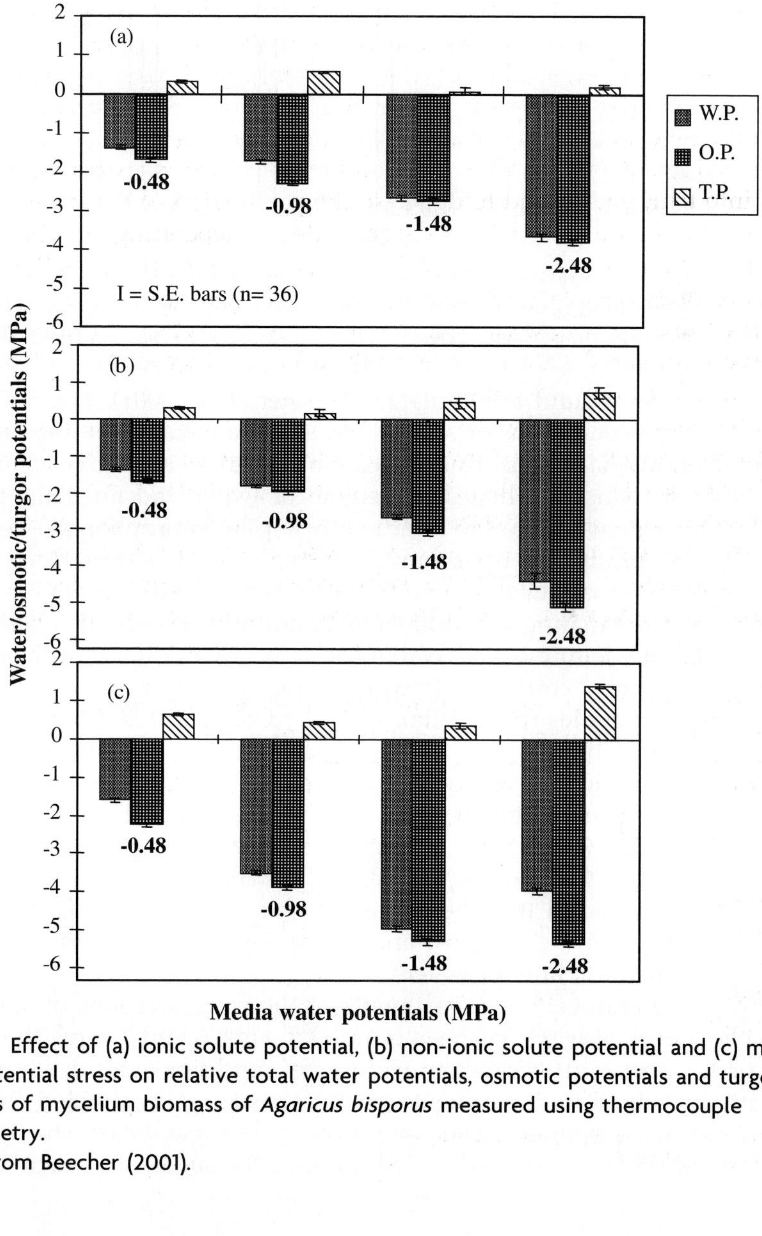

Figure 4 Effect of (a) ionic solute potential, (b) non-ionic solute potential and (c) matric water potential stress on relative total water potentials, osmotic potentials and turgor potentials of mycelium biomass of *Agaricus bisporus* measured using thermocouple psychrometry.
Source: From Beecher (2001).

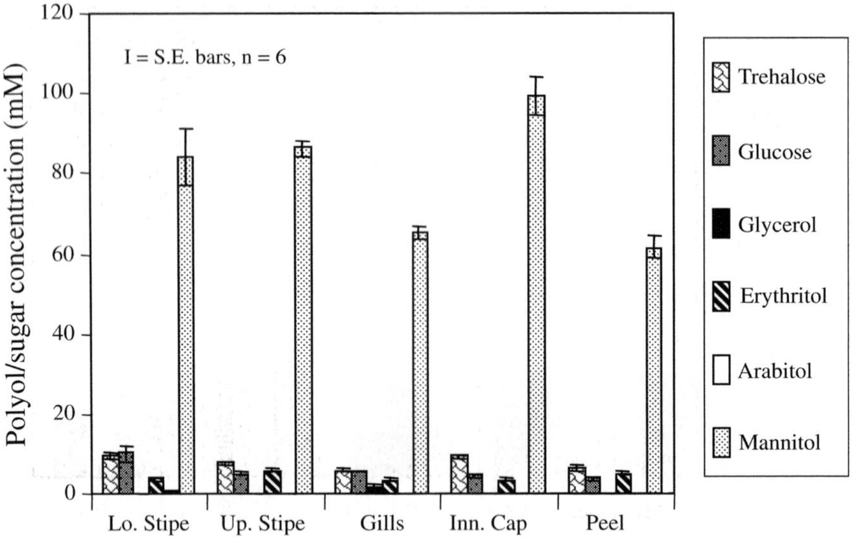

Figure 5 The total concentrations of different sugar alcohols, glucose and trehalose in different tissue of first flush sporophores of the cultivated mushroom *A. bisporus* (Beecher, 2001). Key: Lo., lower; Up., upper; Inn., inner.

most important compound accumulated (Beecher *et al.*, 2001). Fig. 5 shows an example of the concentrations of sugars and sugar alcohols in stipe, pileus and gill tissue in freshly harvested fruit bodies. This accumulation may contribute to a solute gradient which facilitates movement of water and nutrients from the mycelial cords or networks to the rapidly developing sporophores.

Since many saprotrophic basidiomycetes colonising wood produce mycelial cords and rhizomorphs they often move water and essential nutrients over long distances during fluctuating environmental conditions, especially availability of water. For example, Jennings and Watkinson (1982), working with *S. lacrymans*, suggested that mycelial cords facilitated the translocation of water and nutrients to developing fruit bodies (primordia). Work by Coggins *et al.* (1980) on droplet formation at hyphal tips of this species supported the suggestion of Jennings (1974) that translocation along hyphae and cords was due to a bulk flow of solution driven by hydrostatic pressure. This pressure was the result of soluble carbohydrates in the hyphae acting as osmotic solutes causing a flux of water into the hyphae as well as transporting solutes for growth to the hyphal tips. Indeed, Kalberer (1987) suggested that there was a water potential gradient from fruit bodies to mycelium in the compost and casing layer of *A. bisporus* which was responsible for the bulk flow of water.

Occurrence of endogenous adjustments of the concentrations of sugars and sugar alcohols in *A. bisporus* has been examined. Figs. 6 and 7 show the effect of colonies of *A. bisporus* growing from wet to dry conditions and dry to wet conditions respectively. The experiments were carried out in divided 9 cm Petri plates and KCl used to modify the water potential of one sector. The mycelia in each sector and the crossover zone were sampled for sugars and sugar alcohols,

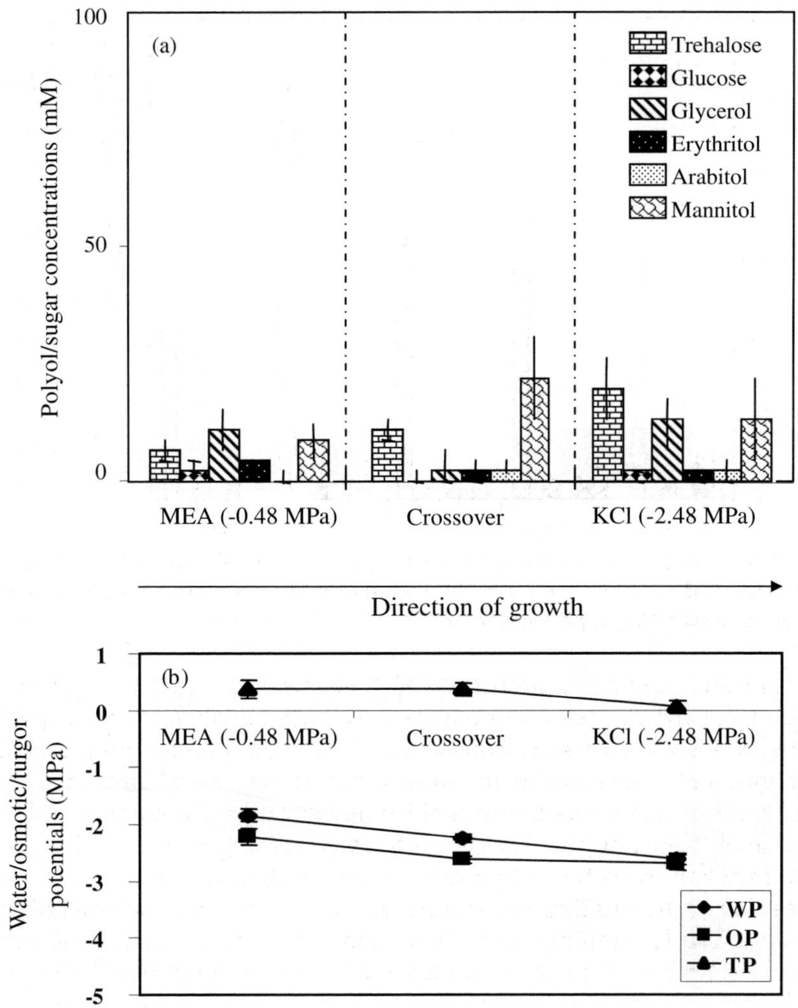

Figure 6 The effect of changing water potential from freely available water to ionic water-stress conditions, in split agar plates, on (a) sugar alcohol and sugar alcohol accumulation in the mycelial biomass in three regions (wet — malt extract agar, MEA, crossover and ionic water stress, −2.48 MPa) and (b) effects on endogenous total water potential, osmotic and turgor potentials of the mycelial regions by *A. bisporus* grown on divided petri plates. *Source*: From Beecher (2001). The fungus was inoculated on the MEA agar side. Standard errors are within the size of the symbol.

and the internal water, turgor and osmotic potentials also measured using thermocouple psychrometry (Beecher, 2001). This showed that there were changes in endogenous synthesis of polyols, reflected by the actual changes in the mycelial turgor and water potentials in each condition and the crossover zone. These studies demonstrated that saprotrophic basidiomycetes can transport water and nutrients over long distances and continue to function under abiotic stress.

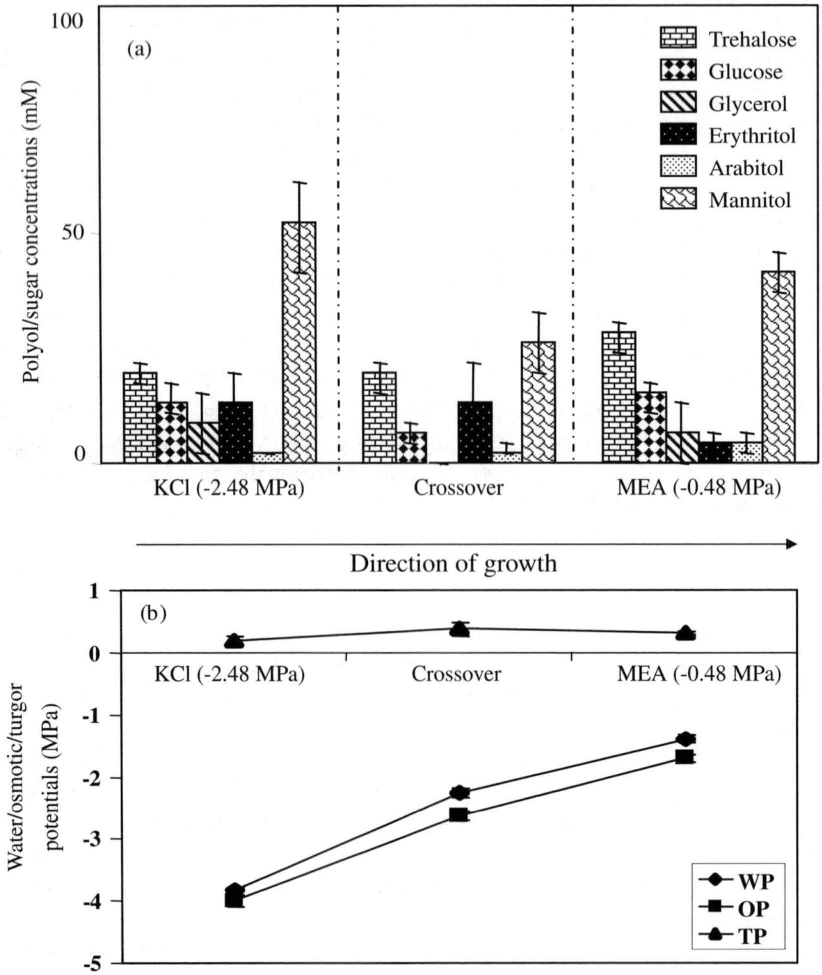

Figure 7 The effect of changing water potential from ionic water stress (−2.48 MPa) to freely available water using an ionic solute on (a) sugar alcohol and sugar alcohol accumulation in the mycelial biomass of *Agaricus bisporus* in three regions (wet — MEA, crossover and ionic water stress, −2.48 MPa) and (b) effects on endogenous total water potential, osmotic and turgor potentials of the mycelial regions. The fungus was inoculated on the −2.48 MPa side. Standard errors are within the size of the symbol.
Source: From Beecher (2001).

3. ENZYME PRODUCTION IN RELATION TO ENVIRONMENTAL FACTORS

The majority of both temperate and tropical basidiomycetes can produce a battery of hydrolytic enzymes to decompose cellulose, hemicellulose and sometimes also lignin (Chapter 2), including cellulases, pectinases, lignin peroxidise and laccases (Ohga and Royse, 2001; Nyanhongo *et al.*, 2002; Pointing *et al.*, 2005).

Temperature stress can affect enzyme production by basidiomycetes. Studies on *Trametes trogii* showed that pectinolytic enzymes such as polymethylgalacturonase, polygalacturanase and pectin lyase were optimally produced over the range 23–28°C. Interestingly while growth under these conditions was optimum at pH 3.3, a pH of 4.5 was best for production of pectinolytic enzymes with production occurring even at 6.2–6.5 (Levin and Forchiassin, 2004). Temperature range and stress has also been shown to have a significant impact on enzyme production. *Abortiporus biennis*, *T. versicolor* and *Cerrena unicolor* have been shown to produce enhanced extracellular peroxidise, superoxidase dismutase and laccase activities following exposure of cultures to marginal high (40°C) and low temperature (<10°C) stress (Fink-Boots *et al.*, 1999). The activation of laccase and peroxidise was rapid (within 96 h) and suggests that biosynthesis of such enzymes is activated rapidly by basidiomycete decay fungi when encountering lignin-based matrices, in both temperate and tropical ecosystems. Studies on a strain of *T. modesta* used a central composite statistical design approach to identify optimum environmental conditions for laccase production (Nyanhongo *et al.*, 2002). This showed that an initial pH of 6.95 and temperature of 30.26°C were optimum for enzyme production. Laccase production was optimal after 6–7 days, while that for other enzymes, such as mannose, was after 3–4 days under optimum temperature conditions. Laccase activity was very stable over a range of temperatures and pH values with optimum at pH 4.5 and 40°C.

Solid organic resources undergoing fungal decomposition may be regarded as a matrix consisting of a solid phase comprised of various lignocellulosic materials and a system of fluid filled voids. Matric potential is commonly a major influence on overall water availability. For a fungus growing within a resource, it is the spatio-temporal distribution of water in relation to the location of hyphae which is important (Boddy, 1986).

Effect of changes in water potential on laccase and cellulase production by *Lentinula edodes* has been quantified using RT-PCR (Ohga and Royse, 2001). Water potential of the resource changed from −0.5 MPa to −1.5 to −2.0 MPa over the fruiting period (Ohga *et al.*, 1998). Cellulase production increased over a 5–10 day period. In contrast, laccase decreased in the drier treatments. Where resource water potential was maintained at a steady-state then both laccase and cellulase activity remained relatively stable over the 15 day cultivation period.

The effect of water stress on production of hydrolytic enzymes has attracted attention because those enzymes are non-specific and can breakdown xenobiotic compounds in soil and compost-based systems, and have been applied in bioremediation systems. A range of enzymes are produced, even under relatively dry soil conditions although sometimes at reduced concentrations (Table 2). Different lignolytic enzymes are produced, depending on soil water potential (Table 3). For example, *Phanerochaete chrysosporium*, often used in bioremediation studies, produced ligninase and cellulases, but not laccase, in soil at both −0.7 and −2.8 MPa (Fragoeiro, 2005). However, very few studies have examined in detail the

Table 2 Production of different extracellular enzymes by *T. versicolor* and *P. chrysosporium* in soil extract broth at two water potentials modified with KCl at 25 °C for 25 days in shaken flask culture in soil extract broth

Species	*T. versicolor*		*P. chrysosporium*	
Water potential (MPa)	−0.7	−2.8	−0.7	−2.8
Enzyme				
Cellulase	2.1	1.6	3.4	4.2
Phosphomonoesterase	9.2	8.9	13.0	22.1
β-glucosidase	14.8	10.2	29.0	9.3
Protease	0	0.3	0.1	6.9
Laccase	91.9	9.3	0	0

Source: From Fragoeiro (2005).

Table 3 Temporal (a) laccase activity (U/g Soil) and (b) total ligninase (% decolourisation of Poly R487) and (c) cellulose (U/g Soil) produced by *T. versicolor*, *Pleurotus ostreatus* and *P. chrysosporium* when added on woodchips to soil microcosms at two water potentials (−0.7 and −2.8 MPa) at 15°C

	Water potential (−MPa)					
	0.7			2.8		
Time (weeks)	6	12	24	6	12	24
(a) Laccase						
T. versicolor	26.7	368.5	4.4	93.3	61.8	0.5
P. ostreatus	96.3	133.3	35.5	272.4	77.0	5.8
P. chrysosporium	0	0	0	0	0	0
(b) Ligninase						
T. versicolor	82.3	74.9	64.4	81.1	84.7	62.4
P. ostreatus	75.5	74.8	63.2	70.3	68.5	55.7
P. chrysosporium	63.5	72.8	64.9	58.9	63.7	66.3
(c) Cellulase						
T. versicolor	33.1	93.8	61.5	28.0	42.7	42.7
P. ostreatus	64.9	76.0	35.4	168.1	167.4	75.3
P. chrysosporium	36.1	52.9	33.9	97.5	61.2	36.9

Note: In all cases the soil microcosms were inoculated with colonised woodchips (5%) and mixed with soil before incubation.
Source: From Fragoeiro (2005).

influence of interacting environmental conditions of water × temperature × pH stress on enzyme production by basidiomycetes.

Biotechnological applications of enzymes produced by saprotrophic basidiomycetes, especially white-rot fungi, have received significant attention (Wessenberg *et al.*, 2003). This is because of the potential for producing copious amounts of extracellular ligninases and laccases in both free fermentation and immobilised systems for industrial applications. Enzymes such as ligninanse and laccases have been successfully applied to the clean up of waste-water streams from the textile industry specifically for the decolourisation and degradation of toxic dyes before discharge into the environment.

REFERENCES

Adaskaveg, J.E. and Gilbertson, R.I. (1989). Cultural studies of four North American species in the *Ganoderma lucidum* complex, with comparison to *G. lucidum* and *G. tsugae*. *Mycological Research*, **92**, 182–191.

Beecher, T.S. (2001). Studies on the water relations of the common cultivated mushroom (*Agaricus bisporus*), PhD Thesis, Cranfield University, Bedford, U.K.

Beecher, T.M., Magan, N. and Burton, K.S. (2000). Osmotic/matric potential affects mycelial growth and endogenous reserves in *Agaricus bisporus*. *Proceedings of 15th International Congress on Science and Cultivation of Edible Fungi* (L.J.L.D. Van Griensven, ed.), pp. 455–461.

Beecher, T.M., Magan, N. and Burton, K.S. (2001). Water potentials and soluble carbohydrate concentrations in tissues of freshly harvested and stored mushrooms. *Postharvest Biology and Technology*, **22**, 121–131.

Boddy, L. (1983). The effect of temperature and water potential on growth rates of wood rotting basidiomycetes. *Transactions of the British Mycological Society*, **80**, 141–149.

Boddy, L. (1986). Water and decomposition process. In: *Water, Fungi and Plants* (P.G. Ayres and L. Boddy, eds.), pp. 27–48. Cambridge University Press, Cambridge.

Boddy, L. and Rayner, A.D.M. (1988). *Fungal decomposition of Wood*. Wiley, New York.

Brown, A.D. (1978). Compatible solutes and extreme water stress in eukaryotic micro-organisms. *Advances in Microbial Physiology*, **17**, 181–242.

Brownell, K.H. and Schneider, R.W. (1983). Roles of matric and osmotic components of water potential and their interaction with temperature in the growth of *Fusarium oxysporum* in synthetic media and soil. *Phytopathology*, **75**, 53–57.

Cartwright, K.St.G. and Findlay, W.P.K. (1958). *Decay of Timber and its Prevention*, 2nd edn. HMSO, London.

Clarke, R.W., Jennings, D.H. and Coggins, C.R. (1980). Growth of *Serpula lachrymans* in relation to water potential of substrate. *Transactions of the British Mycological Society*, **75**, 271–280.

Coggins, C.R., Jennings, D.H. and Clarke, R.W. (1980). Growth of *Serpula lachrymans* in relation to water potential of substrate. *Transactions of the British Mycological Society*, **75**, 271–280.

Dubé, A.J., Dodman, R.L. and Flentje, N.T. (1971). The influence of water activity on the growth of *Rhizoctonia solani*. *Australian Journal of Biological Sciences*, **24**, 57–65.

Eamus, D. and Jennings, D.H. (1986). Water, turgor and osmotic potentials of fungi. In: *Water, Fungi and Plants* (P.G. Ayres and L. Boddy, eds.), pp. 27–48. Cambridge University Press, Cambridge.

Fink-Boots, M., Malarczyk, E. and Leonowicz, A. (1999). Increased enzyme activities and levels of superoxide anion and phenolic compounds in cultures of basidiomycetes after temperature stress. *Acta Biotechnologica*, **19**, 319–330.

Fragoeiro, S. (2005). Use of fungi in bioremediation of pesticides, PhD Thesis, Cranfield University, Bedford, U.K.

Griffin, D.M. (1981). Water and microbial stress. *Advances in Microbial Ecology*, **5**, 91–136.

Harris, R.F. (1981). Effect of water potential on microbial growth and activity. In: *Water Potential Relations in Soil Microbiology* (J.F. Barr, W.R. Gardner and L.F. Elliot, eds.), Soil Science Society of America Special Publication **9**. Soil Science Society of America, Madison, USA, 151 pp.

Hennington, B. (1968). Ecology of decomposition in birch and aspen. In: *Biodeterioration of Materials* (A.H. Walters and J.J. Elphick, eds.), pp. 408–423. Elsevier, London.

Hudson, H. (1986). *Fungal Biology*. Edward Arnold, London.

Jennings, D.H. (1974). Polyol metabolism in fungi. *Advances in Microbial Physiology*, **25**, 149–193.

Jennings, D.H. and Watkinson, S.C. (1982). Structure and development of mycelial strands of *Serpula lachrymans*. *Transactions of the British Mycological Society*, **78**, 465–474.

Kalberer, P.P. (1987). Water potentials of casing and substrate and osmotic potentials of fruit bodies of *Agaricus bisporus*. *Scientia Horticulurae*, **32**, 175–182.
Lacey, J. and Magan, N. (1991). Fungi colonising cereal grain: Their occurrence and water and temperature relationships. In: *Cereal Grain-Mycotoxins, Fungi and Quality in Storage* (J. Chelkowski, ed.), Chapter 5, pp. 77–118. Elsevier, Amsterdam.
Lee, H.B., Magan, N. and Yu, S.H. (2000). Comparison of water relations of three cultivated *Pleurotus* species and *Trichoderma* weed moulds. *The Plant Pathology Journal (Korea)*, **16**, 25–28.
Levin, L. and Forchiassin, F. (2004). Culture conditions for the production of pectinolytic enzymes by the white rot fungus *Trametes trogii* on a laboratory scale. *Acta Biotechnologica*, **18**, 157–166.
Loman, A.A. (1965). The lethal effects of periodic high temperatures on certain lodgepole pine slash decaying basidiomycetes. *Canadian Journal of Botany*, **43**, 334–338.
Luard, E.J. (1982). Effect of osmotic shock on some intracellular solutes in two filamentous fungi. *Journal of General Microbiology*, **128**, 2575–2581.
Luard, E.J. (1983). Activity of isocitrate dehydrogenase from three filamentous fungi in relation to osmotic and solute effects. *Archive of Microbiology*, **134**, 233–237.
Magan, N. (1997). Fungi in extreme environments. In: *Mycota Vol. IV Environmental and Microbial Relationships* (D.T. Wicklow and B. Soderstrom, eds.), Chapter 7. Springer Verlag, Berlin.
Magan, N., Challen, M.P. and Elliot, T.J. (1995). Osmotic, matric potential and temperature effects on vitro growth of *Agaricus bisporus* and *A. bitorquis* strains. In: *Science and Cultivation of Edible Fungi* (T.J. Elliot, ed.), pp. 773–780. Balkema, Rotterdam, NL.
Miller, V.V. and Meyer, E.I. (1934). Effect of desiccation of wood infected by house fungi and their variability. *Zniimod*, **1**, 5–22.
Mswaka, A.Y. and Magan, N. (1998). Cellulase, ligninase and degradation of wood by *Trametes* spp. from the indigenous forests of Zimbabwe. *Mycological Research*, **102**, 1399–1404.
Mswaka, A. and Magan, N. (1999). Water and temperature relations of *Trametes* and related species from the indigenous forests of Zimbabwe. *Mycological Research*, **103**, 1309–1317.
Nyanhongo, G.S., Gomes, J., Gubitz, G., Zvauya, R., Read, J.S. and Steiner, W. (2002). Production of laccase by a newly isolated strain of *Trametes modesta*. *Bioresource Technology*, **84**, 259–263.
Ohga, S., Kato, Y. and Nakaya, M. (1998). Water potential in relation to culture maturity in sawdust-based substrate of *Lentinula edodes*. *Mushroom Science and Technology*, **6**, 65–69.
Ohga, S. and Royse, D.J. (2001). Transcriptional regulation of laccase and cellulose genes during growth and fruiting of *Lentinula edodes* on supplemented sawdust. *FEMS Microbiology Letters*, **201**, 111–115.
Pointing, S.B., Pelling, A.L., Smith, G.J.D., Hyde, K.D. and Reddy, C.A. (2005). Screening of basidiomycetes and xylariaceous fungi for lignin peroxidase and laccase gene-specific sequences. *Mycological Research*, **109**, 115–124.
Rayner, A.D.M. and Boddy, L. (1988). *Fungal Decomposition of Wood*. Wiley, New York.
Tresner, H.D. and Hayes, J.A. (1971). Sodium chloride tolerance of terrestrial fungi. *Applied Microbiology*, **22**, 210–213.
Wessenberg, D., Kyriakides, I. and Agathos, S.N. (2003). White rot fungi and their enzymes for the treatment of industrial dye effluents. *Biotechnology Advances*, **22**, 161–187.

CHAPTER 5

Fruit Bodies: Their Production and Development in Relation to Environment

David Moore, Alan C. Gange, Edward G. Gange and **Lynne Boddy**

Contents

Abstract

Sexual reproduction is important because it generates genetic variation, offers an escape from DNA parasites and provides a means to repair DNA damage. Many fungi exhibit particular patterns of sexual fruit body morphogenesis but the characteristics differ between species. However, it is possible to generalise that within developing fruit body tissues, fungal cells embark on a particular course of differentiation in response to the interaction of their intrinsic genetic programme with external physical signals (light, temperature, gravity, humidity), and/or chemical signals from the environment and other regions of the developing structure. Fruit body morphogenesis is affected by carbon and mineral nutrient availability, and environmental variables including temperature, water availability, CO_2, light and interactions

British Mycological Society Symposia Series
Published by Elsevier Ltd.

with other fungi and bacteria. Changes in the seasonal pattern of fruiting in the UK can be detected from field records made in the last 50 years, and while not all species behave in the same way, mean first fruiting date is now significantly earlier and mean last fruiting date is now significantly later, which results in an extended fruiting season. Significant numbers of species that previously only fruited in autumn now also fruit in spring. Such analyses show that relatively simple field observations of fungi can detect climate change, and that fungal responses are sufficiently sensitive to react to the climate change that has already occurred by adapting their pattern of development. Unfortunately, though it is possible to deduce the decisive steps in development that are open to influence, the molecular controls that normally regulate those steps remain unknown. Extensive genomic analysis shows that sequences crucial to multicellular development in animals or plants do not occur in fungal genomes, so we are ignorant of the basic control processes of fungal multicellular developmental biology.

1. INTRODUCTION

We use the term fruit bodies to encompass all the structures that develop from fungal mycelia to produce and distribute spores or other propagules, including basidiomata—the structures that release sexual spores (meiospores) in Basidiomycota, as well as a range of structures that produce asexual spores (mitospores) and some somatic (vegetative) structures, such as stromata and sclerotia, that can survive adverse conditions. Obviously, the phrase encompasses a very wide range of organs but their common feature is that they are multicellular, and their shape and form emerge as a result of a sequence of developmental adjustments. That is, they exhibit a characteristic pattern of morphogenesis.

1.1 Fungal Morphogenesis

Within the developing tissues of a fruit body, cells embark on a particular course of differentiation in response to the interaction of their intrinsic genetic programme with external physical signals (light, temperature, gravity, humidity), and/or chemical signals from other regions of the developing structure. These chemicals may be termed organisers, inducers or morphogens, and may inhibit or stimulate entry to particular states of determination. Chemical signals may contribute to a morphogenetic field around a structure (cell or organ), which permits continued development of that structure but inhibits formation of another structure of the same type within the field. All of these phenomena contribute to the pattern formation that characterises the 'body plan' created by the particular distribution of differentiated tissues in the multicellular structure. Pattern formation depends on positional information, which prompts or allows the cell to differentiate in a way appropriate to its position in the structure and may be conveyed by concentration gradients of one or more morphogens emitted from one or more spatially distinct organisers. Pattern formation thus involves an instructive process, which provides positional information, and a second interpretive process, in which the receiving cell or tissue responds.

Fungi are 'modular organisms' in which growth is repetitive, and a single individual mycelium will have localised regions at very different stages of development (Andrews, 1995). Consideration of developmental regulatory systems is relevant to the current discussion because *any* effect of the external environment on fruit body development must operate through an influence on the control systems that determine the distribution and growth patterns of the multicellular structure.

The constituent cells of a fungal fruit body are generally considered to be totipotent (able to follow any pathway of differentiation), because a mycelial culture can be produced in vitro from a fragment of a mature, fully differentiated structure, e.g. a fruit body stem. This feature results in a morphogenetic plasticity which surpasses that of other organisms and provides an intellectual challenge in terms of developmental biology, taxonomy and genetics (Watling and Moore, 1994). The only exceptions to totipotency are the meiocytes (the cells within which meiosis occurs), which are committed to sporulation once they have progressed through meiotic prophase (Chiu and Moore, 1988a, 1988b, 1990, 1993; Chiu, 1996). On the other hand, even meiocytes can be 'used' for non-sporulation functions: the hymenium of *Agaricus bisporus* is packed with basidia held in an arrested meiosis and serving a purely structural function (Allen *et al.*, 1992).

Differentiated fungal cells require reinforcement of their differentiation 'instructions'. This reinforcement is part of the context within which they normally develop, but when removed from their normal environment most differentiated hyphae revert to vegetative hyphae. Hyphal differentiation is consequently an unbalanced process in comparison with vegetative hyphal growth. In most hyphal differentiation pathways the balance must be tipped in the direction of 'differentiation' by the *local* microenvironment, which is, presumably, mainly defined by the local population of hyphae.

Another common feature is that morphogenesis is compartmentalised into a collection of distinct developmental processes (called 'subroutines'; Figure 1; Moore, 1998a). These separate (or parallel) subroutines can be recognised at the levels of organs (e.g. cap, stem, veil), tissues (e.g. hymenophore, context, pileipellis), cells (e.g. basidium, paraphysis, cystidium) and cellular components (e.g. uniform wall growth, growth in girth, growth in length, growth in wall thickness). They are distinct genetically and physiologically and may run in parallel or in sequence. When they are played out in their correct arrangement the morphology that is normal to the organism results. If some of the subroutines are disabled (genetically or through physiological stress), the rest may still proceed. This partial execution of developmental subroutines produces an abnormal morphology. The main principles that govern fungal development as deduced from observation, experiment and computer modelling are summarised in Table 1 (from Moore, 2005).

Fungal morphogenesis must be totally different from animals, because fungal cells have walls, and from plants (whose cells also have walls) because hyphae grow only at their tips and hyphal cross-walls form only at right angles to the

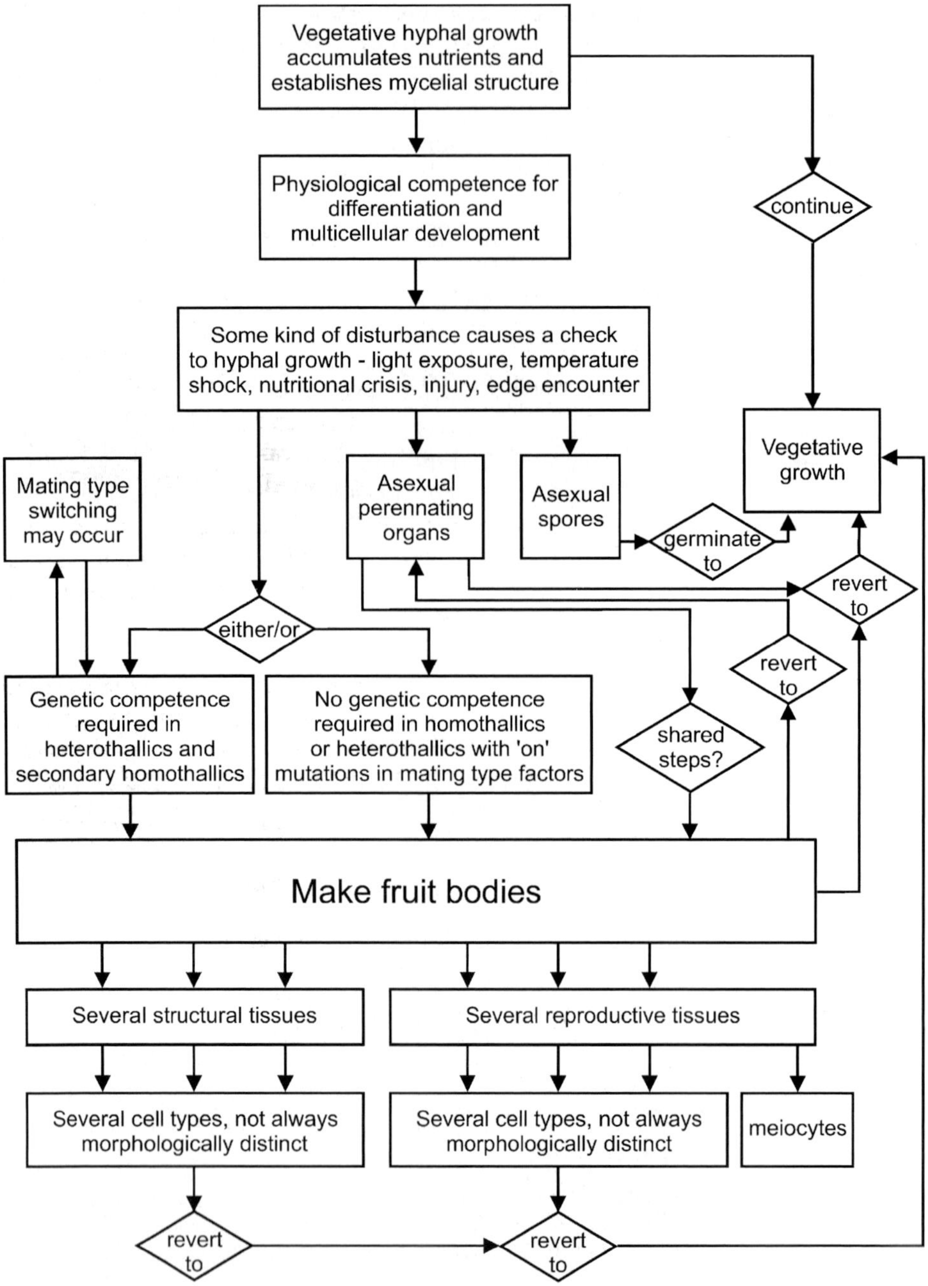

Figure 1 Flowchart showing a simplified view of the processes involved in development of fruit bodies and other multicellular structures in fungi (from Moore, 1998a).

Table 1 The eleven principles that govern fungal development

Principle 1	The fundamental cell biology of fungi on which development depends is that hyphae extend only at their apex, and cross-walls form only at right angles to the long axis of the hypha
Principle 2	Fungal morphogenesis depends on the placement of hyphal branches
Principle 3	The molecular biology of the management of cell-to-cell interactions in fungi is completely different from that found in animals and plants
Principle 4	Fungal morphogenetic programmes are organised into developmental subroutines, which are integrated collections of genetic information that contribute to individual isolated features of the whole programme. Execution of all the developmental subroutines at the right time and in the right place results in a normal structure
Principle 5	Because hyphae grow only at their apex, global change to tropic reactions of all the hyphal tips in a structure is sufficient to generate basic fruit body shapes
Principle 6	Over localised spatial scales coordination is achieved by an inducer hypha regulating the behaviour of a surrounding knot of hyphae and/or branches (these are called Reijnders' hyphal knots)
Principle 7	The response of tissues to tropic signals and the response of Reijnders' hyphal knots to their inducer hyphae, coupled with the absence of lateral contacts between fungal hyphae analogous to the plasmodesmata, gap junctions and cell processes that interconnect neighbouring cells in plant and animal tissues suggest that development in fungi is regulated by morphogens communicated mainly through the extracellular environment
Principle 8	Fungi can show extremes of cell differentiation in adjacent hyphal compartments even when pores in the cross-wall appear to be open (as judged by transmission electron microscopy)
Principle 9	Meiocytes appear to be the only hyphal cells that become committed to their developmental fate. Other highly differentiated cells retain totipotency — the ability to generate vegetative hyphal tips that grow out of the differentiated cell to re-establish a vegetative mycelium
Principle 10	In arriving at a morphogenetic structure and/or a state of differentiation, fungi are tolerant of considerable imprecision (= expression of fuzzy logic), which results in even the most abnormal fruit bodies (caused by errors in execution of the developmental subroutines) being still able

Table 1. (*Continued*)

	to distribute viable spores, and poorly (or wrongly) differentiated cells still serving a useful function
Principle 11	Mechanical interactions influence the form and shape of the whole fruit body as it inflates and matures, and often generate the shape with which we are most familiar

Source: From Moore (2005).

long axis of the hypha. Consequently, fungal morphogenesis depends on the placement of hyphal branches. A hypha must branch to proliferate. To form a multicellular structure, the position at which the branch emerges and its direction of growth must be controlled. A major aspect of that directional control is an autotropism—a tropism to self—in which growth direction of each hyphal branch is influenced by the position of the rest of the mycelium. Exploratory mycelia experience a negative autotropism, which causes them to grow away from the main mycelium and this maintains the outward exploration of the substratum. On the other hand, to create a multicellular structure like a fruit body, positive autotropism is essential to cause hyphae to grow together for hyphal branches to cooperate and coordinate their activities. Tropic reactions imply a signalling system, a signal sensing system and a reaction system. Mathematical models of these systems can be created very successfully (Stočkus and Moore, 1996; Meškauskas *et al.*, 1998, 1999a, 1999b, 2004a, 2004b; Moore *et al.*, 2006), but we know nothing yet about their biochemistry, cell biology or molecular nature. However, it is clear that what mechanisms exist must be *different* to animals and plants because gene sequences known to regulate development in animals and plants do not occur in fungal genomes (Moore *et al.*, 2005; Moore and Meškauskas, 2006).

1.2 Morphogenetic Control Elements

The only major morphogenetic control elements known in fungi are the mating type factors, which regulate pheromone production and pheromone receptors involved in mating, ranging from recognition between sexually competent cells in yeast to governing growth of clamp connections, internuclear recognition and regulation of the distance between the two nuclei in Basidiomycota (Casselton, 2002). However, not all fungi possess mating type factors, and, indeed, even in species that have a well-developed mating type system apparently normal fruit bodies can be formed by haploid cultures, and fruit body formation can usually be separated from other parts of the sexual pathway by mutation (see Chapter 5 in Moore, 1998a).

Generally, vegetative compatibility genes define the individuals of fungal populations, while mating type factors are usually interpreted as favouring the outbreeding of a fungal population (Chiu and Moore, 1999). Consequently, mating type genes contribute to management of the genetics of the population as well as to the sexual development of the individual. Sexual reproduction generates genetic variation, offers an escape from DNA parasites and provides a means to repair DNA damage (Bernstein *et al.*, 1985).

1.3 Importance of Sexual Reproduction

The crucial step in sexual reproduction, which provides the contrast with asexual reproduction, is the fusion of nuclei derived from different individuals. If the individuals involved in a mating have different genotypes, the fusion nucleus will be heterozygous and the products of the meiotic division can be recombinant genotypes. Thus, in one sexual cycle, new combinations of characters can be created in the next generation for selection. Consequently, the most common 'explanation' for sex is that it promotes genetic variability through out-crossing and that variability is needed for the species to evolve to deal with competitors and environmental changes. There is plenty of evidence to show that asexual lineages change little in time and that out-crossing certainly does promote variability in a population, which enables the organism to survive environmental challenges (Hurst and Peck, 1996; Burnett, 2003).

This, though, is a 'group selectionist' interpretation. It argues that variation generated in an *individual* meiosis benefits the *group* or population to which the individual belongs. Yet current theory prefers to emphasise that selection acts on individuals (Carlile, 1987; Dawkins, 1989). A feature that is advantageous in selection must be so because of benefit to the individual itself or its immediate progeny. As noted above, an alternative interpretation of the selective value of a sexual cycle suggests that repair of damaged DNA is the crucial advantage of meiosis (Bernstein *et al.*, 1985). It is argued that bringing together genomes from two different individuals enables DNA damage in one parental chromosome, caused by mutation or faulty replication, to be repaired by comparison and recombination with the normal chromosome provided by the other parent. Genetic fitness would be increased but only when out-crossing ensures heterozygosis. Even an incomplete sexual cycle might be of advantage in this case.

Gene mutations can be recessive and damaging, and different mutations are likely to occur in different mitotically generated cell lines. Just the formation of the diploid (or heterokaryon in most Basidiomycota) by out-crossing will benefit the mated individual if recessive adverse mutations are masked by non-mutant ('wild-type') alleles in the nuclei of the other parent. Out-crossing might also give rise to heterozygous advantage, where the heterozygous phenotype is better than either of its homozygous parents. This has been demonstrated frequently in plants and animals, and also in *Saccharomyces cerevisiae* (James, 1960).

Clearly, the genotype of the parental mycelium makes a crucial contribution to the genetics of the progeny population, but to produce a progeny population the parental mycelium must first produce a crop of fruit bodies and to do that it must grow into and through the substratum to capture, translocate and accumulate sufficient nutrients to support the formation of what can be massive multicellular structures.

2. PHYSIOLOGICAL FACTORS FAVOURING FRUIT BODY PRODUCTION

Fungi enjoy an adaptable and flexible metabolism. It is unlikely that there is a compound, organic or inorganic, on the planet that some fungus cannot utilise, transform, modify or otherwise metabolise (see Chapter 3 in Moore, 1998a). These versatile biochemical capabilities are used in a variety of ways during morphogenesis in fungi and over the past century there have been numerous *in vitro* studies of the nutritional physiology of fruit body production. Nutrients that are inferred to be 'favourable' for fruiting are those that allow the organism to exert its own intrinsic controls over the progress of its metabolism (Hawker, 1950).

2.1 Carbohydrates

An enormous volume of research has been done on this topic (for reviews see Moore-Landecker, 1993; Jennings, 1995; Moore, 1998a), though it is important to remember that conditions in the laboratory are far removed from the natural environment. The crucial insights came from Hawker's (1939, 1947) experiments: simple sugars tend to favour asexual spore production while oligo- and polysaccharides are especially good carbon sources for production of fruit bodies. Glucose often represses fruit body production, even in very low concentrations. The rate with which a fungus can hydrolyse a carbohydrate determines the ability of the carbohydrate to promote fruit body formation (Hawker and Chaudhuri, 1946), so what seems to matter most is the rate of supply and ease of use of substrates as determinants of their value in promoting fruit body formation. It comes as no surprise, therefore, that saprotrophic Basidiomycota on dung fruit more readily than those utilising leaf litter, and in turn than wood decomposers, though, of course, these resources also differ in mineral nutrient content. Likewise, fungi that participate early in community development within a resource fruit more readily than most later colonizers (Cooke and Rayner, 1984; Rayner and Boddy, 1988; Chapter 11), whose carbon sources are more recalcitrant.

2.2 Nitrogen Sources

Similar conclusions are reached when attention turns to the 'best' nitrogen source, which usually proves to be one amino acid or a mixture of amino acids. In most cases inorganic nitrogen and ammonium salts fail to support fruit body development although they may support production of primordia, but amino acids are required to produce the *mature* fruit bodies (reviewed in Moore, 1998a). This suggests that the formation of fruit body initials may be an activity of the vegetative mycelium and it is their further development which constitutes the fundamental 'mode switch' into the fruit body morphogenetic pathway. At least some of the deleterious effects of ammonium salts may be due to their influence on the pH of the medium, though metabolite repression caused by ammonium ions in many Ascomycota may be another cause. Fruit body formation in some fungi is favoured by provision of protein as source of nitrogen. Several basidiomycetes (*A. bisporus*, *Coprinus cinereus* (= *Coprinopsis cinerea*) and *Volvariella volvacea*) are able to use protein as a carbon source as efficiently as they use glucose (Kalisz *et al.*, 1986), so an advantage of protein is that it serves as a source of carbon, nitrogen and sulphur. In more natural conditions, *A. bisporus* and a wide range of other filamentous fungi can utilise dead bacteria as sole source of carbon, nitrogen, sulphur and phosphorus (Fermor and Wood, 1981; Grant *et al.*, 1986).

Higher carbon than nitrogen concentrations are usually required for fruit body production but the optimum C:N ratio varies from ~30:1 to ~5:1 (references in Moore-Landecker, 1993). High concentrations of amino acids tend to delay and/or depress maturation of fruit bodies even in organisms in which fruit body formation is optimal on media containing lower concentrations of amino acids, an effect that may result from the production of large quantities of ammonium as a nitrogen-excretion product on such substrates. When grown on protein as sole carbon source, nitrogen needs to be excreted from the mycelium; when this happens *in vitro* the ammonium concentration of the medium *increases* drastically during mycelial growth. One-third to one-half of the supplied protein-nitrogen was metabolised to ammonia by batch cultures of three saprotrophic basidiomycetes when protein was the sole source of carbon (Kalisz *et al.*, 1986).

2.3 Nutrient Capture

Hyphae absorb sufficient nutrients to support their active vegetative growth *and* to allow accumulation of reserve materials, which may subsequently be translocated to sites of need, including developing fruit bodies. Fruit body primordia may be fairly uniformly dispersed, but locations of enlarging and maturing fruit bodies may be much less evenly spread. For example, in *Coprinus lagopus*, certain favourably placed young fruit bodies may initiate a flow of nutrients in their direction, others that are deprived then fail to mature (Madelin, 1956a, 1956b, 1960). When *C. lagopus* colonies were physically divided in half early in growth, the two halves yielded similar fruit body biomass, whereas the two sides of an intact colony could differ by as much as 10:1, implying that in the latter case the 'minority' half is exporting its nutrients to the 'majority' half (Madelin, 1956b).

Mycelia must have access to sufficient substrates before fruiting is possible. Buller (1931, p. 165) discussed the requirement for a minimum amount of mycelium to support a minimum fruit body in *Coprinus sterquilinus*, arguing that one of the functions of hyphal fusions between (clonal) germlings is to ensure the rapid formation of that minimal size mycelium encompassing a corresponding minimum quantity of substrate. Obviously, the minimum quantity of substrate required varies between species depending on size of the fruit bodies produced. Fungi producing small fruit bodies are able to do so with only a small amount of resource, e.g. minute *Marasmius* and *Mycena* species restricted to leaf petioles, small portions of leaf lamina, beech cupules, etc. (Figure 2). A very large mycelial domain is required to produce the large, perennial brackets of heart-rot fungi (Rayner and Boddy, 1988). It was estimated that all of the nitrogen in 13.6 g of wood would be required to supply 1 g of *Ganoderma applanatum* basidiome, and

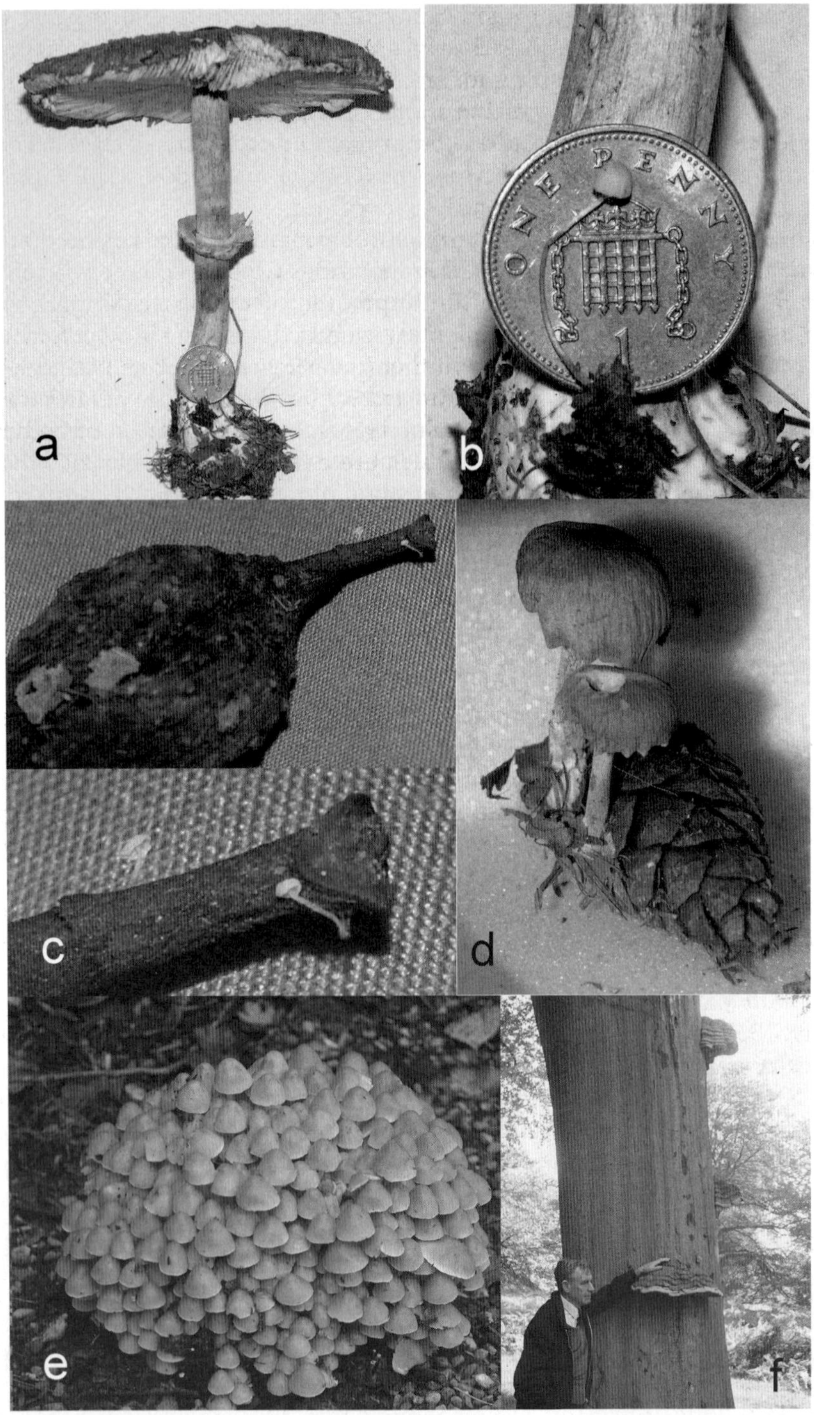
a
b
ONE PENNY
1
c
d
e
f

36.1 g wood to supply 1 g of spores, based on mean nitrogen content of fruit bodies (1.13%), spores (3.05%) and *Betula* sapwood (0.83%; Merrill and Cowling, 1966). Since fruit bodies are commonly 1 kg or more, and several grams of spores are produced each year (*Fomes fomentarius* produced 1.115 g spores in 20 days (Meyer, 1936)), a mycelium would need to draw upon the entire nitrogen content of more than 14 kg wood.

Culture studies indicate that once the minimum substrate size is reached fruit body distribution is governed by a flow of nutrients towards particular developing fruit bodies, rather than localised nutrient depletion or inhibition of development. The generality of this interpretation is based on two consistent observations. First, that many fruit body primordia are generally formed, but only a comparatively small number of them develop into mature fruit bodies; but if fruit body size is related to local nutrient supply, one would expect that all of the primordia on a colony would develop into mature but small fruit bodies, each using those quantities of materials which are available locally and adjusting its size accordingly. Second, a crop consisting of several fruit bodies will often develop as a group, so that any general inhibitory action is unlikely. The concept that nutrients flow towards a favoured centre would permit several neighbouring primordia to mature in a clump, while still withholding nutrients from unfavourably situated primordia. Clearly, different species emphasise different aspects of this physiology in their fruiting behaviour and some are characteristically solitary, e.g. *Phallus impudicus*, while others are caespitose, e.g. *Hypholoma fasciculare* and *Psathyrella multipedata* (Figure 2). Some Basidiomycota, notably Corticiaceae, form fruit bodies over the entire resource surface that they have access to, e.g. *Vuilleminia comedens* on branches in the canopy. Large, skin-like fruit bodies of some Corticiaceae may form at individual sites, subsequently coalescing on contact. Detail is, however, lacking as much less research has been done on these species than on Agarics.

In vitro experiments consistently indicate a general correlation between nutrient exhaustion of the medium and the onset of multicellular morphogenesis; however, reproduction is not an alternative to vegetative hyphal growth but an aspect of the differentiation of vegetative hyphae. Continued growth of the vegetative mycelium is necessary to provide sustenance to its developing fruit bodies. Correlation of fruiting with nutrient exhaustion of the medium does *not* mean that development is prompted by a mycelium that is starving, because the mycelium has accumulated nutrient reserves. Further, the timing of fruiting and the amount of biomass that a fungus commits to fruiting varies with life history

Figure 2 Some fruit bodies of saprotrophic basidiomycota, illustrating a range of sizes and resources: (a) the solitary *Macrolepiota rhacodes* with a coin size marker (20 mm diameter); (b) a fruit body of *Marasmius setosus* with the same coin size marker; (c) even smaller *Marasmius* specimen on the petiole of a beech cupule; (d) *Collybia peronata* on a pine cone; (e) the decidedly caespitose *Psathyrella multipedata*; (f) Terence Ingold posing with *Fomes fomentarius* on a beech tree in Knole Park, Sevenoaks, Kent, 1969 (see Ingold, 2002). Photographs (a)–(e) by David Moore of specimens collected by members of the mid-yorkshire fungus group at Harlow Carr Gardens. **(See Colour Section)**

strategy (Cooke and Rayner, 1984; Rayner and Boddy, 1988; Chapter 11). Rapid and extensive commitment of mycelial biomass is an R-selected (ruderal) characteristic, typical of fungi that rapidly dominate following disturbance. Such fungi are usually not combative and are often rapidly replaced by later arriving, more combative species. They, therefore, must commit to reproduction before they are killed and replaced. By contrast, slower and intermittent commitment to reproduction is characteristic of fungi in stressful environments and/or that are combative, dominating middle stages of community development. Laboratory studies have largely employed species, e.g. *Coprinopsis* spp., *Pleurotus* spp. and *Schizophyllum commune*, that fruit readily in culture, which is a ruderal characteristic; thus, we must be cautious in extrapolating to fungi with other life history strategies.

As we have discussed above, only preconditioned mycelium is capable of undergoing morphogenesis. The preconditioned mycelium must be beyond a particular minimum size, perhaps be of a particular minimum age, and the underlying nature of both these preconditions is that the mycelium has been able to accumulate sufficient supplies of reserve materials to support development of the minimum reproductive structure. For some fungi, exhaustion of a particular metabolite from the medium or substrate may be a signal that prompts morphogenesis in a mycelium that is *not* starving, but is healthy and well provisioned. Exhaustion of one or more constituents of the medium changes the balance of nutrient flow. If the medium is no longer fully supportive, the requirements of active hyphal growth can no longer be met by import from outside the hyphae and the balance must shift from 'reserve material accumulation' to 'reserve material mobilisation'. That change from balanced growth to growth under limitation in external nutrient supply is what signals the onset of morphogenesis. Cellular differentiation leading to fruit body morphogenesis is an expression of unbalanced growth which is precipitated by one or more changes in the balance of metabolism, and itself causes further cycles in which cellular components are re-allocated. Even though nutritional dependence on the external substrate may still be demonstrated, the emphasis shifts towards intramycelial regulation.

While this metabolic change is proceeding there is a change in the behaviour of hyphal branches. For some branches, negative autotropism becomes positive autotropism, so that neighbouring hyphae, often those of the surface or more aerial parts of the mycelium, can interact. They form centres of rapid but self-restricting growth and branching which become the hyphal aggregates or mycelial tufts, perhaps 100–200 μm in diameter, that are the 'initials' of the reproductive structure the organism can produce. Frequently, and especially in culture, these aggregates are formed in great number over the whole surface of the colony. As supplies of nutrients in the medium approach exhaustion repression of the morphogenesis of these hyphal aggregates is lifted and they proceed to develop further. As mentioned above, only a small number of the first-formed hyphal aggregates usually undergo further development and these become the focus for translocation of nutrients, mobilised from the stores in other parts of the colony and transported through the hyphal network to the developing reproductive structures.

Illumination may be required, either to promote further morphogenesis or to direct development into one of a small number of morphogenetic pathways (see below). Particular temperatures may also be required for particular pathways of development. Development usually proceeds in a series of steps that may be coordinated by environmental cues (illumination, temperature, atmosphere) and often involve sweeping re-allocation of cellular components. Within the young fruit body, therefore, new accumulations of 'stored' nutrients arise, and there may be a number of these accumulation–mobilisation–translocation–accumulation cycles during the development of the reproductive structure.

2.4 Non-Nutritional Environmental Variables

As well as carbon and nitrogen nutrition, discussed above, many more environmental variables affect fruit body initiation and development (reviewed by Jennings, 1995; Moore, 1998a; Scrase and Elliott, 1998; Kües and Liu, 2000). Such is the bulk of the literature that we can do little more here than list the major observations.

As the above discussions of metabolism imply, fruit body development requires oxidative metabolism (glycolysis and TCA cycle activity are often amplified) and good aeration is, not surprisingly, associated with successful fruiting. This means not only oxygen but also various volatile metabolites including carbon dioxide. Elevated carbon dioxide concentrations can suppress basidiome initiation in *S. commune* (Raudaskoski and Salonen, 1984). In *Agaricus*, increased elongation of the stem occurs with elevated CO_2, accumulated naturally from respiration, whereas cap and gills expand and spores mature more rapidly when CO_2 is removed (Turner, 1977). It has been argued that the morphogenetic effect on maturation of the fruit body may have ecological advantage: CO_2-enhanced elongation of the stem would raise the gills away from the surface of the substratum where the concentration of CO_2 might be expected to be higher than in the wider atmosphere because of the respiratory activity of microorganisms in the casing soil (Turner, 1977).

High CO_2 levels promote formation of long hyphal compartments in *S. commune*. It has been argued (Raudaskoski and Salonen, 1984) that a wood decomposer like *S. commune* is likely to experience elevated CO_2 within the wood as respiratory CO_2 accumulates. Mycelium that reaches the surface of the wood, however, will be exposed to CO_2 reduced to the atmospheric normal. Such mycelium will be able to form the shortened cells and more compact branching habit, and be predisposed to fruit body formation.

Light has diverse effects on formation of reproductive structures in different basidiomycetes, increasing or decreasing their number, affecting their development or determining whether or not they are produced (Carlile, 1970; Elliott, 1994). In general, the most effective parts of the spectrum are the near-ultraviolet and blue wavelengths, typical of the shaded and litter-covered forest floor. There are indications that the photoreceptor involved in fruit body morphogenesis may be membrane bound. In some fungi levels of intermediary metabolites and coenzymes, and activities of several enzymes respond very rapidly to changes in illumination. The vegetative mycelia of many Ascomycota require exposure to light before they will produce fruit bodies and/or asexual spores, and show specificity not only for particular wavelengths but also for a particular dosage of light radiation. In some Basidiomycota, sequential light exposures are responsible for initiating and programming fruit body morphogenesis, and periods of darkness between illumination events are important. Again, blue (400–520 nm) to near-ultraviolet (320–400 nm) light is the most effective and the work suggests that at least two photosensitive systems operate in fungi, one stimulated by near-ultraviolet and the other by blue light. Because their absorption spectra parallel the action spectra of the blue light photoresponses, carotenes and flavins appear to be the best candidates for photoreceptors.

Production of fruit bodies *in vitro* typically occurs over a more restricted range of temperature than that which will support mycelial growth. Optimum temperatures for fruit body production are generally lower than those most favourable for mycelial growth. In Basidiomycota most information relates to species adopted as laboratory models or for commercial cultivation. Cultivated species frequently need a temperature downshift (by 5–10°C) and lower CO_2 concentrations for fruiting, e.g. *A. bisporus, C. cinereus, Flammulina velutipes, Kuehneromyces mutabilis, Lentinula edodes, Pholiota nameko, Pleurotus ostreatus, Stropharia rugosa-annulata* and *V. volvacea* (Chang and Hayes, 1978; Stamets, 1993). This list includes compost-grown fungi as well as some wood-chip/straw and log-grown wood decomposers, and is not unrepresentative of the wider community of saprotrophic fungi, so it may be that most Basidiomycota require a temperature downshift. A prolonged downshift is not always required; thus, fruit body initiation in *F. velutipes*, which fruits in nature during late autumn to spring, occurs at a continuous regime of 20°C or following 12 h at 15°C (Kinugawa and Furukawa, 1965). Interestingly, the optimum temperature for both mycelial growth and production of fruit body initials by *A. bisporus* is 24°C (Flegg, 1972, 1978a, 1978b). However, temperature downshift is required for further development of initials beyond a cap diameter of ~2 mm.The fruit bodies develop normally when the temperature is lowered to 16°C. So, as with the reaction to nitrogen sources mentioned above, the implication is that formation of fruit body initials/primordia is an aspect of mycelial growth, but their proper development requires a further morphogenetic switch. It is tempting to conclude that these *in vitro* responses reflect the organism's natural response to seasonal changes.

Relative humidity (RH) affects fruit body initiation. Relatively high humidity is usually conducive to initiation of fruiting (Stamets, 1993), though it prevents initiation in *Polyporus ciliatus* (Plunkett, 1956). The water content of the resource may be even more critical. There is a balance between too high a water content that reduces aeration and too low a water potential that provides insufficient water for development (Scrase and Elliott, 1998; Ohga, 1999a; Kashangura *et al.*, 2006). There is variability between strains; *Pleurotus sajor-caju* was able to produce primordia at −2.5 MPa but none at −3.5 MPa even though they were able to grow under these xeric conditions (Kashangura *et al.*, 2006). pH can affect fruit body development, being optimal for several species at 6–7 (Kües and Liu, 2000), but pH 4 for *L. edodes* (Ohga, 1999b).

Physical constraints influence fruit body formation *in vitro*. Sexual reproduction is often initiated when the growing mycelium reaches an obstacle such as the edge of the dish or barriers placed onto the surface of the medium (the 'edge effect' or 'check to growth'). Reproductive structures often arise when mycelial growth had been arrested, by either physical or chemical means (Moore, 1998a). A physical barrier is not absolutely necessary for the 'edge effect', rather the important determining factor is the disturbance in metabolism which results from either encountering the edge of the dish or a major change in nutritional value of the substrate. Thus, different sorts of barrier and different sorts of medium transition are able to disturb the progress of metabolism sufficiently to initiate fruit body formation. The same applies to physical injury to the mycelium, which can stimulate fruit body formation (Leslie and Leonard, 1979a). Fruiting response to mechanical injury in *S. commune* is determined by at least four genes (Leslie and Leonard, 1979a, 1979b), showing that a number of different parallel routes lead to fruit body formation.

Inter- and intraspecific interactions can stimulate reproductive development. In interactions with other fungi this is at least partly a result of damage to vegetative hyphae (Rayner and Boddy, 1988). Many *A. bisporus* strains fruit only when associated with bacteria, e.g. pseudomonads, apparently not due to production of stimulatory compounds but to removal of inhibitory compounds (De Groot *et al.*, 1998). When competing with *C. cinereus* in agar culture, *C. congregatus* fruited from a much smaller resource volume than when growing alone (Schmit, 1999). In contrast, interactions can result in a fungus being confined to territory, e.g. a decay column in wood, that is too small to support fruit body production by that species. Fruit bodies are assembled from contributions of a number of cooperating hyphal systems, usually of the same individual. Hyphal interactions are controlled by the somatic and mating incompatibility systems (Chiu and Moore, 1999) that maintain mycelial individuality. Fruit bodies of somatically compatible Basidiomycota can fuse when the fruit bodies develop in extremely close proximity, as is commonly seen when resupinate fruit bodies meet on wood, and also with stipitate basidiomata, e.g. a fused cap with three stems of *Boletus (Xerocomus) chrysenteron* in Kibby (2006). However, hyphal cooperation is so fundamental that it can even lead to the formation of chimeric fruit bodies. Mixed cultures of two genetically different heterokaryons can produce basidiomata comprising both dikaryons, as seen with *P. nameko* (Babasaki *et al.*, 2003). Even more extreme is the case of fruit bodies of *Coprinus* consisting of two different species, *C. miser* and *C. pellucidus* (Kemp, 1977). The hymenium comprised a mixed population of basidia bearing the distinctive spores of the two species but the chimera extended throughout the fruit body as both species could be recovered by outgrowth from stem segments. All of these features can be interpreted as aspects of the tolerance of imprecision in fungal morphogenesis which has been discussed elsewhere (Moore, 1998a, 1998b, 2005; Moore *et al.*, 1998).

Once fruit bodies have been produced environment, particularly temperature and RH, can affect spore production. For example, in the field spore production by *Hericium erinaceus* is highest at about midday reflecting diurnal temperature and

RH (McCracken, 1970). In the laboratory, at 85–95% RH, spore production increased from a minimum at 0°C to a maximum at 24–27°C, and ceased at 31–33°C. At 20°C, sporulation was greater at 30% RH than at 90% RH (McCracken, 1970).

2.5 Fruiting in the Natural Environment

It is well known that the majority of Basidiomycota fruit in autumn, following mycelial growth and decomposer activity in spring and summer. Temperature and rainfall are considered to be the two main factors affecting productivity (Salerni *et al.*, 2002). In a 21-year fruit body survey of a forest plot in Switzerland, there was considerable variation between years in species richness and productivity, only litter decomposing saprotrophs, *Collybia butyracea* var. *asema* and *C. dryophila*, appearing in all years (Straatsma *et al.*, 2001). *Appearance* of fruit bodies was correlated with July and August temperatures, an increase of 1°C resulting in a delay of fruiting by saprotrophs of ~7 days. In contrast, fruit body *productivity* was correlated with precipitation from June to October (Straatsma *et al.*, 2001), and similar relationships have also been found in Britain and Sweden (Wilkins and Harris, 1946; Wasterlund and Ingelog, 1981).

In a 3-year study of Mediterranean oak forests, there was no evidence for influence of temperature on fruit body species diversity or productivity by most saprotrophs, though there was strong positive correlation between species diversity of wood decay fungi and maximum temperature, and with spring and summer rainfall (Salerni *et al.*, 2002). Temperature and rainfall in the 5 days prior to surveying seemed to have little effect on fruiting, but did so between 10 and 30 days prior to survey.

Climate change has resulted in phenological changes in plants, insects and birds (Parmesan and Yohe, 2003), and this has recently been shown to be the case for fungi (Gange *et al.*, 2007). Analysis of a data set of fruiting records of 200 species of decomposer Basidiomycota in Wiltshire, UK, each of which had been recorded over more than 20 years during 1950–2005, revealed that mean first fruiting date averaged across all species is now significantly earlier, while mean last fruiting date is now significantly later (Figure 3; A.C. Gange, E.G. Gange, T.H. Sparks and L. Boddy, unpublished data). Thus, the fruiting season has been extended since the 1970s. Not all species fruit earlier (47% show an advancement), or produce fruit bodies later into the year (55% continue fruiting later) but of those saprotrophic Basidiomycota that showed significantly earlier fruiting dates ($n = 94$), the average advancement was 7.9 days per decade, while for those with significantly later last fruiting dates ($n = 110$) the delay was 7.2 days per decade. The response differs depending on habitat type: 13% of grassland species fruiting earlier, 48% having later last fruiting; 53% of wood decay fungi fruited earlier, with 20% having later last fruiting. There was a significant relationship between mean fruiting date of those species that normally fruit early in the season (September) and late summer temperature and rainfall (Figure 4). Local July and August mean temperatures have significantly increased (July, $P < 0.05$; August, $P < 0.01$), while rainfall has decreased, though less markedly, over the 56 years of the survey.

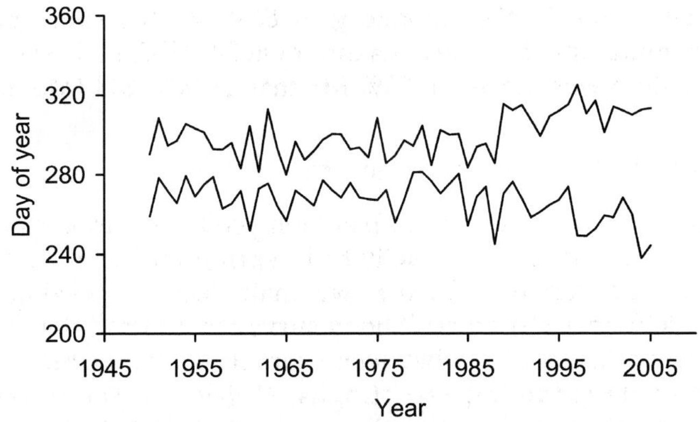

Figure 3 Mean first fruiting date (lower line) and mean last fruiting date (upper line) for 200 saprotrophic basidiomycota over 56 years. Splitting the data into two equal (28 Year) periods reveals no trend in the first half ($P = 0.97$) but a highly significant trend ($P < 0.001$) in the second half (A.C. Gange, E.G. Gange, T.H. Sparks and L. Boddy, unpublished data).

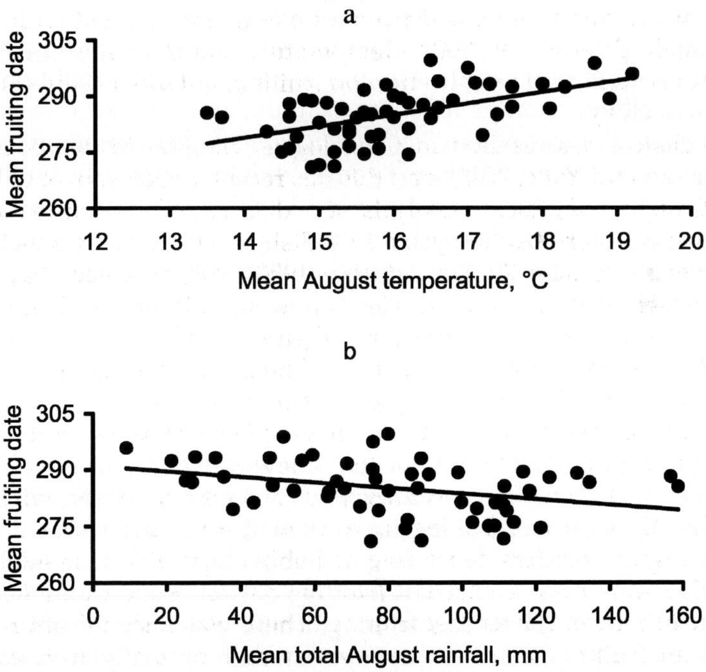

Figure 4 Relationship between mean fruiting date of saprotrophic basidiomycota species that normally fruit early in the season (September) and (a) August temperature ($R^2 = 0.299$, $F_{1,54} = 23.056$, $P = 0.007$) and (b) August rainfall ($R^2 = 0.126$, $F_{1,54} = 7.790$, $P = 0.000$) (A.C. Gange, E.G. Gange, T.H. Sparks and L. Boddy, unpublished data).

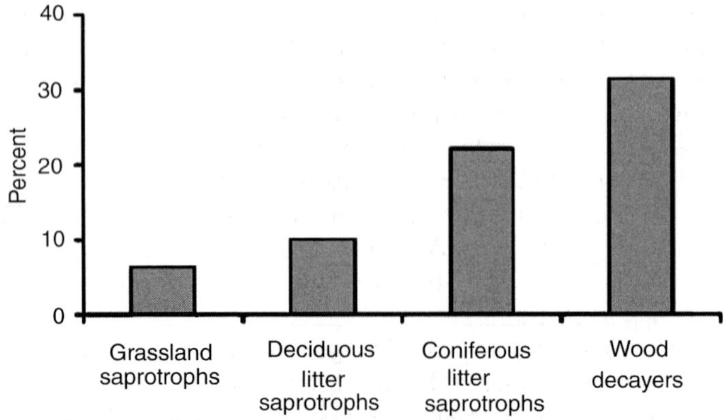

Figure 5 The proportion of saprotrophic basidiomycota in different habitat groups that, before 1975, were not recorded as fruiting in spring, but after this time did so in at least 1 year (A.C. Gange, E.G. Gange, T.H. Sparks and L. Boddy, unpublished data).

As well as changes to autumn fruiting patterns, significant numbers of species that previously only fruited in autumn now also fruit in spring (Figure 5). Since mycelia must be active in uptake of water, nutrients and energy sources before fruit bodies can be produced this suggests that these fungi may now be more active in winter and spring than they were in the past.

Other aspects of the environment can also influence fruiting by affecting microclimate (e.g. ground vegetation and logging waste), providing additional resources or inhibitory compounds. For example, in managed forests: there was lower fruit body biomass where *Pteridium aquilinum* was abundant; in dry years *Mycena* species were more abundant in areas with logging waste, but in wet years they were equally or more abundant in areas without logging waste; fruit body biomass was negatively correlated with grass cover in dry autumns, but positively correlated in wet autumns (Wasterlund and Ingelog, 1981).

3. FRUIT BODY SURVIVAL

As well as the physical size of a fruit body, a significant feature in the ecology of the organism is the length of time that the fruit body remains sufficiently intact to distribute spores. This varies from a few days or weeks for fleshy fungi to several years for perennial brackets, longevity of the latter being associated with structural physical characteristics and production of chemicals that inhibit invertebrate feeding or are toxic to them (Kahlos *et al.*, 1994; Stadler and Sterner, 1998). There appears only to be one detailed study of the lifespan of an agaric, an analysis of the fruit bodies of *A. bisporus* grown in an experimental mushroom farm over 36 days (Umar and Van Griensven, 1997). The fruit bodies remained healthy for 18 days before localised cytological indications of senescence became evident (nuclear and cytoplasmic lysis, permeable cytoplasmic membranes and structural changes to the cell wall). Cells of the fruit body collapsed irregularly and the remnants of the lysed cells aggregated around and between the remaining living hyphal cells. Most of the stem hyphae became empty cylinders. After 36 days, electron microscopy showed that most of the cells throughout the fruit body were severely degenerated and malformed, yet a number of basidia and subhymenial cells remained intact and alive even at 36 days. Interestingly, when mushrooms were cultivated using conventional commercial farming procedures, ~50% of the fruit bodies were infected by *Trichoderma harzianum* and/or *Pseudomonas tolaasi* by 18 days. All such fruit bodies died at 24 days due to generalised severe bacterial and fungal infections leading to tissue necrosis and decay of the caps and stems.

Observations of a wild troop of *Clitocybe nebularis* in a garden in Stockport, Cheshire, began on 21 October 2006, at which time the fruit bodies were young, but close to maturity (5 cm diameter), and continued for 29 days (Figure 6). By 19

Figure 6 Life and death of *Clitocybe nebularis* fruit bodies in a suburban garden in Stockport, Autumn 2006. Observations began on 21 October and continued for 29 days to 19 November. Troops of fruit bodies of *Coprinus micaceus* emerged, matured and decayed ~26 October and November 1 (the latter are illustrated). Some disturbance and grazing (squirrels?) was evident on 10 November, and collapsed fruit bodies by 18 November.

November most of the fruit bodies were beginning to collapse. These basidiomata of *C. nebularis* were still actively releasing spores on 7/8 and 12/13 November, clearly indicating that agarics with large fruit bodies can distribute spores for 3–4 weeks, though viability was not tested. During the observation of *C. nebularis*, two troops of fruit bodies of *Coprinus micaceus* emerged, matured and decayed (~26 October and 1 November), illustrating the alternative (R-selected) strategy of rapid production of short-lived fruit bodies.

The longevity of fruit bodies is obviously important for dispersal, but so also is the period over which spores are actively produced and released, and the viability/germinability of spores produced at different times. While some species retain high germinability of spores produced over several weeks, e.g. *Perenniporia tenuis var. tenuis* and *Coriolopsis gallica*, with others there is a decline, e.g. germinability of *Postia placenta* and *Gloeophyllum trabeum* declined from >94 to 19 and 44%, respectively, 5 weeks after fruiting was initiated in culture (Schmidt and French, 1983).

4. PRINCIPLES OF FUNGAL DEVELOPMENTAL BIOLOGY

Numerous observations show that all aspects of the environment can influence the production and development of fungal fruit bodies. To understand *how* this occurs we need to formalise fruit body development sufficiently to allow recognition of the decisive steps that are open to influence, and we must also identify the molecular controls that normally regulate those steps.

4.1 Underlying Principles

Three generalisations can be extracted from the past century of observations on fruiting physiology. First, the organism internalises nutrients rapidly to gain regulatory control over nutrient access and distribution. By so doing the vegetative mycelium becomes *competent* to produce multicellular structures like fruit bodies. Second, factors that promote fruiting, whether physical or chemical, seem to work by disturbing the normal progress of cellular metabolism. It is the disturbance itself that is the effective factor, overcoming some block to progress and *inducing* the next stage to proceed. Consequently, parallel pathways cover some stages of fruit body development and for these stages different factors seem to be interchangeable (e.g. a particular nutritional state may replace a particular illumination requirement). Third, even relatively simple developmental pathways can be subdivided into stages (at least, initiation, development and maturation) and there seems to be a need for successive signals (successive metabolic disturbances) to maintain progress of the developmental process. Each stage involves *change* in hyphal behaviour and physiology, taking the tissue to a higher order of differentiation.

4.2 Modelling Hyphal Growth and Fruit Body Formation

Hyphal growth is well suited to mathematical modelling, and the recent neighbour-sensing model brings together the basic essentials of hyphal growth kinetics into a vector-based mathematical model that 'grows' a life-like virtual mycelium (or 'cyberfungus') on the user's computer monitor (Meškauskas *et al.*, 2004a, 2004b; Moore *et al.*, 2006). The program has been used in a series of experiments (Meškauskas *et al.*, 2004a, 2004b) to show that complex fungal fruit body shapes can be simulated by applying the same regulatory functions to all of the growth points active in a structure at any specific time. No global control of fruit body geometry is necessary; rather, the shape of the fruit body emerges as the entire population of hyphal tips respond together, in the same way, to the same signals. These computer simulations thus demonstrate that because of the kinetics of hyphal tip growth, very little regulation of cell-to-cell interaction is required to generate fungal fruit body structures. The program includes parameters that can be used to mimic the effects of cell-to-cell signalling and environmental variables. These give the experimenter the opportunity to study the effects of such variables on fungal growth *in silico*.

4.3 Data Mining Fungal Genomes

The notion that control mechanisms of fungal multicellular developmental biology are probably very different from those known in animals and plants that emerges from the work described so far is supported by sequence searches of genomic databases. The unique cell biology of filamentous fungi has clearly caused control of their multicellular development to evolve in a radically different fashion from that in animals and plants. There are no *Wnt*, *Hedgehog*, *Notch*, *TGF*, *p53*, *SINA* or *NAM* sequences in fungi (Moore *et al.*, 2005; Moore and Meškauskas, 2006), but there are presumably analogous or homologous processes in fungal multicellular structures that need to be regulated.

Unfortunately, the demonstration that developmental control sequences of animals and plants lack fungal homologues leaves us knowing nothing about the molecules that *do* govern multicellular development in fungi. Yet these *are* the molecules and mechanisms that generate fungal fruit bodies. The molecular control elements of development *are* the things with which the environment interacts to cause its effects. While we remain ignorant of the basic control processes of fungal developmental biology we will also remain ignorant of the way environment impacts on fungal biology.

REFERENCES

Allen, J.J., Moore, D. and Elliott, T.J. (1992). Persistent meiotic arrest in basidia of *Agaricus bisporus*. *Mycological Research*, **96**, 125–127.

Andrews, J.H. (1995). Fungi and the evolution of growth form. *Canadian Journal of Botany*, **73**, S1206–S1212.

Babasaki, K., Masuno, K. and Murata, H. (2003). Interactions of heterologous mycelia colonized in the substrate govern fruit body production in the cultivated homobasidiomycete *Pholiota nameko*. *Bioscience, Biotechnology, and Biochemistry*, **67**, 100–106.

Bernstein, H., Byerly, H.C., Hopf, F.A. and Michod, R.E. (1985). Genetic damage, mutation and the evolution of sex. *Science*, **229**, 1277–1281.

Buller, A.H.R. (1931). *Researches on Fungi*, Vol. 4. Longmans, Green & Co., London.

Burnett, J. (2003). *Fungal Populations and Species*. Oxford University Press, Oxford.

Carlile, M.J. (1970). The photoresponses of fungi. In: *Photobiology of Microorganisms* (P. Halldal, ed.), pp. 309–344. Wiley Interscience, London.

Carlile, M.J. (1987). Genetic exchange and gene flow: Their promotion and prevention. In: *Evolutionary Biology of the Fungi* (A.D.M. Rayner, C.M. Brasier and D. Moore, eds.), pp. 203–213. Cambridge University Press, Cambridge.

Casselton, L.A. (2002). Mate recognition in fungi. *Heredity*, **88**, 142–147.

Chang, S.T. and Hayes, W.A. (1978). *Biology and Cultivation of Edible Mushrooms*. Academic Press, New York, NY.

Chiu, S.W. (1996). Nuclear changes during fungal development. In: *Patterns in Fungal Development* (S.W. Chiu and D. Moore, eds.), pp. 105–125. Cambridge University Press, Cambridge.

Chiu, S.W. and Moore, D. (1988a). Evidence for developmental commitment in the differentiating fruit body of *Coprinus cinereus*. *Transactions of the British Mycological Society*, **90**, 247–253.

Chiu, S.W. and Moore, D. (1988b). Ammonium ions and glutamine inhibit sporulation of *Coprinus cinereus* basidia assayed *in vitro*. *Cell Biology International Reports*, **12**, 519–526.

Chiu, S.W. and Moore, D. (1990). Sporulation in *Coprinus cinereus*: Use of an *in vitro* assay to establish the major landmarks in differentiation. *Mycological Research*, **94**, 249–253.

Chiu, S.W. and Moore, D. (1993). Cell form, function and lineage in the hymenia of *Coprinus cinereus* and *Volvariella bombycina*. *Mycological Research*, **97**, 221–226.

Chiu, S.W. and Moore, D. (1999). Sexual development of higher fungi. In: *Experimental Fungal Biology* (R.P. Oliver and M. Schweitzer, eds.), pp. 231–271. Cambridge University Press, Cambridge.

Cooke, R.C. and Rayner, A.D.M. (1984). *Ecology of Saprotrophic Fungi*. Longman, London.

Dawkins, R. (1989). *The Selfish Gene*, 2nd edn. Oxford University Press, Oxford.

De Groot, P.W.J., Visser, J., Van Griensven, L.J.L.D. and Schaap, P.J. (1998). Biochemical and molecular aspects of growth and fruiting of the edible mushroom *Agaricus bisporus*. *Mycological Research*, **102**, 1297–1308.

Elliott, C.G. (1994). *Reproduction in Fungi. Genetical and Physiological Aspects*. Chapman & Hall, London.

Fermor, T.R. and Wood, D.A. (1981). Degradation of bacteria by *Agaricus bisporus* and other fungi. *Journal of General Microbiology*, **126**, 377–387.

Flegg, P.B. (1972). Response of the mushroom to temperature with particular reference to the control of cropping. *Mushroom Science*, **8**, 75–84.

Flegg, P.B. (1978a). Effect of temperature on sporophore initiation and development in *Agaricus bisporus*. *Mushroom Science*, **10**, 595–602.

Flegg, P.B. (1978b). Effect of temperature on the development of sporophores of *Agaricus bisporus* beyond a cap diameter of 2 mm. *Mushroom Science*, **10**, 603–609.

Gange, A.C., Gange, E.G., Sparks, T.H. and Boddy, L. (2007). Rapid and recent changes in fungal fruiting patterns. *Science*, **316**, 71.

Grant, W.D., Rhodes, L.L., Prosser, B.A. and Asher, R.A. (1986). Production of bacteriolytic enzymes and degradation of bacteria by filamentous fungi. *Journal of General Microbiology*, **132**, 2353–2358.

Hawker, L.E. (1939). The influence of various sources of carbon on the formation of perithecia by *Melanospora destruens* shear in the presence of accessory growth factors. *Annals of Botany (NS)*, **3**, 455–468.

Hawker, L.E. (1947). Further experiments on growth and fruiting of *Melanospora destruens* shear in the presence of various carbohydrates, with special reference to the effects of glucose and sucrose. *Annals of Botany (NS)*, **11**, 245–259.

Hawker, L.E. (1950). *Physiology of Fungi*. University of London Press, London.

Hawker, L.E. and Chaudhuri, D.D. (1946). Growth and fruiting of certain ascomycetous fungi as influenced by the nature and concentration of the carbohydrate in the medium. *Annals of Botany (NS)*, **9**, 185–194.

Hurst, L.D. and Peck, J.R. (1996). Recent advances in understanding of the evolution and maintenance of sex. *Trends in Ecology and Evolution*, **11**, 79–82.

Ingold, C.T. (2002). A love affair with giant polypores. *Mycologist*, **16**, 162–164.

James, A.P. (1960). The spectrum of severity of mutant effects. II. Heterozygous effects in yeast. *Genetics*, **45**, 1627–1648.

Jennings, D.H. (1995). *The Physiology of Fungal Nutrition*. Cambridge University Press, Cambridge.

Kahlos, K., Kiviranta, J.L.J. and Hiltunen, R.V.K. (1994). Volatile constituents of wild and *in vitro* cultivated *Gloeophyllum odoratum*. *Phytochemistry*, **36**(4), 917–922.

Kalisz, H.M., Moore, D. and Wood, D.A. (1986). Protein utilization by basidiomycete fungi. *Transactions of the British Mycological Society*, **86**, 519–525.

Kashangura, C., Hallsworth, J.E. and Mswaka, A.Y. (2006). Phenotypic diversity amongst strains of *Pleurotus sajor-caju*: Implications for cultivation in arid environments. *Mycological Research*, **110**, 312–317.

Kemp, R.F.O. (1977). Oidial homing and the taxonomy and speciation of basidiomycetes with special reference to the genus *Coprinus*. In: *The Species Concept in Hymenomycetes* (H. Clemençon, ed.), pp. 259–273. J. Cramer, Vaduz.

Kibby, G. (2006). Editorial. *Field Mycology*, **7**, 110.

Kinugawa, K. and Furukawa, H. (1965). The fruit-body formation in *Colybia velutipes* induced by lower temperature treatment of one short duration. *Botanical Magazine Tokyo*, **78**, 240–244.

Kües, U. and Liu, Y. (2000). Fruiting body production in basidiomycetes. *Applied Microbiology and Biotechnology*, **54**, 141–152.

Leslie, J.F. and Leonard, T.J. (1979a). Monokaryotic fruiting in *Schizophyllum commune*: Genetic control of the response to mechanical injury. *Molecular & General Genetics*, **175**, 5–12.

Leslie, J.F. and Leonard, T.J. (1979b). Three independent genetic systems that control initiation of a fungal fruiting body. *Molecular & General Genetics*, **171**, 257–260.

Madelin, M.F. (1956a). The influence of light and temperature on fruiting of *Coprinus lagopus* Fr. in pure culture. *Annals of Botany (NS)*, **20**, 467–480.

Madelin, M.F. (1956b). Studies on the nutrition of *Coprinus lagopus* Fr. especially as affecting fruiting. *Annals of Botany*, **20**, 307–330.

Madelin, M.F. (1960). Visible changes in the vegetative mycelium of *Coprinus lagopus* Fr. at the time of fruiting. *Transactions of the British Mycological Society*, **43**, 105–110.

McCracken, F.I. (1970). Spore production of *Hericium erinaceus*. *Phytopathology*, **60**, 1639–1641.

Merrill, W. and Cowling, E.B. (1966). The role of nitrogen in wood deterioration: Amount and distribution of nitrogen in fungi. *Phytopathology*, **56**, 1085–1090.

Meškauskas, A., Fricker, M.D. and Moore, D. (2004a). Simulating colonial growth of fungi with the neighbour-sensing model of hyphal growth. *Mycological Research*, **108**, 1241–1256.

Meškauskas, A., Jurkoniene, S. and Moore, D. (1999a). Spatial organization of the gravitropic response in plants: Applicability of the revised local curvature distribution model to *Triticum aestivum* coleoptiles. *New Phytologist*, **143**, 401–407.

Meškauskas, A., McNulty, L.J. and Moore, D. (2004b). Concerted regulation of all hyphal tips generates fungal fruit body structures: Experiments with computer visualisations produced by a new mathematical model of hyphal growth. *Mycological Research*, **108**, 341–353.

Meškauskas, A., Moore, D. and Novak Frazer, L. (1998). Mathematical modelling of morphogenesis in fungi. Spatial organization of the gravitropic response in the mushroom stem of *Coprinus cinereus*. *New Phytologist*, **140**, 111–123.

Meškauskas, A., Moore, D. and Novak Frazer, L. (1999b). Mathematical modelling of morphogenesis in fungi. 2. A key role for curvature compensation ('autotropism') in the local curvature distribution model. *New Phytologist*, **143**, 387–399.

Meyer, H. (1936). Spore formation and discharge in *Fomes fomentarius*. *Phytopathology*, **26**, 1155–1156.

Moore, D. (1998a). *Fungal Morphogenesis*. Cambridge University Press, New York, NY.

Moore, D. (1998b). Tolerance of imprecision in fungal morphogenesis. In: *Proceedings of the Fourth Conference on the Genetics and Cellular Biology of Basidiomycetes* (L.J.L.D. Van Griensven and J. Visser, eds.), pp. 13–19. The Mushroom Experimental Station, Horst, The Netherlands.
Moore, D. (2005). Principles of mushroom developmental biology. *International Journal of Medicinal Mushrooms*, **7**, 79–102.
Moore, D., Chiu, S.W., Umar, M.H. and Sánchez, C. (1998). In the midst of death we are in life: Further advances in the study of higher fungi. *Botanical Journal of Scotland*, **50**, 121–135.
Moore, D., McNulty L.J. and Meškauskas, A. (2006). Branching in fungal hyphae and fungal tissues: Growing mycelia in a desktop computer. In: *Branching Morphogenesis* (J. Davies, ed.), pp. 75–90. Landes Bioscience Publishing/Eurekah.com/Springer Science, Georgetown, TX.
Moore, D. and Meškauskas, A. (2006). A comprehensive comparative analysis of the occurrence of developmental sequences in fungal, plant and animal genomes. *Mycological Research*, **110**, 251–256.
Moore, D., Walsh, C. and Robson, G.D. (2005). A search for developmental gene sequences in the genomes of filamentous fungi. In: *Applied Mycology and Biotechnology, Vol. 5, Genes, Genomics and Bioinformatics* (D.K. Arora and R. Berka, eds.), pp. 169–188. Elsevier Science, Amsterdam.
Moore-Landecker, E. (1993). Physiology and biochemistry of ascocarp induction and development. *Mycological Research*, **96**, 705–716.
Ohga, S. (1999a). Effect of water potential on fruit body formation of *Lentinula edodes* in sawdust-based substrate. *Journal of Wood Science*, **45**, 337–342.
Ohga, S. (1999b). Evaluation of maturity by use of pH indicators in sawdust-based cultures of *Lentinula edodes*. *Journal of Wood Science*, **45**, 431–434.
Parmesan, C. and Yohe, G. (2003). A globally coherent fingerprint of climate change impacts across natural systems. *Nature*, **421**, 37–42.
Plunkett, B.E. (1956). The influence of factors of the aeration complex and light upon fruit-body form in pure cultures of an agaric and polypore. *Annals of Botany*, **20**, 563–586.
Raudaskoski, M. and Salonen, M. (1984). Interrelationships between vegetative development and basidiocarp initiation. In: *The Ecology and Physiology of the Fungal Mycelium* (D.H. Jennings and A.D.M. Rayner, eds.), pp. 291–322. Cambridge University Press, Cambridge.
Rayner, A.D.M. and Boddy, L. (1988). *Fungal Decomposition of Wood: Its Biology and Ecology.* Wiley, Chichester.
Salerni, E., Laganà, A., Perini, C., Loppi, S. and De Domonicus, V. (2002). Effects of temperature and rainfall on fruiting of macrofungi in oak forests of the Mediterranean area. *Israel Journal of Plant Sciences*, **50**, 189–198.
Schmidt, E.L. and French, D.W. (1983). Variation in germination percentage of basidiospores collected successively from wood decay fungi in culture. *Canadian Journal of Botany*, **61**, 171–173.
Schmit, J.P. (1999). Resource consumption and competition by unit restricted fungal decomposers of patchy substrates. *Oikos*, **87**, 509–519.
Scrase, R.J. and Elliott, T.J. (1998). Biology and technology of mushroom culture. In: *Microbiology of Fermented Food* (B.J.B. Wood, ed.), Vol. 2, pp. 543–584. Blackie, London.
Stadler, M. and Sterner, O. (1998). Production of bioactive secondary metabolites in the fruitbodies of macrofungi in response to injury. *Phytochemistry*, **49**, 1013–1019.
Stamets, P. (1993). *Growing Gourmet and Medicinal Mushrooms*. Ten Speed Press, Berkeley, CA.
Stočkus, A. and Moore, D. (1996). Comparison of plant and fungal gravitropic responses using imitational modelling. *Plant, Cell & Environment*, **19**, 787–800.
Straatsma, G., Ayer, F. and Egli, S. (2001). Species richness, abundance, and phenology of fungal fruit bodies over 21 years in a Swiss forest plot. *Mycological Research*, **105**, 515–523.
Turner, E.M. (1977). Development of excised sporocarps of *Agaricus bisporus* and its control by CO_2. *Transactions of the British Mycological Society*, **69**, 183–186.
Umar, M.H. and Van Griensven, L.J.L.D. (1997). Morphological studies on the life span, developmental stages, senescence and death of *Agaricus bisporus*. *Mycological Research*, **101**, 1409–1422.
Wasterlund, I. and Ingelog, T. (1981). Fruit body production of larger fungi in some young Swedish forests with special reference to logging waste. *Forest Ecology and Management*, **3**, 269–294.

Watling, R. and Moore, D. (1994). Moulding moulds into mushrooms: Shape and form in the higher fungi. In: *Shape and Form in Plants and Fungi* (D.S. Ingram and A. Hudson, eds.), pp. 270–290. Academic Press, London.

Wilkins, W.H. and Harris, G.C.M. (1946). The ecology of larger fungi. V. An investigation into the influence of rainfall and temperature on the seasonal production of fungi in a beechwood and a pinewood. *Annals of Applied Biology*, **114**, 161–165.

CHAPTER 6

Population Biology of Forest Decomposer Basidiomycetes

Jan Stenlid

Contents

Abstract

This chapter covers aspects of fungal individuality, the size and dynamics of individual mycelia and how its integrity is controlled through somatic incompatibility. It also addresses gene flow and dispersal in wood-decay basidiomycetes.

1. WHAT IS A POPULATION?

A population is a collection of actually or potentially interbreeding individuals of the same species living in a given geographic area. From this follows a need to define the individual and the outer boundaries of the population both in terms of

British Mycological Society Symposia Series
Published by Elsevier Ltd.

geographic distribution and the potential for interbreeding. The following text will address some particular issues related to the concept of the fungal individual, discuss how mycologists have interpreted the population boundaries and give some examples from the biology of basidiomycete wood decomposers.

2. THE FUNCTIONAL FUNGAL INDIVIDUAL

2.1 Nuclear State

Defining the fungal individual has been the topic of several reviews (Rayner *et al.*, 1984; Rayner, 1991a, 1991b; Malik and Vilgalys, 1999; Glass and Dementhon, 2006)). Basically the problems lie in the multitude of genetic constitutions present in Kingdom Fungi and in the fact that fungal vegetative mycelia may fragment, so the genetic entity—the genet—becomes composed of several ramets.

In basidiomycetes, genetically, the vegetative mycelium is typically a dikaryon where two nuclei with dissimilar mating types are maintained in a pair wise manner in each compartment throughout the whole mycelium (Burnett, 2003). In some species, however, the number of copies of each nuclear type is not under strict control. A more inclusive description of the nuclear state is, therefore, to use the term heterokaryon for a mated mycelium and homokaryon for a mycelium harbouring only one nuclear type. There is increasing evidence that the homokaryotic phase in certain cases might be prolonged, especially in protected substrata such as inside the wood of a large log or a living tree (Stenlid, 1994a; Garbelotto *et al.*, 1999; Redfern *et al.*, 2001). Such mycelia are able to act physiologically as functional units. From an evolutionary standpoint, homokaryons can be regarded as dead ends if they do not find a mate. On the other hand, they can function as facilitators for the establishment of subsequent genotypes in the resource and increase the possibilities for a second spore to establish in their vicinity. (Kemp, 1975). The domain occupied by a homokaryon is a selective substratum for conspecific spores capable of mating, but acts to exclude other competing species. In the phase of early colonization of a resource, this can extend and reinforce the spatiotemporal window for spore germination and increase the chances to establish a fertile heterokaryon. Based on this knowledge, species-specific spore traps have been developed using homokaryotic mycelia to catch conspecific spores in the environment (Adams *et al.*, 1984; James and Vilgalys, 2001; Edman *et al.*, 2004a, 2004b).

2.2 The Buller Phenomenon

Heterokaryotic basidiomycete mycelia can deliver one nucleus to unmated homokaryons in its vicinity. This ability has been called the 'Buller phenomenon' (Quintanilha, 1937). Potentially, the Buller phenomenon might lead to complicated networks of mating in a resource. In some basidiomycete species, although successfully mated, homokaryotic sectors can arise within a heterokaryotic mycelium. If such sectors come in contact with another heterokaryon, a remating may occur in line with the Buller phenomenon. Such reassortment of nuclei in somatic incompatibility interaction zones has been recorded in laboratory crossings (Hansen *et al.*, 1993a), and recently also in dense populations under field conditions (Johannesson and Stenlid, 2004). Minute mycelia would normally be out-competed but have the potential of growing away from competitors in the relatively sheltered environment inside wood fibres. An interesting aspect of this is that a nucleus, by combining with a large number of mates, can find optimal combinations for its genetic outfit. Selection can then in essence act on the individual nucleus instead on the heterokaryon level.

2.3 Genetic Limits of the Individual

For practical purposes, the individual has to be defined using a working definition and these units must be recognizable using objective detection criteria. The basidiomycetes themselves recognize self from non-self by an incompatibility system termed somatic incompatibility or vegetative incompatibility (Rayner *et al.*, 1984; Rayner, 1991b; Worrall, 1997; Malik and Vilgalys, 1999). This is an intraspecific postfusion recognition system that normally reduces the exchange of cytoplasm between dissimilar individuals. The system has been shown to be composed of one to three or possibly more genes with multiallelic loci (Hansen *et al.*, 1993b, 1994; Rizzo *et al.*, 1995; Guillaumin, 1998; Malik and Vilgalys, 1999; Marcais *et al.*, 2000; Kauserud *et al.*, 2006; Lind *et al.*, 2007). The exact molecular nature of the interaction is unknown, but recent advances in genome sequencing of basidiomycetes might offer new possibilities to track the actual genes. In Ascomycota, the recognition systems between hyphae have been studied in more detail (Glass *et al.*, 2000). Between 7 and 11 loci have been identified and both allelic (e.g. *Neurospora crassa*, Perkins, 1988) and non-allelic (*Podospora anserina*, Esser and Blaich, 1994) interactions between these have been reported. Protein products from these genes include mating factors, leucine-rich repeats, signal peptides, serine proteases, G-protein, ribonucleotide reductase and a prion analogue (Glass *et al.*, 2000). It is likely that the vegetative incompatibility systems in Asco- and Basidiomycota share some molecular characters but differ in others, for example in some ascomycete species the mating factor is involved in vegetative incompatibility but mating genes have never been shown to be involved in basidiomycete vegetative incompatibility.

The mitochondrial genome is normally uniparentally inherited in basidiomycetes and clonally propagated with genotypic diversity arising from mutation alone (May and Taylor, 1988). However, in the laboratory recombination between mitochondrial genomes can be demonstrated (Matsumoto and FukumasaNakai, 1996). Moreover, Saville *et al.* (1996) provided evidence from field samples of *Armillaria gallica* indicating recombination in the mitochondrial genome in the wild.

Genetic markers such as isozymes, RAPD, AFLP, microsatellites etc. have been used to identify genotypes and delimit individuals (Stenlid, 1985; Smith *et al.*, 1992; Roy *et al.*, 1997; Vainio and Hantula, 2000; Johannesson and Stenlid, 2004). Normally the results of the various methods are congruent from samples from nature (Stenlid, 1985; Guillaumin *et al.*, 1996; Roy *et al.*, 1997; Vasiliauskas and Stenlid, 1999). The power of resolution differs among methods. In outcrossing and outbreeding species, somatic incompatibility normally can detect genets in a local population, yet to reach the same level of discrimination ten or more randomly chosen loci of microsatellites, RAPD or AFLPs may be needed depending on the variation present in the population. One advantage of molecular markers compared to somatic incompatibility tests is that, when studying large numbers of specimens, comparisons can be made directly on banding patterns instead of having to perform a geometrically increasing number of pairing tests. When studying inbred situations with limited genetic differentiation, such as mycelia originating from spore progenies of the same fruit body, a suite of markers might be needed to identify genotypes (Korhonen, 1978a). One recently studied example of a highly inbred wood-decomposer basidiomycete is the dry-rot fungus *Serpula lacrymans* (Kauserud *et al.*, 2007). Here the genetic variation is very low due to several population genetic bottlenecks during the process of adaptation to a life in buildings. Almost no variation is detectable in AFLP (Kauserud *et al.*, 2004), somatic incompatibility (Kauserud, 2004) and microsatellite loci (Högberg *et al.*, 2006) in Europe, while slightly higher variation is found in Asian populations indicating several more recent lines of infestation in Asia.

The evolutionary relevance of maintaining a non-self detection system has been suggested to be a restriction of cytoplasmic virus infection (Milgroom, 1999). Another important aspect of the incompatibility system is that it functions as a way to maintain the integrity of the individual mycelium (Rayner, 1991b). However, the somatic incompatibility barrier is leaky, for example in *Heterobasidion annosum* both dsRNA and mitochondria, as well as nuclei, could move through the barrier (Ihrmark *et al.*, 2002). The leakiness can perhaps be seen as a trade-off between maintaining the resources for a specific nuclear combination and the possibilities offered by reassorting of nuclei mentioned above (Johannesson and Stenlid, 2004).

2.4 Outcrossing and Non-Outcrossing Species

Most basidiomycete species are outcrossing, producing variable offspring from their fruit bodies. However, not all species are outcrossing, and without a fully functional sexual process the spores produced in a basidiome will either contain identical nuclei (homothallic species) or two sister nuclei already mated in the basidium (pseudohomothallic species). Occasionally, both outcrossing and non-outcrossing lines may be present within the same species (Hallenberg, 1991). Similar genotypes can also be the result of mitotic spores, conidia or oida produced on mycelia (Kendrick and Watling, 1979). These spores can, for example be picked up by insects and spread over vast areas. Such clonality can be said to be dispersive while vegetative spread through a resource or set of spatially discontinuous resource gives rise to territorial clones (Anderson and Kohn, 1995).

For a while there were divergent views on whether 'the mycelium acted as an "individual" or as a "social unit"'. The suggestion with the latter was that individual mycelia can pool their resources and collaborate in the colonization and composition of a resource. This was first described by Buller (1931) for *Coprinus sterquilinus* colonizing dung balls (Rayner and Todd, 1979). The idea of the individualistic mycelium, where each genotype builds its reproductive output and fitness by monopolizing resources for its mycelium seemed to be in contrast with the unit mycelium idea, until it was realized that *C. sterquilinus* is a homothallic species (Rayner, 1991b). Thus all spores in the dung ball had the same genotype, hence the cooperation between mycelia originating from different spores can be regarded as fusing of ramets of the same genet.

3. DELIMITING THE POPULATION

3.1 Population Boundaries

One challenge in population biology is how to delimit the assemblage of individuals. Population can be anything from the individuals in a resource to all individuals within a species. The population can also be defined from its geographical borders. Practical limitations for wood-decomposing fungi have often been set by the size of the woodland or forest in which the fungus occurs. Although it might be possible to define the geographic borders they may not represent true borders for actual gene flow, since airborne basidiospores can travel over long distances, for example *H. annosum* spores have been collected on islands more than 300 km from any wood resources (Rishbeth, 1959). From a population genetics point of view, population limits can be defined as when gene flow is below a certain limit. Then a natural approach could be to set the borders at obstacles to gene flow/colonization. This could be either high mountain ranges or vast areas with non-colonizable environment, although what actually constitutes a strong barrier to gene flow is not clear for most fungal species. The frequency of mating then becomes a function of geographic distance, and it is almost impossible to give a fixed limit. Population genetics provides a tool by which we try to understand the range of the actual breeding population.

Also relevant here is the metapopulation approach (Hanski, 1999). In essence a metapopulation is a regional population of local populations each with a certain probability of going extinct. Uncolonized patches can then be recolonized from other populations within the metapopulation system. Hence, the long-term persistence of the species can only occur at the regional or metapopulation level.

The ultimate outer boundary for a population is set by the species limit. The boundary then depends on the species concept, be it morphological, biological or phylogenetic (Harrington and Rizzo, 1999; Taylor *et al.*, 2000; Kohn, 2005; Giraud *et al.*, 2006). While some fungal species defined by morphology show global geographic ranges, when fungal species are defined phylogenetically they typically harbour several to many endemic species (Taylor *et al.*, 2006). It has been further argued that, as a consequence of differences in rates of genetic and morphological change, genetic isolation occurs before a recognizable morphological change. The final step in speciation—reproductive isolation, also follows genetic isolation and may precede morphological change (Taylor *et al.*, 2006). In determining the breeding population both the biological and the phylogenetic species concepts provide basis for experimental work. In *Schizophyllum commune*, sequencing a number of genes followed by phylogenetic analysis showed that regional populations exist in this cosmopolitan species (James *et al.*, 1999). The biological species concept has been widely applied to wood decomposers (e.g. Korhonen, 1978a, 1978b; Chase and Ullrich, 1990; Hallenberg and Larsson, 1992). On many occasions, partial interfertility has been detected, which can be illustrated by the *H. annosum* species complex: European members of the S and F intersterility groups are intersterile, but both groups are interfertile with the North American S group (Capretti *et al.*, 1990; Stenlid and Karlsson, 1991). Intersterility is controlled by at least five genes, and it is necessary for at least one of the loci to have a common plus allele to achieve interfertility between two homokaryotic strains (Chase and Ullrich, 1990). The genes have recently been mapped, which should allow for further characterization of the interacting molecules (Lind *et al.*, 2005). The European S, P and F intersterility groups, with preferences for spruce, pine and fir, respectively, have been given scientific names, *H. parviporum*, *H. annosum* and *H. abietinum*, respectively (Niemelä and Korhonen, 1998).

3.2 Indirect Measures of Gene Flow

Genetic markers can be used to infer indirectly the gene flow between populations (Slatkin, 1985; Excoffier *et al.*, 1992). By studying the distribution of allele frequencies and the degree of heterozygosity in different populations, indexes of subdivision can be established according to the theory of neutral evolution (Wright, 1951). One common measure of population subdivision (Wright's subdivision) Fst, can be translated into gene flow through the formula $Fst = 1/(1+4\,Nm)$ for diploid or dikaryotic organisms, N is the effective population size and m is the fraction of migrants in the population per generation. Provided that the system is in equilibrium, Fst values below 0.1 indicate fairly high-gene flow, 0.1–0.2 low-gene flow and >0.2 indicate isolated populations. However, very large populations will only change gene frequencies extremely slowly through random drift and it might therefore be hard to detect recent subdivision.

Studies of genetic differentiation in wood-decomposing basidiomycetes show that very little subdivision normally occurs in common species such as *Fomitopsis pinicola* (Högberg *et al.*, 1995; Norden, 1997; Högberg and Stenlid, 1999), *H. annosum* (Stenlid *et al.*, 1994), *Phlebiopsis gigantea* (Vainio and Hantula 2000), *Cylindrobasidium evolvens* (Vasiliauskas *et al.*, 1998), *Chondrostereum purpureum* (Gosselin *et al.*, 1999) and *Trichaptum abietinum* (Kauserud and Schumacher, 2003a). In rare species, substantially higher Fst values have been calculated (Högberg *et al.*, 1999; Franzen *et al.*, 2007), indicating that small populations are isolated from each other and the populations might undergo processes leading to inbreeding. To date, studies on inbreeding effects on fitness of wood-decay fungi are scarce, but nevertheless indicate reduced viability of basidiospores (Edman *et al.*, 2004a). However, in the rare species *Fomitopsis rosea* low Fst values have also been reported. The interpretation of the conflicting results in this species can be that founder effects in small populations such as those sampled by Högberg *et al.* (1999) influence the Fst values, while the rather large populations sampled by Kauserud and Schumacher (2003b) probably still reflect the genetic structure of the previously continuous distribution that was present before modern forestry fragmented the habitats.

In *Phlebia centrifuga*, a relatively uncommon decayer of conifer wood in managed forests, populations outside the regions of continuous distribution of suitable forests, show signs of reduced genetic variation. The genetic isolation in the fragmented habitat (Franzen *et al.*, 2007) also matches the measured lack of spore catches on the same geographic scale (Edman *et al.*, 2004a, 2004b).

3.3 Direct Measures of Gene Flow

The establishment of new genets is either by immigration through spore dispersal or recruitment from spore sources within the population. Spore dispersal has the potential to connect populations over a large distance. The process of spore dispersal comprises several steps. First, the spores released into the turbulent air layer, beneath the basidiocarp. If the fruit body is in a forest, then spores will move in the relatively calm air within the canopy of the forest. The majority of spores settle within a few metres of the fruit body. Since sporocarps can produce thousands of spores, a few have the capacity to travel far. Models of the spore dispersal gradient in *H. annosum* suggest that occasionally a spore can disperse and germinate on a suitable resource 500 km away from its source (Stenlid, 1994b). This shows a potential for long-distance spread but it is not necessarily the actual situation for most populations. Spores have been detected 300–1,500 m from the closest known sporocarps in *H. annosum* (Möykkynen *et al.*, 1997) and *P. centrifuga* (Nordén and Larsson, 2000). Within a stand, dispersal seems not to be restricted and, although suitable logs are distributed in a clumped manner, the sporocarps are more randomly scattered among them (Edman and Jonsson, 2001).

In the atmosphere, spores are subject to UV radiation, and there can be 95% loss of vitality within 1–3 days (Kallio, 1973), but pigmented spores survive better than hyaline basidiospores (Burnett, 2003). Basidiospores are relatively small—5–10 μm—and it has been suggested that their movement in air can be compared to that of smoke particles. Schlesinger *et al.* (2006) reported an altered mycoflora on particles carried by dust storms from the Sahara compared to the airborne fungal flora on a clear day. Mims and Mims (2004) showed that dispersal of fungal spores were correlated with smoke particles carried over the Mexican gulf.

Actual spore catches, using species-specific spore trapping techniques, show that spore deposition is highly dependent on spores released within the nearest 3 km of the traps. Interestingly, both in common and uncommon species, spore catches have been minimal >20 km away from known distributions in controlled investigations (Möykkynen and Kontiokari, 2001; Edman *et al.*, 2004a, 2004b). By using spore traps, James and Vilgalys (2001) were able to study the genotypes of wind-borne spores of *S. commune*, and the molecular variation provided no clear evidence for dispersal over large, aquatic barriers within the Caribbean region, but instead suggested that spore-trapping experiments primarily reflect the local, established population (James and Vilgalys, 2001). By contrast, Hallenberg and Kuffer (2001) reported on spore catches of the rare *Peniophora aurantiaca*, during

24 h exposure in the city of Göteborg, Sweden, where the closest known natural occurrence was ~1,000 km away!

Airborne spore dispersal from one continent to another may occasionally happen but the likelihood of surviving the journey and then landing on an appropriate resource makes this scenario very unlikely (Fries, 1987). Cosmopolitan species normally show a distinct genetic structure (James *et al.*, 1999), have intersterility barriers among isolates from different continents (Burnett, 2003), or have clear differentiation of ecological niche (Hallenberg and Larsson, 1992). The conclusion must be that airborne spore establishment from one continent to another of wood-decay basidiomycetes is close to zero.

4. POPULATION DYNAMICS

4.1 Local Structures

Mycelia are modular and can fragment to form spatially disconnected units. Fragmentation of a mycelium might act to change the dynamics of colonization since each fragment might start a new independent colony. The dynamics can be described using exponential rather than linear models, with polycyclic and monocyclic dynamics as in plant pathogenic epidemiology (Van der Plank, 1963). On the other hand, mycelia can be large (see below and Chapter 1), although to what extent they then function as integrated units with concerted energy and nutrient use is not known. Integrated activity provides the potential for increasing competitive strength in foraging and/or in interactions with other organisms. This has implications for nutrient cycling; integrated mycelia are far less prone to nutrient limitations than isolated units (Cairney, 2005). Mycelial systems might transport nutrients over several to many metres (Boddy, 1999; Chapters 1 and 3), and saprotrophic mycelia translocate carbon from woody material to foraging mycelium and even more so to mycelium in connected woody resources (Wells and Boddy, 1995a; Wells *et al.*, 1995).

Interaction studies indicate that mycelial systems with large resources are stronger in combative interactions and in colonization (Holmer and Stenlid, 1993; Wells and Boddy, 1995b; Lindahl *et al.*, 2001). This relates to the concept of inoculum potential—the energy available for a mycelium at the point of infection—often used in plant pathology (Garrett, 1956). Studies so far have indicated that large, ancient (1,000 years) mycelia are genetically stable with no mutations within two studied genets, and concluded that either the mutation rate is lower than in other microorganisms or vegetative expansion of genets requires fewer cell divisions than previously thought (Hodnett and Anderson, 2000).

4.2 Population Demography

The basic entities for population dynamics can be described by

$$P = B + I - D - E$$

where P is the population size, B the birth rate, I the immigration, D the death rate and E the emigration. However, in basidiomycetes, B can be equated to spore dispersal within the population, I to spore recruitment from outside, D is the actual death of mycelia and emigration E must be omitted as the entire biomass of a fungus does not usually move out of an area. An exception is loss of entire resource units, for example logs washed away by rivers. Not surprisingly, there are few demographic studies of fungal populations (Worrall, 1999; Gustafsson, 2002). To perform demographic studies thoroughly, information on the dynamics of mycelial establishment and development, sporocarp dynamics including life-span, actual spore releases, mycelial turnover rates in various types of resources and spore dispersal gradients are needed. The growing interest in fungal conservation biology (Chapter 18) urges this approach.

4.3 Population Age and Size

In young populations, the size of ramets and genets is small, and many spore infections start new individual genets. With time populations tend to consist of a few large individuals, both for mycorrhizal and decay fungi (Dahlberg and Stenlid, 1990; Swedjemark and Stenlid, 1993). Many small individuals die out for stochastic reasons. As a genet grows older and proliferates in the environment it also grows larger. The fact that populations have a discrete distribution of genet sizes indicates that recruitment to a typical population is not continuous, but rather occurs at certain windows of opportunity following disturbance events. For unit restricted species (see Chapter 1) the size of ramets will not have a linear relationship with age, since they will grow in size until the outer border of the resource unit(s) is reached.

Expansion rates for saprotrophic basidiomycetes with non-unit-restricted growth is in the order of 0.3–1.5 m^{-1} $year^{-1}$ (Hansen and Hamelin, 1999). To get a reliable measure of the expansion rate an independent measure of age is needed. This is hard to achieve from markers within the mycelium and has to be sought from known historical events. Normally the maximum age can be calculated but there are also a number of cases where minimum age has been inferred from forestry operations or construction of roads etc. For example, genets of *Armillaria* were estimated to be older than a road that crossed through its distribution (Kile, 1983; Lygis *et al.*, 2005), and the age of *H. annosum* was dated from the onset of thinning operations in previously untouched stands (Bendz-Hellgren *et al.*, 1999). Another possibility may be to use the spike in ^{14}C from nuclear weapon trials date biological material to certain years (Levin and Kromer, 2004). This requires that mycelial structures are built during that period but maybe of use for dating fruit bodies of some perennial species. An indication of fruit-body age can also be obtained from the number of 'growth rings', though these may not always equate to years, as in trees (Niemelä, 1986).

The size distribution of genets within a population represents a footprint of how and when new genets were recruited and how they have expanded over time (Hansen and Hamelin, 1999). Obviously, non-unit-restricted fungi have the potential to grow much larger than those that are restricted to a single resource. Non-unit-restricted mycelia often alter between a foraging phase in their vegetative growth and an exploitative phase in resources encountered (Boddy,

1999; Chapter 1) and can form large territorial clones. Large areas occupied by single genets have been reported for species of *Armillaria* (Korhonen, 1978a; Kile, 1983; Rishbeth, 1991; Smith *et al.*, 1992; Guillaumin *et al.*, 1996; Legrand *et al.*, 1996; Ferguson *et al.*, 2003; Lygis *et al.*, 2005), the largest being 965 ha (Ferguson *et al.*, 2003), with biomasses in the order of 80 tons (Smith *et al.*, 1992). The implication is that spore infection is uncommon and that vegetative growth, mediated by cords, rhizomorphs and root-to-root contact infections of living root systems, is a much more common mode of spread in this genus. The competitive capacity as a saprotroph, as well as the inoculum potential as a pathogen, are both strengthened by the energy provided through the mycelial network as compared to a situation where basidiospores would be the initiating agent. Other species with a similar ability to spread between resource units are the pathogenic root-rot fungi *Phellinus weirii* (Dickman and Cook, 1989), *H. annosum* s.l. (Stenlid, 1985, 1987; Piri, 1996; Piri and Korhonen, 2007), *Inonotus tomentosus* (Lewis and Hansen, 1991), *Phellinus noxius* (Hattori *et al.*, 1996), and the saprotrophic *Mega collybia platyphylla* (Thompson and Rayner, 1982), *Resinicum bicolor* (Kirby *et al.*, 1990), *Phallus impudicus* (Boddy, 1999). *Hymenochaete corrugata* and *H. tabacina* can even move from tree-to-tree via branches glued together by pseudosclerotia (Ainsworth and Rayner, 1990; Stenlid and Holmer, 1991).

Other species have a limited ability to spread vegetatively between resource units and individual mycelia do not frequently extend outside, for example, a single tree. This was the case for *Phaeolous schweinitzii* (Childs, 1937; Barrett and Uscuplic, 1971), and *Collybia fusipes* (Marcais *et al.*, 1998). Inside the bole of a single tree, several genets of *Phellinus tremulae* were detected, indicating multiple establishments by basidiospores (Holmer *et al.*, 1994). Another interesting case of interspecific competition was observed in Douglas fir (*Pseudotsuga mensiezii*). Adams and Roth (1969) reported on a population structure of *Fomitopsis cajanderi* where several genets were present at the site of entrance into the broken tree stems while further down the stem only a few genotypes had expanded at the expense of the others. Typically, multiple spore infections will give rise to small populations of up to tens of decay mycelia in dead stems of conifers (Norden, 1997; Högberg *et al.*, 1999; Kauserud and Schumacher, 2002, 2003a).

Pathogenic root-rot fungi normally require a tight connection between root systems of host plants in order to spread vegetatively. Among the pathogenic species, it is interesting to compare *H. annosum* and *I. tomentosus* both of which have intermediate genet sizes in unmanaged forests, indicating that occasionally they establish via basidiospores in wounds etc. in natural conditions, and thereafter spread by root-to-root contacts (Stenlid, 1985; Lewis and Hansen, 1991). However, *H. annosum*, in contrast to *I. tomentosus*, is able to respond to disturbance in an ecosystem, caused by forestry, by increasing its spore establishment and become a true pest to conifer forestry in the northern hemisphere (Woodward *et al.*, 1998). This is mirrored by a decrease in average genet and ramet sizes (Swedjemark and Stenlid, 1993). Clear responses to forestry activities have also been described for the saprotrophic *R. bicolor* (Kirby *et al.*, 1990). Different size classes in the population indicated pulses of genet growth and resource colonization in response to thinning operations and availability of dead root systems.

Species with unit-restricted mycelia can form populations of genets inside resources. One early study of this was of *Fomitopsis cajanderi* infecting ice broken tops of Douglas fir. Close to the entry point several genets were present in the same cross-section of the tree while further away from the broken top one of the competing genotypes dominated (Adams and Roth, 1969). In a field experimental study on felled beech (*Fagus sylvatica*) logs, Coates and Rayner (1985) showed that in dense competing populations none of the genets may obtain large enough resource pools to support fruiting (see also Chapter 5). This illustrates that interspecific competition can reduce the fitness of all genotypes in an overcrowded population.

The distribution of genotypes in the boles of living trees has also been taken as an indication of the infection biology. Latent infection in beech has, for example been implied by the large-sized genets developing rapidly in trunks and branches after drought (Chapela and Boddy, 1988). Such large mycelia are thought to have been established during earlier phases of tree growth, and dormant propagules distributed extensively in the xylem were triggered to grow as mycelia by the onset of wood drying and increased aeration. Two other examples are provided by *P. tremulae* and *P. pini*. They initially establish in branches, the stubs of which ultimately become incorporated into trunk wood. In due course the sapwood becomes heartwood, again with decreased water content and improved aeration, triggering the development of active decay in the branch stubs buried in the heartwood (Haddow, 1938). The population structure of *P. tremulae* was consistent with this type of establishment (Holmer *et al.*, 1994).

In a set of studies of decay fungi invading bark stripping wounds of Norway spruce, Vasiliauskas *et al.* (1996) illustrated how mating system and infection biology affects the population biology of wood-inhabiting basidiomycetes. Populations of the outcrossing *C. evolvens* (Vasiliauskas *et al.*, 1998a) consisted of distinct genotypes in each tree. The insect vectored basidiomycetes *Amylostereum areolatum* and *A. chailletii* had populations with dispersive clones with ramets of the same genet detected in Sweden and Lithuania (Vasiliauskas *et al.*, 1998; Vasiliauskas and Stenlid, 1999). The pseudohomothallic *Stereum sanguinolentum* also had groups of vegetatively compatible (vc) isolates obtained from widely separated places — Lithuania, Finland, Sweden and Iceland (Vasiliauskas and Stenlid, 1998b; Stenlid and Vasiliauskas, 1998). However, these did not represent true clones as variation occurred inside the vc groups. Although genetic variation between vc groups was high and distributed in concordance with random mating, inside the vc groups the variation was much lower within populations from a single forest stand than between the stands. The pseudohomothallism apparently had lead to highly inbred lines within vc groups.

5. INSECT VECTORS

Some wood-decay fungi are dispersed by insects (Chapter 9), which influences the structure and dynamics of populations of mycelia. The relationship between the insects and fungi varies between being ephemeral to almost total ecological dependence with highly adapted morphological specializations from both partners. An example of low specificity is when wood-boring insects feed on spores of decay fungi and passively carry some spores to the next woody resource visited. A high degree of specialization is seen in the wood wasp–*Amylostereum* relationships (Slippers *et al.*, 2003). Several species of *Sirex* and *Urocerus gigas* are mutualistically associated with *A. areolatum* (*Sirex*) and *A. chailletii* (*Urocerus*). Arthrospores are carried in mycangia inside the body of the females. The spores are inoculated into wood during oviposition. Ramets of a single genet of *A. areolatum* have been isolated from trees scattered in a single forest and from points widely separated geographically in Lithuania, Sweden and Denmark (Vasiliauskas *et al.*, 1998). The dependence of the fungus in northern Europe on insect vectored spread is further illustrated by the fact that fruit bodies are very seldom encountered. There is less emphasis on vertical transmission in *A. chailletii* (Vasiliauskas and Stenlid, 1998). Here ramets are also distributed over large areas, but the higher genotypic variation in the fungus indicates that the insects occasionally pick up new genotypes that have gone through meiosis. The *Sirex–Amylostereum* system can cause significant damages to conifer plantation (Slippers *et al.*, 2003). The life cycles of the *Sirex–Amylostereum* symbionts, where larval development can go on unspotted for 1–3 years, makes the system easy to spread in trade. *S. noctilio–A. aerolatum* has been introduced into the southern hemisphere where they cause massive problems to *Pinus* plantation forestry. The population structure is highly clonal with essentially the same genet being spread to millions of trees in South Africa, Australia, New Zealand, Chile and Uruguay.

6. RAFTING

An interesting possibility for long-distance dispersal is for decay mycelia inhabiting logs that are washed from banks into rivers and transported to the open sea. Such logs are known to be able to drift for long distances from Siberian rivers to the Arctic region, later being washed up on shores hundreds of kilometres away in Iceland and Greenland. This potential route for long-distance migration of decay fungi has not been studied in detail and should attract more attention. It has implications both for understanding evolutionary history and quarantine issues.

7. MAN AS A VECTOR

There is definite evidence for man vectoring fungi over long distances. One well-established case is the movement of European *Armillaria* species to South Africa (Coetzee *et al.*, 2001). Dead and dying oak (*Quercus*) and other woody ornamental trees and shrubs showing signs and symptoms of *Armillaria* root rot were identified in the Company Gardens, Cape Town, South Africa, which were established in the mid-1600s. Molecular markers and pairing tests indicated that the infection centre consisted of a single genet belonging to the European clade of

A. mellea. The size of the genet suggested that the fungus was introduced into the gardens in Cape Town from Europe, perhaps on potted plants, such as grapes or citrus, more than 300 years ago.

Another case was first described by Gonthier *et al.* (2004). Using a phylogenetic approach, *H. annosum* of the North American clade was detected west of Rome, Italy, centred round a military camp established during the Second World War. The fungus subsequently spread to some of the surrounding *Pinus pinea* forests probably by spore infections in thinning operations.

Construction wood is another possibility for establishing decay. The genetic population structure of the dry-rot fungus *S. lacrymans* indicates that it has been spread by western civilization to buildings in temperate and boreal regions throughout the world (Kauserud *et al.*, 2007). One route may have been infected sailing ships that were unsuitable for further sailing and were instead partly used for house construction.

ACKNOWLEDGEMENTS

I am grateful to the Swedish Research Council and the Swedish Research Council for Environment, Agricultural Sciences and Spatial Planning for providing funds for parts of the work. Comments by Lynne Boddy have greatly helped to improve the manuscript.

REFERENCES

Adams, S.H. and Roth, L.F. (1969). Intraspecific competition among genotypes of *Fomes cajenderi* decaying young-growth Douglas-fir. *Forest Science*, **15**, 327–331.

Adams, T.J.H., Williams, E.N.D., Todd, N.K. and Rayner, A.D.M. (1984). A species-specific method of analyzing populations of basidiospores. *Transactions of the British Mycological Society*, **82**, 359–361.

Ainsworth, A.M. and Rayner, A.D.M. (1990). Aerial mycelial transfer by *Hymenochaete corrugate* between stems of hazel and other trees. *Mycological Research*, **94**, 263–288.

Anderson, J.B. and Kohn, L.M. (1995). Clonality in soilborne, plant-pathogenic fungi. *Annual Review of Phytopathology*, **33**, 369–391.

Barrett, D.K. and Uscuplic, M. (1971). The field distribution of interacting strains of *Polyporus schweinitzii* and their origin. *New Phytologist*, **70**, 581–598.

Bendz-Hellgren, M., Brandtberg, P.-O., Johansson, M., Stenlid, J. and Swedjemark, G. (1999). Growth rate of *Heterobasidion annosum* in *Picea abies* established on forest land and arable land. *Scandinavian Journal of Forest Research*, **14**, 402–407.

Boddy, L. (1999). Saprotrophic cord-forming fungi: Meeting the challenge of heterogeneous environments. *Mycologia*, **91**, 13–32.

Buller, A.H.R. (1931). *Researches on Fungi*, Vol. 4. Longmans Green, London.

Burnett, J. (2003). *Fungal Populations and Species*. Oxford University Press, Oxford, UK.

Cairney, J.W.G. (2005). Basidiomycete mycelia in forest soils: Dimensions, dynamics and roles in nutrient distribution. *Mycological Research*, **109**, 7–20.

Capretti, P., Korhonen, K., Mugnai, L. and Romagnoli, C. (1990). An intersterility group of *Heterobasidion annosum*, specialized to *Abies alba*. *European Journal of Forest Pathology*, **20**, 231–240.

Chapela, I.H and Boddy, L. (1988). Fungal colonization of attached beech branches. Spatial and temporal organization of communities arising from latent invaders in bark and functional sapwood, under different moisture regimes. *New Phytologist*, **110**, 47–57.

Chase, T.E. and Ullrich, R.C. (1990). Five genes determining intersterility in *Heterobasidion annosum*. *Mycologia*, **82**, 73–81.

Childs, T.W. (1937). Variability of *Polyporus schweinitzii* in culture. *Phytopathology*, **27**, 29–50.

Coates, D. and Rayner, A.D.M. (1985). Fungal population and community development in cut beech logs. I. Establishment via the aerial cut surface. *New Phytologist*, **101**, 153–171.

Coetzee, M.P.A., Wingfield, B.D., Harrington, T.C., Steimel, O., Coutinho, T.A. and Wingfield, M.J. (2001). The root rot fungus *Armillaria mellea* introduced into South Africa by early Dutch settlers. *Molecular Ecology*, **10**, 387–396.

Dahlberg, A. and Stenlid, J. (1990). Population structure and dynamics in *Suillus bovinus* as reflected by spatial distributions of fungal clones. *New Phytologist*, **115**, 487–493.

Dickman, A. and Cook, S. (1989). Fire and fungus in a mountain hemlock forest. *Canadian Journal of Botany*, **67**, 2005–2016.

Edman, M., Gustafsson, M., Stenlid, J. and Ericson, L. (2004a). Abundance and viability of fungal spores along a forestry gradient: Responses to habitat loss and isolation? *Oikos*, **104**, 35–42.

Edman, M., Gustafsson, M., Stenlid, J., Jonsson, B.G. and Ericson, L. (2004b). Spore deposition of wood-decaying fungi: Importance of landscape composition. *Ecography*, **27**, 103–111.

Edman, M. and Jonsson, B.G. (2001). Spatial pattern of downed logs and wood-decaying fungi in an old-growth *Picea abies* forest. *Journal of Vegetation Science*, **12**, 609–620.

Esser, K. and Blaich, R. (1994). Heterogenic incompatibility in fungi. In: *The Mycota I: Growth, Differentiation and Sexuality* (J.G.H. Wessels and F. Meidhardt, eds.), pp. 211–232. Springer Verlag, Berlin.

Excoffier, L., Smousse, P.E. and Quattro, J.M. (1992). Analysis of molecular variance inferred from metric distances among DNA haplotypes: Application to human mitochondrial DNAS restriction data. *Genetics*, **131**, 479–491.

Ferguson, B.A., Dreisbach, T.A., Parks, C.G., Filip, G.M. and Schmitt, C.L. (2003). Coarse-scale population structure of pathogenic *Armillaria* species in a mixed-conifer forest in the Blue Mountains of northeast Oregon. *Canadian Journal of Forest Research*, **33**, 612–623.

Franzen, F., Vasiliauskas, R., Penttilä, R. and Stenlid, J. (2007). Population genetics of the wood decay fungus *Phlebia centrifuga* P. Karst. in fragmented and continuous habitats. *Molecular Ecology*, in press.

Fries, N. (1987). The third benefactors lecture: Ecological and evolutionary aspects of spore germination in the higher basidiomycetes. *Transactions of the British Mycological Society*, **88**, 1–7.

Garbelotto, M., Cobb, F.W., Bruns, T.D., Otrosina, W.J., Popenuck, T. and Slaughter, G. (1999). Genetic structure of *Heterobasidion annosum* in white fir mortality centers in California. *Phytopathology*, **89**, 546–554.

Garrett, S.D. (1956). *Biology of Root-Infecting Fungi*. Cambridge University Press, London.

Giraud, T., Villareal, L.M.M.A., Austerlitz, F., Le Gac, M. and Lavigne, C. (2006). Importance of the life cycle in sympatric host race formation and speciation of pathogens. *Phytopathology*, **96**, 280–287.

Glass, N.L. and Dementhon, K. (2006). Non-self recognition and programmed cell death in filamentous fungi. *Current Opinion in Microbiology*, **9**, 553–558.

Glass, N.L., Jacobson, D.J. and Shiu, P.K.T. (2000). The genetics of hyphal fusion and vegetative incompatibility in filamentous ascomycete fungi. *Annual Review of Genetics*, **34**, 165–186.

Gonthier, P., Warner, R., Nicolotti, G., Mazzaglia, A. and Garbelotto, M.M. (2004). Pathogen introduction, as a collateral effect of military activity. *Mycological Research*, **108**, 468–470.

Gosselin, L., Jobidon, R. and Bernier, L. (1999). Genetic variability and structure of Canadian populations of *Chondrostereum purpureum*, a potential biophytocide. *Molecular Ecology*, **8**, 113–122.

Guillaumin, J.J. (1998). Genetic control of somatic incompatibility in *Armillaria ostoyae*—A preliminary analysis. In: *Ninth Conference on Root and Butt-Rot of Forest Trees* (C. Delatour, J.J. Guillaumin, B. Lung-Escarmet and B. Marcais, eds.), pp. 31–38. INRA, France.

Guillaumin, J.J., Anderson, J.B., Legrand, P., Ghahari, S. and Berthelay, S. (1996). A comparison of different methods for the identification of genets of *Armillaria* spp. *New Phytologist*, **133**, 333–343.

Gustafsson, M.H. (2002). Distribution and dispersal of wood-decay fungi occurring on Norway spruce logs, Doctorial Thesis, *Sylvestria* 246. Swedish University of Agricultural Sciences.

Haddow, W.R. (1938). The disease caused by *Trametes pini* in white pine (*Pinus strobes*). *Transactions of the Royal Canadian Institute*, **29**, 221–280.

Hallenberg, N. (1991). Speciation and distribution in Corticiaceae (Basidiomycetes). *Plant Systematics and Evolution*, **177**, 93–110.

Hallenberg, N. and Kuffer, N. (2001). Long-distance spore dispersal in wood-inhabiting Basidiomycetes. *Nordic Journal of Botany*, **21**, 431–436.

Hallenberg, N. and Larsson, E. (1992). Mating biology in *Peniophora cinerea* (basidiomycetes). *Canadian Journal of Botany*, **70**, 1758–1764.

Hansen, E.M. and Hamelin, R.C. (1999). Population structure of basidiomycetes. In: *Structure and Dynamics of Fungal Populations* (J.J. Worrall, ed.), pp. 251–282. Kluwer Academic Publishers, Dordrecht.

Hansen, E.M., Stenlid, J. and Johansson, M. (1993a). Somatic incompatibility, partial heterokaryosis, and new nuclear combinations in *Heterobasidion annosum*. *Mycological Research*, **97**, 1223–1228.

Hansen, E.M., Stenlid, J. and Johansson, M. (1993b). Genetic control of somatic incompatibility in the root rotting basidiomycete *Heterobasidion annosum*. *Mycological Research*, **97**, 1229–1233.

Hansen, E.M., Stenlid, J. and Johansson, M. (1994). Somatic incompatibility in *Heterobasidion annosum* and *Phellinus weirii*. In: *Proceedings of the 8th International Conference on Root and Butt Rots* (M. Johansson and J. Stenlid, eds.), pp. 323–333. Swedish University of Agricultural Sciences, Uppsala.

Hanski, I. (1999). *Metapopulation Ecology*. Oxford University Press, New York.

Harrington, T.C. and Rizzo, D.M. (1999). Defining species in Fungi. In: *Structure and Dynamics of Fungal Populations* (J.J. Worrall, ed.), pp. 43–72. Kluwer Academic Publishers, Dordrecht.

Hattori, T., Abe, Y. and Usugi, T. (1996). Distribution of clones of *Phellinus noxius* in a windbreak on Ishigaki Island. *European Journal of Forest Pathology*, **26**, 69–80.

Hodnett, B. and Anderson, J.B. (2000). Genomic stability of two individuals of *Armillaria gallica*. *Mycologia*, **92**, 894–899.

Högberg, N., Holdenrieder, O. and Stenlid, J. (1999). Population structure of the wood decay fungus *Fomitopsis pinicola*. *Heredity*, **83**, 354–360.

Högberg, N. and Stenlid, J. (1999). Population genetics of *Fomitopsis rosea*: A wood decay fungus of the old-growth European taiga. *Molecular Ecology*, **8**, 703–710.

Högberg, N., Stenlid, J. and Karlsson, J.-O. (1995). Genetic differentiation in *Fomitopsis pinicola* (Schwarts: Fr.) Karst. studied by means of arbitrary primed-PCR. *Molecular Ecology*, **4**, 675–680.

Högberg, N., Svegarden, I.B. and Kauserud, H. (2006). Isolation and characterization of 15 polymorphic microsatellite markers for the devastating dry rot fungus, *Serpula lacrymans*. *Molecular Ecology Notes*, **6**, 1022–1024.

Holmer, L., Nitare, L. and Stenlid, J. (1994). Population structure and decay pattern of *Phellinus tremulae* in *Populus tremula* as determined by somatic incompatibility. *Canadian Journal of Botany*, **72**, 1392–1395.

Holmer, L. and Stenlid, J. (1993). The importance of inoculum size for the competitive ability of wood decomposing fungi. *FEMS Microbiology Ecology*, **12**, 169–176.

Ihrmark, K., Johannesson, H., Stenström, E. and Stenlid, J. (2002). Transmission of double-stranded RNA in *Heterobasidion annosum*. *Fungal Genetics and Biology*, **36**, 147–154.

James, T.Y., Porter, D., Hamrick, J.L. and Vilgalys, R. (1999). Evidence for limited intercontinental gene flow in the cosmopolitan mushroom, *Schizophyllum commune*. *Evolution*, **53**, 1665–1677.

James, T.Y. and Vilgalys, R. (2001). Abundance and diversity of *Schizophyllum commune* spore clouds in the Caribbean detected by selective sampling. *Molecular Ecology*, **10**, 471–479.

Johannesson, H. and Stenlid, J. (2004). Nuclear reassortment between vegetative mycelia in natural populations of the basidiomycete *Heterobasidion annosum*. *Fungal Genetics and Biology*, **41**, 563–570.

Kallio, T. (1973). Influence of ultraviolet radiation on the colony formation of *Fomes annosus* diaspores suspended in water. *Karstenia*, **14**, 5–8.

Kauserud, H. (2004). Widespread vegetative compatibility groups in the dry-rot fungus *Serpula lacrymans*. *Mycologia*, **96**, 232–239.

Kauserud, H., Saetre, G.P., Schmidt, O., Decock, C. and Schumacher, T. (2006). Genetics of self/nonself recognition in *Serpula lacrymans*. *Fungal Genetics and Biology*, **43**, 503–510.

Kauserud, H., Schmidt, O., Elfstrand, M. and Högberg, N. (2004). Extremely low AFLP variation in the European dry rot fungus (*Serpula lacrymans*): Implications for self/nonself-recognition. *Mycological Research*, **108**, 1264–1270.

Kauserud, H. and Schumacher, T. (2002). Population structure of the endangered wood decay fungus *Phellinus nigrolimitatus* (Basidiomycota). *Canadian Journal of Botany*, **80**, 597–606.

Kauserud, H. and Schumacher, T. (2003a). Regional and local population structure of the pioneer wood-decay fungus *Trichaptum abietinum*. *Mycologia*, **95**, 416–425.

Kauserud, H. and Schumacher, T. (2003b). Genetic structure of Fennoscandian populations of the threatened wood-decay fungus *Fomitopsis rosea* (Basidiomycota). *Mycological Research*, **107**, 155–163.

Kauserud, H., Svegården, I.B., Saetre, G.P., Knudsen, H., Stensrud, Ö., Schmidt, O., Shuichi, D., Sugiyama, T. and Högberg, N. (2007). Asian origin and rapid global spread of the destructive dry rot fungus *Serpula lacrymans*. *Molecular Ecology*, in press.

Kendrick, B. and Watling, R. (1979). Mitospores in basidiomycetes. In: *The Whole Fungus* (B. Kendrick, ed.), pp. 473–545. National Museum of Natural Sciences, Ottowa, Canada.

Kemp, R.F.O. (1975). Oidia, plasmogamy and speciation in basidiomycetes. *Biological Journal of the Linnean Society*, **7**, 57–69.

Kile, G.A. (1983). Identification of genotypes and the clonal development of *Armillaria-luteobubalina* Watling and Kile in eucalypt forests. *Australian Journal of Botany*, **31**, 657–671.

Kirby, J.J.H., Stenlid, J. and Holdenrieder, O. (1990). Population structure and responses to disturbance of the basidiomycete *Resinicium bicolor*. *Oecologia*, **85**, 178–184.

Kohn, L.M. (2005). Mechanisms of fungal speciation. *Annual Review of Phytopathology*, **43**, 279–308.

Korhonen, K. (1978a). Interfertility and clonal size in the *Armillaria mellea* complex. *Kartenia*, **18**, 31–42.

Korhonen, K. (1978b). Intersterility groups of *Heterobasidion annosum*. *Communicationes Institutionales Forestales Fennica*, **94**, 1–25.

Legrand, P., Ghahari, S. and Guillaumin, J.J. (1996). Occurrence of genets of *Armillaria* spp. in four mountain forests in central France: The colonization strategy of *Armillaria ostoyae*. *New Phytologist*, **133**, 321–332.

Levin, I. and Kromer, B. (2004). The tropospheric $^{14}CO_2$ level in mid-latitudes of the Northern Hemisphere (1959–2003). *Radiocarbon*, **46**, 1261–1272.

Lewis, K.J. and Hansen, E.M. (1991). Vegetative compatibility groups and protein electrophoresis indicate a role for basidiospores in spread of *Inonotus tomentosus* in spruce forests of British Colombia. *Canadian Journal of Botany*, **69**, 1756–1763.

Lind, M., Olson, Å. and Stenlid, J. (2005). An AFLP-markers based genetic linkage map of *Heterobasidion annosum* locating three intersterility genes. *Fungal Genetics and Biology*, **42**, 519–527.

Lind, M., Stenlid, J. and Olson, Å. (2007). Genetics and QTL mapping of somatic incompatibility and intraspecific interactions in basidiomycete *Heterobasidion annosum*. *Fungal Genetics and Biology*, in press.

Lindahl, B.D., Stenlid, J. and Finlay, R.D. (2001). Effects of resource availability on mycelial interactions and P-32 transfer between a saprotrophic and an ectomycorrhizal fungus in soil microcosms. *FEMS Microbiology Ecology*, **38**, 43–52.

Lygis, V., Vasiliauskas, R., Larsson, K.-H. and Stenlid, J. (2005). Wood-inhabiting fungi in stems of *Fraxinus excelsior* in declining ash stands of northern Lithuania, with particular reference to *Armillaria cepestipes*. *Scandinavian Journal of Forest Research*, **20**, 337–346.

Malik, M. and Vilgalys, R. (1999). Somatic incompatibility in fungi. In: *Structure and Dynamics of Fungal Populations* (J.J. Worrall, ed.), pp. 123–138. Kluwer Academic Publishers, Dordrecht.

Marcais, B., Cael, O. and Delatour, C. (2000). Genetics of somatic incompatibility in *Collybia fusipes*. *Mycological Research*, **104**, 304–310.

Marcais, B., Martin, F. and Delatour, C. (1998). Structure of *Collybia fusipes* populations in two infected oak stands. *Mycological Research*, **102**, 361–367.

Matsumoto, T. and FukumasaNakai, Y. (1996). Mitochondrial DNA inheritance in sexual crosses of *Pleurotus ostreatus*. *Current Genetics*, **30**, 549–552.

May, G., Taylor, J.W. (1988). Patterns of mating and mitochondrial DNA inheritance in the agaric basiodiomycete *Coprinus cinereus*. *Genetics*, **118**, 213–220.

Milgroom, M.G. (1999). Viruses in fungal populations. In: *Structure and Dynamics of Fungal Populations* (J.J. Worrall, ed.), pp. 283–305. Kluwer Academic Publishers, Dordrecht.

Mims, S.A. and Mims, F.M. (2004). Fungal spores are transported long distances in smoke from biomass fires. *Atmospheric Environment*, **38**, 651–655.

Möykkynen, T. and Kontiokari, J. (2001). Spore deposition of *Heterobasidion annosum* coll. in *Picea abies* stands of North Karelia, eastern Finland. *Forest Pathology*, **31**, 107–114.
Möykkynen, T., von Weissenberg, K. and Pappinen, A. (1997). Estimation of dispersal gradients of S- and P-type basidiospores of *Heterobasidion annosum*. *European Journal of Forest Pathology*, **27**, 291–300.
Niemelä, T. (1986). Growing old: Twenty years of morphological changes in a polypore. *Windahlia*, **16**, 27–33.
Niemelä, T. and Korhonen, K. (1998). Taxonomy of the genus *Heterobasidion*. In: *Heterobasidion annosum Biology, Ecology, Impact and Control* (S. Woodward, J. Stenlid, R. Karjalainen and A. Hüttermann, eds.), pp. 27–34. CAB International, Wallingford, UK.
Norden, B. (1997). Genetic variation within and among populations of *Fomitopsis pinicola* (Basidiomycetes). *Nordic Journal of Botany*, **17**, 319–329.
Nordén, B. and Larsson, K.-H. (2000). Basidiospore dispersal in the old-growth forest fungus *Phlebia centrifuga* (Basidiomycetes). *Nordic Journal of Botany*, **20**, 215–219.
Perkins, D.D. (1988). Main features of vegetative incompatibility in *Neurospora crassa*. *Fungal Genetics Newsletter*, **35**, 44–46.
Piri, T. (1996). The spreading of the S type of *Heterobasidion annosum* from Norway spruce stumps to the subsequent tree stand. *European Journal of Forest Pathology*, **26**, 193–204.
Piri, T. and Korhonen, K. (2007). Spatial distribution and persistence of *Heterobasidion parviporum* genets on a Norway spruce site. *Forest Pathology*, **37**, 1–8.
Quintanilha, A. (1937). Contribution à l'étude génétique du phénomème de Buller. *Comptes Rendus Hebdomadaires des Séances de l'Académie des Sciences*, **205**, 745.
Rayner, A.D.M. (1991a). The phytopathological significance of mycelial individualism. *Annual Review of Phytopathology*, **29**, 305–323.
Rayner, A.D.M. (1991b). The challenge of the individualistic mycelium. *Mycologia*, **83**, 48–71.
Rayner, A.D.M., Ainsworth, A.M., Adams, T.J.H., Williams, E.N.D. and Todd, N.K. (1984). The biological consequences of the individualistic mycelium. In: *The Ecology and Physiology of the Fungal Mycelium* (D.H. Jennings, ed.), pp. 509–540. Cambridge University Press, Cambridge.
Rayner, A.D.M. and Todd, N.K. (1979). Population and community structure and dynamics of fungi in decaying wood. *Advances in Botanical Research*, **7**, 333–420.
Redfern, D.B., Pratt, J.E., Gregory, S.C. and MacAskill, G.A. (2001). Natural infection of Sitka spruce thinning stumps in Britain by spores of *Heterobasidion annosum* and long-term survival of the fungus. *Forestry*, **74**, 53–71.
Rishbeth, J. (1959). Dispersal of *Fomes annosus* Fr. and *Peniophora gigantea* (Fr.) Massee. *Transactions of the British Mycological Society*, **42**, 243–260.
Rishbeth, J. (1991). *Armillaria* in an ancient broadleaved woodland. *European Journal of Forest Pathology*, **21**, 239–249.
Rizzo, D.M., Rentmeester, R.M. and Burdsall, H.H. (1995). Sexuality and somatic incompatibility in *Phellinus gilvus*. *Mycologia*, **87**, 805–820.
Roy, G., Cormier, M., Hamelin, R.C. and Dessureault, M. (1997). Comparison of RAPD technique and somatic incompatibility tests for the identification of *Phlebiopsis gigantea* strains. *Canadian Journal of Botany*, **75**, 2097–2104.
Saville, B.J., Yoell, H. and Anderson, J.B. (1996). Genetic exchange and recombination in populations of the root-infecting fungus *Armillaria gallica*. *Molecular Ecology*, **5**, 485–497.
Schlesinger, P., Mamane, Y. and Grishkan, I. (2006). Transport of microorganisms to Israel during Saharan dust events. *Aerobiologia*, **22**, 259–273.
Slatkin, M. (1985). Gene flow in natural populations. *Annual Review of Ecology and Systematics*, **16**, 393–430.
Slippers, B., Coutinho, T.A., Wingfield, B.D. and Wingfield, M.J. (2003). A review of the genus Amylostereum and its association with woodwasps. *South African Journal of Science*, **99**, 70–74.
Smith, M.L., Bruhn, J.N. and Anderson, J.B. (1992). The fungus *Armillaria bulbosa* is among the largest and oldest living organisms. *Nature*, **356**, 428–431.
Stenlid, J. (1985). Population structure of *Heterobasidion annosum* as determined by somatic incompatibility, sexual incompatibility, and isoenzyme patterns. *Canadian Journal of Botany*, **63**, 2268–2273.

Stenlid, J. (1987). Controlling and predicting the spread of *Heterobasidion annosum* from infected stumps and trees of *Picea abies*. *Scandinavian Journal of Forest Research*, **2**, 187–198.
Stenlid, J. (1994a). Homokaryotic *Heterobasidion annosum* mycelia in Norway spruce stumps. In: *Proceedings of the 8th International Conference on Root and Butt Rots* (M. Johansson and J. Stenlid, eds.), pp. 249–253. Swedish University of Agricultural Sciences, Uppsala.
Stenlid, J. (1994b). Regional differentiation in *Heterobasidion annosum*. In: *Proceedings of the 8th International Conference on Root and Butt Rots* (M. Johansson and J. Stenlid, eds.), pp. 243–248. Swedish University of Agricultural Sciences, Uppsala.
Stenlid, J. and Holmer, L. (1991). Infection strategy of *Hymenochaete tabacina*. *European Journal of Forest Pathology*, **21**, 313–318.
Stenlid, J. and Karlsson, J.-O. (1991). Partial intersterility in *Heterobasidion annosum*. Intersterility in *Heterobasidion annosum*. *Mycological Research*, **95**, 1153–1159.
Stenlid, J., Karlsson, J.-O. and Högberg, N. (1994). Intraspecific genetic variation in *Heterobasidion annosum* revealed by amplification of minisatellite DNA. *Mycological Research*, **98**, 57–63.
Stenlid, J. and Vasiliauskas, R. (1998). Genetic diversity within and among vegetative compatibility groups of *Stereum sanguinolentum*. *Molecular Ecology*, **7**, 1265–1274.
Swedjemark, G. and Stenlid, J. (1993). Population dynamics of the root rot fungus *Heterobasidion annosum* following thinning of *Picea abies*. *Oikos*, **66**, 247–254.
Taylor, J.W., Jacobson, D.J., Kroken, S., Kasuga, T., Geiser, D.M., Hibbett, D.S. and Fisher, M.C. (2000). Phylogenetic species recognition and species concepts in fungi. *Fungal Genetics and Biology*, **31**, 21–32.
Taylor, J.W., Turner, E., Townsend, J.P., Dettman, J.R. and Jacobson, D. (2006). Eukaryotic microbes, species recognition and the geographic limits of species: Examples from the kingdom Fungi. *Philosophical Transactions of The Royal Society B, Biological Sciences*, **361**, 1947–1963.
Thompson, W. and Rayner, A.D.M. (1982). Spatial structure of a population of *Tricholomopsis platyphylla* in a woodland site. *New Phytologist*, **92**, 103–114.
Vainio, E.J. and Hantula, J. (2000). Genetic differentiation between European and North American populations of *Phlebiopsis gigantea*. *Mycologia*, **92**, 436–446.
Van der Plank, J.E. (1963). *Plant Diseases: Epidemics and Controls*. Academic Press, New York.
Vasiliauskas, R. and Stenlid, J. (1998a). Population structure and genetic variation in *Cylindrobasidium evolvens*. *Mycological Research*, **102**, 1453–1458.
Vasiliauskas, R. and Stenlid, J. (1998b). Influence of spatial scale on population structure of *Stereum sanguinolentum* in Northern Europe. *Mycological Research*, **102**, 93–98.
Vasiliauskas, R. and Stenlid, J. (1999). Vegetative compatibility groups of *Amylostereum areolatum* and *A. chailletii* from Sweden and Lithuania. *Mycological Research*, **103**, 824–829.
Vasiliauskas, R., Stenlid, J. and Johansson, M. (1996). Fungal flora of bark peeling wounds of *Picea abies* in central Sweden. *European Journal of Forest Pathology*, **26**, 285–296.
Vasiliauskas, R., Stenlid, J. and Thomsen, I.M. (1998). Clonality and genetic variation in *Amylostereum aerolatum* and *A. chailletii* from northern Europe. *New Phytologist*, **139**, 751–758.
Wells, J.M. and Boddy, L. (1995a). Phosphorus translocation by saprotrophic basidiomycete cord systems on the floor of a mixed deciduous woodland. *Mycological Research*, **99**, 977–980.
Wells, J.M. and Boddy, L. (1995b). Translocation of soil-derived phosphorus in mycelial cord systems in relation to inoculum resource size. *FEMS Microbiology Ecology*, **17**, 67–75.
Wells, J.M., Boddy, L. and Evans, R. (1995). Carbon translocation in mycelial cord systems of *Phanerochaete velutina* (DC.: Pers.) Parmasto. *New Phytologist*, **129**, 467–476.
Woodward, S., Stenlid, J., Karjalainen, R. and Hüttermann, A. (eds.). (1998). *Heterobasidion annosum Biology, Ecology, Impact and Control*. CAB International, Wallingford, UK.
Worrall, J.J. (1997). Somatic incompatibility in basidiomycetes. *Mycologia*, **89**, 24–36.
Worrall, J.J. (1999). Fungal demography: Mushrooming populations. In: *Structure and Dynamics of Fungal Populations* (J.J. Worrall, ed.), pp. 175–194. Kluwer Academic Publishers, Dordrecht.
Wright, S. (1951). The genetical structure of populations. *Annals of Eugenetics*, **15**, 323–354.

Section 2:
Interactions with other Organisms

CHAPTER 7

Interactions between Saprotrophic Fungi

Steve Woodward and **Lynne Boddy**

Contents

Abstract

Fungal competition for resources can be divided into primary and secondary resource capture. With the former resources have not already been colonized, unlike the latter where combat and antagonistic mechanisms are used to obtain and defend territory. Such interactions can be mediated: (1) at a distance; (2) following contact at the hyphal level; or (3) following contact at the mycelial level. Antagonisms at a distance and at the mycelial level are effected by volatile and diffusible chemicals including enzymes, toxins and other antifungal metabolites. At the hyphal level antagonism is via hyphal interference or parasitism, which again is chemically or enzymatically mediated but on a more localized scale. Interactions have largely been studied on artificial media with the attendant problems of interpretation because of large divergence from field conditions, although soil microcosms have provided a valuable tool for interactions between cord-forming fungi; interactions have also been investigated by inoculating pairs of saprotrophs directly into wood in the field. Microcosm and field-based knowledge of interactions is crucial, not least because of the role of interactions in ecosystem functioning, particularly nutrient cycling and release, their effects on decomposition rates and potential as biological control agents.

British Mycological Society Symposia Series
Published by Elsevier Ltd.

1. INTRODUCTION

Since the publication of *Decomposer Basidiomycetes, their Biology and Ecology* (Frankland *et al.*, 1982), which barely touched on interspecific interactions, considerable research effort has gone into understanding the fungal–fungal interactions that occur during the development of communities in soils and in plant litter. A critical distinction has been made between fungi with primary as opposed to secondary resource capture strategies (Rayner and Boddy, 1988). The former establish in a resource in the absence of competitors, whereas in the latter organisms already present in the resource are replaced. Complete absence of fungi occurs rarely if at all, as fungi are latently present in functional tissues, though in freshly felled wood, for example, their mycelia have not had time to develop extensively and the resource can be considered 'uncolonized' (Chapter 11). In natural ecosystems, resources free of microorganisms are probably rare; in the absence of human interventions, uncolonized woody resources may become available when branches break or trees fall during storms, or following animal damage to trees. In contrast, thinning and clear felling operations in managed forests suddenly expose large quantities of woody resource to colonization, favouring fungi with primary resource capture strategies.

The ability of the interacting fungi to capture and defend territory in resources is achieved through either antagonistic combative mechanisms (e.g. Dowson *et al.*, 1988; Boddy, 2000; Donnelly and Boddy, 2001) or non-selective replacement (Holmer and Stenlid, 1993). Such interactions and replacements result in changes in communities over time (Rayner and Boddy, 1988; Boddy, 2001; Chapter 11). This chapter reviews the significance, mechanisms and approaches to studying interspecific mycelial interactions.

2. MECHANISMS OF INTERACTIONS

Interactions can be mediated: (1) at a distance; (2) following contact at the hyphal level; and (3) following contact at the mycelial level (Boddy, 2000; see below). Antagonism at a distance is effected by volatile and diffusible chemicals, including enzymes, toxins and other antifungal metabolites. Two broad types of antagonism occur at the hyphal level—hyphal interference and parasitism. At the mycelial level—often termed gross mycelial contact—there are probably many different mechanisms, but again they involve the release of enzymes, toxins and other antifungal compounds. Regardless of the mechanism of antagonism, ultimately the outcome can be deadlock where neither species gains headway, replacement where one species wrests territory from the other, partial replacement where one species captures some but not all of the antagonist's territory or mutual replacement where one species takes some of the territory formerly occupied by the other and *vice versa* (Figure 1). Interactions often manifest themselves in natural organic resources as 'interaction zone lines' (Hyppel, 1968; Rayner, 1977; Rayner and Todd, 1979; Figure 1). In wood, these appear as narrow, often dark-coloured, lines in cross-section. They comprise pseudosclerotial plates

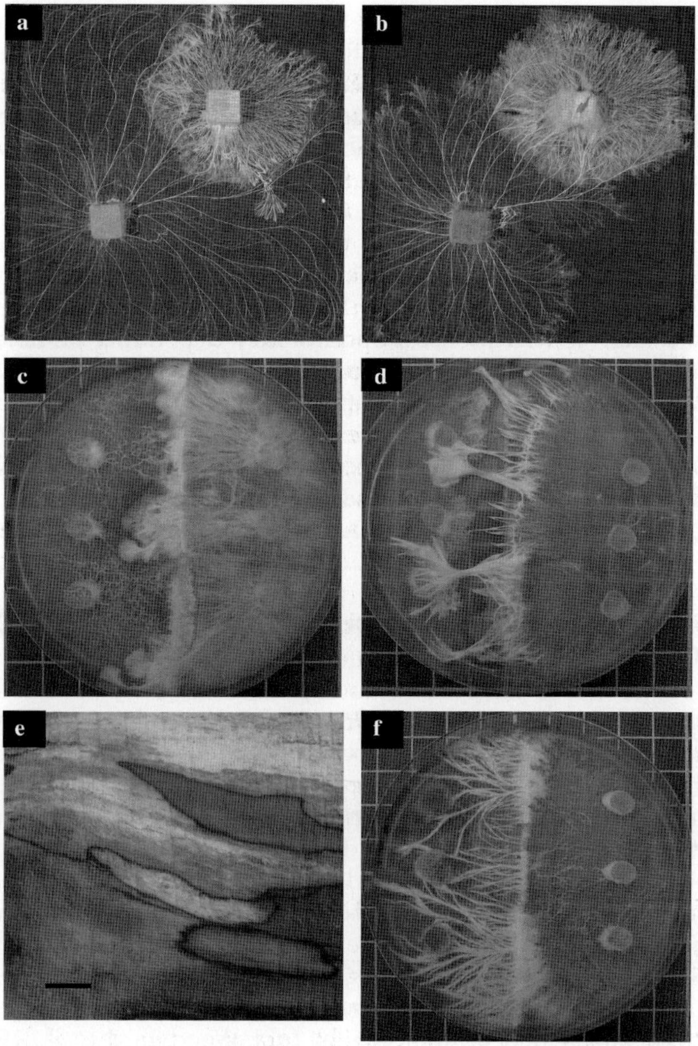

Figure 1 (a and b) Interactions between mycelial systems of *Phanerochaete velutina* (bottom left) and *Hypholoma fasciculare* 12 days after contact. Mycelia are growing from wood inocula across compressed non-sterile soil in 24 × 24 cm trays, incubated at 18–20 °C. (b) Mycelia have been grazed by collembola—80 *Folsomia candida* added 2 days after mycelia made contact. Note the more rapid progress of *P. velutina* across the *H. fasciculare* mycelium in the grazed microcosms, the wood inoculum of the latter being colonized with subsequent replacement. (c) *Resinicium bicolor* (left) being replaced by *H. fasciculare* on malt agar. Note brown pigment production in mycelium of *R. bicolor* in the vicinity of the invading mycelium. (d) *R. bicolor* (left) replacing *Phallus impudicus* on malt agar. Note the zone of lysed *R. bicolor* mycelium, which has subsequently been overgrown by cords of *R. bicolor*. (e) Interaction zone lines in *Fagus sylvatica* logs decaying on the forest floor for 4.5 years. Scale bar 1 cm. (f) *R. bicolor* (right) replacing *P. velutina* as mycelial cords. Digital images (a)–(d) and (f) courtesy of T.D. Rotheray. **(See Colour Section)**

(PSPs), which are often longitudinally extensive and completely surround the territory occupied by a decay fungus. Replacement is often evidenced by 'relic' zone lines (Figure 1f), i.e. PSPs that have been breached and partly decomposed by the invading fungus. Outcome of interactions can be inferred in natural communities, but this is fraught with difficulties, and most studies have therefore been conducted on artificial media or in microcosms in the laboratory (see below).

Which species ultimately dominate is determined by the relative abilities of the antagonists to capture and defend the resource (Rayner and Webber, 1984; Boddy, 1993, 2000). Fungi exhibit an hierarchy of combative ability (e.g. Coates and Rayner, 1985; Holmer and Stenlid, 1997a; Boddy, 2000). However, outcomes between particular combinations of species vary depending on abiotic regime, fungal strain, presence of other microorganisms (Schoeman *et al.*, 1996) and invertebrate grazing (T.D. Rotheray, unpublished; Figure 1a and b) but also can sometimes differ under apparently identical conditions (Boddy, 2000; Woods *et al.*, 2005). The size of the domain occupied by individual competing fungi is also of importance (Holmer and Stenlid, 1993, 1997a), as are the state of decay and other aspects of resource quality and quantity (Boddy, 2000).

3. INTERACTION CHEMISTRY: VOLATILE AND DIFFUSIBLE ORGANIC COMPOUNDS

When two fungi grow in close proximity, changes in both mycelial morphology and secondary compound chemistry occur, resulting in the formation of characteristic 'barrages' and colour changes in the mycelia of both species. These changes are mediated by up-regulation of genes involved in antagonism (Iakovlev *et al.*, 2004; C.A. Eyre, L. Boddy and H.J. Rogers, unpublished), resulting in production of stress compounds, enzymes and low molecular weight secondary metabolites, in the hyphae and into the surrounding environment. The production of such metabolites by one species may have profound effects on other species in the vicinity, mediating antagonism at a distance or following contact, or leading to attraction or stimulation of growth.

Fungi colonizing wood of both conifer (e.g. Woodward *et al.*, 1993; Woods, 1996) and angiosperm (Heilmann-Clausen and Boddy, 2005) trees secrete secondary metabolites in the absence of other fungi, which under natural conditions may function, at least in part, to protect occupied territory from potential invaders. These chemicals may have differential effects on species attempting to colonize, some antagonists being totally inhibited, the growth of others being slowed to a greater or lesser extent, whereas some are apparently unaffected and growth of others stimulated. Moreover, the composition of the secondary metabolites produced by a given species may alter when the presence of a potential competitor is detected (Griffith *et al.*, 1994b, 1994c; Woods, 1996; Hynes *et al.*, 2007). Some chemicals exert their effects indirectly, e.g. by lowering pH, whilst others can be considered to be antibiotics.

Although the production of antibiotic compounds by fungi has long been known (e.g. Fleming, 1929; Brian, 1951), several major questions must be addressed before the function of these compounds in natural environments can be corroborated. With few exceptions, it remains unclear whether such compounds occur in the natural environment in sufficient quantities to impact upon the growth of potential competitors. Other questions include: when and where are the compounds produced, and are they sufficiently persistent in the niche to impact significantly on other organisms?

Certainly, some antimicrobial compounds are persistent; methylbenzoates produced *in vivo* by the brown rot basidiomycete *Sparassis crispa* are present in occupied wood in large quantities, and persist for many years, even following death of the host tree (Woodward *et al.*, 1993). Such high concentrations represent a significant biosynthetic commitment by the fungus, implying considerable benefit to the producing organism. Lipid-soluble antifungal compounds were detected in 85% of fungi isolated from stumps of Sitka spruce (*Picea sitchensis*) when grown in defined liquid medium (Woods, 1996). Activity of these compounds against other fungi varied. For example, three compounds from liquid malt extract cultures of *Stereum sanguinolentum* were antifungal to the bioassay species *Cladosporium cucumerinum* on bioassay-lead thin-layer chromatography plates, but only one of these compounds inhibited mycelial growth of *Hypholoma fasciculare, Heterobasidion annosum* and *Resinicium bicolor.* The same compounds were inactive against *Truncatella* sp., *Melanotus proteus* and *S. sanguinolentum* itself (Figure 2). Active compounds with similar, though not identical, UV spectra were

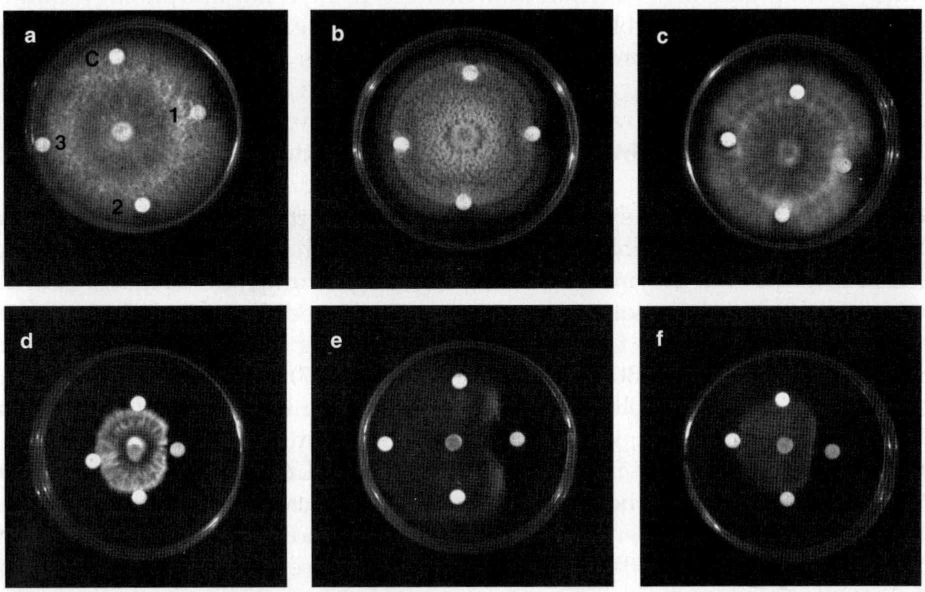

Figure 2 Antifungal activity of organic-phase extracts from 28-day-old 2% malt extract liquid cultures of *Stereum sanguinolentum* towards (a) *S. sanguinolentum*, (b) *Truncatella* sp., (c) *Melanotus proteus*, (d) *Hypholoma fasciculare*, (e) *Heterobasidion annosum* and (f) *Resinicium bicolor*. Antibiotic assay discs were soaked in antifungal compounds 1–3 obtained from *Cladosporium cucumerinum* bioassay-lead preparative TLC plates developed in hexane:ethyl acetate (1:1, v/v). Assay disc C lacked extracts.

also produced by *H. annosum*, *S. sanguinolentum*, *M. proteus* and *R. bicolor* in stumps of Sitka spruce inoculated with individual species. Moreover, additional active compounds were produced during interactions between pairs of fungal species inoculated into stumps.

Beech (*Fagus sylvatica*) wood decayed by a range of fungi contained diffusible metabolites with differential effects on a range of Asco- and Basidiomycota (Heilmann-Clausen and Boddy, 2005). *Stereum hirsutum*-decayed wood reduced growth of most of the fungi tested, whereas that colonized by *Fomes fomentarius* stimulated many species. These results suggested that the strategies utilized by decay fungi in defending an occupied resource may vary considerably *in vivo*: *S. hirsutum* appeared to utilize an antibiosis mechanism in defence; the strategy used by *F. fomentarius*, however, was unknown.

In addition to diffusible secondary metabolites, fungi also produce volatile organic compounds (VOCs; Wheatley, 2002), some of which may impact on potential competitors. VOCs can induce changes in the behaviour of potential competitors at a distance from the producing organism. Both soil and wood are highly porous materials, and the concentrations of VOCs in the lumen of a colonized xylem vessel or in soil pore space may be very high in the absence of rapid diffusion. In *in vitro* interactions between *R. bicolor* and *H. fasciculare* (Figure 1c), changes were noted in the VOCs produced, principally sesquiterpenes, with time (Hynes *et al.*, 2007). These changes appeared to be related temporally with pigment production and changes in mycelial morphology. Similar work on the effect of microbially produced VOCs on growth of *Serpula lacrymans* suggested an impact on protein synthesis (Humphris *et al.*, 2002).

4. INTERACTION CHEMISTRY: ENZYME PRODUCTION

During decomposition, basidiomycetes release many different enzymes (Chapter 2), and there is considerable evidence to suggest a role for up- and down-regulation of particular enzyme groups during fungal–fungal interactions. As with many aspects of interactions, the environmental conditions used during assay can greatly influence up- or down-regulation during experiments (Freitag and Morrell, 1992; White and Boddy, 1992; Lang *et al.*, 1997). Enzymatic production of reactive oxygen species, phenoloxidases and sometimes β-glucosidase increases (Freitag and Morrell, 1992; White and Boddy, 1992; Lang *et al.*, 1997; Iakovlev and Stenlid, 2000; Iakovlev *et al.*, 2003; Baldrian, 2006). Laccase enzymes appear to increase in contact zones between interacting decay-causing species (White and Boddy, 1992), with different temporal effects depending on the species tested (Iakovlev and Stenlid, 2000). Induction of the laccase response was elicited in *Marasmiellus troyanus* by cell-free filtrates of *Marasmius pallescens*, both tropical wood-decomposing basidiomycetes (Ferreira Gregorio *et al.*, 2006), and was accompanied by a concomitant increase in activity of a second group of lignin degrading enzymes—the manganese-dependent lignin peroxidases. Laccase activities also increased in cultures of *Trametes versicolor* and *Pleurotus ostreatus* following challenge with soil microorganisms, including fungi, bacteria and soil or soil extracts (Baldrian, 2004). Laccase does not directly affect decay fungi (Baldrian, 2004); its role is probably defensive, being involved in the production of melanins and similar compounds (Eggert *et al.*, 1995; Baldrian, 2006) which frequently accompany interspecific interactions, although this hypothesis requires further testing.

Fungal chitinases may also play a role in interspecific interactions. Wood decay-causing fungi with secondary resource capture strategies may produce chitinases during the colonization process, causing cell lysis and, therefore, releasing sequestered nitrogen from the thalli of previous colonizers (Lindahl and Finlay, 2006).

5. INTERACTIONS FOLLOWING CONTACT: PARASITISM AND HYPHAL INTERFERENCE

Parasitism between fungi has been widely utilized for practical purposes in the development of mitosporic fungi as biocontrol agents for use in horticulture and agriculture (Whipps, 2001), but has been infrequently studied in saprotrophic Basidiomycota. Host recognition by lectin or agglutinin–carbohydrate interaction is followed by penetration or appression to and growth along and around host hyphae (Chet *et al.*, 1997; Jeffries, 1997; Whipps, 2001). Both enzymes and toxins are produced by the parasitic species acting on the opposing fungal cell wall, enabling penetration of the host by the parasite, or causing lysis of host hyphae (Deane *et al.*, 1998; Howell, 1998; Vàsquez-Garcidueñas *et al.*, 1998). The nutrients exposed and/or released in these processes are then assimilated by the parasitic species for further growth and development. Various enzymes have been implicated in fungal–fungal parasitism, major examples being the β-glucanases and chitinases produced by *Trichoderma* species (Vàsquez-Garcidueñas *et al.*, 1998; Zeiligner *et al.*, 1999).

Potential parasitism by Basidiomycota has rarely been reported, but examples include *Pseudotrametes gibbosa* parasitic on *Bjerkandera* species, and *Lenzites betulina* on *Trametes* species (Rayner *et al.*, 1987). In both cases the parasitism is temporary; when the parasitized fungus has been killed its territory is taken over by the parasite which then operates other antagonistic mechanisms to defend and gain further territory.

Parasitism may be an extended form of the process known as hyphal interference. Hyphal interference occurs when hyphae of two species make contact, leading to the death of the contacted compartment of the susceptible species (Boddy, 2000). Although commonly studied in wood-decaying hymenomycetes (e.g. Ikediugwu *et al.*, 1970), similar processes occur during interactions amongst saprotrophic Ascomycota, and between hymenomycetes and Ascomycota or mitosporic fungi (Woods *et al.*, 2005). In this process, affected cells show vacuolation, loss of opacity, loss of hydrostatic pressure or hyphal swelling and rupture (Ikediugwu *et al.*, 1970; Holdenrieder, 1984; Woods *et al.*, 2005). Competing fungi may also produce thin hyphae compared with those produced in axenic cultures. These impacts on hyphal physiology result from the action of diffusible metabolites released when the interacting species are in close proximity. Interference can be extensive resulting in death; for example, *Phanerochaete magnoliae* overgrows and kills colonies of *Datronia mollis in vitro* (Ainsworth and Rayner, 1991). Small colonies of *H. annosum* in pine stumps and roots may be overgrown and killed by *Phlebiopsis gigantea* (see below).

6. GROSS MYCELIAL CONTACT

Gross mycelial contact is a 'catch-all' term covering interactions that do not involve parasitism, hyphal interference or interaction at a distance. When mycelia meet in agar or soil microcosms, dramatic changes in mycelial morphology occur, both in the vicinity of the antagonist and often elsewhere in the mycelium (Figures 1 and 3; Boddy, 2000; Donnelly and Boddy, 2001). Such changes include production of aerial tufts, barrages, mycelial cords (Figure 1d) and pigment (Figure 1c). Replacement interactions could be divided into those in which the following features are observed: (1) lysis occurs ahead of the advancing mycelium (Figure 1d); (2) overgrowth and through-growth of mycelium with more or less simultaneous death of the weaker antagonist; (3) overgrowth by aggregated mycelial structures with subsequent death of the weaker competitor. Overgrowth should not necessarily be equated with replacement, and must be confirmed, e.g. by isolation. Some fungi are able to overgrow others at the surface (e.g. of agar or compressed soil), but the overgrown fungus remains viable. This is particularly common with *Armillaria* species which may be overgrown but remain viable within a covering of PSPs. Overgrowth can itself sometimes be an important aspect of antagonism. For example, some cord-forming fungi (Chapter 1), which deadlock with opponents on soil, are able to grow over the opposing mycelium to reach the organic resource from which it is growing (Figure 1a and b). Confrontation within the resource can

Figure 3 Effect of the wood decay fungus *Phanerochaete velutina*, growing from a piece of wood (PV), on allocation of carbon to the extra-radical mycelium of the Mycorrhizal fungus *Paxillus involutus* growing in association with *Betula pendula* (a and b), and on *Suillus bovinus* in association with *Pinus sylvestris* in soil microcosms (d and f). (a) and (b) show ectomycorrhizal mycelial growth in the absence of the saprotroph. Plants were pulse-labelled with ^{14}C, and this was quantified in a 20×24 cm below-ground area by digital autoradiography (indicated by a white line in b). Autoradiographs are shown in (b), (e) and (f) with the radioactivity scale indicated in (b). In the interaction with *P. involutus* there were two patches of litter (L) in the microcosm to provide resources for the EM fungus. Note the truncation and browning of the *P. velutina* cords in contact with *P. involutus* (small arrow heads) and the deflection of the growth of the saprotroph on the right of the wood block (large arrow head). The mycorrhizal fungus allocates the carbon it receives from the host plant away from the area of territorial combat, and its growth is locally stopped by contact with *P. velutina*. There was a 60% reduction in ^{14}C allocated to mycelium of *S. bovinus* when interacting with *P. velutina*, up to 30 h after pulse labelling. Presence of ^{14}C (0.03%) was detected in *P. velutina* after 5 days. (a) and (b) Modified from Leake *et al.* (2002) and (c)–(f) modified from Leake *et al.* (2001) with permission from Elsevier and Blackwell. **(See Colour Section)**

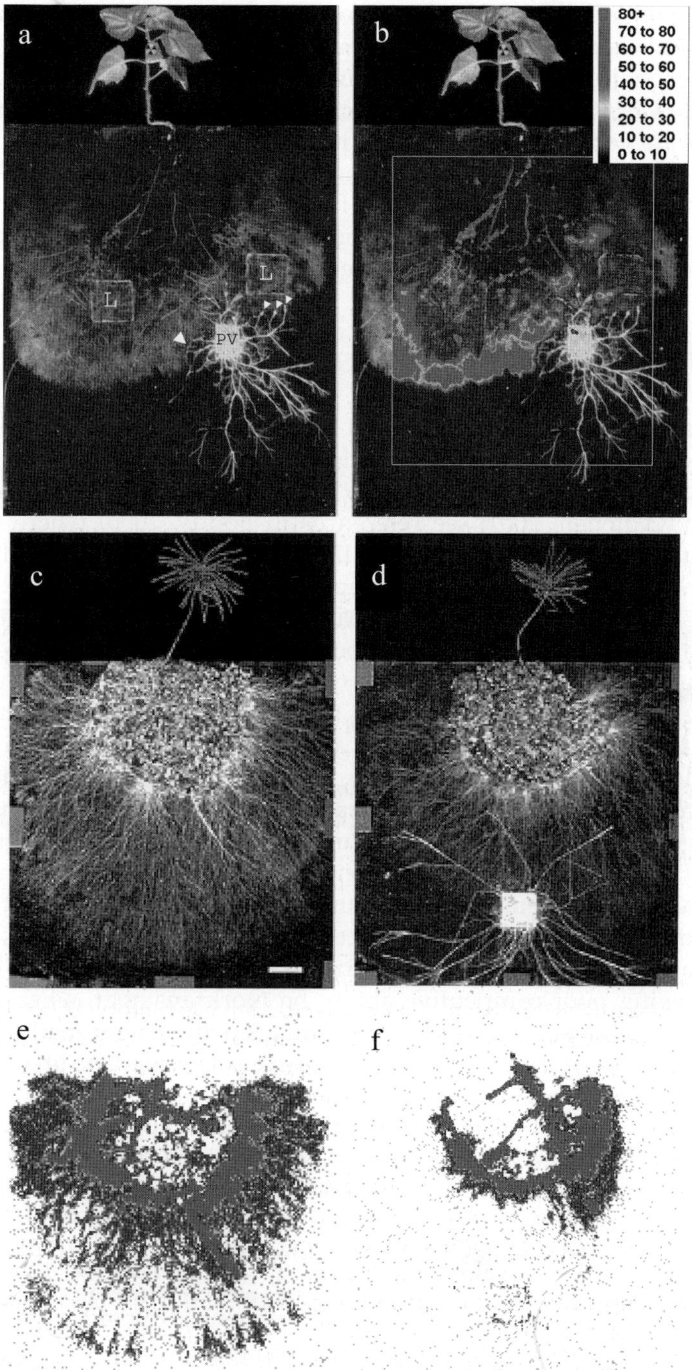
a
L
L
PV
b
80+
70 to 80
60 to 70
50 to 60
40 to 50
30 to 40
20 to 30
10 to 20
0 to 10
c
d
e
f

then result in replacement within the resource and the subsequent demise of extra-resource mycelium.

Since morphological changes are many and varied, it is likely that there is an equally wide variety of antagonistic mechanisms. Morphological changes during interactions are certainly correlated with differences in physiology, enzyme and toxin production (Griffith *et al.*, 1994a, 1994b, 1994c; Rayner *et al.*, 1994; see above). Changes may also occur distant from the site of interaction irrespective of morphology, and have major impacts on nutrient uptake, distribution and loss, and hence on ecosystem functioning (see below).

7. INVESTIGATING INTERACTIONS

Interactions between wood decay fungi have most commonly been investigated in pairs under controlled *in vitro* conditions using agar-based media (Cooke and Rayner, 1984; Rayner and Webber, 1984; Dowson *et al.*, 1988; Pearce, 1990; Murphy and Mitchell, 2001; see references in Boddy, 2000) where they can be clearly visualised (Figure 1e and f). Although it has been argued that valuable information can be gained from this approach (Henningson, 1967; Rayner, 1978; Boddy and Rayner, 1983; Rayner and Webber, 1984), interpretation of the results from such artificial environments may be misleading as responses are not always the same as in more natural systems (Boddy, 2000; Woods *et al.*, 2005, 2006). Contrasting results have been obtained between tests carried out on high sugar media and woody resources (Lundborg and Unestam, 1980; Nicolotti and Varese, 1996; Highley, 1997; Woods *et al.*, 2005). Moreover, differences in outcome often occur between different natural substrata; thus, the outcome of interaction in wood may be different from when mycelia encounter one another in soil (Dowson *et al.*, 1988). In research utilizing fungi isolated from stumps of Sitka spruce, interaction outcomes between fungi differed greatly on the defined, high sugar Norkrans agar compared to spruce-based wood media (Woods *et al.*, 2005). *R. bicolor*, a cord-forming wood colonizer, proved particularly striking in this respect, showing poor competitive ability on Norkrans agar, whereas on substrates based on Sitka spruce wood, including a sawdust–agar medium and autoclaved root blocks, it was the most aggressively competitive of all the species tested. This species rapidly colonizes spruce stumps in the field (Kirby *et al.*, 1990; Holmer and Stenlid, 1997b), out-competing other species of decay-causing hymenomycetes, even in co-inoculations (Woods *et al.*, 2006). In contrast, *Phaeolus schweinitzii*, cause of brown cubical rot in standing conifers, was the most aggressively competitive hymenomycete when tested on Norkrans agar, whereas on the wood-based substrates it was considerably less aggressive (Woods *et al.*, 2005).

Higher levels of complexity, and consequent difficulties in interpreting results, arise when attempting to include more than two species of fungi within an interaction test (e.g. White *et al.*, 1998; Boddy, 2000; Sturrock *et al.*, 2002; Sudin, 2005). Increasing size and complexity in *in vitro* multi-species interaction tests lead to less consistency in the interaction outcomes, and outcomes in three species interactions cannot be predicted based on the relevant two species interactions (White *et al.*, 1998; Boddy, 2000; Sudin, 2005).

In vitro systems based on natural resources may enable better assessment of the *in vivo* nature of interactions between saprotrophic fungi, despite the inherent difficulties in emulating microenvironmental conditions occurring in the field (Rayner and Webber, 1984). Use of trays of compressed, non-sterilized soil maintained under laboratory conditions has been particularly successful for emulating conditions of saprotrophic cord-forming Basidiomycota (Chapter 1) that extend between organic resources in woodlands at the soil litter interface (Donnelly and Boddy, 2001; Figure 1a and b).

When field inoculations into natural resources are utilized to test interspecific interactions, results can be rather difficult to interpret. Uncontrollable variations in environmental conditions, along with differences relating to timing of inoculation, may lead to highly variable results. Despite these possible pitfalls, it is important to consider interactions with respect to natural resources and environmental conditions, to develop a more complete understanding of the processes underlying community development.

Appropriate resources can be directly inoculated with pairs of interacting fungi in the field. Complications can arise, however, for several reasons, but particularly as species which were not deliberately inoculated may colonize the resource after the inoculations, confounding the results. Sitka spruce stumps in Scotland were inoculated with pairs of decay-causing hymenomycetes, as either pre-colonized sawdust inoculum or more discrete pre-colonized woody dowels (Woods *et al.*, 2006). In both inoculation types, many stumps also became colonized by wild strains of the decay-causing agaric *M. proteus*. However, this apparent contamination had little noticeable impact on the types of interactions observed between inoculated species.

8. ECOLOGICAL SIGNIFICANCE OF FUNGAL–FUNGAL INTERACTIONS

Monospecific populations of fungi rarely form within soil and organic resources; thus, interspecific mycelial interactions, with competition for resources, continually occur. These processes affect fungal community dynamics. Since different species effect decomposition, mineralization and nutrient translocation to different extents, outcome of interactions will impact directly on these processes. Moreover, the actual interaction may also affect these processes. For example, decay rate (measured as CO_2 evolution) increased when *Stereum gausapatum* was replaced by *Bjerkandera adusta*, and during deadlock between *L. betulina* and *Flammulina velutipes* in wood (Owens, 1989; Boddy, 2001). Occasionally there was a decrease in CO_2 evolution during interactions, e.g. during replacement of *Chondrostereum purpureum* by *S. hirsutum*. Interactions also affect movement and partitioning of carbon within mycelial systems. ^{14}C monitoring during interactions between saprotrophic cord-forming Basidiomycota in soil provided evidence of a carbon cost of encounter with a competing mycelium (Wells and Boddy, 2002). *P. velutina* switched reliance from the carbon available in the wood from which it grew to that available in the captured wood. It opportunistically utilized carbon previously mobilized by the opponent. In interactions that resulted in deadlock in soil, there was evidence of interspecific carbon exchange, presumably following leakage into soil from damaged hyphae in the interaction zone.

Mineral nutrient uptake, partitioning, movement and release are also affected by interspecific mycelial interactions. ^{32}P-uptake kinetics of cord-forming basidiomycetes was significantly affected by the presence of a competing mycelium (J.M. Wells and L. Boddy, unpublished). Changes in uptake capacity or rate constants were not related to the outcome of the interaction, but probably reflected the ability of species to divert effort preferentially from phosphorus scavenging to territory defence. In some interactions this ability was over-ridden when there was concurrent supply of uncolonized wood resources. There was reciprocal ^{32}P exchange between *R. bicolor, P. velutina* and *H. fasciculare* mycelia in soil which, as with carbon, presumably occurred via leakage in the interaction zone. In contrast, labelled mycelial systems of *P. impudicus* lost ^{32}P only to *R. bicolor,* and this was only detectable 39 days after ^{32}P supply. In addition to nutrient exchange between mycelia, there was movement of nutrients within mycelia during interactions. For example, *P. velutina* and one isolate of *H. fasciculare* preferentially translocated mobilized P to the zone of interaction in soil, whereas a less robust isolate of *H. fasciculare* preferentially translocated mobilized P *away* from the interaction zone (J.M. Wells and L. Boddy, unpublished).

Interspecific mycelial interactions are, together with grazing by invertebrates (Chapter 9), probably the main factor resulting in release of nutrients to soil. For example, interactions between *P. velutina* and *H. fasciculare,* in soil microcosms, resulted in significantly greater losses of ^{32}P to soil than self-pairings (J.M. Wells and L. Boddy, unpublished). Non-self-pairings within a species also resulted in significant losses to soil. In both cases leakage occurred not only in the interaction zone, but also elsewhere.

Many saprotrophic and ectomycorrhizal (EM) Basidiomycota are closely related, and the EM relationship is evolutionarily unstable, having been both gained and lost (Hibbett *et al.,* 2000). It is, therefore, not surprising that many EM hymenomycetes retain key enzyme systems of saprotrophic fungi (Leake and Read, 1997). Since EM hymenomycetes often colonize wood at very late stages of decomposition, and are found in organic soil and litter along with foraging saprotrophic Basidiomycota, there is likely to be intense competition for more labile nitrogen and phosphorus, and for antagonistic mycelial interactions (Lindahl *et al.,* 1999; Leake *et al.,* 2001, 2002, 2004). As with interactions between saprotrophic mycelia in soil, there can be significant transfer of nutrients between the two trophic groups (Lindahl *et al.,* 1999). There were also often marked effects of the saprotrophic *P. velutina* on carbon allocation to extra-radical EM mycelium of *Paxillus involutus* and *Suillus bovinus* (Leake *et al.,* 2001, 2002; Figure 3).

In boreal forest ecosystems, there may be partitioning of niche, reducing interactions between saprotrophic and EM hymenomycetes. Saprotrophs appear to dominate in the upper, organic soil horizons, whereas in lower, mineral layers EM species dominated; these findings suggest that the saprotrophic species lead in the release of carbon from organic materials, whereas the mycorrhizal fungi are the principal agents responsible for mobilization of nitrogen (Lindahl *et al.*, 2007).

In addition to the critical process of nutrient recycling, several hymenomycetes cause serious economically damaging disease in forest trees; although these fungi are often classified as pathogens, they spend much of their lives acting saprotrophically. Examples include species in the genera *Heterobasidion*, *Armillaria* and *Phellinus*. Increased understanding of the dynamics of interspecific competition may lead to the development of additional management intervention techniques for reducing the impact of these pathogens. The application of spore suspensions of *P. gigantea* to stumps of felled conifers, as a biological control of *H. annosum*, is well established (Holdenrieder and Greig, 1998). Other aggressive combatants, e.g. *R. bicolor*, may also be candidates for use as biological control agents (Holmer and Stenlid, 1997b).

9. CONCLUSIONS AND FUTURE PERSPECTIVES

Interactions between saprotrophic fungi clearly are of fundamental importance in ecosystem functioning. Along with EM fungi, these organisms lead in processes of mineralization of nutrients from dead organic sources; their relative abilities to capture a resource and defend it against other species are of paramount importance in the success of these species in a highly competitive environment. Research to date has illustrated the variations possible in interaction types and outcomes; some of the mechanisms involved in the interactions are understood, whilst others remain elusive. In the immediate future, there is much to be gained from the application of molecular methods to interspecific interactions, particularly using gene expression studies. Further research based on gene regulation studies, to include protein expression and changes in the fungal metabolome, will clarify the different mechanisms of interaction.

The use of molecular techniques in the dynamics of fungal ecology, rather than in examining fungal diversity *per se*, is in its infancy but effort is increasing. Questions that require attention include:

- Which fungal genes are up- or down-regulated during secondary colonization of resources, compared with primary resource capture?
- What patterns of gene expression are found during interspecific interactions?
- How does the proteome and metabolome alter following up- and down-regulation of genes during interactions?
- How are carbon and nitrogen (and other nutrients) allocated within the mycelium during interspecific interactions?
- Do the antifungal compounds produced by saprotrophic Basidiomycota function under natural conditions?
- Do changes in the secondary metabolite profiles of interacting fungi differ depending on the species combination?
- What is the site of action of the antifungal secondary metabolites?

Answers to these questions will greatly increase our understanding of the dynamics of saprotrophic Basidiomycota life strategies in ecosystem function.

REFERENCES

Ainsworth, A.M. and Rayner, A.D.M. (1991). Ontogenetic stages from coenocyte to basidiome and their relation to phenoloxidase activity and colonization processes in *Phanaerochaete magnoliae. Mycological Research*, **95**, 1414–1422.

Baldrian, P. (2004). Increase of laccase activity during interspecific interactions of white-rot fungi. *FEMS Microbiology Ecology*, **50**, 245–253.

Baldrian, P. (2006). Fungal laccases — Occurrence and properties. *FEMS Microbiology Reviews*, **30**, 215–242.

Boddy, L. (1993). Saprotrophic cord-forming fungi: Warfare strategies and other ecological aspects. *Mycological Research*, **97**, 641–655.

Boddy, L. (2000). Interspecific combative interactions between wood-decaying basidiomycetes. *FEMS Microbiology Ecology*, **31**, 185–194.

Boddy, L. (2001). Fungal community ecology and wood decomposition processes: From standing tree to complete decay of coarse woody debris. In: *Ecology of Woody Debris in Boreal Forests* (B.G. Jonsson and N. Kruys, eds.). *Ecological Bulletins*, **49**, 43–56.

Boddy, L. and Rayner, A.D.M. (1983). Mycelial interactions, morphogenesis and ecology of *Phlebia radiata* and *P. rufa* from oak. *Transactions of the British Mycological Society*, **80**, 437–448.

Brian, P.W. (1951). Antibiotics produced by fungi. *Botanical Reviews*, **17**, 357–430.

Chet, I., Inbar, J. and Hader, Y. (1997). Fungal antagonists and mycoparasites. In: *The Mycota. IV. Environmental and Microbial Relationships* (D.T. Wicklow and B. Söderström, eds.), pp. 165–184. Springer-Verlag, Heidelberg.

Coates, D. and Rayner, A.D.M. (1985). Fungal population and community development in cut beech logs, III: Spatial dynamics, interactions and strategies. *New Phytologist*, **101**, 183–198.

Cooke, R.C. and Rayner, A.D.M. (1984). *The Ecology of Saprotrophic Fungi*. Longman, London.

Deane, E.E., Whipps, J.M., Lynch, J.M. and Peberdy, J.F. (1998). The purification and characterization of a *Trichoderma harzianum* exochitinase. *Biochimica et Biophysica Acta*, **1383**, 101–110.

Donnelly, D.P. and Boddy, L. (2001). Mycelial dynamics during interactions between *Stropharia caerulea* and other cord-forming, saprotrophic basidiomycetes. *New Phytologist*, **151**, 691–704.

Dowson, C.G., Rayner, A.D.M. and Boddy, L. (1988). The form and outcome of mycelial interactions involving cord-forming decomposer basidiomycetes in homogeneous and heterogeneous environments. *New Phytologist*, **109**, 423–432.

Eggert, C., Temp, U., Dean, J.F.D. and Eriksson, K.E.L. (1995). Laccase-mediated formation of the phenoxazinone derivative, cinnabarinic acid. *FEBS Letters*, **376**, 202–206.

Ferreira Gregorio, A.P., Da Silva, I.R., Sedarati, M.R. and Hedger, J.N. (2006). Changes in production of lignin degrading enzymes during interactions between mycelia of the tropical decomposer basidiomycetes *Marasmiellus troyanus* and *Marasmius pallescens. Mycological Research*, **110**, 161–168.

Fleming, A. (1929). On the antibacterial action of cultures of a *Penicillium*, with special reference to their use in the isolation of *B. influenzae. British Journal of Experimental Pathology*, **10**, 226–236.

Frankland, J.C., Hedger, J.N. and Swift, M.J., eds. (1982). *Decomposer Basidiomycetes, their Biology and Ecology*, p. 335. Cambridge University Press, Cambridge.

Freitag, M. and Morrell, J.J. (1992). Changes in selected enzyme activities during growth of pure and mixed cultures of the white-rot decay fungus *Trametes versicolor* and the potential biocontrol fungus *Trichoderma harzianum. Canadian Journal of Microbiology*, **38**, 317–323.

Griffith, G.S., Rayner, A.D.M. and Wildman, H.G. (1994a). Interspecific interactions and mycelial morphogenesis of *Hypholoma fasciculare* (Agaricaceae). *Nova Hedwigia*, **59**, 47–75.

Griffith, G.S., Rayner, A.D.M. and Wildman, H.G. (1994b). Extracellular metabolites and mycelial morphogenesis of *Hypholoma fasciculare* and *Phlebia radiate* (Hymenomycetes). *Nova Hedwigia*, **59**, 311–329.

Griffith, G.S., Rayner, A.D.M. and Wildman, H.G. (1994c). Interspecific interactions, mycelial morphogenesis and extracellular metabolite production in *Phlebia radiate* (Aphyllophorales). *Nova Hedwigia*, **59**, 331–344.

Heilmann-Clausen, J. and Boddy, L. (2005). Inhibition and stimulation effects in communities of wood decay fungi: Exudates from colonized wood influence growth by other species. *Microbial Ecology*, **49**, 99–406.

Henningson, B. (1967). Interactions between micro-organisms found in birch and aspen pulpwood. *Studia Forestalis Suecica*, **53**, 1–32.

Hibbett, D.S., Gilbert, L.B. and Donoghue, M.J. (2000). Evolutionary instability of ectomycorrhizal symbioses in basidiomycetes. *Nature*, **407**, 506–508.

Highley, T.L. (1997). Control of wood decay by *Trichoderma (Gliocladium) virens*. I. Antagonistic properties. *Material und Organismen*, **31**, 79–89.

Holdenrieder, O. (1984). Untersuchungen zur biologischen Bekämpfung von *Heterobasidion annosum* an Fichte (*Picea abies*) mit antagonistichen Pilzen. II. Interaktionstests auf Holz. *European Journal of Forest Pathology*, **14**, 137–153.

Holdenrieder, O. and Greig, B.J.W. (1998). Biological control. In: *Heterobasidion annosum: Biology, Ecology, Impact and Control* (S. Woodward, J. Stenlid, R. Karjalainen and A. Hüttermann, eds.), pp. 235–258. CAB International, Wallingford.

Holmer, L. and Stenlid, J. (1993). The importance of inoculum size for the competitive ability of wood decomposing fungi. *FEMS Microbiology Ecology*, **12**, 169–176.

Holmer, L. and Stenlid, J. (1997a). Competitive hierarchies of wood decomposing basidiomycetes in artificial systems based on variable inoculum sizes. *Oikos*, **79**, 77–84.

Holmer, L. and Stenlid, J. (1997b). *Resinicium bicolor*; a potential biocontrol agent for *Heterobasidion annosum*. *European Journal of Forest Pathology*, **27**, 159–172.

Howell, C.R. (1998). The role of antibiosis in biocontrol. In: *Trichoderma and Gliocladium. Vol. 2. Enzymes, Biological Control and Commercial Applications* (G.E. Harman and C.P. Kubicek, eds.), pp. 173–184. Taylor & Francis, London.

Humphris, S.N., Bruce, A., Buultjens, E. and Wheatley, R.E. (2002). The effects of volatile microbial secondary metabolites on protein synthesis in *Serpula lachrymans*. *FEMS Microbiology Letters*, **210**, 215–219.

Hynes, J., Müller, C.T., Jones, T.H. and Boddy, L. (2007). Changes in volatile production during the course of fungal mycelial interactions between *Hypholoma fasciculare* and *Resinicium bicolor*. *Journal of Chemical Ecology*, **33**, 43–57.

Hyppel, A. (1968). Antagonistic effects of some soil fungi on *Fomes annosus* in laboratory experiments. *Studia Forestalia Suecica*, **64**, 1–18.

Iakovlev, A., Broberg, A. and Stenlid, J. (2003). Fungal modification of the hydroxyl radical detector coumarin-3-carboxylic acid. *FEMS Microbiology Ecology*, **46**, 197–202.

Iakovlev, A., Olson, Å., Elfstrand, M. and Stenlid, J. (2004). Differential gene expression during interactions between *Heterobasidion annosum* and *Physisporinus sanguinolentus*. *FEMS Microbiology Letters*, **241**, 79–85.

Iakovlev, A. and Stenlid, J. (2000). Spatiotemporal patterns of laccase activity in interacting mycelia of wood-decaying basidiomycete fungi. *Microbial Ecology*, **39**, 236–245.

Ikediugwu, F.E.O., Dennis, C. and Webster, J. (1970). Hyphal interference by *Peniophora gigantea* against *Heterobasidion annosum*. *Transactions of the British Mycological Society*, **54**, 307–309.

Jeffries, P. (1997). Mycoparasitism. In: *The Mycota. IV. Environmental and Microbial Relationships* (D.T. Wicklow and B. Söderstrom, eds.), pp. 149–164. Springer-Verlag, Heidelberg.

Kirby, J.J.H., Stenlid, J. and Holdenrieder, O. (1990). Population structure and responses to disturbance of the basidiomycete *Resinicium bicolor*. *Oikos*, **85**, 178–184.

Lang, E., Eller, G. and Zadrazil, F. (1997). Lignocellulose decomposition and production of lignolytic enzymes during interaction of white rot fungi with soil microorganisms. *Microbial Ecology*, **34**, 1–10.

Leake, J.R., Donnelly, D.P. and Boddy, L. (2002). Interactions between ectomycorrhizal and saprotrophic fungi. In: *Mycorrhizal Ecology* (M.G.A. Van der Heijden and I. Sanders, eds.). Springer-Verlag, Berlin, Heidelberg. *Ecological Studies*, **157**, 346–372.

Leake, J.R., Donnelly, D.P., Saunders, E.M., Boddy, L. and Read, D.J. (2001). Carbon flux to ectomycorrhizal mycelium following ^{14}C pulse labelling of *Pinus sylvestris* L. seedlings: Effects of litter patches and interaction with a wood-decomposer fungus. *Tree Physiology,* **21**, 71–82.

Leake, J.R., Johnson, D., Donnelly, D.P., Muckle, G., Boddy, L. and Read, D.J. (2004). Networks of power and influence: The role of mycorrhizal mycelium in controlling plant communities and agro-ecosystem functioning. *Canadian Journal of Botany,* **82**, 1016–1045.

Leake J.R. and Read D.J. (1997). Mycorrhizal fungi in terrestrial habitats. In: *The Mycota. IV. Environmental and Microbial Relationships* (D.T. Wicklow and B. Söderström, eds.), pp. 281–301. Springer-Verlag, Heidelberg.

Lindahl, B.D. and Finlay, R.D. (2006). Activities of chitinolytic enzymes during primary and secondary colonization of wood by basidiomycetous fungi. *New Phytologist*, **169**, 389–397.

Lindahl, B.D., Ihrmark, K., Boberg, J., Trumbore, S.E., Högberg, P., Stenlid, J. and Finlay, R.D. (2007). Spatial separation of litter decomposition and mycorrhizal nitrogen uptake in a boreal forest. *New Phytologist*, **173**, 611–620.

Lindahl, B., Stenlid, J., Olsson, S. and Finlay, R.D. (1999). Translocation of ^{32}P between interacting mycelia of a wood-decomposing fungus and ectomycorrhizal fungi in microcosm systems. *New Phytologist*, **144**, 183–193.

Lundborg, A. and Unestam, T. (1980). Antagonism against *Fomes annosus*. Comparisons between different methods *in vitro* and *in vivo*. *Mycopathologia*, **70**, 107–115.

Murphy, E.A. and Mitchell, D.T. (2001). Interactions between *Tricholomopsis rutilans* and ectomycorrhizal fungi in paired culture and in association with seedlings of lodgepole pine and Sitka-spruce. *Forest Pathology,* **31**, 331–344.

Nicolotti, G. and Varese, G.C. (1996). Screening of antagonistic fungi against air-borne infection by *Heterobasidion annosum*. *Forest Ecology and Management*, **88**, 249–257.

Owens, E.M. (1989). Fungal community structure and functioning in decomposing wood. PhD Thesis, University of Wales, Cardiff, UK.

Pearce, M.H. (1990). *In vitro* interactions between *Armillaria luteobubalina* and other wood decay fungi. *Mycological Research*, **94**, 753–761.

Rayner, A.D.M. (1977). Fungal colonization of hardwood stumps from natural sources. II. Basidiomycetes. *Transactions of the British Mycological Society,* **69**, 303–312.

Rayner, A.D.M. (1978). Interactions between fungi colonizing hardwood stumps and their possible role in determining patterns of colonization and succession. *Annals of Applied Biology,* **89**, 131–134.

Rayner, A.D.M. and Boddy, L. (1988). *Fungal Decomposition of Wood. Its Biology and Ecology.* Wiley, New York, NY.

Rayner, A.D.M., Boddy, L. and Dowson, C.G. (1987). Temporary parasitism of *Coriolus* spp. by *Lenzites betulina*: A strategy for domain capture in wood decay fungi. *FEMS Microbiology Ecology,* **45**, 53–58.

Rayner, A.D.M., Griffith, G.S. and Wildman, H.G. (1994). Induction of metabolic and morphogenetic changes during mycelial interactions among species of higher fungi. *Biochemical Society Transactions*, **22**, 389–394.

Rayner, A.D.M. and Todd, N.K. (1979). Population and community structure and dynamics of fungi in decaying wood. *Advances in Botanical Research*, **7**, 333–420.

Rayner, A.D.M. and Webber, J.F. (1984). Interspecific mycelial interactions — An overview. In: *The Ecology and Physiology of the Fungal Mycelium* (D.H. Jennings and A.D.M. Rayner, eds.), pp. 383–417. Cambridge University Press, Cambridge, UK.

Schoeman, M.W., Webber, J.F. and Dickinson, D.J. (1996). The effect of diffusible metabolites of *Trichoderma harzianum* on *in vitro* interactions between basidiomycete isolates at two different temperature regimes. *Mycological Research*, **100**, 1454–1458.

Sturrock, C., Ritz, K., Samson, W.B., Brown, J., Staines, H.J., Palfreyman, J.W., Crawford, J. and White, N.A. (2002). The effects of fungal inoculum arrangement (scale and context) on emergent community development in an agar model system. *FEMS Microbiology Ecology,* **39**, 9–16.

Sudin, M. (2005). Development of wood decay in *Parashorea malaanonan* (Blacno) Merr. wounded by logging. PhD Thesis, University of Aberdeen, Aberdeen, Scotland.

Vàsquez-Garcidueñas, S., Leal-Morales, C.A. and Herrera-Estrella, A. (1998). Analysis of the β-1, 3-glucolytic system of the biocontrol agent *Trichoderma harzianum*. *Applied and Environmental Microbiology*, **65**, 2741–2744.

Wells, J.M. and Boddy, L. (2002). Interspecific carbon exchange and cost of interactions between basidiomycete mycelia in soil and wood. *Functional Ecology*, **16**, 153–161.

Wheatley, R.E. (2002). The consequence of volatile organic compound mediated bacterial and fungal interactions. *Anton van Leeuwenhoek*, **81**, 357–364.

Whipps, J.M. (2001). Microbial interactions and biocontrol in the rhizosphere. Journal of Experimental Botany, **52**, 487–511.

White, N.A. and Boddy, L. (1992). Extracellular enzyme location during interspecific fungal interactions. *FEMS Microbiology Letters*, **98**, 75–80.

White, N.A., Sturrock, C., Ritz, K., Samson, W.B., Brown, J., Staines, H.J., Palfreyman, J.W. and Crawford, J. (1998). Interspecific fungal interactions in spatially heterogeneous systems. *FEMS Microbiology Ecology*, **27**, 21–32.

Woods, C.M. (1996). The fungal ecology of Sitka spruce stumps. PhD Thesis, University of Aberdeen, Aberdeen, Scotland.

Woods, C.M., Woodward, S., Pinard, M.A. and Redfern, D.B. (2006). Colonization of Sitka spruce stumps by decay-causing hymenomycetes in paired inoculations. *Mycological Research*, **110**, 854–868.

Woods, C.M., Woodward, S. and Redfern, D.B. (2005). *In vitro* interactions in artificial and wood-based media between fungi colonizing stumps of Sitka spruce. *Forest Pathology*, **35**, 213–229.

Woodward, S., Sultan, Y., Barrett, D.K. and Pearce, R.B. (1993). Two new antifungal metabolites produced by *Sparassis crispa* in culture and in decayed trees. *Journal of General Microbiology*, **139**, 153–159.

Zeiligner, S., Galhaup, C., Payer, K., Woo, S.L., Mach, R.L., Fekete, C., Lorito, M. and Kubicek, C.P. (1999). Chitinase gene expression during mycoparasitic interaction of *Trichoderma harzianum* with its host. *Fungal Genetics and Biology*, **26**, 131–140.

CHAPTER 8

Interactions between Saprotrophic Basidiomycetes and Bacteria

Wietse de Boer and **Annemieke van der Wal**

Contents

Abstract

Bacteria play an important role in the functioning of lignocellulose-degrading basidiomycetes. They can have a negative effect on fungal growth and activity as they are potential competitors for low-molecular weight compounds released by extracellular fungal enzymes. There are also some indications of bacterial mycophagy. On the other hand, basidiomycetes may benefit from the presence of bacteria, in particular with respect to nitrogen supply and detoxification of mycotoxic compounds.

During degradation of wood by basidiomycetes, the environmental conditions become very selective for bacteria because of rapid and strong acidification, production of reactive oxygen species and the presence of toxic fungal secondary metabolites. The bacteria that survive these conditions must have special properties, but research on this is still in its infancy. A better knowledge of the interactions between saprotrophic basidiomycetes and bacteria is not only important from a basic scientific point of view, but will also open up possibilities for new applications in wood conservation and the discovery of metabolites with medical therapeutic value.

British Mycological Society Symposia Series
Published by Elsevier Ltd.

1. INTRODUCTION

Saprotrophic basidiomycetes are continuously confronted with the presence of bacteria in their surroundings. Their interactions with bacteria can range from predatory and highly competitive to mutualistic. In this chapter we present an overview of the current knowledge on these interactions and the consequences for fungal functioning. The focus will be on interactions between wood-decomposing basidiomycetes and bacteria as these interactions have been most studied. Because of the focus on wood decay, we start with a short overview of our knowledge of wood-inhabiting bacteria.

1.1 Wood-Inhabiting Bacteria

Freshly felled wood on the soil surface is rapidly colonized by bacteria (Clausen, 1996). The origin of these bacteria can be the tree itself (endophytes), air, rain water or soil. The early colonizing bacteria are thought to grow on easily degraded substrates like sugars, organic acids, pectin and easily accessed cellulose (Schmidt, 2006). However, there are also bacteria that degrade parts of the lignified cell wall. Based on microscopic observations two types of bacterial wood degradation have been observed namely tunnelling and erosion (Daniel, 2003). Both erosion and tunnelling bacteria probably degrade cellulose (Daniel, 2003). This is a very slow process, and the contribution of these bacteria to wood decay is minor when fungi are present as well (Daniel, 2003). However, under wet conditions they can, together with soft-rot fungi, be the major degraders of wood (Schmidt, 2006).

Actinomycetes can be considered to be the bacterial counterpart of fungi as most of them are hyphal organisms with good capabilities for degrading insoluble organic polymers, for example chitin and cellulose (Goodfellow and Williams, 1983). The importance of actinomycetes in the degradation of wood is, however, still unclear. Some actinomycetes can degrade lignin-model compounds, but the enzymes involved in lignin degradation still remain elusive (Kirby, 2006). A soil isolate of *Streptomyces griseus* did degrade only parts of the cellulose of wood powder, and solubilized only some of the lignin (Arora *et al.*, 2005). If this is the general picture for actinomycetes then their contribution to decay of intact wood is probably minor.

1.2 Competitive and Antagonistic Interactions

Basidiomycetes degrading lignocellulose-rich material exude many different extracellular enzymes and small compounds, mediators, which are involved in the degradation of the different polymers. The oligomers released by the extracellular enzymes form the actual growth substrates for the fungi. These oligomers are, however, also appropriate substrates for most wood- and litter-inhabiting bacteria. Hence, whereas the contribution of bacteria to lignocellulose degradation is probably minor, they can profit from the degradation activities of fungi (Figure 1). Obviously, this could create a situation where the fungus is deprived of a large part of its growth substrates. How do lignocellulolytic basidiomycetes deal with this?

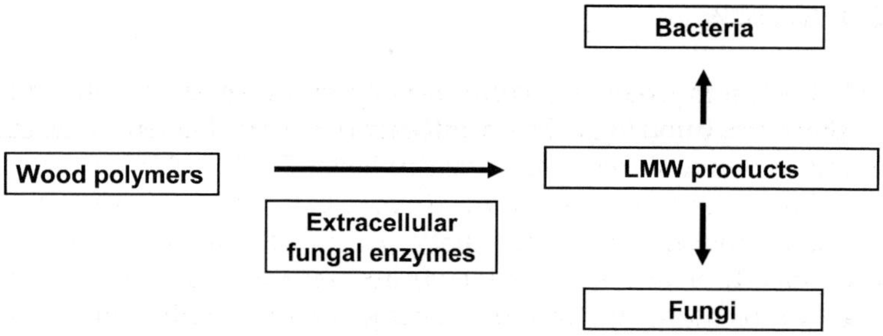

Figure 1 Competition between basidiomycetes and bacteria in decaying wood. The attack of wood polymers is mainly by fungal extracellular enzymes and mediators. The low-molecular weight products (LMW) released by the enzymes are the actual substrates for fungal growth. Bacteria are potentially strong competitors for LMW not for the wood polymers themselves. This situation creates dependency of the bacteria on fungal activity.

It seems that the most obvious strategy is inhibition of surrounding bacteria. Indeed, there are clear indications that such a strategy is employed by several basidiomycetes. The white-rot fungus *Pleurotus ostreatus* drastically reduces the number of bacterial colony-forming units in both soil and straw (Lang *et al.*, 1997; Gramms *et al.*, 1999). Similar findings have been reported for the fairy ring fungus *Marasmius oreades* during the degradation of soil organic matter (Smith, 1980). Gramms (1987) reported the killing of the microflora of timber blocks and sawdust around basidiomycete mycelia.

We have examined the effects of two white-rot fungi, *Hypholoma fasciculare* and *Resinicium bicolor*, on numbers of bacteria near exploratory hyphae (cords) in soil and on numbers of bacteria inhabiting beech wood blocks (Folman, Boddy and de Boer, unpublished results). Culturable bacteria increased slightly, but significantly, in the vicinity of soil mycelia. In contrast, numbers of culturable wood-inhabiting bacteria and total (detected by microscopy) bacteria were considerably reduced after colonization of the blocks by the rot fungi. The fact that not only numbers of colony-forming units but also total microscopic numbers decreased points at a fungal-induced lysis of wood-inhabiting bacteria. In a follow-up study with *H. fasciculare* we observed that decrease in numbers of wood-inhabiting bacteria was already apparent only a few weeks after colonization of the wood blocks by the fungus.

Several mechanisms can be involved in the killing of wood-inhabiting bacteria by basidiomycetes. Decay fungi can rapidly decrease the pH of the resources they colonize by exuding organic acids, for example oxalic acid. A rapid drop in pH is likely to be detrimental to many bacteria, in particular when undissociated forms of weak organic acids are present as well (Booth, 1985). These undissociated organic acids cross the bacteria cell membrane passively by diffusion and cause a drop in the intracellular pH of the bacteria. Bacteria have been shown to adapt to low pH, but this is a gradual process (De Boer *et al.*, 1995).

The production of reactive oxygen species, for example hydroxyl radicals, involved in the modification or degradation of lignin may also be harmful to many bacteria. There is some evidence that decay fungi actually inhibit bacteria in this process, for example production of hydroxyl radicals by the brown-rot fungus *Antrodia vaillantii* increased during interaction with the bacterium *Pseudomonas fluorescens* (Tornberg and Olsson, 2002). Bacteria surviving in wood colonized by decay fungi must have sufficient antioxidative activity to protect themselves from being attacked by free radicals.

Saptrotrophic basidiomycetes are known for their production of secondary metabolites, and these compounds may also play an important role in the inhibition or killing of bacteria (Lorenzen and Anke, 1998; Abraham, 2001; Liu, 2005). It has already been indicated that secondary metabolites can play an important role in the competition between wood-rot fungi (Heilmann-Clausen and Boddy, 2005).

Several basidiomycetes produce organohalogens, including chloroform which is a biocidal compound (Hoekstra *et al.*, 1998; Verhagen *et al.*, 1998). The role of organohalogen production by basidiomycetes is not clear, so far, but it has been suggested that they may also function as mediators in lignin degradation (Öberg *et al.*, 1997). Organohalogens may, however, also suppress competing fungi and bacteria, an aspect that has not yet been examined.

Secondary metabolites may also enhance the competitive position of bacteria against lignocellulolytic basidiomycetes. However, the bacteria are in an incongruous situation as the supply of growth substrates depends on the extracellular enzyme production of the fungus (Figure 1). Hence, strong suppression of the fungus would finally result in starvation of the bacteria. Antagonism of bacteria against fungi appears to be more profitable during the initial stage of wood decay when easily degradable compounds are present. This is a stage where both non-lignocellulolytic fungi and bacteria are present and antagonistic interactions have been indicated (Payne *et al.*, 2000). The bacterial community that develops during this initial decay stage may also inhibit or delay colonization of wood by lignocellulolytic fungi (Greaves, 1970). Hence, for a better understanding of bacterial antagonism against wood-rotting basidiomycetes, it is important to know whether the bacteria are early colonizers of wood or established/maintained during growth and activities of rot fungi.

1.3 Predation, Parasitism and Disease

Penetration of bacterial colonies and subsequent lysis of bacterial cells by cellulolytic and lignolytic basiodiomycetes, for example *P. ostreatus* and *Lentinula edodes*, has been well documented (Barron, 1988; Tsuneda and Thorn, 1994a; Barron, 2003). Nutrient-limiting conditions, in particular nitrogen limitation, seem to trigger this predatory behaviour of the fungi. Therefore, it has been proposed that the bacteria may form a valuable source of nitrogen for the fungi (Barron, 2003). Fungal predation upon bacterial colonies has been observed mostly in artificial media, but one case of fungal consumption of bacterial biomass that had developed on dead nematodes in wood has been reported (Tsuneda and Thorn, 1994b). We observed that two white-rot basidiomycetes, *H. fasiculare* and *R. bicolor*, strongly reduced the number of wood-inhabiting bacteria upon colonization of beech wood blocks (Folman *et al.*, submitted).

Microscopy revealed that the bacteria had actually disappeared, suggesting fungal-induced lysis of the bacteria. Future research should confirm whether the lysed bacterial biomass is actually consumed by the fungi.

Besides being predators, fungi can also be the prey of bacteria. Bacterial mycophagy has, however, been studied much less intensively than fungal mycophagy—mycoparasitism. The scarce reports on bacterial mycophagy, so far, deal almost exclusively with ascomycetes in an experimental setting. Streptomycetes, myxomycetes, paenibacilli, pseudomonads and the recently described genus *Collimonas* have been indicated as potential mycophagous bacteria (De Boer *et al.*, 2005). However, actual proof of importance *in situ* is difficult as all of these bacteria can grow very well on other energy sources, that is they are facultatively mycophagous.. Hence, an equivalent of *Bdellovibrio*—an obligate bacterial predator of bacteria, has not been found so far.

Tsuneda and Thorn (1994b) reported on lytic activities of *Agrobacterium tumefaciens* and *P. tolaasii* against wood-degrading basidiomycetes under laboratory conditions. The potential mycophagous bacterium *Collimonas fungivorans* has been observed in beech wood blocks colonized by the white-rot fungus *R. bicolor* (Folman, Boddy and de Boer, unpublished results).

Bacterial pathogens on fruit bodies of saptrotrophic basidiomycetes may be considered to be a special case of mycophagy. For obvious reasons, these bacterial pathogens have been most intensively studied on commercially grown edible mushrooms, for example *Agaricus bisporus* and *P. ostreatus,* but they have also been found on mushrooms sampled in the field (Bessette, 1984). The best-studied example is that of *P. tolaasii* causing brown blotch disease (Soler-Rivas *et al.*, 1999). When it is present at sufficient density, the bacterium induces disease by producing fungal membrane-disrupting secondary metabolites, in particular the lipodepsipeptide tolaasin. Hence, the bacterium is growing at the expense of hyphal contents that are released due to fungal membrane disruption. Similar strategies have also been found for some other *Pseudomonas* spp. as well as other bacterial species, for example *Janthinobacterium agaracidamnosum* (Lincoln *et al.*, 1999; Godfrey *et al.*, 2001).

1.4 Mutualistic Interactions

Litter- and wood-degrading basidiomycetes have developed strategies for suboptimal concentrations of nitrogen in their substrates. These strategies include recycling of nitrogen from senescent mycelium, reallocation from intracellular stored proteins and uptake and translocation of nitrogen from soil to wood/litter (Cowling and Merrill, 1966; Watkinson *et al.*, 2001; Lindahl and Finlay, 2005). As mentioned in the previous section, it has been proposed that lysis of bacteria may be another strategy in which saprotrophic basidiomycetes obtain nitrogen (Greaves, 1971; Tsuneda and Thorn, 1994b). This hypothesis can be extended to a mutualistic/predatory relationship with nitrogen-fixing bacteria. Nitrogen-fixing bacteria that are adapted to grow in the vicinity of wood-degrading basidiomycetes may provide a continuous source of nitrogen to the fungi. A mutualistic/predatory relationship would imply that only some of the bacteria are being lysed so that the bacterial density can be maintained due to growth of the bacteria on oligomers released by the fungal enzymes.

The importance of nitrogen input by nitrogen-fixing bacteria for decay activities of wood-rot fungi has already been proposed (Cowling and Merrill, 1966). Several reports have indicated the occurrence of nitrogen-fixation in decaying wood (e.g. Jurgensen *et al.*, 1989; Hendrickson, 1991; Brunner and Kimmins, 2003). Brunner and Kimmins (2003) showed that the highest nitrogen-fixation rates were found in the more advanced stages of decay.

Knowledge of the composition of the nitrogen-fixing bacterial communities in decaying wood is very limited. Older studies were largely based on physiological properties to identify culturable bacteria, and conclusions from such studies, for example the absence of the free-living nitrogen-fixing bacterium *Azospirillum* in decaying wood, should be considered with great care (Jurgensen *et al.*, 1989). Obviously, DNA- and/or RNA-based techniques are required to obtain a proper identification of both culturable and non-culturable bacteria that inhabit wood during basidiomycetal decay.

As indicated in the previous section, we observed that colonization of beech wood blocks by the white-rot fungus *H. fasciculare* coincided with a strong reduction in numbers of wood-inhabiting bacteria. A relatively high proportion (25%) of the culturable bacteria that survived the bactericidal effects belonged to the order *Rhizobiales* which are potentially nitrogen-fixing bacteria. These bacteria were not detected in beech wood blocks without *H. fasciculare*, but they may have been masked by the presence of high numbers of *Burkholderia*- and *Xanthomonas* strains. The occurrence of strains related to the methanotrophic bacterium *Methylocapsa acidiphilia*, family *Beijerinckiaceae* of the order *Rhizobiales*, in wood blocks colonized by *H. fasciculare* is of particular interest. *M. acidiphilia* is a nitrogen-fixing methanotrophic bacterium that was isolated from an acidic *Sphagnum* peat bog (Dedysh *et al.*, 2002). It grows with methane and methanol as substrates. Methanol may be the actual substrate for these bacteria, that is they may be methylotrophs rather than methanotrophs, in decaying wood as this is a side-product of fungal ligninolytic activities (Ander and Eriksson, 1984). So, the presence of these putative nitrogen-fixing, methylotrophic strains in wood decayed by white-rot fungi may indicate a mutualistic/predatory interaction namely growth of the bacteria on methanol released by fungal lignolytic activities and supply of nitrogen to the fungus via nitrogen-fixation and cell lysis (Figure 2).

Besides interactions with nitrogen-fixing bacteria, other mutualistic interactions between wood-decomposing basidiomycetes and bacteria have been proposed. Bacteria may provide essential growth factors, for example thiamine (Henningsson, 1967). However, this possibility has not been examined further. Greaves (1971) suggested that bacteria may also stimulate growth and activity of wood-degrading fungi by degrading toxic compounds. In fresh wood several compounds, collectively called wood extractives, occur that may inhibit growth and activity of basidiomycetes. Degradation of such compounds by bacteria has been reported (Burnes *et al.*, 2000; Kallionen *et al.*, 2003). However, this should perhaps be considered as facilitation rather than as mutualism. Only in those cases in which the toxic compounds are produced by the fungi themselves does

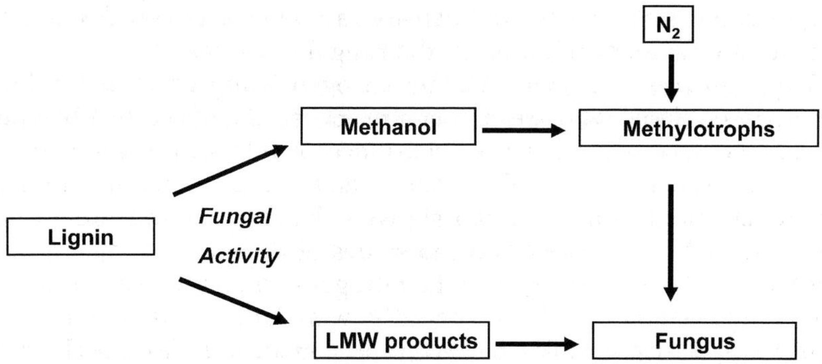

Figure 2 Hypothetical mutualistic interaction between wood-degrading basidiomycetes and nitrogen-fixing, methanol-degrading bacteria (methylotrophs). Methanol, a side-product of fungal lignin degradation, is the growth substrate for nitrogen-fixing bacteria. Part of the biomass of the methylotrophic bacteria is lysed by the fungus and used as nitrogen source. LMW, law molucular weight.

the term mutualism seem to be appropriate, as, for example in the aforementioned removal of methanol, a toxic side-product of fungal lignolysis, by wood-inhabiting methanotrophs.

Bacteria that detoxify fungal cell membrane disrupting compounds produced by bacterial pathogens of fruit bodies (see above) may also be considered as mutualists of basidiomycetes. Tsukamoto *et al.* (2002) isolated several tolaasin-detoxifying strains from wild *Agaricales*. Perhaps the presence of tolaasin-detoxifying strains on wild mushrooms explains why *P. tolaasii* is much more frequently isolated from cultivated mushrooms than from wild ones (Bessette, 1984). More detailed investigations are needed to understand the nutritional requirements of the antagonists of *P. tolaasii*. If they are preferentially selected by the fungus, for example via a resistance to antibacterial compounds and are growing on fungal exudates, this would be true mutualism.

Fruit body formation of several edible mushrooms is dependent on the presence of certain bacteria (Rainey *et al.*, 1990; Cho *et al.*, 2003). Evidence has been presented that this is due to the removal of fungal autoinhibitors by bacteria (Noble *et al.*, 2003). However, the bacteria may also exert a stress on the fungus which triggers fruit body formation.

1.5 Endobacteria

The occurrence of bacteria living in eukaryotic tissues or cells is well known for both plants and animals (Hoffmeister and Martin, 2003). The type of interaction that these endobacteria have with their host ranges from parasitism to mutualism. Much less attention has been given to bacteria living inside fungi. An exception to this, is the research of Bonfante and co-workers on endobacteria of arbuscular mycorrhizal fungi (Bianciotto and Bonfante, 2002). The endobacteria of AM fungi, which belong to the genus *Glomeribacter*, cannot be cultured, indicating an obligate dependence on their hosts.

The work on endobacteria of AM fungi has triggered investigations on the occurrence of endobacteria in other fungi. Recent reports indicate the presence of endobacteria in ectomycorrhizal basidiomycetes (Bertaux *et al.*, 2005; Izumi *et al.*, 2006). However, so far, there are no reports of endobacteria in litter- or wood-degrading basidiomycetes. As Lumini *et al.* (2006) pointed out, actual proof of the possession of endobacteria is not easily given. In particular, in field samples where mycelium can consist of younger and older parts, bacteria may enter damaged hyphae. These bacteria are then inside hyphae, but they are not true endobacteria.

1.6 Specific Associations

Specificity in the associations between saprotrophic basidiomycetes and bacteria may have developed in all the interaction types described. This could be the result of the strong selective environment that is created by lignocellulose-degrading basidiomycetes. Bacteria surviving under these conditions may either be competitors, mutualists or predators. However, so far, information on specificity of associations between saprotrophic basidiomycetes and bacteria is very limited. Seigle-Murandi *et al.* (1996) reported the specific association of three different bacterial strains with hyphae of the white-rot fungus *Phanerochaete chrysosporium*, but this could not be confirmed by others (Janse *et al.*, 1997).

We observed species-specific effects of the white-rot fungi *H. fasciculare* and *R. bicolor* on the bacterial community composition in decaying beech wood blocks, both at the level of bacterial genera and within a bacterial genus (*Burkholderia*) (Folman, Boddy and de Boer, unpublished results). Since both fungi acidify the wood environment and produce lignolytic enzymes (reactive oxygen species), it was to be expected that the specificity was due to differences in secondary metabolite (antibiotics) production.

The surface of fruit bodies of basidiomycetes has been the source for new bacterial species, but it is not known whether the occurrence of these bacteria is really restricted to these fungal environments (Tsukamoto *et al.*, 2001; Lim *et al.*, 2003).

1.7 Practical Applications

It was already recognized by Greaves (1971) that bacteria may have a strong impact on functioning of wood-degrading basidiomycetes. A practical application that was thought of was the use of antagonistic bacteria to prevent wood decay. However, in the following decades few efforts have been undertaken to examine this further (Murray and Woodward, 2003). This is remarkable since there is increasing awareness of the environmentally harmful effects of the use of wood preservatives (Lebow *et al.*, 2004).

Secondary metabolites of basidiomycetes often have antimicrobial properties and are screened for their use as novel medicines (Abraham, 2001; Liu, 2005). Wood-inhabiting bacteria have, so far, not been studied in this context. Yet, they may reveal a new source of antimycotica and antioxidants.

2. CONCLUSIONS AND PERSPECTIVES

Basidiomycetes degrading lignocellulose-rich material can have different types of interactions with bacteria. The knowledge of these interactions is, however, very limited and often based on observations and measurements using experimental conditions. Since bacteria can have both negative and positive effects on the growth and activity of lignocellulolytic fungi, it is important to understand if and how specific associations between bacteria and saprotrophic basidiomycetes have evolved. This is not only of basic scientific interest but it will also open up new possibilities for applications in biological wood conservation and medical therapies.

Both culturable and molecular methods are needed to study the bacteria associated with basidiomycetes. The former will immediately give insight into physiological characteristics and potentially applicable metabolites. The latter may also indicate the presence of unknown bacterial species that are associated with basidiomycetes.

REFERENCES

Abraham, W.R. (2001). Bioactive sesquiterpenes produced by fungi: Are they useful for humans as well? *Current Medical Chemistry*, **8**, 583–606.

Ander, P. and Eriksson, P.-E. (1984). Methanol formation during lignin degradation by *Phanerochaete chrysosporium*. *Applied Microbiology and Biotechnology*, **21**, 96–102.

Arora, A., Nain, L. and Gupta, J.K. (2005). Solid-state fermentation of wood residues by *Streptomyces griseus* B1, a soil isolate, and solubilization of lignins. *World Journal of Microbiology and Biotechnology*, **21**, 303–38.

Barron, G.L. (1988). Microcolonies of bacteria as a nutrient source for lignicolous and other fungi. *Canadian Journal of Botany*, **66**, 2505–2510.

Barron, G.L. (2003). Predatory fungi, wood decay, and the carbon cycle. *Biodiversity*, **4**, 3–9.

Bertaux, J., Schmid, M., Hutzler, P., Hartmann, A., Garbaye, J. and Frey-Klett, P. (2005). Occurrence and distribution of endobacteria in the plant-associated mycelium of the ectomycorrhizal fungus *Laccaria bicolor* S238N. *Environmental Microbiology*, **7**, 1786–1795.

Bessette, A.E. (1984). Distribution of brown blotch bacteria in wild and cultivated species of basidiomycetes. *Applied and Environmental Microbiology*, **48**, 878–880.

Bianciotto, V. and Bonfante, P. (2002). Arbuscular mycorrhizal fungi: A specialized niche for rhizospheric and endocellular bacteria. *Antonie van Leeuwenhoek*, **81**, 365–371.

Booth, I.R. (1985). Regulation of cytoplasmic pH in bacteria. *Microbiological Reviews*, **49**, 359–378.

Brunner, A. and Kimmins, J.P. (2003). Nitrogen fixation in coarse woody debris of *Thuja plicata* and *Tsuga heterophylla* forests on northern Vancouver island. *Canadian Journal of Forest Research*, **33**, 1670–1682.

Burnes, T.A., Blanchette, R.A. and Farrell, R.A. (2000). Bacterial biodegradation of extractives and patterns of bordered pit membrane attack in pine wood. *Applied and Environmental Microbiology*, **66**, 5201–5205.

Cho, Y.-S., Kim, J.-S., Crowley, D.E. and Cho, B.-G. (2003). Growth promotion of the edible fungus *Pleurotus ostreatus* by fluorescent pseudomonads. *FEMS Microbiology Letters*, **218**, 271–276.

Clausen, C.A. (1996). Bacterial associations with decaying wood: A review. *International Biodeterioration and Biodegradation*, **37**, 101–107.

Cowling, E.B. and Merrill, W. (1966). Nitrogen in wood and its role in wood deterioration. *Canadian Journal of Botany*, **44**, 1539–1544.

Daniel, G. (2003). Microview of wood under degradation by bacteria and fungi. In: *Wood Deterioration and Preservation, Advances in Our Changing World* (B. Goodell, D.D. Nicholas and T.P. Schultz, eds.), pp. 34–72. ACS Symposium Series 845.

De Boer, W., Folman, L.B., Summerbell, R.C. and Boddy, L. (2005). Living in a fungal world: Impact of fungi on soil bacterial niche development. *FEMS Microbiology Reviews*, **29**, 795–811.

De Boer, W., Klein Gunnewiek, P.J.A. and Laanbroek, H.J. (1995). Ammonium-oxidation at low pH by a chemolithotrophic bacterium belonging to the genus *Nitrosospira*. *Soil Biology and Biochemistry*, **27**, 127–132.

Dedysh, S.N., Khmelenina, V.N., Suzina, N.E., Trotsenko, Y.A., Semrau, J.D., Liesack, W. and Tiedje, J.M. (2002). *Methylocapsa acidiphila* gen. nov., sp. nov., a novel methane-oxidizing and dinitrogen-fixing acidophilic bacterium from *Sphagnum* bog. *International Journal of Systematic and Evolutionary Microbiology*, **52**, 251–261.

Godfrey, S.A., Harrow, S.A., Marshall, J.W. and Klena, J.D. (2001). Characterization by 16S rRNA sequence analysis of pseudomonads causing blotch disease of cultivated *Agaricus bisporus*. *Applied and Environmental Microbiology*, **67**, 4316–4323.

Goodfellow, M. and Williams, S.T. (1983). Ecology of actinomycetes. *Annual Review of Microbiology*, **37**, 189–216.

Gramms, G. (1987). The influence of the concomitant microflora on the establishment and dieback of decay fungi in standing timber. *Journal of Phytopathology*, **120**, 205–215.

Gramms, G., Voigt, K.-D. and Kirsche, B. (1999). Degradation of polycyclic aromatic hydrocarbons with three to seven aromatic rings by higher fungi in sterile and unsterile soil. *Biodegradation*, **10**, 51–62.

Greaves, H. (1970). The effect of selected bacteria and actinomycetes on the decay capacity of some wood-rotting fungi. *Material und Organismen*, **5**, 265–279.

Greaves, H. (1971). The bacterial factor in wood decay. *Wood Science and Technology*, **5**, 6–16.

Heilmann-Clausen, J. and Boddy, L. (2005). Inhibition and stimulation effects in communities of wood decay fungi: Exudates from colonized wood influence growth by other species. *Microbial Ecology*, **49**, 399–406.

Hendrickson, O.Q. (1991). Abundance and activity of N_2-fixing bacteria in decaying wood. *Canadian Journal of Forest Research*, **21**, 1299–1304.

Henningsson, B. (1967). Interactions between micro-organisms found in birch and aspen pulpwood. *Studia Forestalia Suecica*, **53**, 1–32.

Hoekstra, E.J., Verhagen, F.J.M., Field, J.A., De Leer, E.W.B. and Brinkmans, U.A.T. (1998). Natural production of chloroform by fungi. *Phytochemistry*, **49**, 91–97.

Hoffmeister, M. and Martin, W. (2003). Interspecific evolution: Microbial symbiosis, endosymbiosis and gene transfer. *Environmental Microbiology*, **5**, 641–649.

Izumi, H., Anderson, I.C., Alexander, I.J., Killham, K. and Moore, E.R.B. (2006). Endobacteria in some ectomycorrhiza of Scots Pine (*Pinus sylvestris*). *FEMS Microbiology Ecology*, **56**, 34–43.

Janse, B.J.H., Gaskell, J., Gaskell, J., Cullen, D., Zapanta, L., Dougherty, M.J. and Tien, M. (1997). Are bacteria omnipresent on *Phanerochaete chrysosporium* Burdsall? *Applied and Environmental Microbiology*, **63**, 2913–2914.

Jurgensen, M.F., Larsen, M.J., Wolosiewicz, M. and Harvey, A.E. (1989). A comparison of dinitrogen fixation rates in wood litter decayed by white-rot and brown-rot fungi. *Plant and Soil*, **115**, 117–122.

Kallionen, A., Vaari, A., Rätto, M., Konn, J., Siika-aho, M. and Viikari, L. (2003). Effects of bacterial treatments on wood extractives. *Journal of Biotechnology*, **103**, 67–76.

Kirby, R. (2006). Actinomycetes and lignin degradation. *Advances in Applied Microbiology*, **58**, 125–168.

Lang, E., Kleeberg, I. and Zadrazil, F. (1997). Competition of *Pleurotus* sp. and *Dichomitus squalens* with soil microorganisms during lignocellulose decomposition. *Biosource Technology*, **60**, 95–99.

Lebow, S., Foster, D. and Evans, J. (2004). Long-term soil accumulation of chromium, copper and arsenic adjacent to preservative-treated wood. *Bulletin of Environmental Contamination and Toxicology*, **72**, 225–232.

Lim, Y.W., Baik, K.S., Han, S.K., Kim, S.B. and Bae, K.S. (2003). *Burkholderia sordidicola* sp nov., isolated from the white-rot fungus *Phanerochaete sordida*. *International Journal of Systematic and Evolutionary Microbiology*, **53**, 1631–1636.

Lincoln, S.P., Fermor, T.R. and Tindall, B.J. (1999). *Janthinobacterium agaricidamnosum* sp. nov., a soft rot pathogen of *Agaricus bisporus*. *International Journal of Sytematic Bacteriology*, 49, 1577–1589.

Lindahl, B.D. and Finlay, R.D. (2005). Activities of chitinolytic enzymes during primary and secondary colonization of wood by basidiomycetous fungi. *New Phytologist*, **169**, 389–397.

Liu, J.-K. (2005). N-containing compounds of macromycetes. *Chemical Reviews*, **105**, 2723–2744.

Lorenzen, K. and Anke, T. (1998). Basidiomycetes as a source for new bioactive natural products. *Current Organic Chemistry*, **2**, 329–364.

Lumini, E., Ghignone, S., Bianciotto, V. and Bonfante, P. (2006). Endobacteria or bacterial endosymbionts? To be or not to be. *New Phytologist*, **170**, 205–208.

Murray, A.C. and Woodward, S. (2003). *In vitro* interactions between bacteria isolated from Sitka spruce stumps and *Heterobasidion annosum*. *Forest Pathology*, **33**, 53–67.

Noble, R., Fermor, T.R., Lincoln, S., Dobrovin-Pennington, A., Evered, C. and Mead, A. (2003). Primordia initiation of mushroom (*Agaricus bisporus*) strains on axenic casing materials. *Mycologia*, **95**, 620–629.

Öberg, G., Brunberg, H. and Hjelm, O. (1997). Production of organically-bound chlorine during degradation of birch wood by common white-rot fungi. *Soil Biology and Biochemistry*, **29**, 191–197.

Payne, C., Bruce, A. and Staines, H.J. (2000). Yeast and bacteria as biological control agents against fungal discolouration of *Pinus sylvestris* blocks in laboratory-based tests and the role of antifungal volatiles. *Holzforschung*, **54**, 563–569.

Rainey, P.B., Cole, A.L.J., Fermor, T.R. and Wood, D.A. (1990). A model system for examining involvement of bacteria in basidiome initiation of *Agaricus bisporus*. *Mycological Research*, **94**, 191–195.

Schmidt, O. (2006). *Wood and Tree Fungi. Biology, Damage, Protection, and Use*. Springer, Heidelberg.

Seigle-Murandi, F., Guiraud, P., Croizé, J., Falsen, E. and Eriksson, K.-E. (1996). Bacteria are omnipresent on *Phanerochaete chrysosporium* Burdsall. *Applied and Environmental Microbiology*, **62**, 2477–2481.

Smith, J.D. (1980). Is biological control of *Marasmius oreades* fairy rings possible? *Plant Disease*, **64**, 348–354.

Soler-Rivas, C., Jolivet, S., Arpin, N., Olivier, J.M. and Wichers, H.J. (1999). Biochemical and physiological aspects of brown blotch disease of *Agaricus bisporus*. *FEMS Microbiology Reviews*, **23**, 519–614.

Tornberg, K. and Olsson, S. (2002). Detection of hydroxyl radicals produced by wood-decomposing fungi. *FEMS Microbiology Ecology*, **40**, 13–20.

Tsukamoto, T., Murata, H. and Shirate, A. (2002). Identification of non-pseudomonad bacteria from fruit bodies of wild Agaricales fungi that detoxify tolaasin produced by *Pseudomonas tolaasii*. *Bioscience, Biotechnology and Biochemistry*, **66**, 2201–2208.

Tsukamoto, T., Takeuchi, M., Shida, O., Murata, H. and Shirata, A. (2001). Proposal of *Mycetocola* gen. nov in the family *Microbacteriaceae* and three new species, *Mycetocola saprophilus* sp nov., *Mycetocola tolaasinivorans* sp nov and *Mycetocola lacteus* sp nov., isolated from cultivated mushroom, *Pleurotus ostreatus*. *International Journal of Systematic and Evolutionary Microbiology*, **51**, 937–944.

Tsuneda, A. and Thorn, G. (1994a). Interactions between *Lentinula edodes* and pseudomonads. *Canadian Journal of Microbiology*, **40**, 937–943.

Tsuneda, A. and Thorn, G. (1994b). Interactions of wood decay fungi with other microorganisms, with emphasis on the degradation of cell walls. *Canadian Journal of Botany*, **73**(Suppl. 1), S1325–S1333.

Verhagen, F.J.M., Swarts, H.J., Wijnberg, J.B.P.A. and Field, J.A. (1998). Organohalogen production is a ubiquitous capacity among basidiomycetes. *Chemosphere*, **37**, 2091–2104.

Watkinson, S.C., Burton, K.S. and Wood, D.A. (2001). Characteristics of intracellular peptidase and proteinase activities from the mycelium of a cord-forming wood decay fungus, *Serpula lacrymans*. *Mycological Research*, **105**, 698–704.

CHAPTER 9

Interactions between Basidiomycota and Invertebrates

Lynne Boddy and **T. Hefin Jones**

Contents

Abstract

Basidiomycota are, in most terrestrial habitats, the primary agents of organic matter decomposition. They play a key role in associated ecosystem processes. Inevitably they interact frequently with invertebrates, and these interactions are highly dynamic. They may be direct or indirect, and prove beneficial or detrimental to either or both partners. In this chapter the variety of interactions is explored and the impact of basidiomycetes as a food and habitat resource, and as a predator is assessed. The consequence of fungal–invertebrate interactions on fungal community structure and faunal behaviour is also considered, as are the implications of these effects on nutrient cycling and ecosystem processes such as decomposition and productivity. The sensitivity of the interacting organisms to changes in climate, and general environmental change, with consequential effects on ecosystem activity, is also discussed.

British Mycological Society Symposia Series
Published by Elsevier Ltd.

1. INTRODUCTION

Ground invertebrate biomass differs from one habitat to another, but may often exceed 0.5 tonnes ha^{-1} (Killham, 1994). Numerically, and in terms of ecosystem processes, the most important of these ground-dwelling invertebrates are the oligochaetes (the earthworms and enchytraeid worms), nematodes, arthropods (such as centipedes, millipedes, springtails and mites) and molluscs (including slugs and snails; Gange and Brown, 1997). Saprotrophic fungi, the primary agents of organic matter decomposition in most terrestrial ecosystems (Chapters 2, 3, 10–12, 14–16), may be affected by these invertebrates through a range of mechanisms; many are primarily mycophagous and possess chitinases in their guts for degrading fungal cell walls (Berg *et al.*, 2004).

Invertebrates and fungi are affected by each other in both direct and indirect ways. Direct interactions largely comprise: (1) provision of a fungal food source or habitat for invertebrates (e.g. invertebrates directly grazing on mycelium or fruit bodies) or (2) basidiomycetes killing and then utilizing invertebrate body contents. General insect–fungus interactions, as understood then, were well covered by Carter (1973), Agrios (1980) and Wilding *et al.* (1989). Indirect effects of fungi on invertebrates include: (1) altering behaviour, by attraction, repulsion and arresting activity; and (2) improvement of the nutritional environment by enzymatically softening resources allowing easier feeding, destruction of inhibitory compounds in resources and decreasing carbon:nutrient ratio. Indirect effects of invertebrates on fungi largely relate to: (1) spread, including carriage to, inoculation into and facilitation of colonization of resources; (2) changes to the physical and chemical environment; (3) changes to the microbial community; and also (4) changes to physiology and metabolism. These aspects of fungus–invertebrate interactions, amongst others, were reviewed by Barbosa (1991), de Nooij *et al.* (1992) and Hatcher (1995), highlighting how over many millennia some have evolved into facultative or even obligate symbioses (Table 1). This review highlights the variety of interactions that occur, how these interactions affect ecosystem processes and the possible consequences of global environmental change on these interactions.

2. DIRECT POSITIVE EFFECTS OF BASIDIOMYCOTA ON INVERTEBRATES: FRUIT BODIES AND MYCELIA AS FOOD AND HABITAT

Mycelia are highly nutritious (Swift and Boddy, 1984; see below). Not surprisingly, therefore, many invertebrates use fungi as a food source, either grazing directly on mycelia or fruit bodies, or indirectly by ingesting mycelium within decomposing litter (Maraun *et al.*, 2003). Mycophagy is most prevalent amongst members of the phylum Arthropoda, although there are also many examples within the Mollusca, Enchytraeidae, Annelida, Collembola and Nematoda. For example, nematodes in the genus *Filenchus* fed and reproduced on *Coprinus cinereus, Flammulina velutipes, Rhizoctonia solani* mycelium and even on the

Table 1 Illustrative examples of symbiotic relationships between saprotrophic Basidiomycota and invertebrates

Fungus	Invertebrate	Nature of symbiosis	Effect on fungus	Effect on invertebrate	References
Lepiota, *Leucoagaricus*	Attine ants	Mutualism: ants cultivate the fungus	Provision of plant resources; maintenance of favourable abiotic and biotic environment; carriage of spores to new nests	Nutrition; ingested enzymes	Martin (1992), Baas and Cherrett (1996), Hernandez *et al.* (1999), Ortius-Lechner *et al.* (2000), Mueller (2002)
Termitomyces	Macrotermitinae	Mutualism: ants cultivate the fungus	Provision of plant resources; maintenance of favourable abiotic and biotic environment; carriage of sexual and, in some species, asexual spores to new nests	Nutrition; ingested enzymes	Wood and Thomas (1989), Martin (1992), Aanen *et al.* (2002), Mueller and Gerardo (2002)
Amylostereum areolatum, *A. chailletii* and *Cerrena* (= *Daedalea*) *unicolor*	Siricid wood wasps: *Sirex*, *Urocerus* and *Tremex* spp.	Mutualism: females carry asexual spores to trees; larvae develop in colonized wood	Inoculation into a suitable environment; widespread clonal populations	Softening of wood; improved nutrition; ingested enzymes	Martin (1992), Thomson and Koch (1999)

(Continued)

Table 1 (*Continued*)

Fungus	Invertebrate	Nature of symbiosis	Effect on fungus	Effect on invertebrate	References
Septobasidium	Scale insects	Mutualism and parasitism: scales live within a mycelial mat on the surface of plants	Obligate; nutrition and dispersal	Some scales are parasitized, others benefit from a buffered microclimatic environment and protection from predators	Evans (1989), Henk (2005)
Wood-rotting species, e.g. *Laetiporus sulphureus*, *Fistulina hepatica*, *Phellinus cryptarum*, *Bjerkandera fumosa*, *Trametes versicolor* and *Coniophora puteana*	Death watch beetle *Xestobium rufovilosum*	Larvae burrow in wood colonized by the fungus	May benefit from nutrient input in faeces; harm may accrue from comminution	Softening of wood; improved nutrition	Fisher (1940, 1941)
Phallaceae	Diptera	Mutualism	Spore dispersal	Nutrition	Tuno (1998)
Polypores	Gamasid mites	Feeding within fruit bodies	Decreased reproductive output; spore destruction; spore dispersal	Nutrition and breeding ground	Makarova (2004)
Agarics and polypores	Insecta	Feeding within fruit bodies	Decreased reproductive output; spore destruction; spore dispersal	Nutrition and breeding ground	Hammond and Lawrence (1989), Hanski (1989)

Polypores	Coleoptera	Feeding within fruit bodies	Decreased reproductive output; spore destruction; spore dispersal	Nutrition and breeding ground	Guevara *et al.* (2000a, 2000b, 2000c), Thunes *et al.* (2000), Schigel *et al.* (2004), Orledge and Reynolds (2005)
Agaricus bisporus mycelium	Fungus gnats (Phoridae and Sciaridae)	Mycophagy	Negative	Nutrition and breeding ground	Sheepmaker *et al.* (1996)
Mycelial systems of cord-forming wood decomposers on soil	Collembola	Mycophagy	Morphological changes, increases and decreases in hyphal coverage		Harold *et al.* (2005), Bretherton *et al.* (2006), Tordoff *et al.* (2006), Wood *et al.* (2006)
Inter- and intra-specifically interacting non-compatible mycelia	Fungus gnats (Sciaridae); Collembola	Mycophagy	Outcome of interactions sometimes modified	Nutrition and breeding ground	Boddy *et al.* (1983); Fig. 1
Coprinus cinereus and *Flammulina velutipes* mycelium	Nematoda	Mycophagy			Okada *et al.* (2005)
Coprinus comatus, *Hohenbuehelia*, *Hyphoderma* and *Pleurotus* mycelium	Nematoda	Invertebrate killing	Nutritional	Death	Thorn and Barron (1984), Barron and Thorn (1987), Tzeam and Liou (1993), Luo *et al.* (2004)
Fibularhizoctonia sp.	*Reticulitermes* (Isoptera: Rhinotermitidae)	Sclerotia mimic eggs and are tended by the termites	Protection; transport to a competitor free environment	Enhanced egg survival	Matsuura *et al.* (2000), Matsuura (2005)

nematophagous (see below) *Pleurotus ostreatus* in agar culture and soil (Okada *et al.*, 2005). In another example, the enchytraeid *Enchytraeus crypticus* destroyed colonies of the nematophagous fungus *Hirsutella rhossiliensis*, and reduced biological control of the root-knot nematode *Meloidogyne javanica* (Jaffee, 2004). Mycelia of some species are more palatable than others, and there is sometimes considerable variation between closely related species and strains (Smith *et al.*, 2006a, 2006b), and in different mycelial regions (see below).

Both perennial (e.g. polypore brackets) and ephemeral (e.g. Agaricales) fruit bodies provide food sources and breeding grounds for a diversity of invertebrates, including nematodes (Jaffee, 2006; Li *et al.*, 2006), enchytraeids (Nowak *et al.*, 2005), mites (Makarova, 2004), Collembola (Greenslade *et al.*, 2002; Nakamori and Suzuki, 2005a, 2005b), Coleoptera (beetles; Thunes *et al.*, 2000; Schigel *et al.*, 2004; Orledge and Reynolds, 2005) and Diptera (flies; Jonsell and Nordlander, 2002; Yamashita and Hijii, 2003). Insects that feed on agarics are mostly polyphagic, selective pressures promoting use of a wide variety of species, because the fruit bodies are unpredictable and ephemeral resources, whereas over half of the insects feeding in brackets in Sweden were monophagous (Jonsell and Nordlander, 2002). Ciid beetles can be divided into host-use groups, each ciid taxon only belonging to a single host-use group, but often utilizing several members of that group (Orledge and Reynolds, 2005). The high degree of monophagy in insects in general and the host-use groups of ciids is probably ultimately defined by chemical defences of the host bracket.

Though some invertebrates are able to use living fruit bodies, others are only able to use those that are decaying, and different invertebrate species are found at different stages of decay. Those fungivores that are polyphagous on brackets mainly colonize them after they have been decaying for some time, defence chemicals having presumably decreased by then (Jonsell and Nordlander, 2002). Early colonizers, that tend to be monophagous, have the advantage over later colonizers in that they have so-called priority effects and a more nutritious environment, but they have a narrower choice of fungal resources.

In some cases, the advantage is other than food or habitat resource. Basidiomycete species in the genus *Septobasidium*, for example, form complex relationships with scale insects. These insects are plant parasites and suck plant sap using their long stylet mouthparts. The scale insects are found in the middle layer of the fungus, in chambers that are only slightly larger than the insects themselves and are inter-connected by numerous tunnels. Some of the scale insects are parasitized by the fungus (Dykstra, 1974; Evans, 1989). These parasitized individuals are immobile and usually occupy a chamber where they are attached to the plant via their suctorial tube. The scale insect nourishes itself with the plant sap, and the fungus, attached to the insect by haustoria, is subsequently indirectly nourished by the plant. At first glance, it may appear that this is simply a case of parasitism with the fungus getting all the benefits, but there are benefits to the scale insects. Only some are parasitized, but they are all sheltered from the environment and protected from predators, allowing them to live longer than their free-living brethren. It has been shown that, in at least some cases, the fungal covering over the scale insects is deeper than the reach of the ovipositors of parasitic wasps that seek out scale insects (Morse and Normak, 2006).

Fruit bodies also form an important part of the diet of some vertebrates. Many non-domesticated mammals and birds are opportunistic mycophagists, others, for example, squirrels and several marsupials, are classed as preferential mycophagists concentrating on fruit bodies when they are available, and a few are obligately mycophagous (e.g. two marsupial rat kangaroos—*Potorous longipes* and *P. gilbertii*; Claridge and Trappe, 2005).

3. DIRECT NEGATIVE EFFECTS ON INVERTEBRATES: KILLING BY BASIDIOMYCOTA

As far as we are aware, there are no reports of epizootics of arthropods caused by Basidiomycota, though there are undoubtedly examples of killing. There is certainly evidence that mycelium of *Hypholoma fasciculare* can kill Collembola in some situations (T.D. Rotheray *et al.*, unpublished). Mycelium of certain groups of Basidiomycota kill and subsequently utilize nematodes, often producing specialized killing structures, including *Pleurotus* species (e.g. *P. ostreatus*, *P. cornucopiae* and *P. tuber-regium* which produce droplets of toxin on aerial stalks; Thorn and Barron, 1984; Hibbett and Thorn, 1994); *Hohenbuehelia* species producing adhesive secretory cells on hyphae or conidia (Thorn and Barron, 1984); and some *Hyphoderma* species producing stephanocysts (previously thought to be dispersal propagules) that are trapping devices (Tzean and Liou, 1993) (the stephanocytes exude a chemical that bonds tightly to the nematode cuticle, and in attempting to pull away the nematode may frequently disintegrate (Burdsall, 1969)); and *Coprinus comatus* producing structures looking like spiny balls, which may be the instruments of death (Luo *et al.*, 2004). On the other hand, though *Conocybe lactea* paralyzes by toxic droplets from secretory mycelial appendages, there is currently no evidence of subsequent colonization (Hutchison *et al.*, 1995).

4. DIRECT EFFECTS OF INVERTEBRATES ON FUNGI: MYCOPHAGY

While feeding within fruit bodies, invertebrates inevitably ingest spores. Passage of spores through guts can result in considerable damage resulting in failure to germinate, for example, *Agrocybe* sp. after passage through the gut of earthworms (*Lumbricus terrestris*; Moody *et al.*, 1996). Some Collembola completely break open spores during gut passage, for example, *Hypogastrura* spp. feeding on cultivated *Hypsizygus marmoreus*, though damage varied between Collembola species (Nakamori and Suzuki, 2005a, 2005b). Those species with greater spore-breaking capabilities were more commonly found in the spore-bearing regions of fruit bodies of saprotrophic and mycorrhizal agarics, compared with those with lesser abilities that were found elsewhere in fruit bodies.

The morphology of fruit bodies can be altered as a result of the presence of invertebrates: some Diptera cause galls, enclosing larvae, in long-lived fruit bodies (e.g. *Peniophora cinerea*, *P. limitata* and *Ganoderma applanatum*) and even occasionally in fleshy agarics (e.g. *Panaeolina foenisecii* and *Panaeolus acuminatus* (Hanski, 1989; Spooner, 2003)). Moreover, the reproductive fitness of the fungus can be affected by mycophagy in fruit bodies. Both field surveys and experimental studies indicated that ciid beetles significantly decreased the area of functional hymenium of *Trametes versicolor*: *Octotemnus glabriculus* and *Cis boleti* caused reductions of 58 and 30%, respectively, in experiments (Guevara *et al.*, 2000a, 2000b, 2000c). Production of inhibitory chemicals (Feofilova, 2001), physical structure and phenology (Chapter 5) of fruit bodies are likely to have evolved, at least partly in response to mycophagy.

Mycelia growing across the surface of soil can be completely destroyed by high intensity invertebrate grazing, but less intense grazing can result in dramatic changes in mycelial growth and activity (e.g. Kampichler *et al.*, 2004; Harold *et al.*, 2005; Bretherton *et al.*, 2006; Tordoff *et al.*, 2006; Wood *et al.*, 2006; Figure 1). Despite the plethora of studies on fungal–invertebrate interactions, only recently has attention turned to saprotrophic Basidiomycota, and then largely to cord-formers (Chapter 1) grazed by Collembola. Mycelial morphology and foraging patterns often change dramatically as a result of grazing, changes varying depending on fungal species (Tordoff *et al.*, 2006; Figure 1), resource status (Harold *et al.*, 2005), grazing intensity (density) and invertebrate (Collembola) species (Kampichler *et al.*, 2004). For example, *Folsomia candida* had a large impact, *Proisotoma minuta* often a similar impact, but *Protaphorura armata* often had little impact (G.M. Tordoff *et al.*, unpublished). All Collembola species tested modified *Phanerochaete velutina* and *H. fasciculare* morphology, that of *Resinicium bicolor* was markedly affected by *F. candida*, but not by other Collembola, and effects of grazing on *Phallus impudicus* were negligible. The extra-resource foraging mycelium of *H. fasciculare*, for example, changed from slow, dense exploitative growth to less dense explorative growth during grazing. There were often points of more rapid outgrowth as cords with a fanned margin. The mechanism behind changes in mycelial morphology following invertebrate grazing have not been investigated, but may result from removal of suppression of side branches when apically dominant hyphae are removed (Boddy and Jones, 2006).

Mycelia are preferentially grazed in different regions depending on fungal species. Thus, with *H. fasciculare* there is preferential grazing of hyphal tips at growing margins; with *R. bicolor* there is indiscriminate grazing on cords and fine mycelium, but generally close to wood inoculum; and with *P. velutina* fine mycelium within the colony and hyphal tips at the growing margin (Tordoff *et al.*, 2006). In larger mycelial systems, morphology and hyphal coverage of grazed systems was similar to that of ungrazed systems, apparently because the Collembola grazed on senescing hyphae that would ultimately have died (Wood *et al.*, 2006).

Grazing sometimes results in increased mycelial growth; thus, at very low density grazing by *F. candida* there was 15% greater hyphal coverage by *P. velutina*

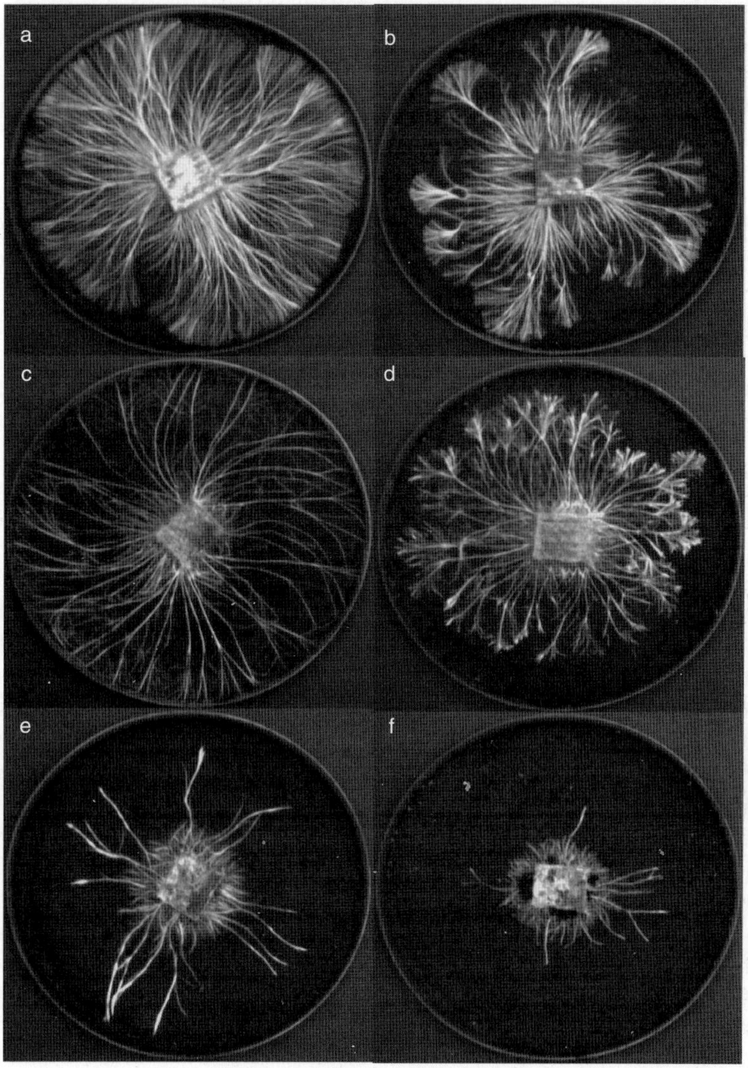

Figure 1 Effect of collembola grazing (*Folsomia candida*) on morphology of mycelia, growing from wood block inocula across compressed non-sterile soil in 13 cm diameter dishes. *Hypholoma fasciculare* ungrazed (a) and grazed by 20 collembola for 10 days (b); *Phanerochaete velutina* ungrazed (c) and grazed by 40 collembola for 10 days (d); *Resinicium bicolor* ungrazed (e) and grazed by 40 collembola for 4 days (f). Mycelia have been allowed to reach 8 cm diameter before collembola were added. From: Tordoff *et al.* (2006).

than in the absence of grazing (Bretherton *et al.*, 2006). In soil systems with high intensity grazing, there was catch-up or even over-compensatory growth following cessation of grazing, in some cases with almost 40% greater hyphal coverage (Bretherton *et al.*, 2006). Such increased growth must be at the cost of more rapid use of the inoculum resource. Similarly, when leaf-cutting ants (that mutualistically cultivated *Leucoagaricus gongylophorus*) were deprived of leaves for 5 days they increased their rate of consumption of fungal staphylae, resulting in increased rates of staphylae production (Bass, 1997). This, in turn, resulted in more rapid substrate exhaustion.

Little is yet known on effects of grazing on fungal physiology. This is obviously crucially important, particularly in terms of the ways and extent of reallocation of nutrients (Chapter 3). There is evidence of mineral nutrient release during grazing of litter colonized by the wood-rotting *T. versicolor* (Anderson *et al.*, 1983). Also, enzymes associated with ligninolysis were switched on by *P. velutina* in the presence of the nematode *Panagrellus redivivus* (Dyer *et al.*, 1992).

Invertebrates can exert direct positive nutritional effects on Basidiomycota mycelia, through nutrient inputs in the faeces. Also, some Basidiomycota can supplement their nutrition by killing and subsequently utilizing invertebrates (see below).

5. INVERTEBRATE EFFECTS ON FUNGAL COMMUNITY STRUCTURE

Invertebrates have differential effects on different species of fungi, resulting from different extents and sites of grazing, secretion of various chemicals including antibiotics, and from physical effects. Consequently, they can affect the species balance within fungal communities. For example, Collembola grazing alters the vertical distribution of *Marasmius androsaceus* and *Mycena galopus* in spruce (*Picea sitchensis*) needle litter (Newell, 1984a, 1984b): *Onychiurus latus* fed preferentially on *M. androsaceus* in laboratory and field tests caused the restriction of this species to the uppermost litter horizon. In inter-specific mycelial confrontations in agar culture and trays of soil in the laboratory, the balance was shifted in favour of one species over another when grazed by the Collembola *F. candida*; for example, in the absence of *F. candida*, *R. bicolor* overgrew *P. velutina*, whereas with Collembola grazing *P. velutina* was able to breach the *R. bicolor* advancing front (T.D. Rotheray *et al.*, unpublished).

In small angiosperm branches decomposing on the forest floor there is a dramatic change in fungal communities following invasion by soil invertebrates: prior to invasion communities are dominated by Basidiomycota, but post-invasion by non-Basidiomycota, presumably resulting from destruction of resident fungi by feeding action, carriage and inoculation of spores of other fungi, and change in microenvironmental conditions (Swift and Boddy, 1984; see below).

The mutualism between *Macrotermitinae* and *Termitomyces* species, in which the former cultivate the latter within their nests, is even more dramatic. *Termitomyces* species are poor competitors and are rapidly over-run if the termites abandon the nest, but are maintained in the fungus comb in active nests in more or less pure culture, despite continual inoculation with other fungi on plant material collected by the termites (Wood and Thomas, 1989; Shinzato *et al.*, 2005). Passage through worker guts reduces germination of non-mutualistic symbiont spores, oral secretions are fungistatic, and nest microclimate (30 °C and elevated CO_2) is optimal for the fungal symbiont. There also seems to be genetic screening of *Termitomyces* strains, either directly via active selection by the termite or indirectly by intra-specific competition on the fungus comb, as evidenced by identical fungal molecular sequences in multiple samples from four different nests of species that have horizontal symbiont transmission (Aanen *et al.*, 2002; see also below).

In the mutualism between attine ants and *Leucoagaricus* spp., *Lepiota* spp. and other Basidiomycota (Mueller, 2002), the fungi are more competitive than in the termite–fungus mutualism, being able to survive for approaching 12 days following abandonment (Bass and Cherrett, 1994). Nonetheless, contaminants are kept out by physical removal of spores by licking, chemical secretions from the ants (Hernandez *et al.*, 1999; Ortius-Lechner *et al.*, 2000), along with antibiotics secreted by actinomycetes (*Streptomyces*) targeted at a virulent Ascomycota mycoparasite (*Escovopsis*; Currie *et al.*, 1999).

Studies in Mexican tropical cloud forest suggest that variability in fungivory (apparent biomass consumed) of understorey Basidiomycota by invertebrate taxa could be explained by apparency (*sensu* Feeny, 1976; "ease of finding")-related characteristics of the above ground structures (colour of pileus, stipe and hymenium; size and aggregation), as has been suggested for plant–herbivore relationships (Crawley, 1983). Considerable inter-specific and inter-taxa variation in fungivory was detected; colour attributes of fruit bodies were not strongly associated with the observed variation of consumption levels, whereas apparent biomass and aggregation size did correlate with the observed variation in fungivory. It was concluded that colouration patterns may not be important for fungivory, whereas genet size and species identity (probably via characteristics unrelated to apparency, such as mycotoxins and nutritional value) seemed to be critical factors (Guevara and Dirzo, 1999). In another example, screening forest soil nematodes for associated fungi showed that the pathogenic fungi *Malassezia restricta* and *M. globosa* were associated with the nematode genus *Malenchus* sp., whereas another nematode, *Tylolaimophorus typicus*, hosted only *M. restricta* (Renker *et al.*, 2003). Preferences have also been shown by lumbricids: *L. terrestris* took far fewer wheat straws colonized by *Agrocybe praecox* than by non-Basidiomycota (Moody *et al.*, 1995).

6. EFFECTS ON INVERTEBRATE BEHAVIOUR

A wide range of volatile organic compounds (VOCs), including alcohols, terpenes, aldehydes, ketones, sesquiterpenes and aromatics, are produced by Basiomycota fruit bodies (Faldt *et al.*, 1999; Rosecke *et al.*, 2000), mycelium (Hynes *et al.*, 2007) and decomposing organic resources (Cole *et al.*, 1989), and often increase in quantity and quality if physically damaged (Stadler and Sterner, 1998; Faldt *et al.*, 1999) or during inter-specific mycelial interactions (Hynes *et al.*, 2007; Chapter 7). Likewise, dissolved organic compounds (DOCs) are also produced (e.g. Su, 2005). Invertebrate responses to VOCs and DOCs include attraction, repulsion, arrestant and antifeeding behaviour (Table 2).

Effects on termites are particularly well documented (see references in Swift and Boddy, 1984; Su, 2005). Mycelium of brown-rot fungi (Chapter 2), wood

Table 2 Examples of effects of basidiomycota on animal behaviour

Fungus/compound	Invertebrate	Nature of effect	References
Agaricus bisporus mycelium	*Megaselia halterata* (Phoridae)	Attraction of gravid females to mycelium	Tibbles *et al.* (2005), Smith *et al.* (2006a)
Agaricus bisporus mycelium	*Lycoriella ingenua* (Sciaridae)		Tibbles *et al.* (2005), Smith *et al.* (2006b)
Perennial fruit bodies	Ciid beetles	Long- and short-range attraction and inter-specific discrimination	Jonsell and Nordlander (1995), Guevara *et al.* (2000a, 2000b), Jonsson *et al.* (2003)
Mating incompatible mycelia of *Stereum* spp. and *Phlebia* spp.	*Bradysia* sp. (Sciaridae)	Apparent attraction to mycelial interaction zone	Boddy *et al.* (1983)
Conocybe lactea mycelium	Nematodes	Activity arrested and antifeedant	Hutchison *et al.* (1995)
Brown-rot fungi, e.g. *Ganoderma applanatum*, *Gloeophyllum trabeum*, *Serpula lacrymans*	*Reticulitermes*, *Heterotermes*, *Coptotermes*, *Kalotermes* spp., Rhinotermitidae	Attraction	Swift and Boddy (1984), Su (2005)
Sesquiterpenes		Antifeedant activity or are ultimately toxic to invertebrates	Kahlos *et al.* (1994), Stadler and Sterner (1998)
Cadinenes, muurolenes and amorphenes	Many	Used insect communication systems	El-Sayed (2005), Hynes *et al.* (2007)

decomposed by them and extractives from such wood are often attractive to termites, and VOCs can stimulate termites to eat more sound wood and build more galleries. White-rot fungi and white-rotted wood are often unattractive and even toxic to termites, though *P. ostreatus* was attractive. White-rot fungal mycelia are, however, attractive to other arthropods. For example, fungus gnats (*Bradysia*; Sciaridae) are highly attracted to and oviposit in interaction zones of mating incompatible mycelia of *Stereum* spp. and *Phlebia* spp. (Boddy *et al.*, 1983; Figure 2a). Collembola are also attracted to and preferentially graze in interaction zones between mycelia growing from woody resources into soil (Figure 2b). These regions are presumably more palatable and leak nutrients, and VOCs are upregulated (Hynes *et al.*, 2007). Sciarids and phorids (Diptera) are attracted to the mycelium and compost of cultivated mushrooms (*Agaricus* species; Grove and Blight, 1983; Tibbles *et al.*, 2005). There were, however, large differences in the sizes of phorid populations emerging from different *Agaricus* species and strains of the same species, which may have resulted from differences in numbers of females choosing to oviposit, as a result of lack of attractants or production of inhibitory chemicals by some strains but not others (Smith *et al.*, 2006a, 2006b).

Not only are direct utilizers of fungi attracted by mycelium and colonized organic resources, but so also are parasitoids of direct users. Thus, *Ibalia leucospoides*, a parasitoid of the wood wasp *Sirex noctilio* that lives in association with the wood decaying *Amylostereum areolatum*, was attracted by volatiles emitted by the fungus (Martínez *et al.*, 2006). The fungal VOCs also appear to elicit increased parasitoid activity, and it has been speculated that they may provide information on the relative densities of host wood wasps (Martínez *et al.*, 2006).

Some fruit bodies produce VOC attractants, the attraction of Diptera to the spore mass of *P. impudicus* being a classic example. Ciid beetles live and breed in the fruit bodies of lignicolous Basidiomycota, and exhibit specific host-use groups (Orledge and Reynolds, 2005). They use VOCs for location of fruit bodies and discrimination of suitable species. For example, *C. boleti* and *O. glabriculus* were attracted to *T. versicolor*, whereas *C. nitidus* was attracted to *Ganoderma* sp., and *Cis bilamellatus* was attracted to both (Guevara *et al.*, 2000b). VOCs are apparently used over long and short distances (Jonsell and Nordlander, 1995; Guevara *et al.*, 2000a, 2000b; Jonsson *et al.*, 2003), though whether the same chemical cues are involved in both instances is not known.

VOCs are also sometimes involved in the regulation of breeding (Guevara *et al.*, 2000a). Both *O. glabricus* and *C. boleti* feed and breed in *T. versicolor* fruit bodies, but *O. glabricus* colonizes young basidiocarps in early spring while *C. boleti* is most abundant in autumn. *O. glabricus* was actually attracted to both old and young fruit bodies but *C. boleti* was only attracted to older fruit bodies. The implication is that changes in VOC emission as fruit bodies age cause a partitioning of resource use by these two ciid beetle species. Differences in attraction in the field may not be as clear-cut as in these laboratory experiments since other abiotic variables may modify effects. For example, the relative attractiveness of two non-Basidiomycota species for oviposition by *Lycoriella inegnua* (Sciaridae) completely changed with CO_2 enrichment of the leaf litter in which they grow (Frouz *et al.*, 2002).

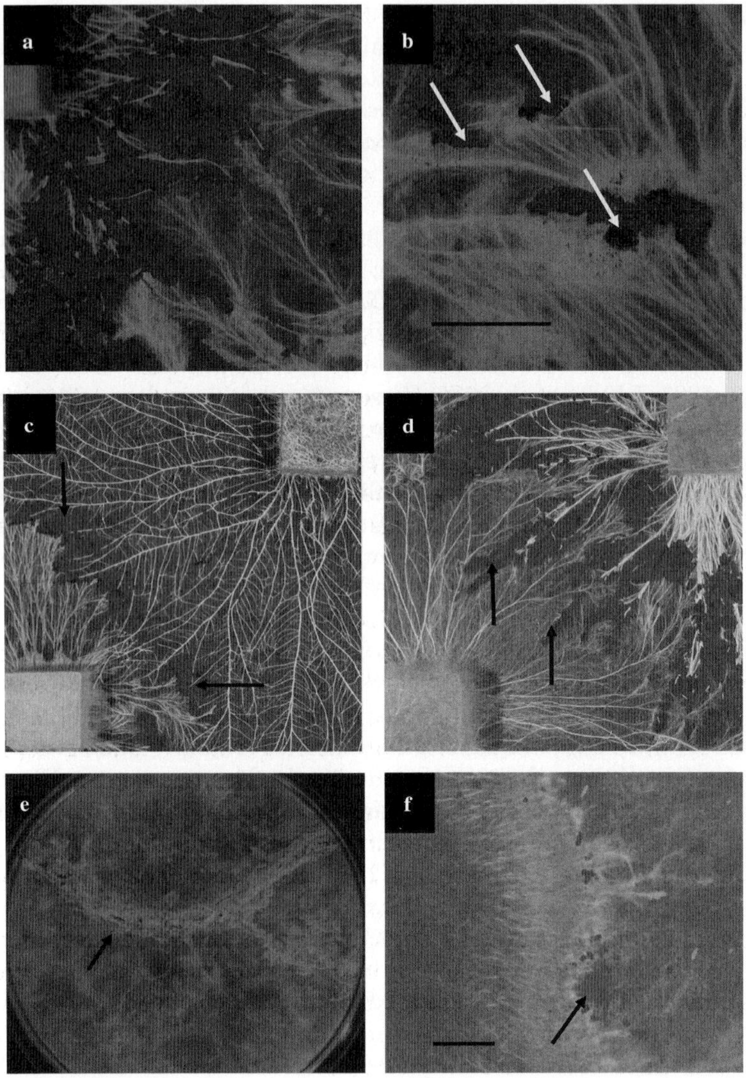

Figure 2 Preferential grazing by invertebrates in interaction zones. (a–e) Interactions between mycelia growing from 2 × 2 × 2 cm wood inocula across compressed non-sterile soil. (a) Considerable grazing by *Folsomia candida* (Collembola) of *Resinicium bicolor* (left) in the region of interaction with *Hypholoma fasciculare*. (b) Localized *F. candida* grazing (arrowed) of *H. fasciculare* near interaction front. (c) Grazing by *F. candida* of both *H. fasciculare* (left) and *R. bicolor* in the interaction zone. (d) *F. candida* grazing of *Phanerochaete velutina* and *R. bicolor*. Note that grazing of *P. velutina* is largely in regions (arrowed) where cords of *R. bicolor* had overgrown *P. velutina*. The *R. bicolor* cords in these regions were completely removed by grazing. (e) Fungus gnat grazing (arrowed) largely in the interaction zone between non-mating compatible homokaryons of *Stereum gausapatum* in agar culture. Note grazed regions (arrowed) in interaction zone. (f) Grazing (arrowed) by *F. candida* in the interaction zone of *P. velutina* (left) and *R. bicolor*, centred largely on the latter, in agar culture. Scale bar 1 cm. (a)–(d) and (f) courtesy of T.D. Rotheray. **(See Colour Section)**

7. IMPROVEMENT OF NUTRITIONAL ENVIRONMENT FOR INVERTEBRATES

The carbon:mineral nutrient ratios vary considerably between different plant litter components, the C:N and C:P ratios of undecayed leaf litter ranging between 25:1 to 100:1 and 450:1 to 1,850:1, respectively, and of undecayed wood between 350:1 to 500:1 and 1,250:1 to >3,500:1 (Dighton and Boddy, 1989). As decomposition proceeds, however, the ratios decrease, providing animals with a resource of higher nutritional quality. Further, fungal mycelium itself has seven times better C:N and C:P ratios than undecayed wood (Swift and Boddy, 1984). Similarly, with external symbiotic relationships, the N and P contents in fungus gardens of *Termitomyces* in the nests of *Macrotermes bellicosus* are greater than in the food initially collected: food 0.28% N, 0.15% P; food store 0.58% N, 0.12% P; fresh fungus comb 0.85% N, 0.19% P; mycotêtes 6.68% N, 0.46% P (see Swift and Boddy, 1984). The attraction of mycophagy is clear: invertebrates need to eat far less fungal mycelium than plant litter to meet their mineral nutrient requirements. Even when invertebrates are feeding on wood and leaf litter a considerable fraction of the nutrients consumed will be within hyphae of Basidiomycota.

The benefits of feeding on fungal-decayed resources is seen with the death watch beetle, *Xestobium rufovillosum* (Anobiidae), that typically attacks wood colonized by the heart-rot (Chapter 11) Basidiomycota *Laetiporus sulphurous*, *Fistulina hepatica* and *Phellinus cryptarum* in oak, and *Bjerkandera fumosa*, *Trametes versicolor* and *Coniophora puteana* in willow and other taxa (Fisher, 1940, 1941). The length of the beetle life cycle is inversely related to the state of decay of the wood: in undecayed wood the larvae developed very slowly or not at all, but in decayed wood the life cycle was completed within 10–17 months, in the laboratory. This is not only a nutritional effect, but also relates to physical factors, the larvae being able to process a larger volume of decayed than undecayed wood, and consequently obtain nitrogen more rapidly (Fisher, 1941; Rayner and Boddy, 1988). Similar nutritional and physical considerations are likely to apply to siricid wood wasps that have a mutualistic symbiotic relationship with *Amylostereum aerolatum* and *A. chailletii* in conifers, and *Cerrena* (=*Daedalea*) *unicolor* in angiosperms (Martin, 1992; Thomson and Koch, 1999; Slippers *et al.*, 2003). The larvae are wood borers that burrow through wood colonized by the fungi; to ensure the presence of the wood-rotting fungus the female carries oidia and inoculates them into the wood during oviposition.

Ingested enzymes, from the fungus, also play an important role in the digestion of plant litter in siricid wood wasp, fungus-growing termite and fungus-growing ant mutualisms (Martin, 1992). Cellulose and hemicellulose are digested during passage through the alimentary tract, predominantly the midgut of *Sirex* spp. larvae, and by the termite and attine ant workers. These acquired enzymes survive gut passage and, in the case of the ants and termites, are concentrated in the faecal droplet which is deposited on fresh plant material (Martin, 1992; Rønhede *et al.*, 2004). This prepares plant material for fungal colonization and increases the initial growth.

Basidiomycota also render palatable wood and leaf litter that is initially repellent or unpalatable to invertebrates due to the presence of allelopathic compounds. Again there are well-documented examples for termites (see references in Swift and Boddy, 1984). There are several examples of trees whose central heartwood is resistant to attack from termites when undecayed, but not once decay has begun. Of course, other aspects of enzyme conditioning (e.g. density reduction) may also play a part. Phenolics are also degraded by the mutualistic fungus partner on the fungus comb within termite nests (Taprab *et al.*, 2005).

8. INVERTEBRATE ASSISTANCE IN DISPERSAL, AND FACILITATION OF COLONIZATION OF RESOURCES

Propagules, be they asexual or sexual spores, hyphal fragments or yeasts, of many Basidiomycota are dispersed by invertebrates on their bodies, passing through their guts, and even in special structures on their bodies. Propagule carriage can thus be incidental (though sometimes with attraction, e.g. Phallales—see above) or obligatory, the latter being most highly developed in the mutualistic associations with termites, ants and wood wasps. Siricidae (wood wasps) carry their symbionts in a pair of pouches, termed mycangia, at the base of the ovipositor, and inoculate both fungus and eggs together into wood (Kajimura, 2000). The importance of wood wasps in dispersing the fungus does, however, appear to be taxon-specific; dispersal through basidiospores is considered of less importance for *A. areolatum*, but common in *A. chailletii* (Kajimura, 2000). The Basidiomycota symbionts of ants fruit only rarely, and normally propagate asexually and are spread by dispersing queens (Mueller *et al.*, 2001). Ants carry, for example, *Attamyces bromatificus* from one leaf-cutter ant colony to a new one in an infra-buccal pocket to ensure successful colonization of new nest material (Cherrett *et al.*, 1989). *Termitomyces* spp. are carried to new colonies by females in the genus *Microtermes* and by kings of *M. bellicosus* (Johnson *et al.*, 1981), but the rest of the Macrotermitinae acquire their symbionts at the nest-founding stage while foraging: sexual spores are consumed and survive passage through the gut, being deposited in a faecal pellet on the fungus comb (Aanen *et al.*, 2002; Mueller and Gerardo, 2002). Spores of some species not involved in tight mutualistic relationships with invertebrates also survive passage through the gut, and sometimes actually require this before readily germinating, e.g. *Ganoderma* sp. passing through the gut of a fly larva (Nuss, 1982).

Invertebrates can also facilitate colonization by breaching outer protective layers of plants providing infection courts, ranging from minute feeding or oviposition sites (e.g. wood wasps—mentioned above) to large wounds allowing colonization by ruderal Basidiomycota (Chapter 11). Tunnels within wood can also facilitate tangential spread.

9. INVERTEBRATE EFFECTS ON THE PHYSICAL AND CHEMICAL ENVIRONMENT

As well as the direct effects already noted, invertebrates can alter the physical and chemical characteristics of organic resources (Swift and Boddy, 1984). For example, temperature and water relations can be altered by loss of bark due to subcortical feeding. Aeration of wood can be considerably improved by tunnelling. On the other hand, comminution provides much smaller particles more favourable to microfungi and bacteria than Basidiomycota, a potential example of ecological engineering (Jones *et al.*, 1994). Invertebrate faeces have an entirely different chemical composition to the resources upon which they feed, for example, they are likely often to be rich in uric acid, and hence nitrogen (Krasnoshchekov and Vishnyakova, 2003).

10. INVERTEBRATE EFFECTS ON FUNGAL PHYSIOLOGY AND METABOLISM

There seems to have been little research performed on effects of invertebrates on Basidiomycota physiology and metabolism, yet such effects are sometimes likely to be large, in view of the dramatic changes to morphology mentioned above. Grazing by *F. candida* altered the partitioning of ^{15}N added to soil close to the wood inoculum (G.M. Tordoff *et al.*, unpublished): less ^{15}N was transferred to new mycelial growth in grazed systems than in ungrazed systems, presumably related to slower growth and altered mycelial morphology. Changes in extracellular enzyme activity have been noted: *P. velutina* and *Stereum hirsutum* exhibited differential responses to the presence of the nematode *Panagrellus redividus* (Dyer *et al.*, 1992). Protease activity increased while esterase and acid phosphotase activities were reduced in *P. velutina*. In contrast, there was increased protease and acid phosphotase activity in *S. hirsutum*.

There is also some evidence of reduction in wood inoculum decay rate when extra-resource mycelia were grazed (Tordoff *et al.*, 2006). On the other hand, a small field study showed an increase in wood decay rate following invertebrate invasion, though others did not (Swift and Boddy, 1984).

11. ECOSYSTEM PROCESSES CONSEQUENCES

While it is clear that many of the effects listed above could potentially lead to quite dramatic effects on various ecosystem processes, there is a dearth of studies which specifically look at this aspect, either qualitatively or quantitatively. Some microcosm studies have demonstrated increased nutrient mineralization from leaf litter colonized by Basidiomycota in the presence of soil fauna (Anderson *et al.*, 1983). Somewhat surprisingly, Collembola grazing did not increase nitrogen mineralization from extra-resource mycelium of *P. velutina* growing from woody resources into the bulk soil (G.M. Tordoff *et al.*, unpublished). Nonetheless, nutrient release from conservative mycelial systems of cord-forming Basidiomycota is likely to occur during grazing (along with during inter-specific mycelial interactions; Chapter 7). Changes in nutrient partitioning within mycelia have the potential to affect the dynamics and spatial heterogeneity of forest floor nutrients.

Basidiomycetes are the primary decomposers of wood and community composition plays an important role in determining the rate of associated ecosystem processes. Toljander *et al.* (2006) assembled 16 species of boreal wood decay fungi (Chapter 12), 8 brown-rot fungi and 8 white-rot fungi in artificial communities and studied the effect of different community species composition on species persistence, wood decomposition and metabolic efficiency. Decomposition was highest at intermediate diversity levels while metabolic efficiency, estimated as the amount of fungal mycelium formed per amount of degraded wood, decreased with increasing community complexity. A similar study exploring the significance of fungal diversity on ecosystem processes associated with the decomposition of different tropical leaf species suggested differential abundance of fungal species affects rates of decomposition dramatically (Santana *et al.*, 2005). As grazing by invertebrates is known to influence, and in some cases modify (see above examples), fungal community structure the implications of faunal activity may be considerable. This area of study is ripe for further investigation!

12. CONSEQUENCES OF GLOBAL ENVIRONMENTAL CHANGE

The effects of global environmental change, including climate change, UV-B radiation, atmospheric and terrestrial pollutants and genetic engineering, on fungi were considered in detail in Frankland *et al.* (1996). In the final section of this chapter we highlight some of the issues pertaining to the effects of current predicted climate change on Basidiomycota and, in particular, on the interactions between these fungi and invertebrates.

With increased frequency of summer drought, damage caused by *Armillaria* spp. is expected to become more prevalent (Rishbeth, 1982; Wargo, 1984). Other Basidiomycota that are dispersed, in some way or another, by invertebrate vectors (see examples above) may see this expansion expedited by the effects of climate. For many invertebrates, particularly insects, geographic ranges are determined largely by climate, while their behaviour, activity and to a large extent abundance are determined by weather variations. Changes in dispersal, migration and distribution patterns may have major consequences.

Invertebrates may exhibit considerable selectivity in their diet preference (Setala *et al.*, 2005). Changes in invertebrate community composition will cascade onto lower trophic levels changing patterns of dominance and composition. Jones *et al.* (1998), in a study in the Ecotron controlled environment facility at Silwood Park (Lawton, 1996), found that the species composition of a springtail (Collembola) community responded strongly to elevated atmospheric carbon dioxide (CO_2). One species, *P. minuta*, dominated communities at present day

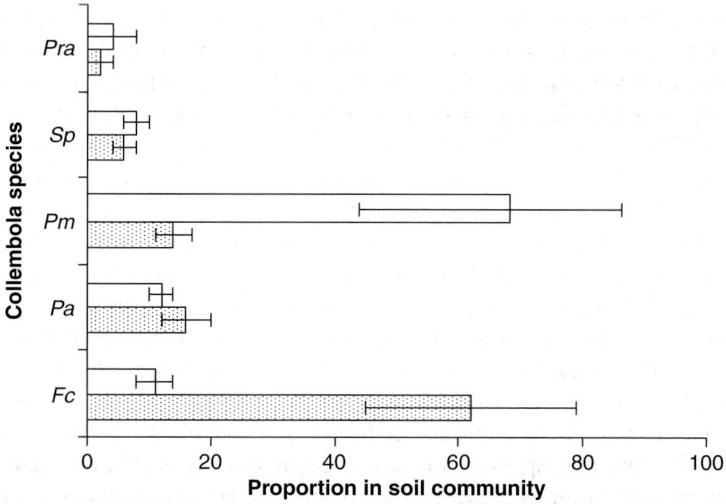

Figure 3 Effect of elevated CO_2 on communities of collembola in soil. Composition of the collembola community at the end (9 months) of ecotron experiments is shown. Microcosm communities were established in soil that was relatively poor in nutrients; They consisted of primary producers, herbivores, secondary consumers (parasitoids) and soil micro- and macroorganisms. All chambers were initiated with the same community; equal densities of each of the five collembola species. For further details see Jones *et al.* (1998). Data (mean $\pm$ SEM) from ambient ($\sim$350 ppm; open bars) and elevated (ambient+200 ppm; stippled) CO_2; $n = 8$ for each treatment. *Pa, Pseudosinella alba; Sp, Sphaeridia pumilis; Pm, Proisotoma minuta; Pra, Protaphorura armata; Fc, Folosmia candida.*

CO_2 concentrations, while another, *F. candida,* dominated when the carbon dioxide concentration was increased (Figure 3). This change in Collembola populations coincided with differences in the species of soil fungi present in the microcosm (in this experiment mainly Ascomycota). Fungal decomposers of cellulose were present at higher biomass in the elevated CO_2 treatments. Only 14 of the 33 species isolated were common to both treatments, whereas 9 and 10 species, respectively, were restricted to present day and elevated CO_2 treatments, a pattern unlikely to occur by chance alone. Elevated CO_2 in this study appeared to have major effects on the fungal and springtail soil decomposer components of the food chain. The results were very similar in a Swiss grassland study using a Free-Air CO_2 Enrichment (FACE) facility (Jones *et al.,* 2000). No comparable study has been carried out centred on Basidiomycota but with the degree of invertebrate grazing known to occur it would seem that increases in CO_2, and other associated changes, will undoubtedly modify biotic community composition.

The yellow dunes, where marram grass (*Ammophila arenaria*) is more or less completely dominant, are characterized by a group of saprotrophic fungi able to thrive at this early stage of succession. They appear to be confined to sand dunes and to require continuous input of newly deposited sand, often using marram or buried rabbit dung as a resource. They are rarely or never found inland. Species include *Psathyrella ammophila, Phallus hadriani* and the rare taxa *Coprinus ammophilae* and *Hohenbuehelia culmicola* (Rotheroe, 1996). Yellow dune, along with dune slack, are the two low lying habitats most likely to be affected by rising sea-levels in Britain; as well as losing rare fungal communities associated invertebrate populations may also decline.

13. CONCLUSION

The interactions between Basidiomycota and invertebrates take various forms. Direct benefits may be associated with food and habitat resources, and in some cases, protection; the fungi may also kill some fauna. More indirect effects relate to changes in nutrient distribution, rate of decomposition and effects of both fungal and invertebrate community structures. Chemicals emanating from Basidiomycota may also have marked influence on invertebrate foraging strategy and behaviour. Modern techniques (e.g. image analysis, stable isotope studies) allow us to study these interactions in some detail; there remains, however, a dearth of studies which specifically explore the consequences of these interactions on a number of ecosystem processes. This is in marked contrast to our knowledge and understanding of the role, for example, of invertebrate herbivory on plant productivity, transpiration and decomposition rates (Crawley, 1983; Strong *et al.*, 1984). Similarly, although limited speculation is possible on how these interactions, and the consequential effects on nutrient cycling, decomposition and soil fertility, will be affected by predicted climate change scenarios, it remains an area demanding scientific attention.

REFERENCES

Aanen, D.K., Eggleton, P. and Rouland-Lefèvre, C. (2002). The evolution of fungus-growing termites and their mutualistic fungal symbionts. *Proceedings of the National Academy of Sciences of the United States of America*, **99**, 14887–14892.

Agrios, G.N. (1980). Insect involvement in the transmission of fungal pathogens. In: *Vectors of Plant Pathogens* (K.F. Harris and K. Maramorosch, eds.), pp. 293–324. Academic Press, New York, NY.

Anderson, J.M., Ineson, P. and Huish, S.A. (1983). Nitrogen and cation mobilization by soil fauna feeding on leaf litter and soil organic matter from deciduous woodlands. *Soil Biology and Biochemistry*, **15**, 463–467.

Barbosa, P. (1991). Plant pathogens and nonvector herbivores. In: *Microbial Mediation of Plant–Herbivore Interactions* (P. Barbosa, V.A. Krischik and C.G. Jones, eds.), pp. 341–382. Wiley, New York, NY.

Barron, G.L. and Thorn, R.G. (1987). Destruction of nematodes by species of *Pleurotus*. *Canadian Journal of Botany*, **65**, 774–778.

Bass, M. (1997). The effects of leaf deprivation on leaf-cutting ants and their mutualistic fungus. *Ecological Entomology*, **22**, 384–389.

Bass, M. and Cherrett, J.M. (1994). The role of leaf-cutting ant workers (Hymenoptera: Formicidae (in fungus garden maintenance)). *Ecological Entomology*, **19**, 215–220.

Bass, M. and Cherrett, J.M. (1996). Fungus garden structure in the leaf-cutting ant *Atta sexdens* (Formicidae, Attini). *Symbiosis*, **21**, 9–24.

Berg, M.P., Stoffer, M. and van den Heuvel, H.H. (2004). Feeding guilds in Collembola based on digestive enzymes. *Pedobiologia*, **48**, 589–601.

Boddy, L., Coates, D. and Rayner, A.D.M. (1983). Attraction of fungus gnats to zones of intraspecific antagonism on agar plates. *Transactions of the British Mycological Society*, **81**, 149–151.

Boddy, L. and Jones, T.H. (2006). Mycelial responses in heterogeneous environments: Parallels with macroorganisms. In: *Fungi in the Environment* (G.M. Gadd, S.C. Watkinson and P. Dyer, eds.), pp. 112–140. Cambridge University Press, Cambridge.

Bretherton, S., Tordoff, G.M., Jones, T.H. and Boddy, L. (2006). Compensatory growth of *Phanerochaete velutina* mycelial systems grazed by *Folsomia candida* (Collembola). *FEMS Microbiology Ecology,* **58**, 33–44.

Burdsall, H.H., Jr. (1969). Stephanocysts: Unique structures in the Basidiomycetes. *Mycologia,* **61**, 915–923.

Carter, W. (1973). *Insects in Relation to Plant Diseases*. Wiley, New York, NY.

Cherret, J.M., Powell, R. and Stradling, D.J. (1989). The mutualism between leaf - cutting ants and the fungus. In: *Insect–Fungus Interactions* (M. Wilding, M.M. Collins, P.M. Hammond and J.F. Webber, eds.), pp. 25–68. Academic Press, London.

Claridge, A.W. and Trappe, J.M. (2005). Sporocarp mycophagy: Nutritional, behavioural, evolutionary, and physiological aspects. In: *The Fungal Community: Its Organization and Role in the Ecosystem* (J. Dighton, J.F. White and P. Oudemaus, eds.), pp. 599–611. Taylor & Francis, Boca Raton.

Cole, R.A., Jones, T.H. and Finch, S. (1989). Deterrent effects of carboxylic acids on cabbage root fly oviposition. *Annals of Applied Biology,* **119**, 35–44.

Crawley, M.J. (1983). *Hebivory. The Dynamics of Animal–Plant Interactions*. Blackwell Science, Oxford.

Currie, C.R., Scott, J.A., Summerbell, R.C. and Malloch, D. (1999). Fungus-growing ants use antibiotic-producing bacteria to control garden parasites. *Nature,* **398**, 701–704.

de Nooij, M.P., Biere, A. and Linders, E.G.A. (1992). Interaction of pests and pathogens through host predisposition. In: *Pests and Pathogens: Plant Responses to Foliar Attack* (P.G. Ayres, ed.), pp. 143–160. BIOS, Oxford.

Dighton, J. and Boddy, L. (1989). The role of fungi in N, P, S cycling in forests. In: Nitrogen, Phosphorous and Sulphur utilization by fungi (L. Boddy, R. Marchant and D.J. Read, eds.), pp. 269–298. Cambridge University Press, Cambridge.

Dyer, H.C., Boddy, L. and Preston-Meek, C.M. (1992). Effect of the nematode *Panagrellus redivivus* on growth and enzyme production by *Phanerochaete velutina* and *Stereum hirsutum. Mycological Research,* **96**, 1019–1028.

Dykstra, M.J. (1974). Some ultrastructural features in genus *Septobasidium. Canadian Journal of Botany,* **52**, 971–978.

El-Sayed, A.M. (2005). *The pherobase: Database of insect pheromones and semiochemicals.* www.pherobase.net (accessed March 2007).

Evans, H.C. (1989). Mycopathogens of insects of epigeal and aerial habitats. In: *Insect–Fungus Interactions* (M. Wilding, M.M. Collins, P.M. Hammond and J.F. Webber, eds.), pp. 205–238. Academic Press, London.

Faldt, J., Jonsell, M., Norlander, G. and Borg-Karlson, A.K. (1999). Volatiles of bracket fungi *Fomitopsis pinicola* and *Fomes fomentarius* and their functions as insect attractants. *Journal of Chemical Ecology,* **25**, 567–590.

Feeny, P. (1976). Plant apparency and chemical defense. *Recent Advances in Phytochemistry,* **10**, 1–40.

Feofilova, E.P. (2001). The kingdom Fungi: Heterogeneity of physiological and biochemical properties and relationships with plants, animals, and prokaryotes. *Applied Biochemistry and Microbiology,* **37**, 124–137.

Fisher, R.C. (1940). Studies of the biology of the death watch beetle (*Xestobium rufovillosum* de Greer). III. Fungal decay in timber in relation to the occurrence and rate of development of the insect. *Annals of Applied Biology,* **27**, 545–557.

Fisher, R.C. (1941). Studies of the biology of the death watch beetle (*Xestobium rufovillosum* de Greer). IV. The effect of type and extent of fungal decay upon the rate of development of insects. *Annals of Applied Biology,* **28**, 244–260.

Frankland, J.C, Magan, N. and Gadd, G.M. (eds.) (1996). *Fungi and Environmental Change.* Cambridge University Press, Cambridge.

Frouz, J., Novakova, A. and Jones, T.H. (2002). The potential effect of high atmospheric CO_2 on soil fungi–invertebrate interactions. *Global Change Biology,* **8**(4), 339–344.

Gange, A.C. and Brown, V.K. (1997). *Multitrophic Interactions in Terrestrial Systems*. Blackwell Science, Oxford.

Greenslade, P., Simpson, J.A. and Grgurinovic, C.A. (2002). Collembola associated with fungal fruit-bodies in Australia. *Pedobiologia*, **46**, 345–352.

Grove, J.F. and Blight, M.M. (1983). The oviposition attractant for the mushroom phorid *Megaselia halterata* — The identification of volatiles present in mushroom house air. *Journal of the Science of Food and Agriculture*, **34**, 181–185.

Guevara, R. and Dirzo, R. (1999). Consumption of macro-fungi by invertebrates in a Mexican tropical cloud forest: Do fruit body characteristics matter? *Journal of Tropical Ecology*, **15**, 603–617.

Guevara, R., Hutcheson, K.A., Mee, A.C., Rayner, A.D.M. and Reynolds, S.E. (2000a). Resource partitioning of the host fungus *Coriolus versicolor* by two ciid beetles: The role of odour compounds and host ageing. *Oikos*, **91**, 184–194.

Guevara, R., Rayner, A.D.M. and Reynolds, S.E. (2000b). Orientation of specialist and generalist fungivorous ciid beetles to host and non-host odours. *Physiological Entomology*, **25**, 288–295.

Guevara, R., Rayner, A.D.M. and Reynolds, S.E. (2000c). Effects of fungivory by two specialist ciid beetles (*Octotemnus glabriculus* and *Cis boleti*) on the reproductive fitness of their host fungus *Coriolus versicolor*. *New Phytologist*, **145**, 137–144.

Hammond, P.M. and Lawrence, J.F. (1989). Appendix: Mycophage in insects: A summary. In: *Insect–Fungus Interactions* (M. Wilding, M.M. Collins, P.M. Hammond and J.F. Webber, eds.), pp. 275–324. Academic Press, London.

Hanski, I. (1989). Fungivory: Fungi, insects and ecology. In: *Insect–Fungus Interactions* (M. Wilding, M.M. Collins, P.M. Hammond and J.F. Webber, eds.), pp. 25–68. Academic Press, London.

Harold, S., Tordoff, G.M., Jones, T.H. and Boddy, L. (2005). Mycelial responses of *Hypholoma fasciculare* to collembola grazing: Effect of inoculum age, nutrient status and resource quality. *Mycological Research*, **109**, 927–935.

Hatcher, P.E. (1995). Three-way interaction between plant pathogenic fungi, herbivorous insects and their host plants. *Biological Reviews*, **70**, 639–694.

Henk, D.A. (2005). Co-evolutionary diversification in *Septobasidium*, a fungal symbiont of scale insects, PhD Thesis, Duke University, Durham, NC.

Hernandez, J.V., Cabrera, A. and Jaffe, K. (1999). Mandibular gland secretion in different castes of the leaf-cutter ant *Atta laevigata*. *Journal of Chemical Ecology*, **25**, 2433–2444.

Hibbett, D.S. and Thorn, R.G. (1994). Nematode-trapping in *Pleurotus tubberregium*. *Mycologia*, **86**, 696–699.

Hutchison, L.J., Madzia, S.C. and Barron, G.L. (1995). The presence and antifeedant function of toxin-producing secretory cells on hyphae of the lawn-inhabiting agaric *Conocybe lactea*. *Canadian Journal of Botany*, **74**, 431–434.

Hynes, J., Mueller, C.T., Jones, T.H. and Boddy, L. (2007). Changes in volatile production during the course of fungal mycelial interactions between *Hypholoma fasciculare* and *Resinicium bicolor*. *Journal of Chemical Ecology*, **33**, 43–57.

Jaffee, B.A. (2004). Wood, nematodes, and the nematode-trapping fungus *Arthrobotrys oligospora*. *Soil Biology and Biochemistry*, **36**, 1171–1178.

Jaffee, B.A. (2006). Interactions among a soil organic amendment, nematodes, and the nematode-trapping fungus *Dactylellina candidum*. *Phytopathology*, **96**, 1388–1396.

Johnson, R.A., Thomas, R.J., Wood, T.G. and Swift, M.J. (1981). The inoculation of the fungus comb in newly founded colonies of some species of the Macrotermitinae (Isoptera) from Nigeria. *Journal of Natural History*, **15**, 751–756.

Jones, C.G., Lawton, J.H. and Shachak, M. (1994). Organisms as ecosystem engineers. *Oikos*, **69**, 373–386.

Jones, T.H., Bezemer, T.M., Korner, C., Lawton, J.H. and Thompson, L.J. (2000). Comparing studies of artificial and natural ecosystem responses to CO_2 enrichment. *Biotronics*, **29**, 1–7.

Jones, T.H., Thompson, L.J., Lawton, J.H., Bezemer, T.M., Bardgett, R.D., Blackburn, T.M., Bruce, K.D., Cannon, P.F., Hall, G.S., Hartley, S.E., Howson, G., Jones, C.G., Kampichler, C., Kandeler, E. and Ritchie, D.A. (1998). Impacts of rising atmospheric carbon dioxide on model terrestrial ecosystems. *Science*, **280**, 441–443.

Jonsell, M. and Nordlander, G. (1995). Field attraction of Coleoptera to odours of the wood-decaying polypores *Fomitopsis pinicola* and *Fomes fomentarius*. *Annales Zoologici Fennici*, **32**, 391–402.

Jonsell, M. and Nordlander, G. (2002). Insects in polypore fungi as indicator species: A comparison between forest sites differing in amounts and continuity of dead wood. *Forest Ecology and Management*, **157**, 101–118.

Jonsson, M., Kindvall, O., Jonsell, M. and Nordlander, G. (2003). Modelling mating success of saproxylic beetles in relation to search behaviour, population density and substrate abundance. *Animal Behaviour*, **65**, 1069–1076.

Kahlos, K., Kiviranta, J.L.J. and Hiltunun, R.V.K. (1994). Volatile constituents of wild and *in vitro* cultivated *Gloeophyllum odoratum*. *Phytochemistry*, **36**, 917–922.

Kajimura, H. (2000). Discovery of mycangia and mucus in adult female xiphydriid woodwasps (Hymenoptera: Xiphydriidae) in Japan. *Annals of the Entomological Society of America*, **93**, 312–317.

Kampichler, C., Rolschewski, J., Donnelly, D.P. and Boddy, L. (2004). Collembolan grazing affects the growth strategy of the cord-forming fungus *Hypholoma fasciculare*. *Soil Biology and Biochemistry*, **36**, 591–599.

Killham, K. (1994). *Soil Ecology*. Cambridge University Press, Cambridge.

Krasnoshchekov, Y.N. and Vishnyakova, Z.V. (2003). Changes in properties of soils in foci of Siberian moth (*Dendrolimus sibiricus*) outbreaks. *Eurasia Soil Science*, **36**, 1298–1306.

Lawton, J.H. (1996). The Ecotron facility at Silwood Park: The value of "big bottle" experiments. *Ecology*, **77**, 665–669.

Li, Y., Jeewon, R., Hyde, K.D., Mo, M.H. and Zhang, K.Q. (2006). Two new species of nematode-trapping fungi: Relationships inferred from morphology, rDNA and protein gene sequence analyses. *Mycological Research*, **110**, 790–800.

Luo, H., Mo, M., Huang, X., Li, X. and Zhang, K. (2004). *Coprinus comatus*: A basidiomycete fungus forms novel spiny structures and infects nematodes. *Mycologia*, **96**, 1218–1225.

Makarova, O.L. (2004). Gamasid mites (Parasitiformes, Mesostigmata), dwellers of bracket fungi, from the Pechora-Ilychskii Reserve, Republic of Komi. *Entomological Review*, **84**, 667–672.

Maraun, M., Martens, H., Migge, S., Theenhaus, A. and Scheu, S. (2003). Adding to 'the enigma of soil animal diversity': Fungal feeders and saprophagous soil invertebrates prefer similar food substrates. *European Journal of Soil Biology*, **39**, 85–95.

Martin, N.M. (1992). The evolution of insect–fungus associations: From contact to stable symbiosis. *American Zoologist*, **32**, 593–605.

Martínez, A.S., Fernández-Arhex, V. and Corley, J.C. (2006). Chemical information from the fungus *Amylostereum areolatum* and host-foraging behaviour in the parasitoid *Ibalia leucospoides*. *Physiological Entomology*, **31**, 336–340.

Matsuura, K. (2005). Distribution of termite egg-mimicking fungi ("termite balls") in *Reticulitermes* spp. (Isoptera: Rhinotermitidae) nests in Japan and the United States. *Applied Entomological Zoology*, **40**, 53–61.

Matsuura, K., Tanaka, C. and Nishida, T. (2000). Symbiosis of the termite and the sclerotium-forming fungus: Sclerotia mimic termite eggs. *Ecological Research*, **15**, 405–414.

Moody, S.A., Briones, M.J.I., Piearce, T.G. and Dighton, J. (1995). Selective consumption of decomposing wheat straw by earthworms. *Soil Biology and Biochemistry*, **27**, 1209–1213.

Moody, S.A., Piearce, T.G. and Dighton, J. (1996). Fate of some fungal spores associated with wheat, straw, decomposition on passage through the guts of *Lumbricus terrestris* and *Aporrectodea longa*. *Soil Biology and Biochemistry*, **28**, 533–537.

Morse, G.E. and Normak, B.B. (2006). A molecular phylogenetic study of armoured scale insects (Hemiptera: Diaspididae). *Systemic Entomology*, **31**, 338–349.

Mueller, U.G. (2002). Ant versus fungus versus mutualism: Anti-cultivar conflict and a deconstruction of the attine ant fungus symbiosis. *American Naturalist*, **160**, S67–S98.

Mueller, U.G. and Gerardo, N. (2002). Fungus-farming insects: Multiple origins and diverse evolutionary histories. *Proceedings of the National Academy of Sciences of the United States of America*, **99**, 15247–15249.

Mueller, U.G., Schultz, T.R., Currie, C.R., Adams, R.M.M. and Malloch, D. (2001). The origin of the attine ant-fungus mutualism. *Quarterly Review of Biology*, **76**, 169–197.

Nakamori, T. and Suzuki, A. (2005a). Spore-breaking capabilities of collembolans and their feeding habit within sporocarps. *Pedobiologia*, **49**, 261–267.
Nakamori, T. and Suzuki, A. (2005b). Preference of three collembolan species for fruit-bodies of three species of basidiomycete fungi. *Pedobiologia*, **49**, 119–125.
Newell, K. (1984a). Interaction between two decomposer Basidiomycetes and a collembolan under Sitka spruce: Distribution, abundance and selective grazing. *Soil Biology and Biochemistry*, **16**, 227–233.
Newell, K. (1984b). Interaction between two decomposer Basidiomycetes and a collembolan under Sitka spruce: Grazing and its potential effects on fungal distribution and litter decomposition. *Soil Biology and Biochemistry*, **16**, 235–239.
Nowak, E., Piotrowska-Seget, Z. and Chmielewski, K. (2005). Response of enchytraeid community (Oligochaeta, Enchytraeidae) to manipulation of microbial biomass. *Polish Journal of Ecology*, **53**, 53–63.
Nuss, I. (1982). Die Bedeutung der proterosporen: Schluβfolgerungen aus untersuchungen an *Ganoderma* (Basidiomycetes). *Plant Systematics and Evolution*, **141**, 53–79.
Okada, H., Harada, H. and Kadota, I. (2005). Fungal-feeding habits of six nematode isolates in the genus *Filenchus*. *Soil Biology and Biochemistry*, **37**, 1113–1120.
Orledge, G.M. and Reynolds, S.E. (2005). Fungivore post-use groups for cluster analysis: Patterns of utilisation of fungal fruiting bodies by ciid beetles. *Ecological Entomology*, **30**, 620–641.
Ortius-Lechner, D., Maile, R., Morgan, E.D. and Boomsma, J.J. (2000). Metaplural gland secretion of the leaf-cutter ant *Acromyrmex octospinosus*: New compounds and their functional significance. *Journal of Chemical Ecology*, **26**(7), 1667–1683.
Rayner, A.D.M. and Boddy, L. (1988). Fungal communities in the decay of wood. *Advances in Microbial Ecology*, **10**, 115–166.
Renker, C., Alphei, J. and Buscot, F. (2003). Soil nematodes associated with the mammal pathogenic fungal genus *Malassezia* (Basidiomycota: Ustilaginomycetes) in central European forests. *Biology and Fertility of Soils*, **37**, 70–72.
Rishbeth, J. (1982). Species of *Armillaria* in southern England. *Plant Pathology*, **31**, 239–249.
Rosecke, J. Pietsch, M. and König, W.A. (2000). Volatile constituents of wood-rotting basidiomycetes. *Phytochemistry*, **54**, 747–750.
Rotheroe, M. (1996). Implications of global warming and rising sea-levels for macrofungi in UK dune systems. In: *Fungi and Environmental Change* (J.C. Frankland, N. Magan and G.M. Gadd, eds.), pp. 51–60. Cambridge University Press, Cambridge.
Rønhede, S., Boomsma, J.J. and Rosendahl, S. (2004). Fungal enzymes transferred by leaf-cutting ants in their fungus gardens. *Mycological Research*, **108**, 101–106.
Santana, M.E., Lodge, D.J. and Lebow, P. (2005). Relationship of host recurrence in fungi to rates of tropical leaf decomposition. *Pedobiologia*, **49**, 549–564.
Schigel, D.S., Niemelä, T., Similä, M., Kinnunen, J. and Manninen, O. (2004). Polypores and associated beetles of the North Karelian Biosphere Reserve, eastern Finland. *Karstenia*, **44**, 35–56.
Setala, H., Berg, M.P. and Jones, T.H. (2005). Trophic structure and functional redundancy in soil communities. In: *Biological Diversity and Function in Soils* (R.D. Bardgett, M.B. Usher and D.W. Hopkins, eds.), pp. 236–249. Cambridge University Press, Cambridge.
Sheepmaker, J.W.A., Geels, F.P., Griensven Van, L.J.L.D. and Smits, P.H. (1996). Substrate dependent larval development and emergence of the mushroom pests *Lycoriella auripila* and *Megaselia halterata*. *Entomologia Experimentalis et Applicata*, **79**, 329–334.
Shinzato, N., Muramatsu, M., Watanabe, Y. and Matsui, T. (2005). Termite-regulated fungal monoculture in fungus combs of a macrotermitine termite *Odontotermes formosanus*. *Zoological Society of Japan*, **22**, 917–922.
Slippers, B., Coutinho, T.A., Wingfield, B.D. and Wingfield, M.J. (2003). A review of the genus *Amylostereum* and its association with woodwasps. *South African Journal of Science*, **99**, 70–74.
Smith, J.E., Challen, M.P., White, P.F., Edmondson, R.A. and Chandler, D. (2006a). Differential effect of *Agaracus* host species on the population development of *Megaselia* and *halterata* (Diptera: Phoridae). *Bulletin of Entomological Research*, **96**, 1–7.

Smith, J.E., White, P.F., Edmondson, R.A. and Chandler, D. (2006b). Effects of different *Agaracus* species on the development of the mushroom sciarid fly *Lycoriella ingenua. Entomologia Experimentalis et Applicata,* **120**, 63–69.

Spooner, B. (2003). A dipteran gall of *Panaeolina* and *Panaeolus. Field Mycology,* **4**, 91–92.

Stadler, M. and Sterner, O. (1998). Production of bioactive secondary metabolites in the fruit bodies of macrofungi as a response to injury. *Phytochemistry,* **49**, 1013–1019.

Strong, D.R., Jr., Lawton, J.H. and Southwood, T.R.E. (1984). *Insects on Plants: Community Patterns and Mechanisms.* Blackwell Science, Oxford.

Su, N.Y. (2005). Directional change in tunnelling of subterranean termites (Isoptera: Rhinotermitidae) in response to decayed wood attractants. *Journal of Economic Entomology,* **98**, 471–475.

Swift, M.J. and Boddy, L. (1984). Animal–microbial interactions during wood decomposition. In: *Invertebrate–Microbe Interactions* (J.M. Anderson, A.D.M. Rayner and D.W.H. Walton, eds.), pp. 89–131. Cambridge University Press, Cambridge.

Taprab, Y., Johjima, T., Maeda, Y., Moriya, S., Trakulnaleamsai, S., Noparatnaraporn, N., Ohkuma, M. and Kudo, T. (2005). Symbiotic fungi produce laccases potentially involved in phenol degradation in fungus combs of fungus-growing termites in Thailand. *Applied and Environmental Microbiology,* **71**, 7696–7704.

Thomson, I.M. and Koch, J. (1999). Somatic compatibility in *Amylostereum areolatum* and *A. chailletii* as a consequence of symbiosis with siricid wood wasps. *Mycological Research,* **103**, 817–823.

Thorn, R.G. and Barron, G.L. (1984). Carnivorous mushrooms. *Science,* **224**, 76–78.

Thunes, K.H., Midtgaard, F. and Gjerde, I. (2000). Diversity of Coleoptera of the bracket fungus *Fomitopsis pinicola* in a Norwegian spruce forest. *Biodiversity and Conservation,* **9**, 833–852.

Tibbles, L.L., Chandler, D., Mead, A., Jervis, M.A. and Boddy, L. (2005). Behavioural response of the flies *Megaselia halterata* and *Lycoriella castanescens* to different mushroom cultivation substrates. *Entomologia Experimentalis et Applicata,* **116**, 73–81.

Toljander, Y.K., Lindahl, B.D., Holmer, L. and Hogberg, N.O.S. (2006). Environmental fluctuations facilitate species co-existence and increase decomposition in communities of wood decay fungi. *Oecologia,* **148**, 625–631.

Tordoff, G.M., Jones, T.H. and Boddy, L. (2006). Grazing by *Folsomia candida* (Collembola) differentially affects mycelial morphology of the cord-forming basidiomycetes *Hypholoma fasciculare, Phanerochaete velutina* and *Resinicium bicolor. Mycological Research,* **110**, 335–345.

Tuno, N. (1998). Spore dispersal of *Dictyophora* fungi (Phallaceae) by flies. *Ecological Research,* **13**, 7–15.

Tzean, S.S. and Liou, J.Y. (1993). Nematophagous resupinate basidiomycete fungi. *Phytopathology,* **83**, 1015–1020.

Wargo, P.M. (1984). How stress predisposes trees to attack by *Armillaria Mellea* - a hypothesis. In : Proceedings of the 6th International Conference on Root and Butt Rots of Forest Trees (G.A. Kile, ed). IUFRO working party S2. 06. 01., CSIRO, Melbourne, Australia pp. 115–122.

Wilding, N., Collin, N.M., Hammond, P.M. and Webber, J.F., eds. (1989). *Insect–Fungus Interactions.* Academic Press, London.

Wood, T.G. and Thomas, R.J. (1989). The mutualistic association between macrotermitinae and *Termitomyces.* In: *Insect–Fungus Interactions* (M. Wilding, M.M. Collins, P.M. Hammond and J.F. Webber, eds.), pp. 69–92. Academic Press, London.

Wood, J., Tordoff, G.M., Jones, T.H. and Boddy, L. (2006). Reorganization of mycelial networks of *Phanerochaete velutina* in response to new woody resources and collembola grazing. *Mycological Research,* **110**, 985–993.

Yamashita, S. and Hijii, N. (2003). Effects of mushroom size on the structure of a mycophagous arthropod community: Comparison between infra communities with different types of resource utilisation. *Ecological Research,* **18**, 131–143.

Section 3:
Basidiomycete Communities: Structure and Function

CHAPTER 10

Distribution and Function of Litter Basidiomycetes in Coniferous Forests

Björn Lindahl and **Johanna Boberg**

Contents

Abstract

The spatio-temporal distribution of basidiomycetes colonising coniferous litter, and their impact on carbon and nitrogen cycling is reviewed. After a brief phase of colonisation by phyllosphere ascomycetes and ephemeral basidiomycetes, litter is colonised by fungi with higher capacity for litter decomposition, such as *Mycena* species. This change in fungal community composition involves a shift in substrate utilisation from relatively readily available sugars to cellulose. The fungi overcome the nitrogen deficiency experienced during litter colonisation by translocating nitrogen from older parts of the mycelium. To retrieve nitrogen from well-decomposed litter, the fungi presumably use carbohydrates translocated from fresh litter as a co-substrate to attack polyphenol–nitrogen complexes. After a few years of decomposition, the saprotrophic fungal community is out-competed by mycorrhizal fungi. Their direct access to photo-assimilates is likely to be a major competitive advantage in highly recalcitrant and cellulose-depleted resources. Decomposition is continued by the mycorrhizal fungi, although at a very slow rate. It is not until this phase that net losses of nitrogen from the litter occur, as the mycorrhizal fungi forage for nitrogen to support

British Mycological Society Symposia Series
Published by Elsevier Ltd.

themselves and their host plants. In low-productivity coniferous forests, saprotrophic fungi, thus, re-cycle litter carbon to the atmosphere, whereas re-cycling of litter nitrogen to growing plants is performed largely by mycorrhizal fungi.

1. INTRODUCTION

Litter decomposition is central in discussions and models of global carbon balances, and soil respiration represents a major input of carbon dioxide to the atmosphere (Schlesinger and Andrews, 2000). Northern forest ecosystems have been identified as major global carbon sinks (Myneni *et al.*, 2001). The carbon and nitrogen cycles are tightly interlinked, and the effects of nitrogen deposition on litter decomposition are of particular interest, but the relationships are complex and observations often contradictory (Neff *et al.*, 2002). Basidiomycetes occupy a central role as colonisers of litter, humus and soil in forests, but have not been studied extensively. Considering the vast literature on the ecosystem processes that these organisms carry out, the lack of knowledge of their identity, ecology and physiology is particularly striking. Most process-oriented research on nutrient cycling does not take the specific ecophysiology of litter fungi into account, but treats all microorganisms as a single functional entity. In the first part of this chapter, we discuss the spatio-temporal distribution of different functional groups of fungi, and review results from some recent studies that have used molecular identification techniques to investigate fungal succession in coniferous litter. In the second part of the chapter, the role of basidiomycetes in re-cycling of litter-bound nitrogen and carbon is examined. The nitrogen and carbon dynamics of decomposing litter is related to resource translocation in fungal mycelia.

2. FUNGAL SUCCESSION AND INTERACTIONS IN FOREST LITTER: A MOLECULAR APPROACH

One of the major challenges in studying fungal ecology is identification of the fungi present in a particular substrate. Traditionally, litter fungi have been studied primarily through isolation of fungi growing out from litter components placed on artificial media (Frankland *et al.*, 1990). During the last 15 years, molecular methods have been developed for fungal identification and have been used extensively to analyse communities of mycorrhizal fungi (Horton and Bruns, 2001). Only recently have techniques been developed that enable DNA-based identification of mixed fungal communities directly from natural substrata (Landeweert *et al.*, 2003), and the first molecular studies of fungal communities in forest litter have been published (O'Brien *et al.*, 2005; Lindahl *et al.*, 2007). The major advantage of DNA-based methods over identification of isolated fungi is that analysis of extracted DNA does not discriminate against non-culturable and slow-growing taxa. Molecular methods may, in fact, provide very different views of fungal communities than isolation-based methods (Allmér *et al.*, 2006). In addition, sequence comparisons with databases may enable identification of non-sporulating taxa, including basidiomycetes. The major drawbacks of molecular identification techniques are the large costs involved and the required access to a molecular biology laboratory. In addition, molecular techniques suffer from the same problems as most other methods, when attempting to analyse complex microbial communities from limited numbers of samples (Taylor, 2002).

Briefly, the methods are based on polymerase chain reaction (PCR) amplification of a small part of the fungal genome (usually the ITS region) using fungal-specific primers (White *et al.*, 1990) and DNA extracted from field samples as template. The PCR products, containing a mixture of DNA from many different fungal taxa, are incorporated into plasmid vectors and cloned into bacteria. When the transformed bacteria are cultured on agar plates, each bacterial colony originates from a single bacterial cell and thus contains DNA from a single fungal taxon. Bacterial clones (e.g. 25–250 per sample) are collected and the cloned fungal DNA is re-amplified and sequenced (Landeweert *et al.*, 2003). Taxonomic identities may be assigned by comparisons with reference sequences obtained from identified sporocarps or sporulating cultures. Cloning is costly and involves a large amount of laboratory work, and if large numbers of samples are analysed, the cloning approach has to be combined with a community fingerprinting method, such as terminal restriction fragment length polymorphism (TRFLP). We have conducted a series of studies where DNA was extracted from needle litter, moss litter, humus or soil from Swedish coniferous forests, and the total fungal community was analysed by cloning, sequencing and TRFLP analysis. Some of the results are outlined below (Lindahl *et al.*, 2007 and unpublished data).

2.1 Early Colonisers

Communities of litter fungi are often considered in terms of successional colonisation patterns, and the established picture is that ascomycetes dominate at early stages of litter decomposition, whereas basidiomycetes dominate later (Frankland, 1998). When plant litter arrives at the forest floor, needles and leaves are already colonised by fungi. Phyllosphere ascomycetes colonise living plant tissues as biotrophic parasites or sometimes necrotrophic pathogens, but a significant fraction of these fungi persist in dead litter, and some are more active as litter saprotrophs than as biotrophic endophytes (Osono, 2006). During the first year of pine needle decomposition, primarily soluble sugars and less recalcitrant substrates are lost from the litter (Figure 1; Berg *et al.*, 1982), implying that fungi without explicit cellulolytic capacity may also be active in the litter. In *Pinus sylvestris* needles at early decomposition stages, *Lophodermium pinastri* (Rhytismataceae) dominated together with *Sydowia polyspora* (Dothioraceae), a Lachnoid species (Hyaloscyphaceae) and another helotialean taxon. These species declined rapidly in abundance during the initial year on the forest floor. *Scleroconidioma sphagnicola* (Dothioraceae) also occurred abundantly during the first year but was able to persist for up to 2 years, confirming the suggestion by Koukol *et al.* (2006) that this species has a high capacity to endure competition.

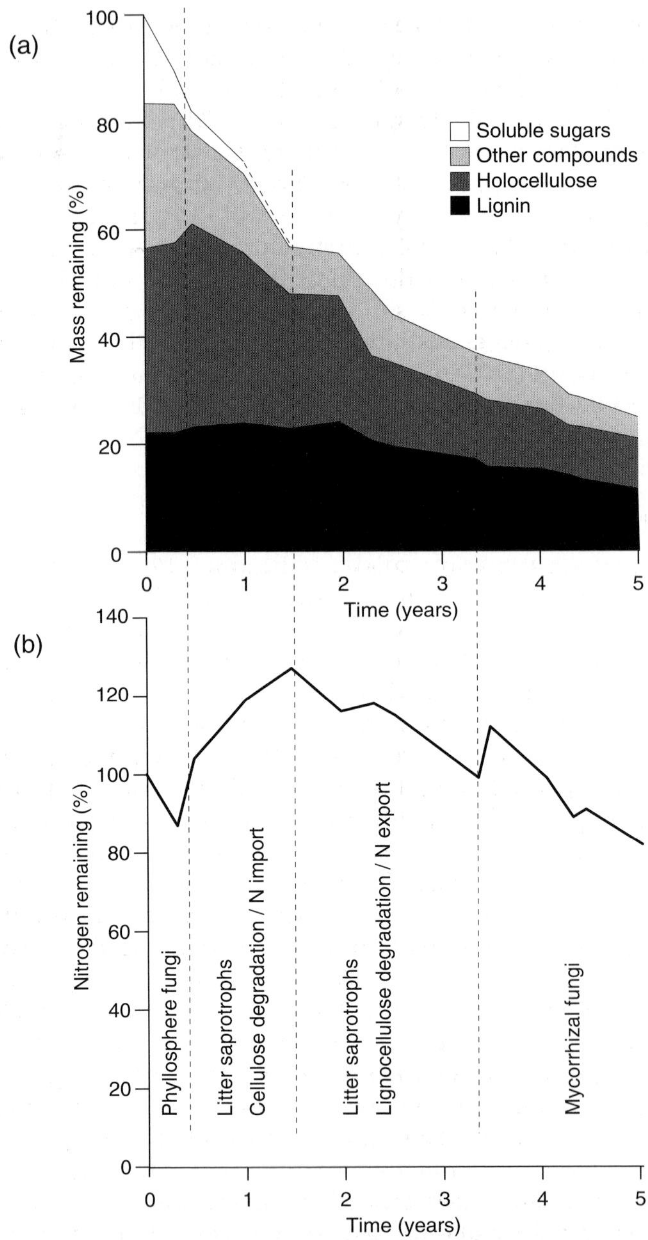

Figure 1 Data from a litterbag experiment (berg *et al.*, 1982) in which scots pine needles were incubated on the forest floor for 5 years. Samples were retrieved at regular intervals and analysed for chemical composition. The plots show (a) change in major chemical components of the litter, and (b) absolute amounts of nitrogen in the litter. In both plots, the chemical content is relative to the initial litter content. In (a), 'Lignin' refers to acid insoluble material (klason lignin), 'holocellulose' includes hemicellulose as well as chitin (fungal production of chitin may explain the initial increase in this fraction), 'soluble sugars' are mainly fructose, glucose and pinitol, and 'other compounds' are mainly lipids and terpenes. Data on 'soluble sugars' were available for the first year of decomposition only.

In addition to these ascomycetes, basidiomycetes within the genera *Athelia* and *Sistotrema* were detected after a few months but were absent after 1 year, indicating a ruderal strategy (Cooke and Rayner, 1984). *Athelia* and *Sistotrema* species form delicate resupinate sporocarps consisting of basidia on a thin subiculum directly on the substratum (Larsson *et al.*, 2004) and are therefore likely to be overlooked in sporocarp inventories. The simple sporocarps, producing large amounts of spores with a relatively small investment of biomass, could reflect the ruderal strategy. Records of DNA from *Athelia* species in decaying litter of *Pinus taeda* as well as in mixed deciduous litter in eastern USA (O'Brien *et al.*, 2005), and from *Sistotrema* species in pine and spruce litter from three different Swedish locations, suggest that these fungi are common and widely spread members of saprotrophic litter communities. *Marasmius androsaceus*, known to form sporocarps on relatively fresh litter (Frankland, 1984), was also detected in needles a few months after abscission.

2.2 The Cellulose Decomposition Phase

After the early decomposition phase, a community of secondary colonisers replaces the primary colonisers. In our studies of pine litter colonisation, and also in an American study of mixed deciduous litter (O'Brien *et al.*, 2005), *Mycena* species dominated the Basidiomycota DNA pool. The ecophysiology of *Mycena galopus* has been well investigated by Frankland and co-workers (reviewed by Frankland, 1998). In our studies, identities of the *Mycena* species remain unknown, due to lack of database references. More detailed studies of the molecular phylogeny of litter basidiomycetes in general, and of the genus *Mycena* in particular, are required. Throughout the cellulose decomposition phase, the *Mycena* species and other basidiomycetes were accompanied by ascomycetes. In litter of *Pinus sylvestris, P. taeda, Picea abies* and mixed deciduous species, ascomycetes accounted for 50–70% of the fungal taxa. The distribution of clones more or less reflected the distribution of taxa, suggesting that ascomycetes constituted a major fraction of the mycelial biomass in the litter. Almost all of the ascomycetes recorded as DNA in the litter were either Leotiomycetes (primarily within the Hyaloscyphaceae, Helotiales) or Dothideomycetes. The fact that ascomycetes constituted a major fraction of the litter-colonising fungi does not, however, necessarily imply that they were responsible for a major part of the decomposition. Osono and Takeda (2002) tested the litter-degrading capacity of 60 different species over 21 months. They found that, apart from Xylariaceae, ascomycetes caused little mass loss of the litter, and their litter-decomposing capacity was exceeded by that of basidiomycetes by an order of magnitude. In the study of Osono and Takeda (2002), however, Leotiomycetes were represented by one species only. In the pine forest study of Lindahl *et al.* (2007), each litter sample (taken with a 28 mm diameter corer) contained on average 1.6 basidiomycete taxa, ranging up to 5. In 15% of the samples, no basidiomycetes were detected. These samples did not diverge from the rest in decomposition, suggesting either that some ascomycetes may cause significant decomposition, and/or that there is a high degree of temporal variation in basidiomycete colonisation.

2.3 Mycorrhizal Fungi Take over at Late Decomposition Stages

In our pine forest study, after 2–4 years of decomposition, when the litter began to lose its structural integrity and eventually became humus, the community of saprotrophic litter fungi was replaced by a totally different community dominated by mycorrhizal species (Figure 2). In well-decomposed litter and humus samples, ectomycorrhizal basidiomycetes, mainly within the genera *Piloderma* and *Cortinarius*, dominated the DNA pool together with ascomycetes, again from the Leotiomycetes and Dothideomycetes. Helotialean ascomycetes continued to play an important role at later stages of decomposition, but these taxa were different from the ones in less decomposed litter. Mycorrhizal symbiosis with conifers or ericaceous plants appears to be widespread throughout the Helotiales (Vrålstad *et al.*, 2002), and it is likely that the late helotialean colonisers were mycorrhizal. The Dothideomycetes recorded at later decomposition stages

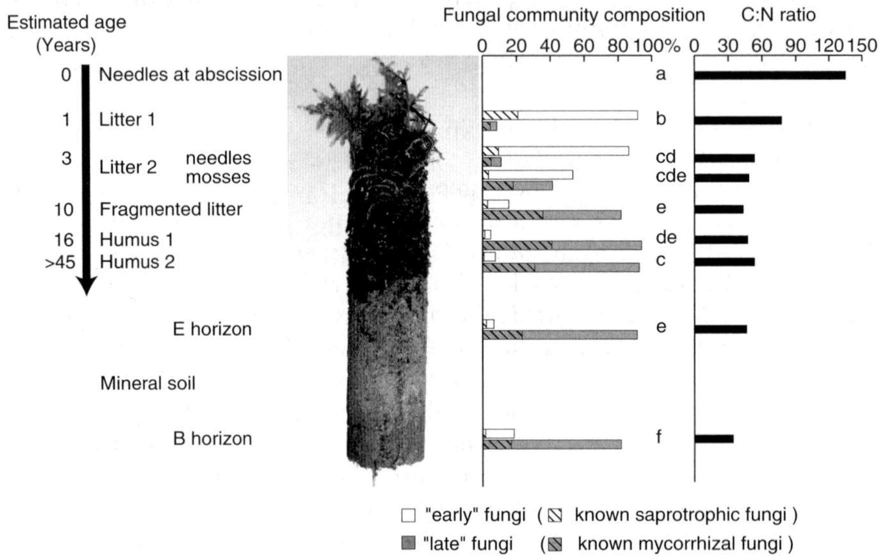

Figure 2 The graph shows a 28 mm diameter core taken through the forest floor and upper mineral soil of a swedish *P. sylvestris* forest. In a study by Lindahl *et al.* (2007), 27 such cores were collected and subdivided into vertical horizons. The fungal communities in the different horizons were analysed using ITS-PCR and TRFLP in combination with cloning and sequencing. The results indicate that, as decomposition progresses, an 'early' community (here defined as taxa more abundant in litter samples than in humus samples) is replaced by a 'late' community (here defined as taxa more abundant in humus samples than in litter samples). Age estimates represent time since abscission, as determined by analysis of the content of ^{14}C (from thermonuclear bomb tests). Fungal community composition in each horizon is related to carbon:nitrogen ratios. During the early, presumably saprotrophic phase, C:N decreases progressively with litter age. During the later, presumably mycorrhizal phase, C:N increases with age of the organic matter. In the mineral soil C:N ratios are slightly lower. Different letters in the C:N ratio bar diagram Indicate significant differences ($P < 0.05$). Reproduced with publisher's permission from Lindahl *et al.* (2007). **(See Colour Section)**

belonged to the genus *Capronia*, in which taxa have been identified as mycorrhizal symbionts of ericaceous plants (Allen *et al.*, 2003). In addition, we recovered DNA from a group of recently discovered fungi, only distantly related to other ascomycetes, which hitherto have been described only in terms of their DNA sequence (Schadt *et al.*, 2003). DNA sequences attributed to this intriguing group have also been obtained from mycorrhizal pine roots (Rosling *et al.*, 2003), suggesting that they are root-associated. Thus, almost all the fungi detected in the litter-humus after a few years of saprotrophic decomposition were recognised as, or presumed to be, mycorrhizal. In soil microcosms, we demonstrated antagonistic interactions between a saprotrophic basidiomycete and certain ectomycorrhizal fungi (Lindahl *et al.*, 1999). The outcome of the combative interactions was strongly affected by the amount of resources available to the saprotroph (Lindahl *et al.*, 2001). Together these observations suggest that litter saprotrophs, when their cellulose supply is depleted, are out-competed by mycorrhizal fungi that have direct access to plant-derived photo-assimilates. The spatial separation of the two functional groups of fungi may simply reflect the fact that saprotrophs obtain their carbon predominantly from litter, deposited from above, whereas mycorrhizal fungi obtain their carbon from roots that enter the forest floor from below.

Particularly noteworthy is the absence of sequences from typical 'soil fungi', such as *Trichoderma* and *Penicillium*, in the clone-libraries. According to conventional isolation-based studies, these genera usually dominate at later stages of litter decomposition (e.g. Widden and Parkinson, 1973; Söderström and Bååth, 1978; Frankland, 1998; Virzo de Santo *et al.*, 2002), and their absence in the DNA pool highlights the practical problems involved in isolating representative fungi from litter and soil. Most mycorrhizal fungi grow very slowly on agar plates, and many do not grow at all. When the dominant fungi in a sample fail to grow out on agar, opportunistic fungi, that initially constitute a very small fraction of the biomass in the sample or are present only as spores, may rapidly take over the agar plates. For example, Virzo de Santo *et al.* (2002) studied fungal colonisation of coniferous litter and observed dominance of basidiomycete mycelium on the litter from after 6 months and further throughout the entire 4-year experiment. However, when litter fragments were placed on malt agar plates, basidiomycetes were isolated only during the first 2 years, and were subsequently replaced by *Trichoderma*, *Penicillium* and *Mucor* species. These apparently conflicting observations are consistent with colonisation of well-decomposed litter primarily by slower growing mycorrhizal basidiomycetes, which escape detection using conventional isolation methods.

3. NITROGEN AND CARBON CYCLING BY LITTER SAPROTROPHS

Although litter is colonised by fungi when it reaches the forest floor, the phyllosphere fungi are soon replaced by other fungi colonising from adjacent, decaying litter. Dowson *et al.* (1989) described how the fairy-ring forming fungus *Clitocybe nebularis* in deciduous forest colonised fresh litter by horizontal growth-sweeps across the forest floor. Mature rings typically consisted of a 30–40 cm wide band with the mycelial front extending into fresh litter and the inner edge leaving bleached litter behind (Chapter 1). In contrast, Frankland *et al.* (1995) described a 'sit-and-wait' strategy, where *Mycena galopus* in a coniferous forest formed sporocarps at the same position year after year, presumably reflecting the location of the mycelium. In such cases, it appears that the fungi forage for resources mostly vertically, extending out from more decayed litter below into freshly fallen litter deposited on top. Both examples illustrate the polar growth of litter-degrading mycelia, and how they constantly advance into new resource-units from older, depleted ones. The dynamics of resource colonisation by saprotrophic mycelia has been thoroughly investigated in wood-decomposing fungi. A wide range of studies describe how saprotrophic basidiomycetes in heterogeneous environments optimise resource utilisation and maximise their competitive strength by translocating carbohydrates and nutrients throughout their mycelia (reviewed by Boddy, 1999; Chapter 1). Translocation implies that the whole mycelium functions as a single entity where the nutritional status of one part of the mycelium affects distant parts. Resources are translocated within the mycelium from areas with a surplus of a specific resource to areas with high demand, according to source–sink relationships (Lindahl and Olsson, 2004; Chapter 3). In coniferous ecosystems, the amount of nitrogen available to organisms is generally low, and decomposition rates are tightly linked to nitrogen availability. Therefore, in the following sections, we focus on the dynamics of this particular element within decomposing litter. During decomposition of litter, nitrogen is enzymatically released from decomposing plant tissues, but immediately absorbed by mycelium in the litter. The two pools are difficult to separate, and in the following text the nitrogen dynamics of the entire litter–mycelium complex is discussed.

3.1 Nitrogen Import During Early Decomposition Stages

Additions of nitrogen increased both fungal respiration and biomass (chitin assay) as well as weight loss of Scots pine needles inoculated with *M. epipterygia* in laboratory systems (Boberg and Lindahl, unpublished data), indicating that the needle decomposer was constrained by low nitrogen availability. Furthermore, cellulase activity increased in response to nitrogen additions to litter in the field (Carreiro *et al.*, 2000). When litter-filled mesh bags were incubated in the field, the nitrogen content of leaf and wood litter increased during the first years of decomposition, not only in relative concentration, but also in absolute amounts (Figure 1; Berg *et al.*, 1982; Fahey *et al.*, 1985; Dighton and Boddy, 1989). Using ^{15}N tracer isotope, Hart and Firestone (1990) and Frey *et al.* (2000) demonstrated vertical translocation of nitrogen into surface litter in a forest and an agricultural soil. Thus, fungi appear to translocate nitrogen into fresh, nitrogen-poor litter, and thereby promote their growth and activity. This supposition was supported by a positive correlation between the amount of imported nitrogen and fungal biomass increase (Berg and Söderström, 1979) as well as litter weight loss

(Berg and Staaf, 1981). Furthermore, nitrogen import has been shown to be more pronounced in litter with low initial nitrogen content (Berg and Staaf, 1981).

3.2 Nitrogen Export During Late Saprotrophic Decomposition Stages

During the second year of pine needle decomposition, the absolute amounts of nitrogen in litter peak and subsequently decrease (Figure 1; Berg *et al.*, 1982). As decomposition progresses, the quality of litter as a carbon and energy source decreases, fungal growth declines, and decomposing litter changes from a sink for nitrogen to a source. Thus, the import of nitrogen into freshly colonised litter is likely to be covered by export from more degraded litter. By re-allocating nitrogen to growing mycelium, fungi minimise losses during mycelial senescence within depleted resources. A similar pattern has been demonstrated in the redistribution of ^{32}P radiotracer in *Phanerochaete velutina* mycelia colonising wood blocks at different stages of decomposition (Wells *et al.*, 1998). Only after 4 years of decomposition has the absolute amount of nitrogen in decomposing needles decreased to the initial level, after which net loss of nitrogen eventually takes place (Figure 1; Berg *et al.*, 1982). Presumably, at this stage, the fungal community has changed and is dominated by mycorrhizal rather than saprotrophic fungi. There, thus, appears to be no net release of nitrogen from the litter–mycelium complex during decomposition by saprotrophs, only redistribution from mycelium in more degraded litter into freshly colonised needles.

Traditional nitrogen cycling theory is based around the mineralisation of organic nitrogen and release of ammonium during decomposition. Release of inorganic nitrogen occurs when microorganisms experience carbohydrate deficiency and therefore utilise organic, nitrogen-containing compounds as a source of energy, leaving ammonium as a by-product (Myrold, 1998). This concept was developed for unicellular microorganisms, which are restricted to resources in their immediate vicinity and may therefore easily experience carbon deficiency. Filamentous fungi, on the other hand, may circulate resources throughout their entire mycelia, and mycelium experiencing local carbon deficiency may be supported from more or less distant resources (Chapter 3). In coniferous forest ecosystems, litter input is often more continuous than in other ecosystems, and needles may constitute a high-quality source of cellulose for at least 2 years. Thus, carbon limitation and subsequent nitrogen mineralisation is unlikely to occur in litter-decomposing fungi. We have studied the effect of translocation on nitrogen mineralisation by *Marasmius androsaceus* and *Mycena epipterygia* in laboratory microcosms. Culturing the fungi on an amino acid as a sole source of both nitrogen and carbon resulted in the release of ammonium to the medium. In cultures where part of the mycelium colonised Scots pine needles, however, production of ammonium in the compartment containing the amino acid was much lower, indicating that the fungi were able to translocate carbohydrates to avoid local carbon limitation (Boberg and Lindahl, unpublished data). By avoiding carbon limitation and maintaining a strong sink for nitrogen within freshly colonised litter, fungi may conserve acquired nitrogen within their mycelia rather than releasing it as ammonium. In fact, the concentration of mineral nitrogen in undisturbed coniferous forest soils is generally very low (Persson *et al.*, 2000).

3.3 Lignin Decomposition by Saprotrophs

As cellulose decomposition proceeds, the concentration of the more recalcitrant lignin increases (Figure 1; Berg *et al.*, 1982). At later stages of decomposition, decay correlate well with lignin concentration in the litter (McClaugherty and Berg, 1987). Polyphenolic compounds, either tannins present in the fresh litter or products of lignin decomposition, form recalcitrant complexes with nitrogen-containing compounds, such as proteins and chitin (Kelley and Stevenson, 1995). As a result, nitrogen progressively becomes incorporated into the highly recalcitrant, polyphenolic litter fraction during decomposition (Berg, 1988). In highly decomposed coniferous forest humus, more than half of the nitrogen was found in the acid insoluble (i.e. polyphenolic) fraction (Johnsson *et al.*, 1999). Basidiomycetes have been highlighted as the main organisms responsible for lignin degradation, using elaborate oxidative enzyme systems (Rayner and Boddy, 1988; Chapter 2). There are, however, large energy costs associated with the synthesis and activation of these enzymes. No organism has been found to use macromolecular lignin as a sole carbon source, and lignin degradation is believed to be a co-metabolic process requiring other substrates, such as cellulose, as energy sources (Kirk and Farrel, 1987; Hatakka, 2001). Most literature claims that the major benefit of lignin decomposition is increased access to ligno-cellulose. However, the well-developed ligninolytic enzyme systems of litter fungi may also be used to decompose humified polyphenolic compounds (Steffen *et al.*, 2002), increasing the availability of nitrogen rather than carbohydrates. Given a resource characterised by low nitrogen availability, such as coniferous litter, the energy investment associated with the production of ligninolytic enzymes could be rewarded by increased nitrogen availability. Hypothetically, litter fungi could thus translocate carbohydrates from fresh litter to forage for polyphenolic-bound nitrogen in well-decomposed litter. This hypothesis is supported by the observations that nitrogen additions decreased the activity of ligninolytic enzymes both in pure cultures of *P. chrysosporium* in the lab (Keyser *et al.*, 1978; Kirk and Fenn, 1982) and in field trials (Carreiro *et al.*, 2000; Frey *et al.*, 2004; Sinsabaugh *et al.*, 2005). In line with the reduced enzyme activities, increasing exogenous nitrogen hampered decomposition of more recalcitrant material, that is high in lignin and polyphenolic compounds (Fog, 1988; Berg and Matzner, 1997). The decomposition-stimulating effect of nitrogen during early decomposition stages thus contrasts with the inhibition of degradation at later stages (McClaugherty and Berg, 1987).

3.4 Very Late Decomposition Stages: Humus and Mycorrhizal Fungi

In the humus layer, saprotrophic litter fungi seem to be out-competed by mycorrhizal fungi (Lindahl *et al.*, 2007). Ectomycorrhizal symbiosis with tree roots has evolved on several independent occasions during evolution of basidiomycetes, and a major fraction of ground-living basidiomycetes are ectomycorrhizal. Phylogenetic relationships suggest that reversions from biotrophic symbiosis back to saprotrophy have also occurred on several occasions during evolution, and most of the saprotrophic litter-degrading basidiomycetes of today, for example genera in the euagaric clade such as *Mycena, Marasmius* and *Galerina*, are likely to have evolved from mycorrhizal ancestors (Hibbett *et al.*, 2000). The evolutionary instability of mycorrhizal symbiosis and the sometimes relatively close genetic relationship between ectomycorrhizal fungi and litter saprotrophs suggest that these two groups may not be as functionally distinct as previously thought. Most ectomycorrhizal fungi have some saprotrophic capabilities and produce enzymes such as proteases and polyphenol oxidases that enable mobilisation of nutrients in complex organic forms (Read and Perez-Moreno, 2003; Lindahl *et al.*, 2005). Overall, most evidence suggests that free-living saprotrophic fungi have a much higher capacity to degrade fresh litter than ectomycorrhizal fungi (e.g. Colpaert and van Tichelen, 1996). However, as decomposition progresses, the competitive advantage endowed by labile carbon supplied by a plant host may compensate for the modest decomposition capacity.

The shift in community composition from saprotrophic taxa at initial stages of litter decomposition to mycorrhizal taxa at later stages coincides with a simultaneous shift in the carbon and nitrogen dynamics of the litter (Figure 2; Lindahl *et al.*, 2007). During the saprotrophic phase, C:N ratios decreased with time, presumably due to consumption of litter carbon by fungal respiration, and retention of nitrogen in fungal biomass. In the mycorrhizal phase, C:N ratios increased slightly with time, presumably due to mobilisation and subsequent translocation of organic nitrogen to the plant roots in combination with respiration of host-derived carbohydrates rather than litter carbohydrates.

Throughout this discussion, nitrogen circulation within the forest floor has been described as a closed system. This is, of course, an over-simplification, as the nitrogen pools are continuously replenished by fixation and deposition of atmospheric nitrogen, and drained through loss of nitrogen-containing compounds to the ground water. In most coniferous forest ecosystems, throughput of nitrogen is, however, likely to be relatively small compared to the internal circulation.

ACKNOWLEDGEMENTS

Funding from The Swedish Research Council for Environment, Agricultural Sciences and Spatial Planning (FORMAS) as well as The Swedish University of Agricultural Sciences (SLU) is gratefully acknowledged. Thanks are also due to Dr. Katarina Ihrmark and Professors Roger Finlay, Jan Stenlid, Peter Högberg, Torgny Näsholm, Tryggve Persson and Björn Berg, for research collaboration and Dr. Andy Taylor, Dr. Petra Fransson and Professor Lynne Boddy and Dr. Juliet Frankland, for critical reading of the manuscript.

REFERENCES

Allen, T.R., Millar, T., Berch, S.M. and Berbee, M.L. (2003). Culturing and direct DNA extraction find different fungi from the same ericoid mycorrhizal roots. *New Phytologist*, **160**, 255–272.

Allmér, J., Vasiliauskas, R., Ihrmark, K., Stenlid, J. and Dahlberg, A. (2006). Wood-inhabiting fungal communities in woody debris of Norway spruce (*Picea abies* (L.) Karst.), as reflected by sporocarps, mycelial isolations and T-RFLP identification. *FEMS Microbiology Ecology*, **55**, 57–67.

Berg, B. (1988). Dynamics of nitrogen (N-15) in decomposing Scots pine (*Pinus sylvestris*) needle litter: Long-term decomposition in a Scots pine forest 6. *Canadian Journal of Botany*, **66**, 1539–1546.

Berg, B., Hannus, K., Popoff, T. and Theander, O. (1982). Changes in organic chemical components of needle litter during decomposition. Long-term decomposition in a Scots pine forest I. *Canadian Journal of Botany*, **60**, 1310–1319.

Berg, B. and Matzner, E. (1997). Effect of N deposition on decomposition of plant litter and soil organic matter in forest systems. *Environmental Review*, **5**, 1–25.

Berg, B. and Söderström, B. (1979). Fungal biomass and nitrogen in decomposing Scots pine needle litter. *Soil Biology and Biochemistry*, **11**, 339–341.

Berg, B, and Staaf, H. (1981). Leaching, accumulation and release of nitrogen in decomposing forest litter. In: *Terrestrial Nitrogen Cycles* (E. Clark and T. Rosswall, eds.), pp. 163–178. Swedish Natural Science Research Council, Stockholm, Sweden.

Boddy, L. (1999). Saprotrophic cord-forming fungi: Meeting the challenge of heterogeneous environments. *Mycologia*, **91**, 13–32.

Carreiro, M.M., Sinsabaugh, R.L., Repert, D.A. and Parkhurst, D.F. (2000). Microbial enzyme shifts explain litter decay responses to simulated nitrogen deposition. *Ecology*, **81**, 2359–2365.

Colpaert, J.V. and van Tichelen, K.K. (1996). Decomposition, nitrogen and phosphorus mineralization from beech leaf litter colonized by ectomycorrhizal or litter decomposing basidiomycetes. *New Phytologist*, **134**, 123–132.

Cooke, R.C., Rayner, A.D.M. (1984). *Ecology of Saprotrophic Fungi*. Longman, London, UK.

Dighton, J. and Boddy, L. (1989). Role of fungi in nitrogen, phosphorus and sulphur cycling in temperate forest ecosystems. In: *Nitrogen, Phosphorus and Sulphur Utilization by Fungi* (L. Boddy, R. Marchant and D.J. Read, eds.), pp. 269–298. Cambridge University Press, Cambridge.

Dowson, C.G., Rayner, A.D.M. and Boddy, L. (1989). Spatial dynamics and interactions of woodland fairy ring fungus, *Clitocybe nebularis*. *New Phytologist*, **111**, 699–705.

Fahey, T.L., Yavitt, J.B., Pearson, J.A. and Knight, D.H. (1985). The nitrogen cycle in lodgepole pine forests, southeastern Wyoming. *Biogeochemistry*, **1**, 257–275.

Fog, K. (1988). The effect of added nitrogen on the rate of decomposition of organic-matter. *Biological Reviews of the Cambridge Philosophical Society*, **63**, 433–462.

Frankland, J.C. (1984). Autecology and the mycelium of a woodland litter decomposer. In: *The Ecology and Physiology of the Fungal Mycelium* (D.H. Jennings and A.D.M. Rayner, eds.), pp. 241–260. Cambridge University Press, Cambridge, UK.

Frankland, J.C. (1998). Fungal succession: Unravelling the unpredictable. *Mycological Research*, **102**, 1–15.

Frankland, J.C., Dighton, J. and Boddy, L. (1990). Methods for studying fungi in soil and forest litter. *Methods in Microbiology*, **22**, 343–404.

Frankland, J.C., Poskitt, J.M. and Howard, D.M. (1995). Spatial development of populations of a decomposer fungus, *Mycena galopus*. *Canadian Journal of Botany*, **73**, S1399–S1406.

Frey, S.D., Elliott, E.T., Paustian, K. and Peterson, G.A. (2000). Fungal translocation as a mechanism for soil nitrogen inputs to surface residue decomposition in a no-tillage agroecosystem. *Soil Biology and Biochemistry*, **32**, 689–698.

Frey, S.D., Knorr, M., Parrent, J.L. and Simpson, R.T. (2004). Chronic nitrogen enrichment affects the structure and function of the soil microbial community in temperate hardwood and pine forests. *Forest Ecology and Management*, **196**, 159–171.

Hart, S.C. and Firestone, M.K. (1990). Forest floor–mineral soil interactions in the internal nitrogen cycle of an old-growth forest. *Biogeochemistry*, **12**, 73–97.

Hatakka, A. (2001). Biodegradation of lignin. In: *Lignin, Humic Substances and Coal* (M. Hofrichter and A. Steinbüchel, eds.), Vol. 1., pp. 129–180. Wiley-VCH, Weinheim, Germany.

Hibbett, D.S., Gilbert, L.B. and Donoghue, M.J. (2000). Evolutionary instability of ectomycorrhizal symbioses in basidiomycetes. *Nature*, **407**, 506–508.
Horton, T.R. and Bruns, T.D. (2001). The molecular revolution in ectomycorrhizal ecology: Peeking into the black-box. *Molecular Ecology*, **10**, 1855–1871.
Johnsson, L., Berggren, D. and Kårén, O. (1999). Content and bioavailability of organic forms of nitrogen in the O horizon of a podzol. *European Journal of Soil Science*, **50**, 591–600.
Kelley, K.R. and Stevenson, F.J. (1995). Forms and nature of organic N in soil. *Fertilizer Research*, **42**, 1–11.
Keyser, P., Kirk, K. and Zeikus, J.G. (1978). Ligninolytic enzyme system of *Phanerochaete chrysosporium*: Synthesized in the absence of lignin in response to nitrogen starvation. *Journal of Bacteriology*, **135**, 790–797.
Kirk, T.K. and Farrell, R.L. (1987). Enzymatic combustion: The microbial degradation of lignin. *Annual Review of Microbiology*, **41**, 465–505.
Kirk, T.K. and Fenn, P. (1982). Formation and action of the ligninolytic system in basidiomycetes. In: *Decomposer Basidiomycetes: Their Biology and Ecology* (J.C. Frankland, J.N. Hedger and M.J. Swift, eds.), pp. 67–90. British Mycological Society Symposium **4**. Press Syndicate of the University of Cambridge, Cambridge.
Koukol, O., Mrnka, L., Kulhankova, A. and Vosatka, M. (2006). Competition of *Scleroconidioma sphagnicola* with fungi decomposing spruce litter needles. *Canadian Journal of Botany*, **84**, 469–476.
Landeweert, R., Leeflang, P., Kuyper, T.W., Hoffland, E., Rosling, A., Wernars, K. and Smit, E. (2003). Molecular identification of ectomycorrhizal mycelium in soil horizons. *Applied and Environmental Microbiology*, **69**, 327–333.
Larsson, K.H., Larsson, E. and Koljalg, U. (2004). High phylogenetic diversity among corticioid homobasidiomycetes. *Mycological Research*, **108**, 983–1002.
Lindahl, B.D., Finlay, R.D. and Cairney, J.W.G. (2005). Enzymatic activities of mycelia in mycorrhizal fungal communities. In: *The Fungal Community: Its Organization and Role in the Ecosystem* (J. Dighton, J.F. White and P. Oudemans, eds.), 3rd edn., pp. 331–348. CRC Press, Boca Raton, FL, USA.
Lindahl, B.D., Ihrmark, K., Boberg, J., Trumbore, S.E., Högberg, P., Stenlid, J. and Finlay, R. (2007). Spatial separation of litter decomposition and mycorrhizal nitrogen uptake in a boreal forest. *New Phytologist*, **173**, 611–620.
Lindahl, B.D. and Olsson, S. (2004). Fungal translocation: Creating and responding to environmental heterogeneity. *Mycologist*, **18**, 79–88.
Lindahl, B., Stenlid, J. and Finlay, R. (2001). Effects of resource availability on mycelial interactions and ^{32}P-transfer between a saprotrophic and an ectomycorrhizal fungus in soil microcosms. *FEMS Microbiology Ecology*, **38**, 43–52.
Lindahl, B., Stenlid, J., Olsson, S. and Finlay, R. (1999). Translocation of ^{32}P between interacting mycelia of a wood decomposing fungus and ectomycorrhizal fungi in microcosm systems. *New Phytologist*, **144**, 183–193.
McClaugherty, C. and Berg, B. (1987). Cellulose, lignin and nitrogen concentrations as rate regulating factors in late stages of forest litter decomposition. *Pedobiologia*, **30**, 101–112.
Myneni, R.B., Dong, J., Tucker, C.J., Kaufmann, R.K., Kauppi, P.E., Liski, J., Zhou, L., Alexeyev, V. and Hughes, M.K. (2001). A large carbon sink in the woody biomass of Northern forests. *Proceedings of the National Academy of Science of the United States of America*, **98**, 14784–14789.
Myrold, D.D. (1998). Transformations of Nitrogen. In: *Principles and Application of Soil Microbiology* (D.M. Sylvia, J.J. Fuhrmann, P.G. Hartel and D.A. Zuberer, eds.), pp. 259–294. Prentice-Hall, Inc., London, UK.
Neff, J.C., Townsend, A.R., Gleixner, G., Lehman, S.J., Turbull, J. and Bowman, W.D. (2002). Variable effects of nitrogen additions on the stability and turnover of soil carbon. *Nature*, **419**, 915–917.
O'Brien, H.E., Parrent, J.L., Jackson, J.A., Moncalvo, J.-M. and Vilgalys, R. (2005). Fungal community analysis by large-scale sequencing of environmental samples. *Applied and Environmental Microbiology*, **71**, 5544–5550.

Osono, T. (2006). Role of phyllosphere fungi of forest trees in the development of decomposer fungal communities and decomposition processes of leaf litter. *Canadian Journal of Microbiology*, **52**, 701–716.

Osono, T. and Takeda, H. (2002). Comparison of litter decomposing ability among diverse fungi in a cool temperate deciduous forest in Japan. *Mycologia*, **94**, 421–427.

Persson, T., Rudbeck, A., Jussy, J.H., Colin-Belgrand, M., Priemé, A., Dambrine, E., Karlsson, P.S. and Sjöberg, R.M. (2000). Soil nitrogen turnover: Mineralisation, nitrification and denitrification in European forest soils. In: *Carbon and Nitrogen Cycling in European Forest Ecosystems* (E.-D. Schulze, ed.), pp. 297–331. Springer Verlag, Heidelberg.

Rayner, A.D.M. and Boddy, L. (1988). *Fungal Decomposition of Wood: Its Biology and Ecology.* Wiley, Chichester, UK.

Read, D.J. and Perez-Moreno, J. (2003). Mycorrhizas and nutrient cycling in ecosystems: A journey towards relevance? *New Phytologist*, **157**, 475–492.

Rosling, A., Landeweert, R., Lindahl, B.D., Larsson, K.-H., Kuyper, T.W., Taylor, A.F.S. and Finlay, R.D. (2003). Vertical distribution of ectomycorrhizal fungal taxa in a podzol soil profile. *New Phytologist*, **159**, 775–783.

Schadt, C.W., Martin, A.P., Lipson, D.A. and Schmidt, S.K. (2003). Seasonal dynamics of previously unknown fungal lineages in tundra soils. *Science*, **301**, 1359–1361.

Schlesinger, W.H. and Andrews, J.A. (2000). Soil respiration and the global carbon cycle. *Biogeochemistry*, **48**, 7–20.

Sinsabaugh, R.L., Gallo, M.E., Lauber, C., Waldrop, M.P. and Zak, D.R. (2005). Extracellular enzyme activities and soil organic matter dynamics for northern hardwood forests receiving simulated nitrogen deposition. *Biogeochemistry*, **75**, 201–215.

Söderström, B.E. and Bååth, E. (1978). Soil microfungi in three Swedish coniferous forests. *Holarctic Ecology*, **1**, 62–72.

Steffen, K.T., Hatakka, A. and Hofrichter, M. (2002). Degradation of humic acids by the litter-decomposing basidiomycete *Collybia dryophila*. *Applied and Environmental Microbiology*, **68**, 3442–3448.

Taylor, A.F.S. (2002). Fungal diversity in ectomycorrhizal communities: Sampling effort and species detection. *Plant and Soil*, **244**, 19–28.

Wells, J.M., Harris, M.J. and Boddy, L. (1998). Temporary phosphorus partitioning in mycelial systems of the cord-forming basidiomycete *Phanerochaete velutina*. *New Phytologist*, **140**, 283–293.

White, T.J., Bruns, T., Lee, S. and Taylor, J. (1990). Amplification and direct sequencing of fungal ribosomal RNA genes for phylogenetics. In: *PCR Protocols: A Guide to Methods and Applications* (M.A. Innis, D.H. Gelfland, J.J. Sninsky and T.J. White, eds.), pp. 315–322. Academic Press, San Diego, CA, USA.

Widden, P. and Parkinson, D. (1973). Fungi from Canadian coniferous forest soils. *Canadian Journal of Botany*, **51**, 2275–2290.

Virzo de Santo, A., Rutigliano, F.A., Berg, B., Fioretto, A., Puppi, G. and Alfani, A. (2002). Fungal mycelium and decomposition of needle litter in three contrasting coniferous forests. *Acta Oecologica*, **23**, 247–259.

Vrålstad, T., Myhre, E. and Schumacher, T. (2002). Molecular diversity and phylogenetic affinities of symbiotic root-associated ascomycetes of the Helotiales in burnt and metal polluted habitats. *New Phytologist*, **155**, 131–148.

CHAPTER 11

Distribution and Role of Mat-Forming Saprobic Basidiomycetes in a Tropical Forest* † ‡

D. Jean Lodge, William H. McDowell, Jordan Macy, Sarah Katherine Ward, Rachel Leisso, Karla Claudio-Campos and **Kerstin Kühnert**

Contents

Abstract

This chapter provides a brief synopsis of previous studies on the ecology of agaric decomposers that form litter 'mats' in tropical forests, augmented by data from temperate forest studies. Description of several experiments in tropical forests of the Luquillo Mountains in Puerto Rico is included. These studies showed higher rates of mass loss in leaves that were decomposed on basidiomycete mycelia (i.e., white-rot) than in the absence of

* This manuscript was prepared by a USA government employee on official time, so the information is in the public domain and not subject to copyright.
† All work by the first author, except experiments on basidiomycete responses to moisture and preparation of this manuscript, was done as part of the Center for Forest Mycology Research, Forest Products Laboratory, USDA Forest Service, which is located at the University of Wisconsin in Madison, WI, but the author was stationed at the Sabana Field Research Station of the International Institute of Tropical Forestry in Puerto Rico.
‡ The use of trade or firm names in this publication is for reader information and does not imply endorsement by the U.S. Department of Agriculture of any product or service.

British Mycological Society Symposia Series
Published by Elsevier Ltd.

basidiomycetes. The density of litter mats that were bound by basidiomycetes decreased with elevation and increased with slope. Addition of nitrogen inhibited mycelial growth. Capture of new litter by basidiomycetes was inhibited by nitrogen at high elevation. Litter-binding basidiomycetes exhibited differential responses to moisture, associated with full and partial shades. *Micromphale bevipes* was the only species that grew better in partial than in full shade. *Marasmius crinis-equi* had the highest rates of new attachments to litter in both full and partial shade, and was considered most suitable for use in restoration of steep road cuts and landslides to reduce erosion.

1. INTRODUCTION

Colonies of saprobic basidiomycetes in the Agaricales commonly bind leaf litter together in both tropical and temperate forests. These fungi regulate nutrient cycling and rates of decomposition, and their litter-binding activity reduces erosion on steep slopes in forests where they are abundant. Published research on the ecology of basidiomycete litter mats is summarized here. While most of the previous work on decomposer basidiomycete ecology has been conducted in temperate zones, the principles are essentially the same in the tropics—it is primarily the relative strengths of various ecosystem drivers and flux rates that differ between temperate and tropical forests.

A series of experiments that examined the ecology of basidiomycete litter decomposers in a montane forest on the Caribbean island of Puerto Rico are described: (1) comparison of early leaf decomposition rates with different levels of basidiomycete-induced white-rot; (2) a survey of basidiomycete litter mat abundance along an elevation gradient in the Luquillo Mountains; (3) mycelial mat movements in relation to slope; (4) basidiomycete mat growth and litter trapping rate in response to nitrogen addition; and (5) effects of shading and associated moisture on four litter-binding basidiomycetes, that was part of an experiment testing use of fungal mats to control erosion.

2. THE ROLE OF BASIDIOMYCETE LITTER DECOMPOSERS IN NUTRIENT CYCLING

In temperate and boreal forests, fungi play an important role in ecosystems by decomposing organic matter, thereby releasing nutrients that then become available to plants (Swift *et al.*, 1979; Beare *et al.*, 1992). Unlike most other fungi, basidiomycetes have enzymes that enable them to delignify 'low-quality' litter with high lignin and low nutrient content (Hintikka, 1970; Carreiro *et al.*, 2000; see Chapters 3 and 10). Non-unit-restricted basidiomycetes (i.e., those that grow from one resource in search of others; Chapter 1) have the potential to colonize and degrade low-quality resources more rapidly than unit-restricted fungi, by translocating nutrients from partly decomposed resources, enabling them to build biomass in new resources that are deficient in nutrients (Swift, 1977;

Frankland, 1982; Watkinson, 1984; Chapters 2, 10 and 11). This is equally the case in tropical systems (Lodge, 1993; Lodge *et al.*, 1994).

3. BASIDIOMYCETE EFFECTS ON DECOMPOSITION RATES

Most studies on effects of basidiomycetes on leaf decomposition have been from temperate zones, specifically in relation to the inhibitory effects of exogenous nitrogen inputs on rates of decomposition of high lignin resources (Magill and Aber, 1988; Berg and Matzner, 1997; Carreiro *et al.*, 2000; Hobbie and Vitousek, 2000) or on enzymes involved in delignification (Fenn and Kirk, 1981; Kirk and Farrell, 1987; Carreiro *et al.*, 2000; Waldrop *et al.*, 2004; Chapter 2 and 10). Several temperate zone studies examined effects of fungal enzyme systems on rates of leaf decomposition (Cromack and Caldwell, 1992; Osono and Takeda, 2002; Chapter 10), but there have been few studies on tropical basidiomycete (Urairuj *et al.*, 2003; Chapter 2). Santana *et al.* (2005) found that basidiomycetes significantly increased rates of decomposition in low-quality leaf litter by 22% beyond that caused by microfungi alone in a microcosm experiment in Puerto Rico.

Similarly, white-rot basidiomycetes significantly increased the rate of mass loss during a field experiment in a secondary wet subtropical forest at Sabana in the Luquillo Mountains of Puerto Rico (D.J. Lodge *et al.*, unpublished). A natural mixture of freshly fallen leaves (10 g fresh mass = 4.0 g oven dry mass) was allowed to decompose for 3 months beginning in mid-June 2004 (rainy season). In paired comparisons, after 3 months, decay of litter placed on white-rot litter basidiomycete mats was 8.4% greater than on adjacent (< 50 cm away) forest floor lacking mats (12 pairs; Figure 1).

The acceleration of early leaf decomposition by some basidiomycetes may be partly attributed to their capacity to colonize rapidly and translocate nutrients, via rhizomorphs and cords, which allows them to build biomass quickly in new

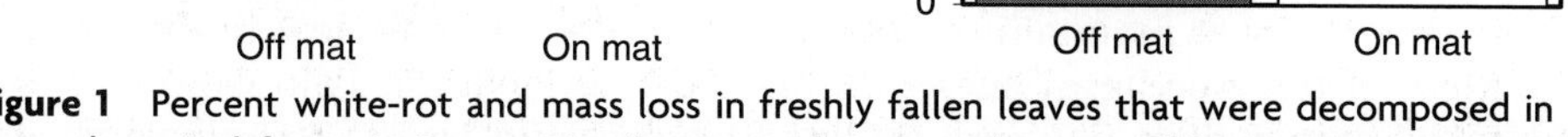

Figure 1 Percent white-rot and mass loss in freshly fallen leaves that were decomposed in wet subtropical forest in Puerto Rico for 3 months situated on or off (< 50 cm away) litter mats formed by white-rot basidiomycetes. Percent of leaf area with white-rot was estimated visually twice using calibrated leaf models. (a) Percent of leaf area affected by white-rot on and off of basidiomycete litter mats. (b) Percent mass loss on and off of basidiomycete litter mats. The difference in rate of mass loss between treatments was significant (paired *t*-test, $P = 0.0028$).

resources where nutrients are in short supply (Hintikka, 1970; Lodge, 1993; Carreiro *et al.*, 2000; Chapters 1, 3 and 10). Nutrient translocation by basidiomycetes may be largely responsible for early increases in nutrient content beyond 100% of the initial content in some tropical studies (Lodge, 1993; Lodge *et al.*, 1994). In temperate leaf litter nitrogen is generally in short supply, whereas phosphorus is more limiting in low elevation tropical forests, and both N and P may be cycled tightly in tropical montane forests (Vitousek, 1984).

At least part of the increased rate of decomposition we observed when litter was placed on basidiomycete mats could be attributed to the greater ligninolytic capacities of basidiomycetes compared to microfungi (Cromack and Caldwell, 1992; Chapter 2). Santana *et al.* (2005) found a similar increase in leaf mass loss beyond that caused by microfungi alone in microcosms that contained ligninolytic basidiomycetes. Similarly, fungi with ligninolytic capabilities caused the greatest mass loss in a comparison of litter fungi from a temperate deciduous forest in Japan (Osono and Takeda, 2002), and in a comparison of tropical endophytic fungi (Urairuj *et al.*, 2003). Most microfungi present in litter are anamorphic states of ascomycetes that lack ligninolytic enzymes, but species of Xylariaceae do have the ability to decompose lignocellulose (Osono and Takeda, 2002; Urairuj *et al.*, 2003).

4. LITTER MAT ABUNDANCE ALONG AN ELEVATION GRADIENT

Percent of ground covered by basidiomycete litter mats, and extent of individual mats, was assessed along an elevation gradient in the Luquillo Experimental Forest (from 8 June to 4 July 1994; rainy season), by examining leaves along transects to determine if they were attached to surrounding litter by hyphal strands, cords, rhizomorphs or holdfasts (D.J. Lodge *et al.*, unpublished). The percentage of ground covered by litter mats generally increased with slope and decreased with elevation (Figure 2).

The decrease in mat cover with elevation was not constant, however, and there was a strong dip associated with the cloud condensation level at ca. 600 m above sea level (asl) in the Palo Colorado forest type, independent of slope (Figure 2).

Inputs from leaf litter fall decreased sharply in elfin forest at high elevation in the Luquillo Mountains (Weaver *et al.*, 1986; Lodge *et al.*, 1991). The litter layer in high-elevation forests is discontinuous and thin (Weaver *et al.*, 1986)—a condition unfavorable to basidiomycetes that require a nearly continuous layer of litter on the forest floor to maintain their mycelia and incorporate new resources rapidly. Most of the basidiomycetes that formed litter mats in Palo Colorado and elfin forest bound litter that was in direct contact, and lacked rhizomorphs and cords. Adding litter to the forest floor significantly increased the number of rhizomorphs and cords in a forest at lower elevation in the Luquillo Mountains (D.J. Lodge, unpublished data), indicating that litter depth and supply rate influence mat-forming basidiomycetes. It is not known why a dip in litter mat cover occurred ~600 m asl, but there are corresponding boundaries in plant

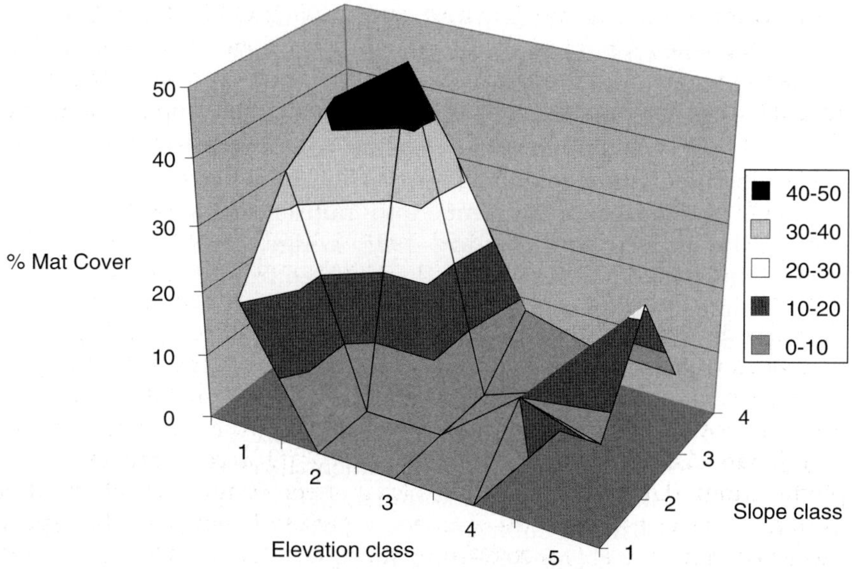

Figure 2 Percent of land area covered by decomposer basidiomycete litter mats as a function of slope (percent) and elevation along three 5–30 m transects, in the Luquillo Mountains of Puerto Rico, laid from ridge crest to stream bank at 50 m intervals along an elevation gradient from 150 to 1,000 m asl. Samples were divided into the following classes, with sample sizes (n) in parentheses. Elevation classes were: (1) 150–300 m ($n = 5$); (2) 301–450 m ($n = 8$); (3) 451–600 ($n = 3$); (4) 601–750 m ($n = 8$); (5) 751–1,000 m ($n = 10$). Slope classes were: (1) 0–10%; (2) 11–20%; (3) 21–30%; (4) 31–40%; (5) 41–50%.

species ranges at this elevation (J. Barone, J.K. Zimmermann, J. Thomlinson, N.L. Brokaw and P. Anglada, unpublished data).

The increase in litter mat cover with increasing slope (Figure 2) was probably influenced by two factors. First, many of the shallower slopes were near stream channels that overflowed during storms, disrupting the mats. Second, litter accumulated on the upslope sides of mats located on steep slopes, providing a concentration of new resources (see below).

5. EFFECTS OF NITROGEN ADDITION ON LITTER BASIDIOMYCETES

Forests are increasingly affected by nitrogen inputs from air pollutants. Several studies have focused on the effects of nitrogen additions from fertilizer or actual or simulated air pollutants on litter decomposition in tropical (Hobbie and Vitousek, 2000) as well as temperate forests (Magill and Aber, 1988; Berg and Matzner, 1997; Carreiro *et al.*, 2000; Schröter *et al.*, 2003; Gallo *et al.*, 2004; Waldrop *et al.*, 2004; Chapter 10). High nitrogen concentrations can have paradoxical effects on litter decomposition rates, accelerating decay of labile components while inhibiting decay of recalcitrant highly lignified components in both temperate (Fog, 1988; Carreiro *et al.*, 2000; Wardle, 2000) and tropical forests (Hobbie and Vitousek, 2000). Nitrogen additions often inhibit basidiomycete production or activity of ligninolytic enzymes during decomposition of refractory organic material (Kirk and Farrell, 1987; Cromack and Caldwell, 1992; Carreiro *et al.*, 2000; Gallo *et al.*, 2004; Waldrop *et al.*, 2004). Shifts in the composition of a basidiomycete decomposer community in response to nitrogen additions may reflect reduced competitive advantage of fungi that produce hyphal cords and rhizomorphs able to translocate nutrients into nutrient-depauperate food bases (Cromack and Caldwell, 1992; Boddy, 1993; Lodge, 1993). Cords and rhizomorphs of basidiomycete decomposers disappeared from the litter layer of plots receiving complete fertilization (macro- and micronutrients) in Puerto Rico (Lodge, 1993). Such losses of cord and rhizomorph systems may accelerate losses of soil organic matter and nutrients on steep slopes (Lodge and Asbury, 1988; Lodge, 1993; Miller and Lodge, 1997).

The effects of nitrogen addition on basidiomycete litter decomposers were studied at two sites in the Luquillo Experimental Forest in Puerto Rico (D.J. Lodge *et al.*, unpublished). The Bisley watershed (18°18′58″N, 65°44′10″W) was a low-elevation (250–300 m asl) subtropical wet forest that received ~3,500 mm of rain per year. This was a late secondary forest in which the native forest was characterized by tabonuco trees (*Dacryodes excelsa*), but there was an abundance of nitrogen-fixing trees in the Fabaceae (*Inga vera*) that had been planted as overstorey for a coffee plantation. The Icacos site (18°16′32″N, 65°47′4″W) was a high-elevation (620 m asl) lower montane rainforest characterized by Palo Colorado trees (*Cyrilla antillana*) that received 4,000–4,500 mm of rain per year.

At each location, three pairs of plots (nitrogen addition and control) were matched for slope, dominant vegetation, elevation and watershed type. The nitrogen plots had received 50 kg N ha^{-1} $year^{-1}$ since January 2002 (25 kg N ha^{-1} in two applications per year) as ammonium nitrate pellets (as in Hall and Matson, 1999; Magill *et al.*, 2004). The control plots did not receive fertilizer. The location and maximum extent of three to five discrete basidiomycete litter mats was marked with flags in each of the plots at Icacos and Bisley before the first fertilization in January 2002, and again in June 2003. Mean fungal mat sizes increased over the 18-month study because they were originally marked during the dry season and remeasured during the rainy season. However, mat sizes increased more in control than in nitrogen treatment plots, though not quite significantly (one-sided test, $P = 0.062$), and were more marked for plots at higher elevation in Icacos than those at lower elevation in Bisley (Figure 3).

Changes in leaf attachment rates in response to nitrogen addition at Bisley and Icacos plots were determined by tethering numbered, freshly fallen leaves to marker flags with 1 m lengths of nylon line and placing them on basidiomycete mats (Table 1). The marker flags were placed downslope of the attached leaves, so the leaves were free to move. The leaves were checked at 8 days for the presence or absence of fungal attachments to the litter mat. At Icacos there was suggestion of correlation between nitrogen addition and decreased leaf attachment rate, after 8 days ($P = 0.0625$). Rate of leaf attachment decreased after fertilization but increased in control plots (Figure 4). In contrast, there was no relationship between nitrogen fertilization and leaf attachment at Bisley (Figure 4).

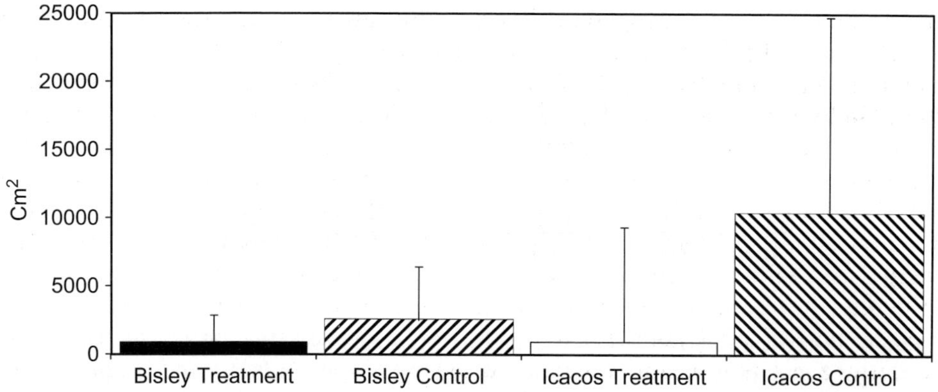

Figure 3 Mean increases in decomposer basidiomycete litter mat size from the early dry season (January 2002, just prior to nitrogen additions to treatment plots) to the early wet season (July) 18 months later. Treatment plots received 25 kg N ha^{-1} each January and June in the form of ammonium nitrate pellets; paired control plots had no nitrogen additions. Forests were located at Bisley (250–300 m asl) and Icacos (620 m asl) in the Luquillo Mountains of Puerto Rico (three pairs of replicate plots at high and low elevations). Data on mean mycelial mat area for the five pairs of plots were pooled among sites. Pair-wise differences in mat area were analyzed using a one-sided Wilcoxon Signed Ranks Test ($P = 0.0625$; Hollander and Wolfe, 1973).

Table 1 Numbers of tethered, freshly fallen leaves of *Dacryodes excelsa* and *Clusia krugiana* placed on litter mats in paired nitrogen loading and control plots before and after nitrogen fertilization in June 2003

	Bisley	Icacos
Plots	3 pairs, +N and control	3 pairs, +N and control
2 weeks before fertilization	10 *D. excelsa* leaves per plot (1 per mat)	6 *C. krugiana* leaves per plot (1 per mat)
1 week after fertilization	20 *D. excelsa* leaves per plot (1 per mat)	20 *C. krugiana* and 20 *D. excelsa* (2 each per mat)

Note: Rate of leaf attachment to the mats by basidiomycetes was determined at 8 days. Paired comparisons between treatments within forest type were made using Wilcoxon Signed Ranks tests (Hollander and Wolfe, 1973).

The decreased growth and inhibition of new resource capture by white-rot basidiomycetes following nitrogen addition at Icacos fits with previous studies that showed nitrogen inhibition of ligninolytic enzyme production by white-rot fungi (Kirk and Farrell, 1987; Cromack and Caldwell, 1992; Carreiro *et al.*, 2000; Gallo *et al.*, 2004; Waldrop *et al.*, 2004), which probably affects their competitive abilities (Carreiro *et al.*, 2000). The lack of response to nitrogen addition in the lower elevation forest at Bisley may have been related to the natural abundance of nitrogen there. Lowland tropical forests are often saturated with nitrogen (Silver *et al.*, 2005). The low-elevation forest at Bisley had more nitrogen cycling via fine litter than forests at higher elevation in the Luquillo Mountains (Lodge *et al.*, 1991).

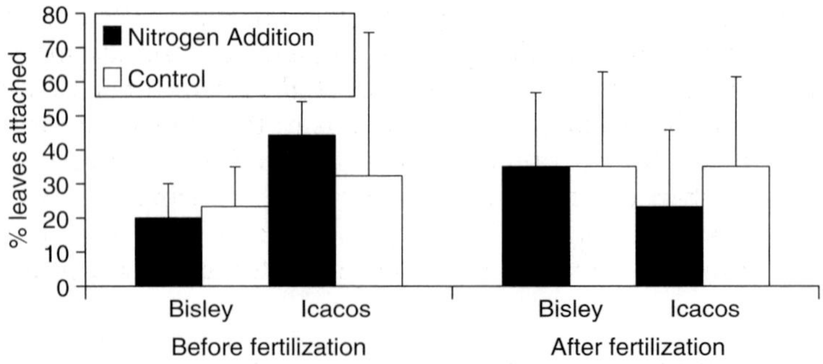

Figure 4 Percent of marked leaves attached to litter mats by decomposer basidiomycete fungi after 8 days in subtropical forest at low and high elevations (Bisley and Icacos, respectively) in the Luquillo Mountains of Puerto Rico, immediately before and after a nitrogen addition. Suppression of basidiomycete leaf attachment rates was suggestive in the high-elevation forest at Icacos ($P = 0.0625$) but not at low elevation in bisley.

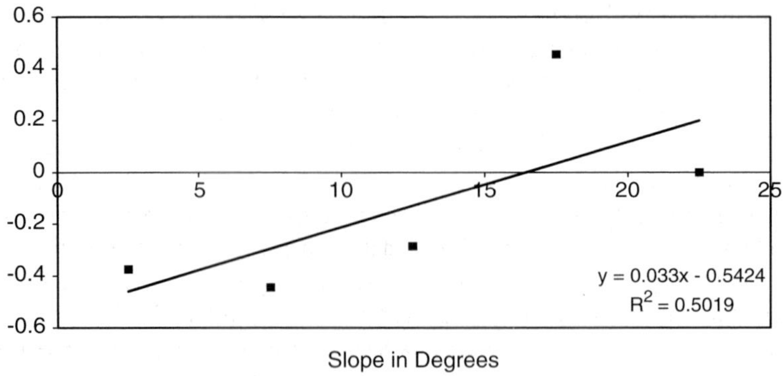

Figure 5 Mean frequency of decomposer basidiomycete litter mats moving up- or downslope as a function of degree of slope. Movements of mats over 18 months were recorded as +1 for upslope, −1 for downslope and 0 for no movement relative to slope axis. The slopes were determined using a level and plumb line. Data were pooled from two forests at low and high elevations in the Luquillo Mountains of Puerto Rico (Bisley and Icacos, respectively) and divided among five slope classes at 5° intervals: 0–5, >5–10, >10–15, >15–20 and >20–25°. The effect of slope on the mean frequency of mat movement was analyzed using regression analysis in excel (Version 9.0.3821, Microsoft Corp.).

6. MOVEMENT OF BASIDIOMYCETE LITTER MATS

There was a trend (though not significant, $R^2 = 0.5019$, $P = 0.18$) of movement of mats proportional to slope, over the 18-month study at Bisley and Icacos, mats on shallow slopes more frequently moving downslope while those on steeper slopes more frequently moving uphill (Figure 5). Two factors probably contributed to this pattern. First, as noted above, many of the shallower slopes in Bisley were in valleys subjected to overland flow during torrential storms (Weaver *et al.*, 1987). Thus, litter and soil organic matter that has been exported from slopes is probably exported in streams during high rainfall events (Weaver *et al.*, 1986; Lodge and Asbury, 1988), resulting in the highest soil organic matter contents being located on ridges and slopes (Weaver *et al.*, 1986) rather than in bottomlands, as occurs in the Appalachian Mountains of the USA (Orndorff and Lang, 1984). Mortality and damage to basidiomycete mycelia from overland flow was higher in valleys than on slopes in the Bisley plots during this study. Second, litter mats acted as terrestrial debris dams, and the basidiomycetes then grew into litter that had tumbled downslopes and accumulated on their upslope side.

7. BASIDIOMYCETE RESPONSES TO MOISTURE

Opening of a forest canopy by natural or anthropogenic disturbance causes dramatic environmental changes on the forest floor. Following canopy damage in Puerto Rico caused by Hurricane Hugo, the litter layer experienced more rapid drying as a result of greater exposure to solar irradiation and wind (Lodge *et al.*, 1991, 1994), resulting on ridges in higher mortality of *Collybia johnstonii*—a species that forms superficial mycelial fans on leaf surfaces, and was the dominant litter fungus in tabonuco forest prior to the hurricane (Lodge and Cantrell, 1995; Lodge, 1996). Higher stress tolerance of some marasmioid species allowed them to replace *C. johnstonii* on ridges (Lodge and Cantrell, 1995).

Effect of moisture on attachment rate of litter by mycelial mats of *Micromphale brevipes*, *Marasmius guyanensis* (both terrestrial species), *Marasmius crinis-equi* and *Micromphale* sp. (both aerial species), from Luquillo Experimental Forest, was studied by dividing mats and placing them in separate baskets (18 cm $\times$ 25 cm, with mesh bottoms) and placing in either shade or partial shade, moistened with 2 l of stream water on alternate days (except on rainy days), and resupplied with litter weekly. Evaporation potential was measured in the mat source environments, and the shade and partial shade propagation areas (using 3 ml vials of distilled water into which a 2 mm diameter cotton wick was inserted with 5 mm of the wick exposed to the air). Mean evaporation potentials were higher in partial (2.5 mm week^{-1}) than in full shade (0.75 mm week^{-1}). There were significant ($P = 0.009$) differences in attachment rates among species; *M. crinis-equi* formed significantly more attachments than the other species (Figure 6). Differences between attachment rates were not significant between light environments ($P = 0.548$), though two of the four species had better attachment rates in full shade, but *M. brevipes* had better development in partial shade (Figure 5). *M. brevipes* also had the driest source environment.

8. EFFECTS OF BASIDIOMYCETE LITTER MATS ON EROSION

Litter mat formation by basidiomycete decomposers benefits tropical forest ecosystems as a whole, both by conserving nutrients against leaching losses (Lodge, 1993; Lodge *et al.*, 1994) and by reducing erosion (Lodge and Asbury, 1988; Lodge *et al.*, 1994). The presence of basidiomycetes reduced downhill litter

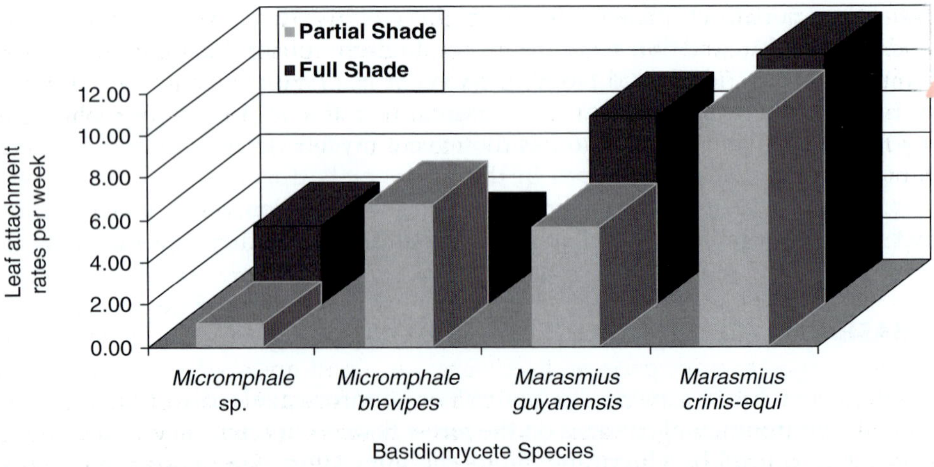

Figure 6 Mean weekly rates of leaf attachment by mycelia of four decomposer basidiomycetes in full and partial shades at El Verde in the Luquillo Mountains of Puerto Rico. Mycelia were grown in baskets (18 cm × 25 cm) filled with a thin layer of old fallen leaves or twigs lacking basidiomycetes, then the mats followed by freshly fallen leaves or more twigs on the top. Control treatments were constructed similarly, but without the mats (these remained free of fungal attachments). Every week during June 2006, the number of attached and unattached leaves or twigs in each basket were counted, followed by addition of freshly fallen leaves and twigs. Mean weekly attachment rates were compared using one-way ANOVA followed by Tukey's adjusted LSD comparisons. The overall two-way ANOVA for attachment rates was significant ($P = 0.0025$) and differences were found among species ($P = 0.009$), but not light environments ($P = 0.548$).

loss on steep slopes of a subtropical wet forest in Puerto Rico by up to 41% (Lodge and Asbury, 1988). Furthermore, the fungal litter mats protected soil surfaces from erosion losses of soil organic matter and nutrients, thereby maintaining soil fertility (Lodge and Asbury, 1988).

9. SUMMARY AND FUTURE WORK

White-rot basidiomycetes that formed litter mats significantly increased rates of mass loss during the first 3 months of decomposition in field experiments in a Puerto Rican tropical wet forest. Litter mats generally decreased in density with increasing elevation and decreasing slope. The decrease in mat density with increasing elevation was correlated with reduced litter fall rates at high elevation, but more studies are needed to confirm this relationship. While abundance of white-rot litter decomposers is expected to decrease with increasing litter nutrient concentrations, this aspect has not been explored in tropical forests.

Litter mats tended to move upslope on steep inclines, probably because they grew into the litter that accumulated on their upslope side, but downslope on more gentle gradients associated with overland flows in valleys. The mobile basidiomycete mats in tropical montane forests are thus less likely to leave distinctive signatures in soils, such as the patchwork of mull and mor soils observed in Finland in association with decomposer basidiomycete mats (Hintikka, 1970). Studies of the abundance of litter mats formed by decomposer basidiomycetes are needed from tropical forests with less topographic relief and from non-insular areas in order to determine their general importance in the tropics.

Nitrogen additions had a significant negative effect on litter mat size and leaf trapping rates in high-elevation forest that had limited nitrogen cycling, but little or no effect on basidiomycete mats in a lower elevation forest where nitrogen in litter fall was more abundant. This raises the possibility of partial resilience among decomposer basidiomycetes to nitrogen loading. Studies of ligninolytic enzyme production in response to nitrogen addition are needed to explore this possibility. As nutrient cycling in lowland tropical forests is more often limited by phosphorus than nitrogen, and decomposer basidiomycetes play a key role in recycling of phosphorus in the litter layer (Lodge, 1993, 1996), negative effects of nitrogen from air pollutants on decomposer basidiomycetes could significantly slow leaf decomposition in tropical forests.

Basidiomycetes that formed litter mats and subcanopy nets were tested for moisture preferences. The fungi tested indicated differential responses to shading and associated moisture. *M. crinis-equi* (a subcanopy species) produced most attachments, and has potential for reclamation of steep eroding slopes. Further applied studies are needed to develop protocols for using litter-binding basidiomycetes in reclamation of road cuts and landslides, and to determine their effectiveness.

ACKNOWLEDGMENTS

This work was done in cooperation with the University of Puerto Rico as part of the USA National Science Foundation Long-Term Ecological Research Program at the Luquillo Experimental Forest (Grants DBI-0218039 and DEB-0080538) to the Terrestrial Ecology Division, University of Puerto Rico, and the International Institute of Tropical Forestry (IITF), US Department of Agriculture, Forest Service. The first author was supported by Forest Service, Center for Forest Mycology Research and the International Institute of Tropical Forestry. Housing for Dr. W.H. McDowell and K. Kühnert was provided at the Sabana Field Research Station, IITF. Housing, training and stipends for S. Ward and K. Claudio-Campos were provided by a USA National Science Foundation Grant for Research Experience for Undergraduate Students to the University of Puerto Rico. We thank Dr. F.N. Scatena and Mrs. E. Treviño (both previously with IITF) for arranging the international research internship for K. Kühnert, and Dr. G. Stanoz at the Department of Plant Pathology, University of Wisconsin, for arranging R. Leisso's research internship.

REFERENCES

Beare, M.H., Parmelee, R.W., Hendrix, P.F., Cheng, W., Coleman, D.C. and Crossley, D.A. (1992). Microbial and faunal interactions and effects on litter nitrogen and decomposition in agroecosystems. *Ecological Monographs*, **62**, 569–591.

Berg, B. and Matzner, E. (1997). Effect of N deposition on decomposition of plant litter and soil organic matter in forest systems. *Environmental Review*, **5**, 1–25.

Boddy, L. (1993). Cord-forming fungi: Warfare strategies and other ecological aspects. *Mycological Research*, **97**, 641–655.
Carreiro, M.M., Sinsabaugh, R.L., Repert, D.A. and Parkhurst, D.F. (2000). Microbial enzyme shifts explain litter decay responses to simulated nitrogen deposition. *Ecology*, **81**, 2359–2365.
Cromack, K., Jr. and Caldwell, B.A. (1992). The role of fungi in litter decomposition and nutrient cycling. In: *The Fungal Community: Its Organization and Role in the Ecosystem* (G.C. Carroll and D.T. Wicklow, eds.), 2nd edn., pp. 653–668. Marcel Dekker, New York, NY.
Fenn, P. and Kirk, T.K. (1981). Relationship of nitrogen to the onset and suppression of ligninolytic activity and secondary metabolism in *Phanerochaete chrysosporium*. *Archives of Microbiology*, **130**, 59–65.
Fog, K. (1988). The effect of added nitrogen on the rate of decomposition of organic matter. *Biological Review*, **63**, 433–462.
Frankland, J.C. (1982). Biomass and nutrient cycling by decomposer basidiomycetes. In: *Decomposer Basidiomycetes* (J.C. Frankland, J.N. Hedger and M.J. Swift, eds.), British Mycological Society Symposium, Vol. 4, pp. 241–261. Cambridge University Press, Cambridge.
Gallo, M., Amonette, R., Lauber, C., Sinsabaugh, R.L. and Zak, D.R. (2004). Microbial community structure and oxidative enzyme activity in nitrogen-amended north temperate forest soils. *Microbial Ecology*, **48**, 218–229.
Hall, S.J. and Matson, P.A. (1999). Nitrogen oxide emissions after nitrogen additions in tropical forests. *Nature*, **400**, 152–155.
Hintikka, V. (1970). Studies on white-rot humus formed by higher fungi in forest soils. *Communicationes Instituti Forestalis Fenniae*, **69**, 1–68.
Hobbie, S.E. and Vitousek, P.M. (2000). Nutrient limitation of decomposition in Hawaiian forests. *Ecology*, **81**, 1867–1877.
Hollander, M. and Wolfe, D.A. (1973). *Nonparametric Statistical Methods*. Wiley, New York, NY.
Kirk, K.T. and Farrell, R.L. (1987). Enzymatic combustion: The microbial degradation of lignin. *Annual Review of Microbiology*, **41**, 465–505.
Lodge, D.J. (1993). Nutrient cycling by fungi in wet tropical forests. In: *Aspects of Tropical Mycology* (S. Isaac, J.C. Frankland, R. Watling and A.J.S. Whalley, eds.), British Mycological Society Symposium Series 19, pp. 37–57. Cambridge University Press, Cambridge.
Lodge, D.J. (1996). Microorganisms. In: *The Food Web of a Tropical Rain Forest* (D. Reagan and R. Waide, eds.), pp. 54–80. The University of Chicago Press, Chicago, IL.
Lodge, D.J. and Asbury, C.E. (1988). Basidiomycetes reduce export of organic matter from forests slopes. *Mycologia*, **80**, 888–890.
Lodge, D.J. and Cantrell, S. (1995). Fungal communities in wet tropical forests: Variation in time and space. *Canadian Journal of Botany*, **73**, s1391–s1397.
Lodge, D.J., McDowell, W.H. and McSwiney, C.P. (1994). The importance of nutrient pulses in tropical forests. *Trends in Ecology and Evolution*, **9**, 384–387.
Lodge, D.J., Scatena, F.N., Asbury, C.E. and Sanchez, M.J. (1991). Fine litterfall and related nutrient inputs resulting from Hurricane Hugo in subtropical wet and lower montane rain forests of Puerto Rico. *Biotropica*, **23**, 364–372.
Magill, A.H. and Aber, J.D. (1988). Long-term effects of experimental nitrogen additions on foliar litter decay and humus formation in forest ecosystems. *Plant and Soil*, **203**, 301–311.
Magill, A.H., Aber, J.D., Currie, W.S., Nadelhoffer, K.J., Martin, M.E., McDowell, W.H., Mellillo, J.M. and Steudler, P. (2004). Ecosystem response to 15 years of chronic nitrogen additions at the Harvard Forest LTER, Massachusetts, USA. *Forest Ecology and Management*, **196**, 7–28.
Miller, R.M. and Lodge, D.J. (1997). Fungal responses to disturbance: Agriculture and forestry. In: *Environmental and Microbial Relationships* (D. Wicklow and B. Söderström, eds.), pp. 65–84. Springer-Verlag, Berlin.
Orndorff, K.A. and Lang, G.E. (1984). Leaf litter redistribution in a West Virginia hardwood forest. *Journal of Ecology*, **69**, 225–235.
Osono, T. and Takeda, H. (2002). Comparison of litter decomposing ability among diverse fungi in a cool temperate deciduous forest in Japan. *Mycologia*, **94**, 421–427.
Santana, M., Lodge, D.J. and Lebow, P. (2005). Relationship of host recurrence in fungi to rates of tropical leaf decomposition. *Pedobiologia*, **49**, 549–564.

Schröter, D., Woters, V. and De Ruiter, P.C. (2003). C and N mineralization in the decomposer food webs of a European forest transect. *Oikos*, **102**, 294–308.

Silver, W.L., Thompson, W., Reich, A., Ewel, J.J. and Firestone, M.K. (2005). Nitrogen cycling in tropical plantation forest: Potential control on nitrogen retention. *Ecological Applications*, **15**, 1604–1614.

Swift, M.J. (1977). The ecology of wood decomposition. *Science Progress, Oxford*, **64**, 175–199.

Swift, M.J., Heal, O.W. and Anderson, J.M. (1979). *Studies in Ecology, Vol. 5. Decomposition in Terrestrial Ecosystems*. Blackwell Scientific, Oxford.

Urairuj, C., Kanongnuch, C. and Lumyong, S. (2003). Ligninolytic enzymes from tropical Xylariaceae. *Fungal Diversity*, **13**, 209–219.

Vitousek, P.M. (1984). Litterfall, nutrient cycling, and nutrient limitation in tropical forests. *Ecology*, **65**, 285–298.

Waldrop, M.P., Zak, D.R. and Sinsabaugh, R.L. (2004). Microbial community response to nitrogen deposition in northern forest ecosystems. *Soil Biology and Biochemistry*, **36**, 1443–1451.

Wardle, D. (2000). *Communities and Ecosystems. Linking the Above and Belowground Components*. Princeton University Press, Princeton, NJ.

Watkinson, S.C. (1984). Morphogenesis of the *Serpula lacimans* colony in relation to its functions in nature. In: *The Ecology and Physiology of the Fungal Mycelium* (D.H. Jennings and A.D.M. Rayner, eds.), pp. 151–184. Cambridge University Press, Cambridge.

Weaver, P.L., Birdsey, R.A. and Lugo, A.E. (1987). Soil organic matter in secondary forests in Puerto Rico. *Biotropica*, **19**, 17–23.

Weaver, P.L., Medina, E., Pool, D., Dugger, K., Gonzalez-Liboy, J. and Cuevas, E. (1986). Ecological observations in the dwarf forest of the Luquillo Mountains of Puerto Rico. *Biotropica*, **18**, 79–85.

CHAPTER 12

Basidiomycete Community Development in Temperate Angiosperm Wood

Lynne Boddy and **Jacob Heilmann-Clausen**

Contents

Abstract

The wide variety of dead wood habitats supports a wide variety of specialized fungi, which globally may exceed 100,000 species. Of these the majority of known taxa are Basidiomycota. They exhibit a wide variety of strategies to gain and hold territory within wood, defined by their mode of dispersal, establishment, competitive ability and adaptation to disturbance and stress factors. Many habitat factors affect community composition and development, both exogenous, e.g. microclimatic regime, and endogenous, e.g. interspecific interactions. Initial microenvironmental factors—at one extreme high stress and at the other extreme absence of abiotic stress—are major determinants of the communities that establish. Following initial

British Mycological Society Symposia Series
Published by Elsevier Ltd.

establishment, community development is influenced by four main driving forces: stress aggravation (worsening of abiotic environmental conditions), stress alleviation (improvement in abiotic conditions), disturbance and combat (interspecific competition for space rather than directly for nutrients). The ecological strategies adopted by wood decay Basidiomycota, habitat factors influencing community development and community development pathways are discussed in relation to angiosperm wood.

1. INTRODUCTION

Dead wood is a very important arena for growth and sporulation of saprotrophic basidiomycetes, supporting thousands of different species with variable adaptations to the wood environment. Fallen trunks and branches often constitute the main bulk of dead wood in long unmanaged deciduous forests (Christensen *et al.*, 2005). Heartwood of ancient trees is another major source of dead wood, though less common in plantations managed for wood production, as trees are often harvested long before substantial heartwood develops. Attached dead branches, stumps and buried roots are other important dead wood habitats, which may constitute an important fraction of the total dead wood volume, not least in managed or formerly managed stands (Nordén *et al.*, 2004a). The balance between different dead wood types varies depending on forest types and dominating tree species. In a survey of unmanaged forest stands in Lithuania dead standing wood predominated in alder stands (*Alnus* spp.), but not in birch (*Betula* spp.), aspen (*Populus tremuloides*) and oak (*Quercus* spp.) (Vasiliauskas *et al.*, 2004). Similarly, the relative proportion of standing dead wood was considerably higher in montane European beech forest reserves (41–47% of total dead wood volume) compared with lowland reserves (23–29% of total dead wood volume) (Christensen *et al.*, 2005), probably reflecting a combination of higher windstorm damage in lowland areas in NW Europe and the presence of silver fir (*Abies alba*) in most evaluated montane stands. Even the structure of fungal communities may have a marked impact on the frequency of snags (natural snags). In the beech forests of Halland, Sweden, *Fomes fomentarius* is dominant in the primary decay community and due to its decay, usually concentrated in the mid-stem section, almost all trees experience tops breaking and falling to the floor, while uprooted trees, common in most other beech forest zones, are scarce (Heilmann-Clausen, 2005).

The wide variety of dead wood habitats is known to support a vast variety of more or less specialized fungi, but the global number of fungal species involved in wood decay is unknown. In Sweden alone, more than 2,500 species have been recorded as primarily associated with dead wood (Dahlberg and Stokland, 2004), corresponding to ~20% of an estimated total of 12,000 species. With an estimated 1.5 million fungal species worldwide (Hawksworth, 2001) the total number of wood-inhabiting fungi on the global scale is likely to exceed 100,000 species, even if the wood-inhabiting fraction is considerably lower than in the Swedish case.

Fungal community development in dead wood has been investigated intensively, especially since the 1980s, and a general understanding of principles and factors affecting the process is beginning to emerge. Here we first review ecological strategies in wood decay fungi, in terms of dispersal, life-history and colonization strategies. Second, important habitat factors influencing community development are considered. Third, community development pathways and the ways in which environmental factors determine community composition are discussed.

2. FUNGAL STRATEGIES IN DECAYING WOOD

Wood-inhabiting fungi show a vide variety of strategies to gain and hold territories in wood, defined not least by their mode of dispersal and establishment, their competitive ability and by individual adaptation to the various disturbance and stress factors influencing life in decaying wood. In addition, passive or active interactions with other wood-inhabiting organisms may play an important role for some species depending on insect vectors, mycoparasitic relations or other successor–predecessor relations. Table 1 summarizes these strategies/ecological roles, with examples from angiosperms in temperate Europe, and below we review the importance of dispersal, life-history strategies and successor–predecessor relationships.

2.1 Modes of Dispersal

As a habitat dead wood is characterized by: (i) the limited duration of each habitat patch (e.g. a fallen trunk) and (ii) an unpredictable and heterogeneous distribution of habitat patches in space and time, influenced by twigs and branches falling, natural tree mortality factors (e.g. storms, drought) and in recent centuries by forest management cycles and landscape fragmentation. To persist in this highly dynamic context wood decay fungi need to be able to spread between sinking and rising habitat patches (Jonsson *et al.*, 2005). They do so in two main ways, depending on whether or not they are confined within the woody resource—'resource-unit-restricted' or 'non-resource-unit-restricted', respectively. Resource-unit-restricted fungi are disseminated in space and time via spores, whereas non-resource-unit-restricted fungi can spread both via spores and as mycelium growing out of the wood in search of new resources.

Fungi able to spread as mycelium from one resource in search of another have considerable advantages. They operate a variety of search strategies to optimize discovery of new resources (Boddy, 1993, 1999; Boddy and Jones, 2006; Chapter 1). On arrival at a new resource they can exert considerable inoculum potential—their carbon and mineral nutrient requirements are met by translocation through the established mycelium (Chapter 3), and they can attack resident fungi over a large front. Moreover, when new resources have been located and colonized, the fungi can reallocate biomass that is growing through soil in non-successful areas of search (Boddy, 1999; Boddy and Jones, 2006). Fungi that extend outside resources can also operate a 'sit and wait' strategy, whereby a

Table 1 Ecological strategies and life-history traits of fungal colonizers of angiosperm wood, through pioneer, early, late and final stages of decomposition

General category	Ecological strategy	Subtype	Site and mode of establishment	Stage of succession	Life-history strategy[a]	Typical lifespan of active mycelia	Examples	References
Primary: Involved in lignocellulose decomposition prior to colonization by other fungi	Heart rot agents	Top rot	Wounds and broken branches exposing heartwood, sapwood of dead branches, branch stubs, etc., engulfed as tree grows outwards	Pioneer to late stage	S	Many years	*Fistulina hepatica*, *Laetiporus sulphureus*, *Pholiota aurivellus*, *Ossicaulis lignatilis*, *Volvariella bombycina*	Phillips and Burdekin (1982), Rayner and Boddy (1988), personal observations
		Butt rot	Roots; latently present	Pioneer to late stage	S	Many years	*Armillaria* spp., *Ganoderma adspersum*, *Heterobasidion annosum*, *Inonotus dryadeus*, *Pholiota squarrosa*, *Kretzschmaria deusta*[b]	Phillips and Burdekin (1982), Rayner and Boddy (1988), Butin (1995), Woodward *et al.* (1998)
	Pathogens	Root pathogens	Spores and mycelium via wounded roots, root grafts, subcortical growth and direct penetration	Pioneer to late stage	C, S	Many years	*Armillaria* spp., *Collybia fusipes*, *Grifola frondosa*, *Heterobasidion annosum*, *Meripilus giganteus*	Rayner and Boddy (1988), Butin (1995), Woodward *et al.* (1998), Fox (2000), Piou *et al.* (2002)
		Wound parasites	Wounds	Pioneer	R, S	2–5 years	*Chondrostereum purpureum*, *Trametes* spp., *Stereum sanguinolentum*	Rayner and Boddy (1988), Butin (1995)

	Latent colonizers/ endophytes of sapwood	Natural pruners of branches and twigs	Sites of entry unconfirmed but probably various; propagules remain latent in functional sapwood for many years	Pioneer	R, S	Few months to few years	*Vuilleminia comedens*, *Colpoma quercinum*[b], *Peniophora* spp., *Biscogniauxia nummularia*[b], *Hypoxylon fragiforme*[b], *Stereum gausapatum*	Chapela and Boddy (1988a), Rayner and Boddy (1988), Fisher and Petrini (1990), Griffith and Boddy (1990), Hirst (1995), Boddy (2001)
		Trunk colonizers	As above	Pioneer to late stage	S	Up to several years	*Biscogniauxia nummularia*[b], *Eutypa spinosa*[b], *Piptoporus betulinus*, *Fomes fomentarius*	Chapela (1989), Hendry *et al.* (1998, 2002), Danby (2000), Baum *et al.* (2003)
		Canker formers	As above	Pioneer to late stage	S	Few months to several years	*Biscogniauxia nummularia*[b], *Eutypa spinosa*[b], *Hypoxylon mammatum*, *Neonectria* spp.[b]	Rayner and Boddy (1988), Pinon and Manion (1991), Hendry *et al.* (1998, 2002)
	Ruderal primary colonizers	Wound colonizers	Freshly exposed wood as spores; some via insect vectors	Pioneer	R	Few months to few years	Mostly non-Basidiomycota	Rayner and Boddy (1988)
		Felled wood and stumps					*Chondrostereum purpureum*, *Cylindrobasidium evolvens*	Coates and Rayner (1985a, 1985b, 1985c)
Secondary: Involved in lignocellulose decomposition after colonization by primary and other secondary colonizers	Resource unit restricted	Combative invaders	Exposed wood, via spores	Early to late stages	C, R	2–5 years	*Bjerkandera adusta*, *Phlebia radiata*, *Stereum hirsutum*, *Trametes* spp.	Boddy and Rayner (1983), Coates and Rayner (1985a, 1985b, 1985c)

(*Continued*)

Table 1 *(Continued)*

General category	Ecological strategy	Subtype	Site and mode of establishment	Stage of succession	Life-history strategy[a]	Typical lifespan of active mycelia	Examples	References
		Temporary mycoparasites	Initially via specific fungal hosts	Early to late stages	C	2–5 years	*Antrodiella* spp., *Lenzites betulina*, *Trametes gibbosa*	Rayner *et al.* (1987), Chapela *et al.* (1988), Niemelä *et al.* (1995)
		Defensive	Exposed wood via spores	Early to late stages	S, C, R	Many years	*Xylaria hypoxylon*[b]	Coates and Rayner (1985a, 1985b, 1985c), Chapela *et al.* (1988)
		Desiccation tolerant	Dead wood prone to desiccation, via spores	Early to late stages	S, R	Several years	*Auricularia auricula-judae*, many corticoid fungi, *Neolentinus lepideus*	Theden (1961), Boddy and Rayner (1983), Boddy (2001), Heilmann-Clausen (2001)
		Heat and desiccation tolerant	Sun-baked wood, presumably via spores	Mid to late stages	S	Many years	*Gloeophyllum* spp., *Trametes hirsuta*, *Pycnoporus cinnabarinus*, *Schizophyllum commune*	Heilmann-Clausen (2001), Huckfeldt *et al.* (2005)
		Acid and polyphenol tolerant	Heartwood rich in tannins, via spores	Mid to late stages	S	Many years	*Daedalea quercina*, *Hymenochaete rubiginosa*	Rayner and Boddy (1988)
		Late stage polypores and agarics	Well-decayed wood, presumably via spores	Late to final stages	C, S, R	Several years	*Lentinellus cochleatus*, *Ceriporia* spp., *Mycena* spp., *Pluteus* spp.	Chapela *et al.* (1988), Rayner and Boddy (1988), Heilmann-Clausen (2001)

	Non-resource restricted	Cord-formers	Felled and fallen wood, and stumps, via mycelia cords and spores	Early to final stages	C	Several years	*Hypholoma fasciculare*, *Lycoperdon* spp., *Megacollybia platyphylla*, *Phallus impudicus*, *Phanerochaete* spp.	Chapter 1; Thompson (1984), Coates and Rayner (1985b, 1985c), Boddy (1993, 1999), Heilmann-Clausen (2001)
Tertiary: Not primarily involved in lignocellulose decomposition	Soil and litter fungi	Litter and humus colonizers	All wood via mycelium, and invertebrate-carried spores	Late to final stages	S ± C ± R	Many years, though at most several years in very decayed wood	*Clitocybe* spp., *Collybia* spp., *Marasmius* spp., *Lepiota* spp., *Trechispora* spp., non-Basidiomycota	Personal observations, Swift and Boddy (1984)
		Ectomycorrhiza formers	All wood via mycelium	Late to final stages	S ± C ± R	Few to many years, though at most several years in very decayed wood	*Byssocorticium* spp., *Lactarius subdulcis*, *Tomentella* spp., *Piloderma* spp.	Tedersoo *et al.* (2003)
	Mycoparasites		All wood, presumably via spores	Early to final stages		1–3 years	*Achroomyces* spp., *Tremella* spp., *Hypomyces* spp.[b]	Helfer (1991), Zugmaier *et al.* (1994), Hansen and Knudsen (1997)

Note: The classification into ecological strategies is not rigid, and categories sometimes overlap. For example, *Piptoporus betulimus* causes heart rot in *Betula* being latently present.

[a]Most common characteristics, though not necessarily for all examples.

[b]Ascomycota.

large mycelial network waits for resources to land on it and then actively colonizes those resources, often with responses occurring elsewhere in the system (Boddy and Jones, 2006; Chapter 1). On the down-side, in the quest for new resources, non-resource-unit-restricted fungi risk loss of a large amount of biomass as a result of invertebrate grazing (Chapter 9), killing by other microorganisms and death in a harsh environment.

Sexual spores, as well as allowing genetic recombination, provide a means of long-distance spread, exceeding hundreds and possibly thousands of kilometres (e.g. Vilgalys and Sun, 1994; Hallenberg, 1995; Hallenberg and Küffer, 2001). Sexual and asexual spores may also provide a means of dispersal in time as has been described in ectomycorrhizal communities (Baar *et al.*, 1999; Kjøller and Bruns, 2003). Little is known of the role of resting spores in wood decay fungi, but several basidiomycetes, e.g. *Piptoporus quercinus* (Roberts, 2002), and members of the corticoid genera *Botryobasidium* and *Trechispora* (Eriksson *et al.*, 1973–1988), are known to form thick-walled spores which most likely function as resting spores. In one species, *Hyphodontia paradoxa*, chlamydospores have been described as part of a drought resistant strategy, allowing the species to maintain its position in the fungal community of attached *Quercus* twigs and branches during desiccating conditions (Boddy and Rayner, 1983).

A major disadvantage of spatial dispersal by spores is that it is usually highly haphazard with small chance of landing on appropriate resources and encountering locally favourable conditions. In addition, the majority of liberated spores settle within a very short distance of the sporocarp (e.g. Nordén and Larsson, 2000). Wind-borne spores have no active means of locating and settling on appropriate substrata, though variations in spore size and shape and the presence of protuberances may reflect adaptation to various establishment environments or animal vectors (Hallenberg and Parmasto, 1998; Nordén *et al.*, 1999). Some wood decay fungi have, however, evolved mutualistic relationships with invertebrates in which spores are transported to an appropriate location (Chapter 9). More haphazard dispersal over shorter distances is believed to be common in wood decay fungi, with various invertebrates and less commonly vertebrates functioning as dispersal vectors between resource units (Rayner and Boddy, 1988).

Germinating spores are faced with several obstacles before a reproductive mycelium can be established. They only have a small endogenous food reserve available for colonization (i.e. low inoculum potential), and gaining access to the resource often involves competitive/combative interactions (Chapter 7) with already established mycelia (Boddy, 2000). Hyphae developing from spores that are genetically identical (which is most likely if they are asexually derived) may act synergistically, but when they are genetically different somatic incompatibility is likely to result in competition rather than synergy (Coates and Rayner, 1985c).

2.2 Life-History Strategies

Wood decay fungi have evolved different life-history strategies to fit them for the many and varied niches available in different types of woody resource and at different times in the decay process. Based on ideas from plant ecology (Grime, 1977, 1979) fungi have been described to be influenced by three major environmental determinants: (i) incidence of competitors; (ii) stress—an abiotic factor that limits the production of biomass by the *majority* of organisms, e.g. extremes of pH, temperature and water content; and (iii) disturbance—a sudden event that provides new resources with or without partial or complete destruction of the resident organisms, termed, respectively, destructive and enrichment disturbances (Pugh, 1980; Cooke and Rayner, 1984; Rayner and Boddy, 1988; Andrews, 1991). This subsequently led to the formulation of three primary fungal life-history strategies: (i) competitive (C-selected); (ii) stress-tolerant (S-selected); and (iii) ruderal (R-selected). Many fungi, however, are subjected to combinations of these environmental determinants and hence secondary (combination of two) or tertiary (combination of three) strategies have evolved.

Applying ecological strategies developed for plants to fungi is not without difficulties since they have very different lifestyles (Ohtonen *et al.*, 1997). Plants compete for light and nutrients in two different spaces (above and below ground), while wood decay fungi compete for nutrients in a single space (inside a piece of wood), though cord-forming basidiomycetes may transport nutrients between units. In solid bulky substrata such as wood competition is for space; occupancy of territory allows the fungus to access the nutrients held therein as and when necessary. The term combat is, therefore, in most situations more appropriate for wood decay fungi than the term competition (Cooke and Rayner, 1984; Rayner and Boddy, 1988), though competition for nutrients in the classical sense does occur when mycelia grow out through soil. Competition/combat between wood decay fungi often occurs between mycelia meeting along a clearly defined border (Boddy, 2000), while competition in plant communities is spatially diffuse. Also, wood is a temporary resource that is continually diminishing and physically and chemically changing. Thus, conditions during initial mycelial establishment may be very different from those experienced by the established mycelium. Further, wood decay fungi need to move to new resources before the old ones disappear.

Bearing in mind these reservations the classification of wood decay fungi in terms of R-, S- and C-selection is still very useful. These strategies can be used to define the behaviour of an organism in a particular context, but not to classify organisms *per se*, since different behavioural characteristics may be adopted at different stages in the life-cycle, when the mycelium is in different physiological/biochemical modes, and under different environmental regimes. For example, the coenocytic margin of *Phlebia radiata* mycelium extends rapidly, utilizes simple organic compounds and does not respond aggressively to other fungi (R-selected characteristics—see below), while the more mature, septate mycelium is able to use lignocellulose and is antagonistic to other mycelia (C-selected) (Boddy and Rayner, 1983).

Fungi with a lot of R-selected characteristics are favoured in the relative absence of stress in uncrowded environments resulting from disturbance. Characteristics include: the ability to reproduce rapidly; effective dispersal; narrow enzymic ability, hence utilization of simple easily available organic substrates;

and rapid growth. Fungi with C-selected characteristics are favoured in relatively non-stressed, undisturbed conditions; the domains of fungi colonizing organic resources increasingly overlap and competition/combat ensues. Other C-selected characteristics include long-life expectancy and wide enzymic ability. S-selected characters include: ability to cope with a particular abiotic stress or set of stresses; slow or intermittent commitment to reproduction; and often slow growth.

2.3 Predecessor–Successor Relationships

All species involved in wood decomposition alter the wood, though changes vary depending on species and conditions. For example, structural compounds, e.g. cellulose, hemicellulose and lignin, are removed to different extents, resulting in different types of rot, e.g. brown rot, white rot and soft rot (Rayner and Boddy, 1988; Chapter 2). Also, stimulatory and inhibitory secondary metabolites are produced, that can remain active even after death of the fungus (Heilmann-Clausen and Boddy, 2005). Not surprisingly, therefore, earlier decay agents may influence subsequent community development. Some specific early decay fungi of coniferous wood appear to have a large influence on which later stage decay fungi become established (Renvall, 1995; Holmer *et al.*, 1997; Chapter 12), and likewise predecessor–successor relationships seem to occur in angiosperm wood. For example, *Antrodiella hoehnelii* almost always follows *Inonotus nodulosus* and *I. radiatus*, while *Hericium coralloides* is typically a successor of *Inonotus obliquus*, *I. cuticularis* or *F. fomentarius* in Scandinavia and Central Europe (Niemelä *et al.*, 1995; Heilmann-Clausen and Christensen, 2004). In most cases the mechanisms underlying predecessor–successor relationships are poorly known, but for *Lenzites betulina* following *Trametes versicolor* a temporary parasitic relationship occurs (Rayner *et al.*, 1987; Chapter 7). More permanent parasitic relationships occur in several *Tremella* species parasitizing various wood-inhabiting fungi (e.g. Zugmaier *et al.*, 1994).

3. DEAD WOOD AS A HABITAT FOR WOOD DECAY FUNGI

A number of habitat factors have marked impact on species composition and development in communities of wood decay fungi. Some are exogenous, e.g. soil type and microclimatic regimes, and typically influence fungal communities on a gross scale. Others are closely linked with fungal activity itself or to tree architecture, and may hence vary dynamically in time or over short distances.

3.1 Microclimate

Microclimatic conditions, notably water content and temperature (Boddy, 1984, 1986), are major determinants of fungal community development in wood (see reviews in Rayner and Boddy, 1988; Boddy, 2001). Low water availability impedes metabolic functioning, while high water contents impose poor aeration and hence restrict aerobic processes, including fungal wood decay. Both extremes typically result in mycelial death if prolonged (Chapter 4). Similarly, low temperatures decrease the rate of metabolic activity, while high temperatures are detrimental to enzyme function, and can result in death (Chapter 4). Temperature and water regime also affect the balance in combative interactions (Boddy, 2000; Chapter 7).

Some wood-inhabiting fungi have evolved to cope with some of these extremes and predominate under appropriate conditions, i.e. they are S-selected. For example, the mycelium of many pyrenomycetes is tolerant of very dry conditions (Boddy *et al.*, 1985, 1989), whereas the mycelium of Basidiomycota tends not to be, though some are desiccation tolerant by virtue of resting spores, e.g. *H. paradoxa* produces chlamydospres, with rapid resumption of mycelial growth when conditions improve. Some polypores associated with sun-exposed logs are able to resist very high temperatures, e.g. *Gloeophyllum abietinum*, *G. sepiarium* and *G. trabeum* 4 h at 95 °C (Huckfeldt *et al.*, 2005). The majority of Basidiomycota can probably survive freezing, and some certainly remain active at temperatures approaching freezing, several corticoid basidiomycetes producing actively sporulating sporocarps under snow (Gilbertson, 1973); the agarics *Pleurotus ostreatus* and *Flammulina velutipes* produce sporocarps resistant to subzero temperatures, reinitiating active growth during milder periods (Yakovlev *et al.*, 2000).

3.2 Factors Influencing Microclimatic Regimes in Dead Wood

Microclimate varies spatially and temporally (Boddy, 1984, 1986) over a range of scales. At the forest scale two important microclimatic gradients are evident: a horizontal gradient from forest interior to forest edge, clearcut or natural canopy gap and a vertical gradient from forest floor to canopy. Forest interiors and lower levels in stands tend to have higher air humidity, lower wind speed, lower maximum and higher minimum temperatures compared to gaps, forest edges and open land, and gradients run from stable to variable microclimatic conditions (e.g. Chen *et al.*, 1993; Morecroft *et al.*, 1998; Ritter *et al.*, 2005). Microclimatic stress is hence low in wood decomposing on the forest floor in closed forests, while fungi in dead wood in the canopy or on the floor of exposed forest edges are subject to stressful conditions. In addition, environmental conditions vary vastly between functional sapwood (with its high water content and low aeration), dysfunctional sapwood and heartwood (both of which are drier and better aerated), and according to wood diameter (Boddy, 1984, 1986; Rayner and Boddy, 1988).

The microclimatic regime in dead wood may even be affected biotically by fungi living in the wood. Thus, some fungi are able to regulate the moisture contents of occupied wood. Pyrenomycetes, e.g. *Xylaria hypoxylon*, seem to maintain wood drier than ambient conditions (Boddy *et al.*, 1989; Heilmann-Clausen, 2001), while *Armillaria* species occupy wood wetter than ambient, by virtue of the pseudosclerotial plates that they produce (Lopez-Real and Swift, 1975; Chapela, 1987). The strategy seems to be to create an environment stressful for the majority of fungi, thereby avoiding combative exclusion from more competitive species. Microclimatic gradients may also be created and maintained by mosses growing on dead wood (Bader *et al.*, 1995; Chlebicki *et al.*, 1996; Heilmann-Clausen *et al.*, 2005), and ground vegetation around and above it (Heilmann-Clausen, 2001).

Temporal microclimatic fluctuations occur daily and annually, and over the long term—global climate changes. Fluctuations in temperature and water content over the year are probably one factor promoting high fungal diversity in dead wood, by creating a natural variation in the relative fitness of competing species. Indeed, fluctuations in temperature increased the number of inoculated fungal species reisolated after 6 months in a wood-chip microcosm experiment (Toljander *et al.*, 2006).

3.3 Tree Species and Composition of Fungal Communities

Many wood-inhabiting fungi appear, based on fruit bodies, to have a preference for certain tree species. In some cases host preferences relate to interspecific differences in chemical composition, e.g. pH, presence of allelopathic compounds, bark and wood morphology (Rayner and Hedges, 1982; Rayner and Boddy, 1988). The heartwood of *Quercus robur*, for instance, is well known for its high durability and distinctive mycota owing to its low pH and high content of tannins (Wald *et al.*, 2004; Heilmann-Clausen *et al.*, 2005). In other cases intimate interactions between living host tissue and fungi able to infect functional sapwood seem to play a key role (e.g. Hřib and Rypáček, 1981; Chapela *et al.*, 1991; Hendry *et al.*, 1993).

The percentage of apparently specific and selective species in different fungal groups is variable, tending to be higher among polypores and pyrenomycetes than corticoids and agarics (Figure 1). These differences reflect differences in dominating ecological strategies among these groups. Fungal species interacting with live hosts, i.e. latent decay fungi (see below) and heart rot agents, are well represented among pyrenomycetes and polypores, but less so by agarics and corticoid species. Based on incidence of fruit bodies, heart rot fungi often appear selective for individual or closely related tree species, e.g. *F. hepatica* and *P. quercinus* for *Quercus* spp. Other fungi are apparently strongly selective for a particular tree genus, but sometimes occur albeit infrequently on other trees, e.g. *Grifola frondosa* and *Inonotus dryadeus* on *Quercus* (Ryvarden and Gilbertson, 1992). Further, some appear to be selective for unrelated tree species, e.g. *Laetiporus sulphureus* on *Quercus*, *Prunus*, *Salix* and *Taxus* spp., though in North America there is some indication, based on fruit body morphology, cultural characteristics and molecular evidence, that there may be five taxa of *Laetiporus* (Banik and Burdsall, 2000).

Among latent decay fungi host selectivity is prominent and in some genera strict host specificity seems to occur, at least based on sporocarp observations, e.g. some *Peniophora* species (Table 2).

3.4 Effects of Soil Type on Wood Decay Fungi

As with plant community composition, the occurrence of terrestrial fungi is strongly influenced by soil type and chemistry (Tyler, 1985; Arnolds *et al.*, 1994), but little is known of the effects on wood-inhabiting fungi. It is obvious that non-unit-restricted fungi, e.g. cord-formers, are directly influenced by the soil and litter environment through which they grow in search of new wood resources (e.g. Abdalla and Boddy, 1996; Donnelly and Boddy, 1998). Also, for

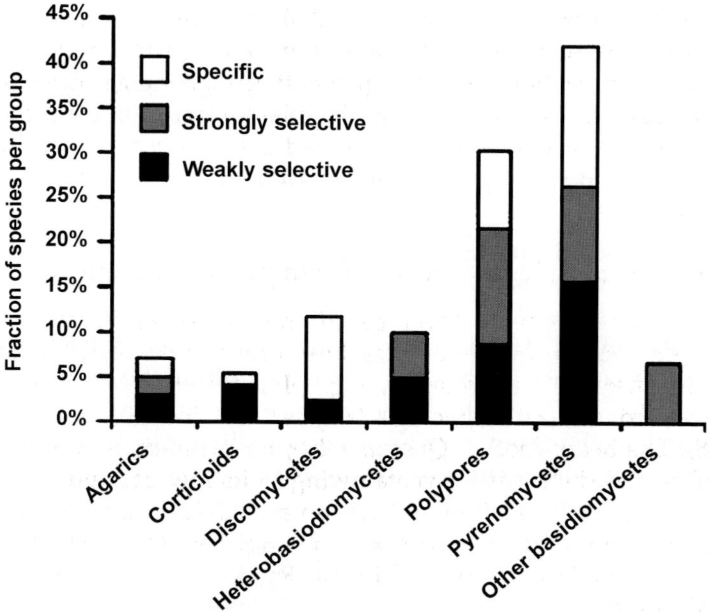

Figure 1 Proportion of wood decay fungi that can be considered weakly or strongly selective or specific for host tree species, according to sporocarp surveys (from Heilmann-Clausen, 2003). Information was extracted from a dataset obtained from Draved Skov, Denmark, during 1999 and 2000. The site contained seven main angiosperm tree species.

Table 2 Host ranges in *Peniophora* species growing on deciduous hosts in scandinavia

Apparently host specific	Host selective	Broad host range
P. eriksonii (*Alnus*)	*P. cinerea* (mostly *Fagus*)	*P. incarnata* (deciduous, less often coniferous hosts)
P. hydnoidea (*Carpinus*)	*P. laurentii* (*Populus tremula* and *Betula*)	*P. lycii* (deciduous hosts)
P. lilacea (*Ulmus carpinifolia*)	*P. limitata* (mostly *Fraxinus*, *Syringa* and *Ligustrum*)	*P. nuda* (deciduous hosts)
P. polygonia (*Populus tremula*)	*P. quercina* (*Fagus* and *Quercus*)	*P. suecica* (deciduous hosts)
P. rufa (*Populus tremula*)	*P. violaceolivida* (mostly *Salix* and *Populus*)	
P. rufomarginata (*Tilia*)		

Source: Based on data in Hansen and Knudsen (1997).

soil-borne tree pathogens, e.g. *Armillaria* spp., *Heterobasidion annosum* and *Collybia fusipes*, soil conditions influence infection incidence (Camy *et al.*, 2003; Thor *et al.*, 2005). However, the extent to which unit-restricted wood decay fungi are affected by soil type and chemistry is uncertain. Effects would be largely indirect; wood chemistry and structure (e.g. year ring widths) are affected by soil chemistry and general growth conditions (Sundberg *et al.*, 1993), and a recent study from coniferous wood has indicated that these might affect fungal species composition (Edman *et al.*, 2006).

4. COMMUNITY DEVELOPMENT PATHWAYS

Fungi colonizing wood are first faced with the problem of arrival at and entry into a suitable resource. Subsequently the priority is establishment within that resource, which at the earliest stages of community development may be uncolonized by other fungi, but is most commonly against a backdrop of well-established mycelia. Once established, remaining within the resource depends on defence of territory held. Eventually, as the wood resource rots away, decay fungi need to exit in search of new resources, and the life-cycle continues.

Community development can begin under highly stressful conditions, under conditions where abiotic stresses are completely absent, or under conditions somewhere between these extremes. Such microenvironmental factors are strong determinants of the communities that develop (see below). Following initial establishment, community development is influenced by four main driving forces: stress aggravation (worsening of abiotic environmental conditions), stress alleviation (improvement in abiotic conditions), disturbance and combat (interspecific competition for space rather than directly for nutrients) (Figure 2; Cooke and Rayner, 1984; Rayner and Webber, 1984; Rayner and Boddy, 1988; Boddy, 2001; Heilmann-Clausen, 2001). Stress aggravation and alleviation is often brought about externally, e.g. improvement or worsening of water regime by wetting or drying, though the organisms themselves can also effect changes, e.g. translocation of water into dry wood. Alleviation of stress allows fungi with a preponderance of R- and/or C-selected characteristics to predominate, whereas stress aggravation leads to predominance of species tolerant of that particular stress. As colonization proceeds leading to a closed community with no uncolonized territory available, organisms with more C-selected characteristics begin to dominate, and during these middle stages of decomposition, community change is mainly effected by the fungi themselves. As decomposition proceeds the wood resource is used up, and an increasing proportion of the nutrients are contained within living hyphae, bacteria and mesofauna. Little is known of the ecological forces most important during these stages of decay, but the well-developed micro- and mesofauna probably play a crucial role, both as grazers on mycelia and also as a potential food source for wood decay fungi. In small branches, there can be dramatic changes in the fungal community following invasion by invertebrates, away from domination by Basidiomycota (Swift and Boddy, 1984).

4.1 Colonization Strategies in Living Trees

Initially the wood cylinder is protected from fungal colonization by physical and chemical barriers of the bark, host resistance of living cells and the high water

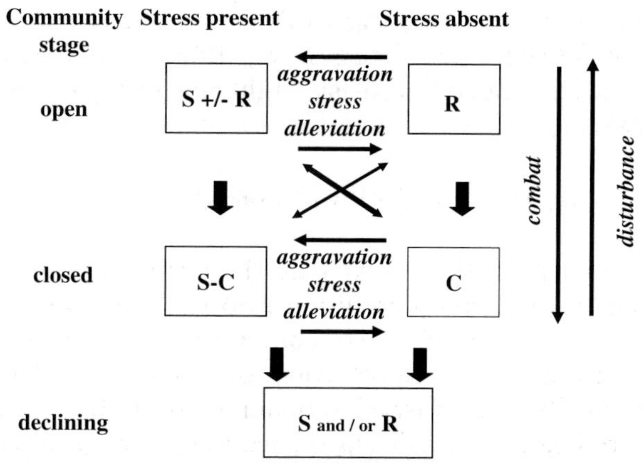

Figure 2 Schema showing potential fungal community development pathways following colonization of newly available wood (top of diagram), through an open community stage in which there is still unoccupied territory, to a closed community in which all territory is occupied (i.e. colonization must be by secondary resource capture), and finally to communities developing in well-decayed wood, characterized by substrate modification and invasion by soil invertebrates. Major ecological characteristics of dominant organisms are indicated in boxes: S, stress tolerant; C, combative; R, ruderal. Driving forces for change are indicated in italic. Direction in which the driving force pushes the community is indicated by arrows. Aggravation of stresses will push the community to the left, whereas alleviation will push the community to the right. As the community moves from open to closed, combat will be the main driving force for change. In contrast, destructive disturbance will push the community towards species with more ruderal characteristics (Adapted from Boddy, 2001; Rayner and Webber, 1984).

content of functional sapwood. Nevertheless, fungal colonization of wood tissue in living trees is common. Functional sapwood is water saturated, and contains virtually no oxygen and many living cells, while the inner core—heartwood—often contains large quantities of allelopathic compounds, but has an ameliorated gaseous regime. Though heartwood is inimical to fungal growth it is slightly less hostile than functional sapwood; hence, in the standing tree establishment of actively growing mycelia occurs predominantly in the former, and decay is most rapid and extensive there. Paradoxically, in felled and fallen trunks sapwood decays considerably more quickly than does heartwood. This is because sapwood is no longer functional in water conductivity and cells are dead; hence, the gaseous regime improves dramatically, cells are not in possession of host defence mechanisms and nutrients are more easily accessible.

Fungi may gain access to heartwood in living trees by different routes (Rayner and Boddy, 1988), but most commonly via wounds, e.g. after breakage of large branches. Volumes of sapwood can become dysfunctional, e.g. branches may die as a result of drought or light suppression and trunks and branches may experience wounding due to forestry activity, storm damage, mammal activity, invertebrate penetration or fire. Conditions within the sapwood are then ameliorated allowing development of actively growing fungal mycelia. Finally, fungal infection of intact wood may follow direct inoculation during oviposition by an insect vector (Chapter 9), and by pathogenesis, e.g. *H. annosum* that colonizes via contact of healthy roots with those of trees and stumps that are already colonized, and *Armillaria* species that arrive as rhizomorphs. In sapwood, primary colonizers are usually followed by more competitive secondary decay fungi prior to fall.

In some cases whole trees may die standing. Standing wood and attached branches are highly subject to microclimatic variation, being particularly prone to desiccation and to high internal temperatures that may occur in exposed standing trunks during sunny days. Community development in these dead wood habitats is hence characterized by many species with S- or R-selective traits.

4.2 Decay in Attached Branches

The most detailed studies on fungal community development in standing trees have focused on the sapwood of attached ash and oak branches (Boddy and Rayner, 1983; Boddy *et al.*, 1987). In southern Britain the most common primary colonizers of oak branches were *Peniophora quercina, Stereum gausapatum, Vuilleminia comedens, Phellinus ferreus* and *Phlebia rufa* (Boddy and Rayner, 1983). Latent presence has been demonstrated for the first three (Hirst, 1995; Boddy, 2001), but presumably they all have S- and R-selected characters allowing them to exist latently in functional sapwood, and then to develop overtly as mycelia as soon as the high water content/poor aeration regime is alleviated. *T. versicolor, P. radiata* and *Stereum hirsutum* were identified as combative secondary colonizers whose establishment depends on conditions having ameliorated sufficiently to allow their growth, and whether they are better combatants under the conditions obtaining. Other secondary colonizers, e.g. *Hyphoderma setigerum* and *H. paradoxa*, were less combative but tolerant of desiccation and insect activity. By virtue of this they sometimes replaced both primary and combative secondary colonizers. In the same way *Peniophora lycii* persisted in terminal branch regions. Analogous communities seem to develop in attached branches of many other deciduous tree species (Butin and Kowalski, 1986; Boddy *et al.*, 1987; Unterseher and Tal, 2006), though the primary colonizers are predominantly Ascomycota in some species, e.g. in *Fraxinus* (Boddy *et al.*, 1987). The amount of decay occurring in the canopy prior to fall is variable. Complete decay of sapwood can occur in the canopy if it is supported by heartwood, but if not, branches and twigs will fall to the floor, where decay continues.

4.3 Decay in Standing Dead Trunks

Knowledge of community development in standing dead trunks is limited. In boreal Fennoscandia dead pine trees, known as kelo-trees, may stand for centuries and have been reported to support highly specialized fungal communities (Niemelä *et al.*, 2002). Data from managed Danish beech forests similarly indicate natural snags have a distinctive mycoflora based on sporocarp observations (Heilmann-Clausen and Aude, 2006). For instance, the heterobasidiomycetes *Phleogena faginea* was very prominent, occurring on 29% of investigated snags, but only on 2% of the fallen logs. When long standing snags or standing dead trees fall to the ground they support fungal communities different from logs in the same decay stage which have decomposed on the ground. Several polypores and tooth fungi of conservation interest seem to fruit preferentially on such fallen dry trees or snags (Heilmann-Clausen and Christensen, 2004), but statistically supporting data are lacking.

Sun-exposed wood in standing trunks will experience extreme temperature fluctuations. Interestingly, some isolates of the rare heart rot agent *P. quercinus*, usually found in oak trees in wood pasture, grow well at 30 °C (Wald *et al.*, 2004), and we hypothesize that tolerance to high temperatures and/or desiccation is common among species living in standing dead wood. In this respect it is noteworthy that a compiled list of wood decay fungi from road side trees in Germany is remarkably rich in heart rot fungi and other species associated with decay in living trees (Kreisel, 2000). Several of the reported species, e.g. *Inonotus nidus-pici, I. hispidus, Ganoderma adspersum* and *Sarcodontia crocea*, are very sparsely recorded from forest environments in the same region and it seems likely that many of them are tolerant of heat-stress with low competitive ability in more shaded conditions.

4.4 Colonization Strategies in Fallen Wood

Fallen wood is often already colonized or even well decayed by the time it arrives on the forest floor (e.g. Boddy and Swift, 1983; Harmon *et al.*, 1986). This is especially the case for branches falling due to natural pruning and for old trees dying gradually in natural forests. Following arrival on the forest floor established fungal communities are faced with a swift change in microclimatic conditions, and with competition from fungi with well-established mycelia in the forest floor, and from combative fungi, arriving as spores, that were not tolerant of the microclimate of the aerial environment. Unfortunately, to date, there have been no studies documenting the changes in communities that developed while wood was attached to a standing tree once the wood has fallen to the floor.

Wood lacking a well-established decay community does sometimes reach the floor, e.g. younger trees toppling over or tree tops breaking from larger trees in heavy storms. Community development in these situations is likely to be similar to that in felled logs (below).

4.5 Decay in Felled Logs and Large Branches

Fungal community development in bulky wood on the forest floor has been studied in a number of cases, especially in *Fagus* spp., which we describe here as a model system for wood decay in angiosperms. Decay community development in other deciduous tree species seems to follow similar pathways (Gricius *et al.*, 1999; Hood *et al.*, 2004; Lindhe *et al.*, 2004), though some differences are evident reflecting differences in bark and wood morphology and wood chemistry. In many *Betula* spp. the bark is usually intact until final decay stages (J. Heilmann-Clausen, personal observation), making access for secondary colonizers more difficult. Similarly, species which possess true heartwood, e.g. *Quercus* spp., present widely different decay environments, where decay proceeds following different pathways in sapwood and heartwood.

There have been detailed studies on fungal community structure of felled beech logs, at the mycelial level, during the first 4.5–5 years of decay (Coates and Rayner, 1985a, 1985b, 1985c; Chapela and Boddy, 1988b; Chapela *et al.*, 1988; Boddy *et al.*, 1989; Willig and Schlechte, 1995), and of fruit bodies to very late stages of decay (Ueyama, 1966; Siepmann, 1973; Lange, 1992; Heilmann-Clausen, 2001). The different studies revealed similar patterns of community structure and development though species composition sometimes differed. Primary colonizers derived from those latently present in functional sapwood were prominent during early stages of decay, including a number of pyrenomycetes (Chapela and Boddy, 1988a, 1988b; Heilmann-Clausen, 2001; Hendry *et al.*, 2002), but also basidiomycetes, especially *F. fomentarius* (Baum *et al.*, 2003), and possibly *I. nodulosus* and *Exidia plana* (Heilmann-Clausen, 2001). After a few weeks secondary Basidiomycota, including *Bjerkandera adusta*, *T. versicolor* and *S. hirsutum*, arrived by spores at cut surfaces, and their decay columns became clearly resolved from 24 weeks (Chapela and Boddy, 1988b) (Figure 3). On some sites, though not others, the Ascomycota, *X. hypoxylon*, became increasingly evident between

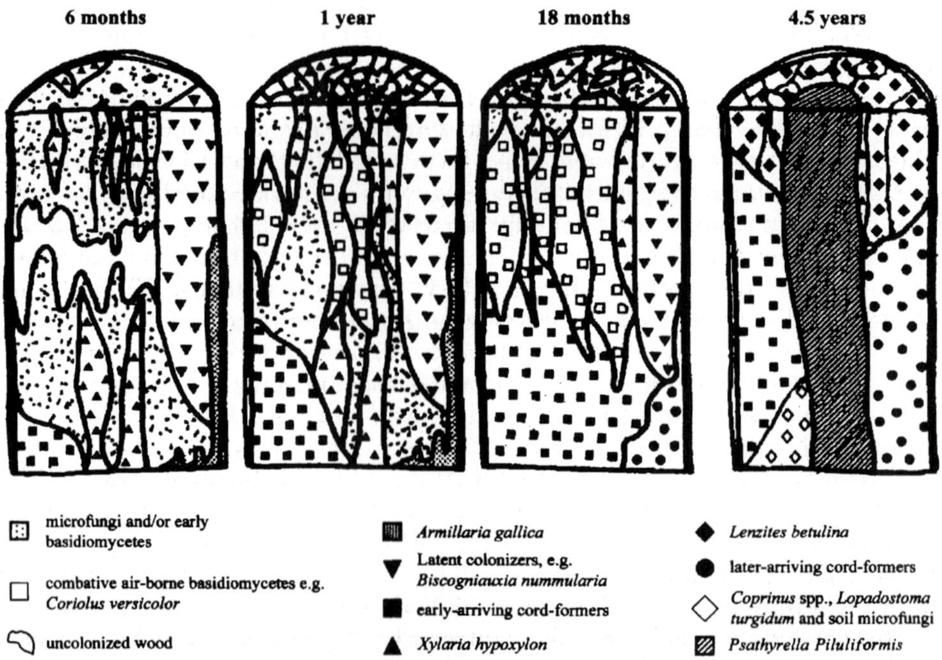

Figure 3 Diagram illustrating the way in which a hypothetical community might develop in a single beech log, placed upright with the base in contact with soil, over time. No two communities will ever be identical because differences in fungi arriving, and in environmental conditions, at every time point will result in different patterns of development. Nonetheless, species general temporal patterns are discernable (from Boddy, 2001; 6–18 months adapted from Coates and Rayner, 1985c).

12 and 52 weeks. In large logs the same species dominated by 3–9 years after felling (Willig and Schlechte, 1995; Heilmann-Clausen, 2001) probably following establishment in foci on the bark surface or in small crevices exposing sapwood. In sun-exposed beech logs the bark often loosens within 1 year and communities of stress-tolerant fungi quickly develop, including secondary invaders of which *Schizophyllum commune* and *Trametes hirsuta* are the most prominent (Willig and Schlechte, 1995; Heilmann-Clausen, 2001).

In small logs, cord-forming saprotrophs, e.g. *Hypholoma fasciculare*, *Megacollybia platyphylla*, *Phallus impudicus* and *Phanerochaete velutina*, and the rhizomorphic *Armillaria gallica* were colonizing by 6 months, the latter usually subcortically, and then in peripheral regions (Coates and Rayner, 1985a, 1985b, 1985c; Boddy, 2001). Eventually, the combative cord-formers occupied substantial decay columns after replacing earlier colonizers (Figure 3). After 1–1.5 years many of the early colonizers had declined, though *S. hirsutum* persisted in a few logs, and *T. versicolor* and *X. hypoxylon* (Ascomycota) were still present in at least 45% of logs by 4.5 years (Chapela *et al.*, 1988; Boddy, 2001). Cord-forming Basidiomycota dominated many logs by 4.5 years though other Basidiomycota, *Coprinus* spp., *L. betulina* (invading by temporary mycoparasitism of *T. versicolor*; Rayner *et al.*, 1987), *Psathyrella piluliformis*, and the Ascomycota, *Lopadostoma turgidum* and various soil Mucorales and Deuteromycota were also evident by then. The presence of the latter groups generally reflected disturbance by invading soil invertebrates.

In large logs, colonization by cord-forming Basidiomycota and other combative secondary decay fungi seems to progress slower than in smaller logs (Willig and Schlechte, 1995; Heilmann-Clausen, 2001), probably because access is more difficult due to the protective bark layer and longer distance from cut ends. This allows more time for latent decay fungi to develop large mycelia, and especially *E. spinosa* (Ascomycota) and *F. fomentarius* are able to sustain sporulating mycelia for decades (Lange, 1992; Heilmann-Clausen, 2001). In the Ascomycota *E. spinosa*, and *X. hypoxylon* in smaller logs, this seems to reflect a defensive strategy in which these species maintain occupied wood drier than its surroundings (Boddy *et al.*, 1989; Heilmann-Clausen, 2001).

Little is known of community development during late stages of wood decay, but sporocarp based studies indicate that agarics, especially *Mycena* and *Pluteus* spp., become increasingly dominant (Lange, 1992; Heilmann-Clausen, 2001), together with cord-formers, which often persist throughout the decay process. In addition, ectomycorrhizal species and soil/litter saprotrophs are often present, as evidenced by sporocarps (Lange, 1992; Heilmann-Clausen, 2001) and ectomycorrhizal roots (Harvey *et al.*, 1976; Tedersoo *et al.*, 2003).

The nutrient content in well-decayed wood is higher per unit volume than in undecayed wood, due to loss of carbon as respiratory CO_2 and import of mineral nutrients via mycelia of non-resource-restricted fungi (Swift and Boddy, 1984; Boddy and Watkinson, 1995; Ódor and Standovár, 2003; Laiho and Prescott, 2004; Chapters 1 and 3). Most nutrients are, however, bound in living fungal hyphae, bacteria and fauna, while the wood tissue *per se* is depleted of readily available nutrients. Some of the fungi inhabiting wood at late stages of decay may accordingly depend much more on alternative sources of nutrition. There is a dearth of information in this area, but it is well known that many wood-inhabiting fungi can utilize invertebrates, bacteria and algae (Barron, 2003; Chapters 8 and 9), and some, including not only saprotrophs (Niemelä *et al.*, 1995) but also ectomycorrhizal species (Lindahl *et al.*, 1999; Leake *et al.*, 2001), are able to retrieve nutrients from mycelia of other fungi, due to combative or mycoparasitic interactions. Competition/combat for space is then no longer the main driving force for community change. Rather, drivers are likely to be nutrient stress, stress relating to shifting microenvironmental regimes and disturbance by larger animals destroying dead wood and mycelia, and grazing by invertebrates. Grazing results in dramatic morphological and presumably also physiological changes in mycelia of Basidiomycota, and comminution (Chapter 9). At very late stages the composition of the remaining wood components becomes increasingly similar to soil, and R-selected soil fungi and bacteria often dominate (Swift and Boddy, 1984), though ectomycorrhizal fungi may also become established (Tedersoo *et al.*, 2003).

4.6 Decay in Fallen Twigs and Small Branches

Fungal communities in fallen twigs and small branches have not been well studied, but based on sporocarps there appear to be large differences in species composition compared with more bulky resources (Nordén *et al.*, 2004b). In early decay stages differences relate to the pre-fall infection history. At fall twigs are typically dominated by Ascomycota and corticoid Basidiomycota several of which developed from latent propagules, e.g. *Hypoxylon* spp. (Ascomycota), *V. comedens* and *Peniophora* spp. (Griffith and Boddy, 1988, 1990; Boddy and Griffith, 1989).

Small diameter dead wood provides much less stable microclimatic conditions than larger wood, being highly prone to desiccation and temperature fluctuations (Boddy, 1983), which will affect both species composition and decomposition rate. In a microcosm study, fluctuating microclimatic regime enhanced both species diversity and decay rate (Toljander *et al.*, 2006). In the field, the decay rate of small diameter wood is faster than large diameter wood (Stone *et al.*, 1998; Tarasov and Birdsey, 2001), which could indicate that overall conditions for decay are more favourable in small diameter wood, reflecting better aeration and less allelopathic chemicals than in larger wood, and/or may relate to differences in species composition. Thus, fallen twigs are often rapidly colonized by cord-forming Basidiomycota (Boddy, 1993), which are known as fast and effective decayers. Due to the relatively fast decomposition of small diameter wood, complex interactions between decay fungi, ectomycorrhizal fungi and invertebrates appear to be less distinct or at least to occur for shorter time than in bulky resources, and many late stage specialists reported from logs and large branches seem to be absent from twigs and small branches, based on sporocarp evidence (Kruys *et al.*, 1999; Sippola and Renvall, 1999).

4.7 Decay in Stumps, Buried and Submerged Wood

Cut stumps represent a rather special habitat for fungi. The aerial cut surface allows rapid establishment of fungi with air-borne propagules, but also cord-forming basidiomycetes have easy access via dead wood underground (Rayner, 1977a, 1977b; Rayner and Boddy, 1988; Pearce and Malajczuk, 1990). In addition, a variety of fungi that were probably present before felling, e.g. heart rotters such as *Ganoderma* spp., and latent invaders such as the Ascomycota *D. concentrica* (on ash) are frequent. Close soil contact buffers environmental fluctuations, making stumps a more stable environment than most other dead wood types. Most fungi causing decay in stumps are equally or more frequent on other dead wood types, but some species seem to take advantage of the special conditions for establishment and growth. The success of *H. annosum* as a pathogen in conifer plantations is thus partly due to its superior ability for establishment and spread via cut stumps (Woodward *et al.*, 1998), but there are also examples of fungi occurring on deciduous wood which are more common on stumps compared to natural dead wood types (Heilmann-Clausen and Aude, 2006). Buried roots and other subsoil types of dead wood also have a number of Basidiomycota which are very rarely seen on non-buried wood. Some, e.g. *Meripilus giganteus*, *C. fusipes* and *G. frondosa*, seem to be primary decay agents that establish before tree death/felling, while others, e.g. *Xerula radicata* and *Hydropus subalpinus*, are likely to be secondary invaders, establishing in dead wood.

Freshwater is present at some forest sites as permanent or temporary ponds and streams. In alder swamp forests, for example, standing water is normal, especially during winter, and a major fraction of the dead wood pool may be inundated for part of the year. Several species seem to be more or less specific to temporarily inundated wood, e.g. *Phlebia subochracea* and *Bulbillomyces farinosus* (Winterhoff, 1993). The latter produces vegetative dispersal structures (bulbils) which are probably adapted to dispersal by water. Also some polypores, e.g. *Physisporinus vitreus* and *P. sanguineus*, seem to be well adapted to growth under very wet conditions (Schmidt *et al.*, 1997; Aude *et al.*, 2006). Though there are a few aquatic Basidiomycota (Chapter 17), decay of permanently submerged wood is effected almost entirely by Ascomycota and bacteria (Rayner and Boddy, 1988; Kim and Singh, 2000).

REFERENCES

Abdalla, S.H.M. and Boddy, L. (1996). Effect of soil and litter on outgrowth patterns of mycelial systems of *Phanerochaete velutina*. *FEMS Microbiology Ecology*, **20**, 195–204.

Andrews, J.H. (1991). *Comparative Ecology of Microorganisms and Macroorganisms*. Springer, New York, NY.

Arnolds, E., Opdam, A., van Steenis, W. and de Vries, B. (1994). Mycocoenology of stands of *Fagus sylvatica* L. in the northeastern Netherlands. *Phytocoenologia*, **24**, 507–530.

Aude, E., Heilmann-Clausen, J. and Bennet, T.V. (2006). Er Urørthed en Trussel mod Naturen i Ellesumpe? *HabitatVision, Rapport* **06-02**.

Baar, J., Horton, T.R., Kretzer, A.M. and Bruns, T.D. (1999). Mycorrhizal colonization of *Pinus muricata* from resistant propagules after a stand-replacing wildfire. *New Phytologist*, **143**, 409–418.

Bader, P., Jansson, S. and Jonsson, B.G. (1995). Wood-inhabiting fungi and substratum decline in selectively logged boreal spruce forests. *Biological Conservation*, **72**, 355–362.

Banik M.T. and Burdsall, H.H. (2000). Incompatibility groups among North American populations of *Laetiporus sulphureus* sensu lato. *Mycologia*, **92**, 649–655.

Barron, G.L. (2003). Predatory fungi, wood decay, and the carbon cycle. *Biodiversity*, **4**, 3–9.

Baum, S., Sieber, T.N., Schwarze, F.W.M.R. and Fink, S. (2003). Latent infections of *Fomes fomentarius* in the xylem of European beech (*Fagus sylvatica*). *Mycological Progress*, **2**, 141–148.

Boddy, L. (1983). Effect of temperature and water potential on growth rate of wood-rotting basidiomycetes. *Transactions of the British Mycological Society*, **80**, 141–149.

Boddy, L. (1984). The micro-environment of basidiomycete mycelia in temperate deciduous woodlands. In: *Ecology and Physiology of the Fungal Mycelium* (D.H. Jennings and A.D.M. Rayner, eds.), pp. 261–289. Cambridge University Press, Cambridge.

Boddy, L. (1986). Water and decomposition processes in terrestrial ecosystems. In: *Water, Plants and Fungi* (P.G. Ayres and L. Boddy, eds.), pp. 375–398. Cambridge University Press, Cambridge.

Boddy, L. (1993). Saprotrophic cord-forming fungi: Warfare strategies and other ecological aspects. *Mycological Research*, **97**, 641–655.

Boddy, L. (1999) Saprotrophic cord-forming fungi: Meeting the challenge of heterogeneous environments. *Mycologia*, **91**, 13–32.

Boddy, L. (2000). Interspecific combative interactions between wood-decaying basidiomycetes. *FEMS Microbiology Ecology*, **31**, 185–194.

Boddy, L. (2001). Fungal community ecology and wood decomposition processes in angiosperms: From standing tree to complete decay of coarse woody debris. *Ecological Bulletins*, **49**, 43–56.

Boddy, L., Bardsley, D.W. and Gibbon, O.M. (1987). Fungal communities in attached ash branches. *New Phytologist*, **107**, 143–154.

Boddy, L., Gibbon, O.M. and Grundy, M.A. (1985). Ecology of *Daldinia concentrica*: Effect of abiotic variables on mycelial extensions and interspecific interactions. *Transactions of the British Mycological Society*, **85**, 201–211.

Boddy, L. and Griffith, G.S. (1989). Role of endophytes and latent invasion in the development of decay communities in sapwood of angiospermous trees. *Sydowia*, **41**, 41–73.

Boddy, L. and Jones, T.H. (2006). Mycelial responses in heterogeneous environments: Parallels with macroorganisms. In: *Fungi in the Environment* (G.M. Gadd, S.C. Watkinson and P. Dyer, eds.), pp. 112–140. Cambridge University Press, Cambridge.

Boddy, L., Owens, E.M. and Chapela, I.H. (1989). Small scale variation in decay rate within logs one year after felling: Effects of fungal community structure and moisture content. *FEMS microbiology Ecology*, **62**, 173–184.

Boddy, L. and Rayner, A.D.M. (1983). Ecological roles of basidiomycetes forming decay communities in attached oak branches. *New Phytologist*, **93**, 77–88.

Boddy, L. and Swift, M.J. (1983). Wood decomposition in an abandoned beech and oak coppiced woodland in south-east England. I. Patterns of wood-litter fall. *Holaractic Ecology*, **6**, 320–332.

Boddy, L. and Watkinson, S.C. (1995). Wood decomposition, higher fungi and their role in nutrient cycling. *Canadian Journal of Botany*, **73**(Suppl. 1), S1377–S1383.

Butin, H. (1995). *Tree Diseases and Disorders*. Oxford University Press, Oxford.

Butin, H. and Kowalski, T. (1986). Die natürliche Astreinigung und ihre biologischen Voraussetzungen. III. Die Pilzflora von Ahorn, Erle, Birke, Hainbuche und Esche. *European Journal of Forest Pathology*, **16**, 129–138.

Camy, C., de Villebonne, D., Delatour, C. and Marçais, B. (2003). Soil factors associated with infection by *Collybia fusipes* and decline of oaks. *Forest Pathology*, **33**, 253–266.

Chapela, I.H. (1987). Structure and development of fungal communities in attached beech (*Fagus sylvatica* L.) branches, PhD Thesis, University of Wales, Cardiff, UK.

Chapela, I.H. (1989). Fungi in healthy stems and branches of American beech and aspen: A comparative study. *New Phytologist*, **113**, 65–75.

Chapela, I.H. and Boddy, L. (1988a). Fungal colonization of attached beech branches. II. Spatial and temporal organization of communities arising from latent invaders in bark and functional sapwood, under different moisture regimes. *New Phytologist*, **110**, 47–57.

Chapela, I.H. and Boddy, L. (1988b). The fate of early fungal colonizers in beech branches decomposing on the forest floor. *FEMS Microbiology Ecology*, **53**, 273–284.

Chapela, I.H., Boddy, L. and Rayner, A.D.M. (1988). Structure and development of fungal communities in beech logs four and a half year after felling. *FEMS Microbiology Ecology*, **53**, 59–70.

Chapela, I.H., Petrini, O. and Hagmann, L. (1991). Monolignol glucosides as specific recognition messengers in fungus–plant symbioses. *Physiological and Molecular Plant Pathology*, **39**, 289–298.
Chen, J., Franklin, J.F. and Spies, T.A. (1993). Contrasting microclimates among clearcut, edge, and interior of old-growth Douglas-fir forest. *Agricultural and Forest Meteorology*, **63**, 219–237.
Chlebicki, A., Żarnowiec, J., Cieśliński, S., Klama, H., Bujakiewicz, A. and Załuski, T. (1996). Epixylites, lignicolous fungi and their links with different kinds of wood. *Archivum Geobotanicum*, **6**, 75–133.
Christensen, M., Hahn, K., Mountford, E., Ódor, P., Standovár, T., Rozenbergar, D., Diaci, J., Wijdeven, S., Meyer, P., Winter, S. and Vrska, T. (2005). Dead wood in European beech forest reserves. *Forest Ecology and Management*, **210**, 267–282.
Coates, D. and Rayner, A.D.M. (1985a). Fungal population and community development in cut beech logs, I. Establishment via the aerial cut surface. *New Phytologist*, **101**, 153–171.
Coates, D. and Rayner, A.D.M. (1985b). Fungal population and community development in cut beech logs, II. Establishment via the buried cut surface. *New Phytologist*, **101**, 173–181.
Coates, D. and Rayner, A.D.M. (1985c). Fungal population and community development in cut beech logs, III: Spatial dynamics, interactions and strategies. *New Phytologist*, **101**, 183–198.
Cooke, R.C. and Rayner, A.D.M. (1984). *Ecology of Saprotrophic Fungi*. Longman, London.
Dahlberg, A. and Stokland, J.N. (2004). Vedlevande Arters Krav på Substrat—Sammanställning och Analys av 3.600 arter. *Rapport* **7**, Skogsstyrelsen.
Danby, A.J. (2000). Latent endophytic fungi in sapwood of *Betula pendula* and *B. pubescens*, PhD Thesis, University of Wales, Cardiff, UK.
Donnelly, D.P. and Boddy, L. (1998). Developmental and morphological responses of mycelial systems of *Stropharia caerulea* and *Phanerochaete velutina* to soil nutrient enrichment. *New Phytologist*, **138**, 519–531.
Edman, M., Möller, R. and Ericsson, L. (2006). Effects of enhanced tree growth rate on the decay capacities of three saprotrophic wood-fungi. *Forest Ecology and Management*, **232**, 12–18.
Eriksson, J., Hjortstam, K., Larsson, K.-H. and Ryvarden, L. (1973–1988). *The Corticiaceae of North Europe*, Vol. 1–8. Fungiflora, Oslo, Norway.
Fisher, P.J. and Petrini, O. (1990). A comparative study of fungal endophytes in xylem and bark of *Alnus* species in England and Switzerland. *Mycological Research*, **94**, 313–319.
Fox, R.T.V. (2000). *Armillaria Root Rot: Biology and Control of Honey Fungus*. Intercept Press, Andover.
Gilbertson, R.L. (1973). Notes on some corticoid fungi associated with snow banks in southern Arizona. *Persoonia*, **7**, 171–182.
Gricius, A., Urbonas, V., Kutorga, E. and Matelis, A. (1999). Succession of fungi on dead timber of various trees. *Botanica Lithuanica*, **5**, 61–70.
Griffith, G.S. and Boddy, L. (1988). Fungal communities in attached ash (*Fraxinus excelsior*) twigs. *Transactions of the British Mycological Society*, **91**, 599–606.
Griffith, G.S. and Boddy, L. (1990). Fungal decomposition of attached angiosperm twigs. I. Decay community development in ash, beech and oak. *New Phytologist*, **116**, 407–415.
Grime, J.P. (1977). Evidence for the existence of three primary strategies in plants and its relevance to ecological and evolutionary theory. *American Naturalist*, **111**, 1169–1194.
Grime, J.P. (1979). *Plant Strategies and Vegetation Processes*. Wiley, Chichester.
Hallenberg, N. (1995). Dispersal abilities and distributional patterns in Aphyllophorales, with emphasis on corticoid fungi. *Acta Universitatis Upsaliensis, Symbolae Botanicae Upsaliensis*, **30**(3), 95–100.
Hallenberg, N. and Küffer, N. (2001). Long-distance spore dispersal in wood-inhabiting Basidiomycetes. *Nordic Journal of Botany*, **21**, 431–436.
Hallenberg, N. and Parmasto, E. (1998). Phylogenetic studies in species of Corticiaceae growing on branches. *Mycologia*, **90**, 640–654.
Hansen, L. and Knudsen, H. (1997). *Nordic Macromycetes. Vol. 3, Heterobasidioid, Aphyllophoroid and Gatromycetoid Basidiomycetes*. Nordsvamp, Copenhagen.
Harmon, M.E., Franklin, J.F., Swanson, F.J., Sollins, P., Gregory, S.V., Lattin, J.D., Anderson, N.H., Cline, S.P., Aumen, N.G., Sedell, J.R., Lienkaemper, G.W., Cromack, K., Jr. and Cummins, K.W. (1986). Role of coarse woody debris in temperate ecosystems. *Advances in Ecological Research*, **15**, 133–302.

Harvey, A.E., Larsen, M.J. and Jurgensen, M.F. (1976). Distribution of ectomycorrhizae in a mature Douglas-fir/larch forest in western Montana. *Forest Science*, **22**, 393–398.

Hawksworth, D.L. (2001). The magnitude of fungal diversity: The 1.5 million species estimate revisited. *Mycological Research*, **105**, 1422–1432.

Heilmann-Clausen, J. (2001). A gradient analysis of communities of macrofungi and slime moulds on decaying beech logs. *Mycological Research*, **105**, 575–596.

Heilmann-Clausen, J. (2003). Wood-inhabiting fungi in Danish deciduous forests-diversity, habitat preferences and conservation, PhD Thesis, Danish Research Academy/Danish Forest and Landscape Research Institute.

Heilmann-Clausen, J. (2005). Diversity of saproxylic fungi on decaying beech wood in protected forests in the county of Halland. *Information från Länsstyrelsen Halland, Meddelande* **7**.

Heilmann-Clausen, J. and Aude, E. (2006). Forvaltning af dødt ved i naturnære bøgebevoksninger—til gavn for biodiversiteten? *HabitatVision, Rapport* **06-01**.

Heilmann-Clausen, J., Aude, E. and Christensen, M. (2005). Cryptogam communities on decaying deciduous wood—Does tree species diversity matter? *Biodiversity and Conservation*, **14**, 2061–2078.

Heilmann-Clausen, J. and Boddy, L. (2005). Inhibition and stimulation effects in communities of wood decay fungi: Exudates from colonized wood influence growth by other species. *Microbial Ecology*, **49**, 1–8.

Heilmann-Clausen, J. and Christensen, M. (2004). På jagt efter urskovssvampe. *Svampe*, **49**, 45–52.

Helfer, W. (1991). *Pilze auf Fruchtkörpern—Untersuchungen zur Ökologie, Systematik und Chemie*. Libri Botanici, Vol. 1, Eching.

Hendry, S.J., Boddy, L. and Lonsdale, D. (1993). Interactions between callus cultures of European beech, indigenous ascomycetes and derived fungal extracts. *New Phytologist*, **123**, 421–428.

Hendry, S.J., Boddy, L. and Lonsdale, D. (2002). Abiotic variables effect differential expression of latent infections in beech (*Fagus sylvatica* L.). *New Phytologist*, **155**, 449–460.

Hendry, S.J., Lonsdale, D. and Boddy, L. (1998). Strip-cankering of beech (*Fagus sylvatica*): Pathology and distribution of symptomatic trees. *New Phytologist*, **140**, 549–565.

Hirst, J.E. (1995). The ecology and physiology of endophytes of angiosperm stems, PhD Thesis, University of Wales, Cardiff, UK.

Holmer, L., Renvall, P. and Stenlid, J. (1997) Selective replacement between species of wood-rotting basidiomycetes, a laboratory study. *Mycological Research*, **101**, 714–720.

Hood, I.A., Beets, P.N., Kimberley, M.O., Gardner, J.F., Oliver, G.R. and Pearce, S. (2004). Colonisation of podocarp coarse woody debris by decomposer basidiomycete fungi in an indigenous forest in the central North Island of New Zealand. *Forest Ecology and Management*, **196**, 311–325.

Hřib, J. and Rypáček, V. (1981). A simple callus test to determine the aggressiveness of wood-destroying fungi. *European Journal of Forest Pathology*, **11**, 270–274.

Huckfeldt, T., Schmidt, H. and Quader, O. (2005). Ökologische untersuchungen am Echten Hausschwamm und weiteren Hausfaülepilzen. *Holz als Roh- und Werkstoff*, **63**, 209–219.

Jonsson, B.G., Kruys, N. and Ranius, T. (2005). Ecology of species living on dead wood—Lessons for dead wood management. *Silva Fennica*, **39**, 289–309.

Kim, Y.S. and Singh, A.P. (2000). Micromorphological characteristics of wood biodegradation in wet environments: A review. *IAWA Journal*, **21**, 135–155.

Kjøller, R. and Bruns, T.D. (2003). Rhizopogon spore bank communities within and among California pine forests. *Mycologia*, **95**, 603–613.

Kresiel, H. (2000). Pilze an Strassenbäumen in Ostdeutschland. *Hoppea*, **61**, 169–182.

Kruys, N., Fries, C., Jonsson, B.G., Lämås, T. and Ståhl, G. (1999). Wood-inhabiting cryptogams on dead Norway spruce (*Picea abies*) trees in managed Swedish boreal forests. *Canadian Journal of Forest Research*, **29**, 178–186.

Laiho, R. and Prescott, C.E. (2004). Decay and nutrient dynamics of coarse woody debris in northern coniferous forests: A synthesis. *Canadian Journal of Forest Research*, **34**, 763–777.

Lange, M. (1992). Sequence of Macromycetes on decaying beech logs. *Persoonia*, **14**, 449–456.

Leake, J.R., Donnelly, D.P., Saunders, E.M., Boddy, L. and Read, D.J. (2001). Rates and quantities of carbon flux to ectomycorrhizal mycelium following ^{14}C pulse labelling of *Pinus sylvestris* seedlings:

Effects of litter patches and interactions with a wood-decomposer fungus. *Tree Physiology*, **144**, 183–193.

Lindahl, B., Stenlid, J., Olsson, S. and Finlay, R. (1999). Translocation of ^{32}P between interacting mycelia of a wood-decomposing fungus and ectomycorrhizal fungi in microcosm systems. *New Phytologist*, **144**, 183–193.

Lindhe, A., Asenblad, N. and Toresson, H.G. (2004). Cut logs and high stumps of spruce, birch, aspen and oak—Nine years of saproxylic fungi succession. *Biological Conservation*, **119**, 443–454.

Lopez-Real, J.M. and Swift, M.J. (1975). The formation of pseudosclerotia ('zone lines') in wood decayed by *Armillaria mellea* and *Stereum hirsutum*. II. Formation in relation to moisture content of wood. *Transactions of the British Mycological Society*, **64**, 465–471.

Morecroft, M.D., Taylor, M.E. and Oliver, H.R. (1998). Air and soil microclimates of deciduous woodland compared to an open site. *Agricultural and Forest Meteorology*, **90**, 141–156.

Niemelä, T., Renvall, P. and Pentillä, R. (1995). Interactions of fungi at late stages of wood decomposition. *Annales Botanici Fennici*, **32**, 141–152.

Niemelä, T., Wakkenius, T. and Kotiranta, H. (2002). The kelo tree, a vanishing substrate of specified wood-inhabiting fungi. *Polish Botanical Journal*, **47**, 91–101.

Nordén, B., Appelquist, T., Lindahl, B. and Henningsson, M. (1999). Cubic rot fungi—Corticoid fungi in highly brown rotted spruce stumps. *Mycologia Helvetica*, **10**, 13–24.

Nordén, B., Götmark, F., Tonnberg, M. and Ryberg, M. (2004a). Dead wood in semi-natural temperate broadleaved woodland: Contribution of coarse and fine dead wood, attached dead wood and stumps. *Forest Ecology and Management*, **194**, 235–248.

Nordén, B. and Larsson, K.-H. (2000). Basidiospore dispersal in the old-growth forest fungus *Phlebia centrifuga* (Basidiomycetes). *Nordic Journal of Botany*, **20**, 215–219.

Nordén, B., Ryberg, M., Götmark, F. and Olausson, B. (2004b). Relative importance of coarse and fine woody debris for the diversity of wood-inhabiting fungi in temperate broadleaf forests. *Biological Conservation*, **117**, 1–10.

Ódor, P. and Standovár, T. (2003). *Changes of Physical and Chemical Properties of Dead Wood during Decay.* NAT-MAN Working Report 23, http://www.flec.kvl.dk/natman.

Ohtonen, R., Aikio, S. and Vare, H. (1997). Ecological theories in soil biology. *Soil Biology & Biochemistry*, **29**, 1613–1619.

Pearce, M.H. and Malajczuk, N. (1990). Stump colonization by *Armillaria luteobubalina* and other wood decay fungi in an age series of cut-over stumps in karri (*Eucalyptus diversicolor*) regrowth forests in south-western Australia. *New Phytologist*, **115**, 129–138.

Phillips, D.A. and Burdekin, D.A. (1982). *Diseases of Forest and Ornamental Trees*. Macmillan, London.

Pinon, J. and Manion, P.D. (1991). *Hypoxylon mammatum* and its toxins—Recent advances in understanding their relationships with canker disease of poplar. *European Journal of Forest Pathology*, **21**, 202–209.

Piou, D., Delatour, C. and Marçais, B. (2002). Hosts and distribution of *Collybia fusipes* in France. *Forest Pathology*, **32**, 29–41.

Pugh, G.J.F. (1980). Strategies in fungal ecology. *Transactions of the British Mycological Society*, **75**, 1–14.

Rayner, A.D.M. (1977a). Fungal colonization of hardwood stumps from natural sources. I. Non-basidiomycetes. *Transactions of the British Mycological Society*, **69**, 292–302.

Rayner, A.D.M. (1977b). Fungal colonization of hardwood stumps from natural sources. II. Non-basidiomycetes. *Transactions of the British Mycological Society*, **69**, 303–312.

Rayner, A.D.M. and Boddy, L. (1988). *Fungal Decomposition of Wood; Its Biology and Ecology.* Wiley, Chichester.

Rayner, A.D.M., Boddy, L. and Dowson, C.G. (1987). Temporary parasitism of *Coriolus* spp. by *Lenzites betulina*: A strategy for domain capture in wood decay fungi. *FEMS Microbiology Ecology*, **45**, 53–58.

Rayner, A.D.M. and Hedges, M.J. (1982). Observations on the specificity and ecological role of basidiomycetes colonizing dead elm wood. *Transactions of the British Mycological Society*, **78**, 370–373.

Rayner, A.D.M. and Webber, J.F. (1984). Interspecific mycelial interactions—An overview. In: *Ecology and Physiology of the Fungal Mycelium* (D.H. Jennings and A.D.M. Rayner, eds.), pp. 383–417. Cambridge University Press, Cambridge.

Renvall, P. (1995). Community structure and dynamics of wood-rotting Basidiomycetes on decomposing conifer trunks in northern Finland. *Karstenia*, **35**, 1–51.

Ritter, E., Dalsgaard, L. and Einhorn, K.S. (2005). Light, temperature and soil moisture regimes following gap formation in a semi-natural beech-dominated forest in Denmark. *Forest Ecology and Management*, **206**, 15–33.

Roberts, P. (2002). *Report on the oak polypore Piptoporus quercinus (syn. Buglossoporus quercinus), a UK BAP priority species and Schedule 8 species*. English Nature Research Report No. 458, Peterborough.

Ryvarden, L. and Gilbertson, R.L. (1992). *European Polypores. I.* Fungiflora, Oslo.

Schmidt, O., Schmitt, U., Moreth, U. and Potsch, T. (1997). Wood decay by the white-rotting basidiomycete *Physisporinus vitreus* from a cooling tower. *Holzforschung*, **51**, 193–200.

Siepmann, R. (1973). Isolierung und bestimmung von Basidiomyceten und anderden pilzen aus waldlagernedm Buchenholz (*Fagus sylvatica*). *Material und Organismen*, **8**, 271–294.

Sippola, A.-L. and Renvall, P. (1999). Wood-decomposing fungi and seed-tree cutting: A 40 year perspective. *Forest Ecology and Management*, **115**, 183–201.

Stone, J.N., MacKinnon, A., Parminter, J.V. and Lertzman, K.P. (1998). Coarse woody debris decomposition documented over 65 years on southern Vancouver Island. *Canadian Journal of Forest Research*, **28**, 788–793.

Sundberg, B., Ericsson, A., Little, C.H.A., Näsholm, T. and Gref, R. (1993). The relationship between crown size and ring width in *Pinus sylvestris* L. stems: Dependence on indole-3-acetic acid, carbohydrates and nitrogen in the cambial region. *Tree Physiology*, **12**, 347–362.

Swift, M.J. and Boddy, L. (1984). Animal–microbial interactions during wood decomposition. In: *Invertebrate–Microbe Interactions* (J.M. Anderson, A.D.M. Rayner and D.W.H. Walton, eds.), pp. 89–131. Cambridge University Press, Cambridge.

Tarasov, M.E. and Birdsey, R.A. (2001). Decay rate and potential storage of coarse woody debris in the Leningrad Region. *Ecological Bulletin*, **49**, 137–147.

Tedersoo, L., Kõljalg, U., Hallenberg, N. and Larsson, K.-H. (2003). Fine scale distribution of ectomycorrhizal fungi and roots across substrate layers including coarse woody debris in a mixed forest. *New Phytologist*, **159**, 153–166.

Theden, G. (1961). Untersuchungen über die Fähigkeit holzzerstörender pilze zur Trockenstarre. *Angewandte Botanik*, **35**, 131–145.

Thompson, W. (1984). Distribution, development and functioning of mycelial cord systems of decomposer basidiomycetes of the deciduous woodland floor. In: *The Ecology and Physiology of the Fungal Mycelium* (D.H. Jennings and A.D.M. Rayner, eds.), pp. 185–215. Cambridge University Press, Cambridge.

Thor, M., Ståhl, G. and Stenlid, J. (2005). Modelling root rot incidence in Sweden using tree, site and stand variables. *Scandinavian Journal of Forest Research*, **20**, 165–176.

Toljander, Y.K., Lindahl, B.D., Holmer, L. and Högberg, N.O.S. (2006). Environmental fluctuations facilitate species co-existence and increase decomposition in communities of wood decay fungi. *Oecologia*, **148**, 625–631.

Tyler, G. (1985). Macrofungal flora of Swedish beech forest related to soil organic matter and acidity characteristics. *Forest Ecology and Management*, **10**, 13–29.

Ueyama, A. (1966). Studies on the succession of higher fungi on felled beech logs (*Fagus crenata*) in Japan. *Material und Organismen*, **1**, 325–332.

Unterseher, M. and Tal, O. (2006). Influence of small scale conditions on the diversity of wood decay fungi in a temperate, mixed deciduous forest canopy. *Mycological Research*, **110**, 169–178.

Vasiliauskas, R., Vasiliauskas, A., Stenlid, J. and Matelis, A. (2004). Dead trees and protected polypores in unmanaged north-temperate forest stands of Lithuania. *Forest Ecology and Management*, **193**, 355–370.

Vilgalys, R. and Sun, B.L. (1994). Assessment of species distributions in *Pleurotus* based on trapping of airborne basidiospores. *Mycologia*, **86**, 270–274.

Wald, P., Crockatt, M., Gray, V. and Boddy, L. (2004). Growth and interspecific interactions of the rare oak polypore *Piptoporus quercinus*. *Mycological Research*, **108**, 189–197, 465.

Willig, J. and Schlechte, G.B. (1995). Fungal succession on wood after windthrow in a natural beech-forest reserve. *Allgemeine Forst Zeitschrift*, **15**, 814–818.

Winterhoff, W. (1993). Die Großpilzflora von Erlenbruchwäldern und deren Kontaktgesellschaften in der nordbadischen Oberrheinebene. *Beihefte zu den Veröffentlichungen für Naturschutz und Landschaftspflege in Baden-Württemberg*, **74**, 1–100.

Woodward, S., Stenlid, J., Karjalainen, R. and Hütterman, A.A. (1998). *Heterobasidion annosum: Biology, Ecology, Impact and Control*. CABI, Wallingford.

Yakovlev, A.Y., Borovskii, G.B., Penzina, T.A., Petrov, A.N. and Voinikov, V.K. (2000). The influence of subzero temperatures on mycelial growth and fruit bodies vitality of some higher xylotrophic basidiomycetes. *Mikologiya i Fitopatologiya*, **34**, 56–62.

Zugmaier, W., Bauer, R. and Oberwinkler, F. (1994). Mycoparasitism of some *Tremella* species. *Mycologia*, **86**, 49–56.

CHAPTER 13

Wood-Decay Basidiomycetes in Boreal Forests: Distribution and Community Development

Jan Stenlid, Reijo Penttilä and **Anders Dahlberg**

Contents

Abstract

Dead wood and wood-inhabiting fungi are of key importance for biodiversity in boreal forests, and also for global CO_2 dynamics. Of more than 10,000 non-lichenised fungal species in Fennoscandia, over 2,500 are wood-inhabiting. Anthropogenic influences such as forest harvesting and fire suppression have reduced the availability of dead wood in forests, resulting in many wood-decay fungi being considered threatened. Classic inventory approaches have been complemented by pure culture studies of mycelia and recently by molecular detection methods. Nutrient cycling and interspecific interactions play important parts in the development of fungal communities. Boreal fungal communities are, in general, less diverse and more similar on a global scale compared to communities from the temperate regions.

1. INTRODUCTION

Awareness of the key importance of dead wood and wood-inhabiting fungi for biodiversity in boreal forests has increased substantially during the last two decades, as a result of extensive research in northern Europe and North America

British Mycological Society Symposia Series
Published by Elsevier Ltd.

(Harmon *et al.*, 1990; Dahlberg and Stokland, 2004; Jonsson *et al.*, 2005; Harper *et al.*, 2006). This improved understanding has also been communicated to decision makers in politics and forestry and resulted in information, guidelines and rules for practical management to enhance biodiversity associated with dead wood.

Of the estimated 10,000 plus non-lichenised fungal species in Fennoscandinavia, more than 2,500 are wood-inhabiting (Siitonen, 2001; Gärdenfors *et al.*, 2003; Dahlberg and Stokland, 2004). Roughly 1,500 of these are basidiomycetes and 1,000 are ascomycetes. These estimates are predominantly based on observations of sporocarps, and the contribution of ascomycetes and other species with inconspicuous sporocarps or rarely fruiting species are probably both qualitatively and quantitatively underestimated. Among the 1,500 wood-inhabiting basidiomycetes, at least 550 are corticoid, 200 polypores, 150 agarics and 100 heterobasidiomycetes. One challenge is to understand the presence of this high diversity given the relatively few tree species present in boreal forests. Niche separation can be the result of spatial, temporal and qualitative differences in the environment (Table 1).

Dead organic matter is of paramount importance as the energy source for the major part of the biological diversity in forest ecosystems. Less than 10% of the

Table 1 The number of potential combinations of factors that affect the quality of dead wood and thus the conditions for the occurrence of wood-inhabiting fungi is almost unlimited

Factors affecting the quality of the wood	Examples	Number of types
Tree species	*Picea abies, Pinus sylvestris, Betula spp, Populus tremula*	>10
Dimensions of the wood	Fine woody debris and coarse woody debris	>2
Decay class	Recently dead, initial decay, intermediate decay, advanced decay, very decayed	5
Part of tree	Branches, trunk, root	>3
Part of the wood	Bark, sapwood, heartwood	>3
The wood's cause of death	Weakened and slowly dying due to age, insects or fungal pathogens or directly killed after storm, fire or cutting	>3
The increment speed of the wood	Slow to rapid	>2
Microenvironment around the wood	Dry, damp, wet, sunk in water, sunexposed, shaded	>5
Other factors	Associated species	>5
Number of potential combination possibilities with this example of factors >100,000		

approximately 20,000 multicellular species in Fennoscandian boreal forests have photosynthetic ability while the majority are either part of the decomposer or herbivore food web. Fungi are the key drivers in the decomposer food web as they are mainly decomposers of cellulose and lignin (Rayner and Boddy, 1988). Most species associated with dead wood do not live on the wood tissue directly but on species that decompose it. For example, a large number of insect species and other invertebrates feed on fungal mycelium.

2. THE IMPORTANCE OF BOREAL WOOD

Boreal forests are important as a major sink for CO_2 (Bonan and Vancleve, 1992; Gower *et al.*, 2001; Goodale *et al.*, 2002; Liski *et al.*, 2006). However, this is true mainly for forests with a high net production of biomass, while in regeneration and declining phases, forests will function as sources of respired CO_2 due to the decomposition of wood and other organic material. Models of wood decomposition have usually not explicitly included variation in the fungal community (Mäkinen *et al.*, 2006), rather the decomposer community is treated as a 'black box' (Hyvönen *et al.*, 2005; Montes and Canellas, 2006). However, there are substantial differences in decomposition rate depending on the species involved (Tanesaka *et al.*, 1993; Osono and Takeda, 2002; Urcelay and Robledo, 2004) and the complexity of the community. Intermediate complexity might be the most efficient in decomposing wood. In an experimental study with combinations of 1–16 wood-decay species, the highest community wood decomposition rate was in species combinations with intermediate complexity (Toljander *et al.*, 2006). This was particularly true when the temperature was fluctuating between 7 and 31 °C. Presumably, competition suppressed decomposition in the most complex communities, while the intermediate complexity allowed for adaptation to fluctuating environmental conditions.

The decay community is influenced by the amount and quality of resource present. Wood from fast growing trees decays faster (Edman *et al.*, 2006), but Heijari *et al.* (2005) showed no consistent relationship between fertilization and durability of *Coniophora puteana* in pine or spruce wood following long-term fertilization. Managed forests contain only 5–40% of the quantity of dead wood present in natural forests and as a consequence harbour a much lower diversity of species (e.g. Lindblad, 1998; Fridman and Walheim, 2000). When trees die they may remain standing or as dead snags for a time, during which decomposition is slower than for fallen logs. In particular, dead pine (*Pinus*) trees may often remain standing for extended periods (even up to 200 years) as 'kelo trees' before falling (Niemelä *et al.*, 2002; Rouvinen *et al.*, 2002). The time spent as snags has been implicated as a major factor responsible for the variation in decomposition rate of dead wood. Vanderwel *et al.* (2006) reported that snags remain for 90 years while fallen trees remained for only 55–60 years in white and red pine (*Pinus strobus* and *P. resinosa*). Krankina and Harmon (1995) showed a higher decay rate for birch (*Betula* spp.) with an average time to disappearance of 28 years, and 20% of standing volume in dead wood. Siitonen *et al.* (2000) indicated even higher proportions of dead wood in old growth—33% of the total volume of wood of which snags made up 30%. Decay rate is also affected by temperature; Kueppers *et al.* (2004) found decomposition rate doubled in sites that were on average 3 °C warmer. Since primary production is less sensitive than decomposition to changes in temperature, boreal forests have a tendency to accumulate more dead organic matter than temperate forests.

Regeneration of seedlings is greater and growth is better among decaying wood than in the surrounding soil, particularly in coniferous forests (Szewczyk and Szwagrzyk, 1996; Gray and Spies, 1997; Hörnberg *et al.*, 1997; Takahashi *et al.*, 2000; Kuuluvainen and Kalmari, 2003; Narukawa *et al.*, 2003; Baier *et al.*, 2006). Decaying logs may provide better microclimatic conditions including a more balanced water regime, and better nutrient status due to reduced competition with large established root systems. Decomposing logs may also be an important site for interaction between wood decomposers and mycorrhizal fungi (Lindahl *et al.*, 1999, 2001; Leake *et al.*, 2001; Tedersoo *et al.*, 2003), or for multifunctional fungal species (Vasiliauskas *et al.*, 2007). Brown-rot fungi play an essential role in the functioning of coniferous (boreal) forest ecosystems (Gilbertson, 1980, 1981). For example, brown-rot residues, which may make up to 30% of the upper 30 cm of soil in coniferous forests (e.g. Harvey *et al.*, 1976), increase the soil water holding capacity and provide a favourable environment for seed germination and seedling establishment (Jurgenson *et al.*, 1977). Brown-rot residues provide a main site of ectomycorrhizal development (Harvey *et al.*, 1976), and also a site of nitrogen fixation in coniferous forest soils (Larsen *et al.*, 1979).

3. THE DECAY COMMUNITY COMPLEX

Traditionally the approach to the study of wood-decay fungi has been by survey of trees and logs for externally visible fruit bodies. This is, however, only the 'tip of the iceberg' and many organisms remain unaccounted for. The decomposer community consists of fungi, bacteria and arthropods etc. Mammals can play an important part in fragmenting the woody material, thereby making it more accessible to the microorganisms that possess the enzymatic capacity for breakdown of the lignocellulose. Bacteria and fungi can break down cellulose, but it is mainly white-rot basidiomycetes and xylariacious ascomycetes that break down lignin (Rayner and Boddy, 1988; Chapter 2). In undecayed wood there is a surplus of carbon and decomposers have to cope with a limited supply of nitrogen and other nutrients. During decomposition, coarse wood is initially a sink for nitrogen and becomes in later stages a source (Holub *et al.*, 2001; Laiho and Prescott, 2004). However, patterns are complex and in some tree species dead boles function as nutrient sources throughout their decomposition history (Brais *et al.*, 2006). At least four different aspects are important for the nitrogen economy of decomposer basidiomycetes and their capacity to maintain nutrients. (1) Fungi have adapted a nutrient reallocation approach. Thus, for example nitrogen is reused in the mycelium rather than recycled within the environment (Boddy and Watkinson, 1995; Lindahl and Finlay, 2006). (2) During early stage decomposition there is a net import of nutrients into logs. Logs can become a sink for nutrients (Zimmerman *et al.*, 1995; Romero *et al.*, 2005). Foraging and translocation by mycelial aggregates, rhizomorphs or mycelial cord systems, probably play important roles in the nutrient dynamics in fallen logs and stumps (Rayner and Boddy, 1988; Kirby *et al.*, 1990; Chapters 1 and 3). (3) Mycelia may capture small animals and special structures, such as nematode trapping rings, may help in this process (Tunlid *et al.*, 1992; Chapter 9). (4) Bacteria in wood may fix significant amounts of nitrogen; Brunner and Kimmins (2003) reported up to 1–2 kg/ha/year from Vancouver Island. A similar range of nitrogen-fixation was reported for the symbiosis between a cyanobacterium (*Nostoc* sp.) and the ubiquitous feather moss, *Pleurozium schreberi* that alone fixes between 1.5 and 2.0 kg/ha/year in mid to late-successional forests of northern Fennoscandia (DeLuca *et al.*, 2002).

4. GEOGRAPHIC DISTRIBUTION ZONES

The boreal zone is climatically represented by short summers and long winters. It is also climatically divided into oceanic parts with higher precipitation and continental areas with lower precipitation, for example western parts of Fennoscandia have an oceanic climate, while towards the east the climate becomes more continental (e.g. Ahti *et al.*, 1968; Tuhkanen, 1980). The boreal zone has also been divided latitudinally, that is hemiboreal, southern boreal, middle boreal and northern boreal vegetation zones (Ahti *et al.*, 1968). The hemiboreal zone is characterized by temperate broad-leaved tree species such as oak (*Quercus robur* in Europe or *Q. mongolica* in the Far East) and ash (*Fraxinus excelsior* in Europe or *F. mandschurica* in the Far East) mainly on better soils, and spruce (*Picea*) or pine on poorer soils (Esseen *et al.*, 1997; Dai and Penttilä, 2006). All the other zones are characterized by the dominance of coniferous tree species, together with deciduous tree species typically belonging to the genera *Betula, Populus* or *Salix*. Due to the dominance of coniferous tree species the boreal zone is also often called the northern coniferous forest region (e.g. Gilbertson, 1980). In the boreal zone, the host tree species have wide distributions, and although they are not circumpolar, the species in different continents are closely related both taxonomically and ecologically (e.g. Hallenberg, 1991). This, in addition to the rather continuous forest belt across the boreal zone and the ability of fungi to spread long distances via spores, contributes to the fact that a very large proportion of boreal wood-decay basidiomycetes seem to be widely distributed or circumpolar (e.g. Gilbertson, 1980; Hallenberg, 1991, 1998; Chapter 13).

Gilbertson (1980) analysed the distribution of wood-decay polypores and corticoid basidiomycetes shared between North America and other continents and found that the highest number was from the boreal zone (80–90% shared species) and that the fraction of shared species decreases in warmer regions. For example, less than 60% of the polypores found in temperate forests of East Asia are also present in North America (Nunez and Stokland, 2000) while only 30% of the polypores species found in New Zealand are shared with North

America (Gilbertson, 1980). We conducted an analysis of the distribution of polypores in the northern hemisphere using two datasets collected from different zones of boreal East Asia (Table 2). The number of shared polypore species between different continents and the number of circumpolar species were approximately 10% higher in the more northern Kamchatka and Magadan areas in Russian Far East than in the hemiboreal zone in north-eastern China (Table 2). The lower number of shared and circumpolar species in the hemiboreal zone in north-eastern China can largely be explained by the high number of endemic species (21 species; Dai and Penttilä, 2006) in comparison with the Kamchatka and Magadan areas (5 endemic species; data collected by Parmasto, Kotiranta and Mukhin, personal communication). A similar trend is also present for vascular plants, especially trees (Adams and Woodward, 1989; Latham and Ricklefs, 1993; Qian and Ricklefs, 2000). The lack of Pleistocene glaciations compared to other parts of Eurasia or North America, including Kamchatka and Magadan (Zhang *et al.*, 2006), favourable geography (continuous dispersal routes between East Siberia and South-East Asia for millions of years allowing coexistence of boreal, temperate and tropical species) and high numbers of host tree species are the likely explanations for the higher species diversity of polypores in north-east Asia (Dai, 1996; Nunez and Stokland, 2000; Dai and Penttilä, 2006).

A comparison of the climatic specialization of wood-inhabiting fungi is presented in Table 3. Although records are far from complete, very few polypore species are predominantly boreal and have a restricted distribution. In the temperate zone the number of endemic polypore species is definitely higher in East Asia than in Europe or North America, probably because of limited spread of glaciers in East Asia during the Ice Ages (Dai, 1996; Nunez and Stokland, 2000). Species richness of polypores and the number of endemic polypore species are higher in the hemiboreal zone of East Asia than, for example in the boreal Fennoscandia (Dai and Penttilä, 2006). However, we need more information on the species distribution from boreal East Asia and North America to verify whether the boreal zone in East Asia is more species rich and has more endemic species than Europe or North America.

In Europe, approximately 50% of the polypore species can be considered as strictly or predominantly temperate or Mediterranean, and up to 30% are equally common both in the boreal and temperate/Mediterranean zone (based on data in Ryvarden and Gilbertson 1993; 1994). Only 5% of European polypore species are totally restricted to the boreal zone and approximately 15% are predominantly boreal with scattered occurrences mainly in the mountains of central and southern Europe. Species that predominantly occur in the boreal region, and especially in the continental parts, have been called taiga (Russian expression for boreal conifer forests) species (Eriksson, 1958; Ryvarden, 1993a, 1993b). The majority of taiga species prefer coniferous hosts—which primarily occur in the boreal and mountainous regions—and a large number of them are also brown-rot fungi (Gilbertson, 1980, 1981). Brown-rot fungi are more efficient in obtaining energy from wood for growth and reproduction than white-rot fungi and, hence, better suited for cold and dry habitats with a short growing season (Gilbertson, 1980). Ryvarden (1993a, 1993b) speculated that the restricted distribution of 'taiga-species'

Table 2 Examples of boreal and boreal/temperate polypore species showing restricted distribution patterns

Distribution of species	Examples of species
Endemic to Europe	*Oligoporus septentrionalis*
	Oxyporus borealis
	Phellinus populicola
	Postia luteocaesia
	Sistotrema albolutea
Endemic to Asia	*Anomoporia vesiculosa* (temperate)
	Antrodiella gypsea (temperate)
	A. ussurii (temperate)
	Castanoporus castaneus (temperate)
	Ceriporiopsis cremea
	Daedaleopsis sinensis (temperate)
	Nigroporus ussuriensis (temperate)
	Oxyporus bucholtzii (temperate)
	Perenniporia maackiae (temperate)
	Polyporus choseniae
	Poriodontia subvinosa (temperate)
	Rigidoporus camschadalica
	Spongipellis sibirica
Endemic to North America	*Echinodontium tinctorium*
	Ganoderma oregonense
	Hapalopilus mutans (temperate)
	Inonotus glomeratus
	Junghuhnia zonata
	Oligoporus anguliporus
	Oxyporus nobilissimus
	Oxyporus similis
	Perenniporiu ellisiana
	Phellinus repandus
	Trichaptum subchartaceum
	Tyromyces subgiganteus (temperate)
Eurasian species	*Antrodia mellita*
	Antrodia pulvinascens
	Antrodiella citrinella
	Ceriporiopsis resinascens
	Diplomitoporus flavescens
	Junghuhnia pseudozilingiana
	Piloporia sajanensis
	Skeletocutis jelicii
	S. papyraceae
European–North American species	*Heterobasidion annosum sensu stricto*
	Lindtneria trachyspora (temperate)
	Oligoporus mappus

Table 2. (*Continued*)

Distribution of species	Examples of species
	Oligoporus ptychogaster
North American–North East Asian species	*Anomoporia flavissima*
	Auriporia aurea
	Cryptoporus volvatus (temperate)
	Ganoderma tsugae (temperate)
	Oxyporus cuneatus (temperate)
	P. variegate (temperate)
	Phellinus vaninii
	Trametes conchifer Pil. (temperate)

Source: Gilbertson and Ryvarden (1986, 1987), Ryvarden and Gilbertson (1993, 1994), Nunez and Ryvarden (2000, 2001), Dai (2000) and distributional information from some other relevant literature were used as sources in estimating the distributional pattern of the species. Nomenclature of the species is according to the above sources (excluding Dai, 2000) and Niemelä (2005). Predominantly temperate species with scattered occurrences in the boreal zone are indicated.

Table 3 The number of shared and circumpolar polypore species from two datasets collected from Russian Far East (Kamchatka peninsula, Parmasto, 1963; Magadan Area, Kotiranta and Mukhin, unpublished data) and north-eastern China (Fenglin Nature Reserve, Hemiboreal Region, Dai and Penttilä, 2006)

Area	Number of species	Percentage in North America	Percentage in Europe	Circumpolar sp. (%)
Russian Far East	118	87	94	86
North-eastern China	153	80	83	76

Note: A few species in the datasets (e.g. recently described new species in the genera *Antrodiella* and *Skeletocutis*) were omitted from the comparison, since their existence especially in North America could not be reliably identified.

is caused by an adaptation to a lower optimum growth temperature which makes them less competitive, and thus often non-existent, in areas with a warmer climate. According to Ryvarden (1993a) coniferous-living *Amylocystis lapponica* and *Antrodia albobrunnea* among polypores, and corticoid *Laurilia sulcata* and *Phlebia centrifuga* are good examples of the so-called taiga species. Other polypores showing a restricted distribution in coniferous forest regions with a more continental climate include *Antrodia crassa, Diplomitoporus crustulinus, Gloeophyllum protractum, Inonotopsis subiculosus, Piloporia sajanensis, Pycnoporellus alboluteus, Skeletocutis stellae* and *Trichaptum laricinum* on coniferous hosts, and *Daedaleopsis septentrionalis, Haploporus odorus, Phellinus populicola* and *Polyporus pseudobetulinus* on deciduous hosts (Gilbertson and Ryvarden, 1986, 1987; Ryvarden and Gilbertson, 1993, 1994).

5. APPROACHES FOR STUDYING THE COMMUNITY

Many studies of wood-decay fungi have been based on what is fruiting on the surface. The advantage of such observations is that a large number of woody units can be surveyed in a relatively short time. Further, fruiting reflects the reproductive output and thereby represents an important aspect of fitness. However, there are several reasons why studies solely based on fruiting structures will only give a partial view of the composition, activity and importance of the fungal community involved in wood decomposition. One reason is that this gives a heavy bias towards species with large conspicuous fruit bodies, in particular basidiomycetes, predominantly belonging to the polyporacae, agaricales and corticiacae. Organisms such as bacteria, arthropods and fungal species lacking large fruit bodies will remain undetected by this approach. Other drawbacks are associated with the timing of fruiting. Inevitably, there is a lag phase between establishment of a mycelium and formation of a fruit body. The time required to build enough resources within the mycelium for fruiting can be from a few weeks up to several years depending on species, resource quality and environmental conditions (Chapter 5). For example, a species that in spite of being established in early stages of decomposition only produces fruit bodies at late stages of the development, will erroneously appear to be a late successional. Other species might have a much shorter lag before fruiting and the interpretation might again be misleading. Furthermore, certain cues might be needed for fruiting, such as specific temperature regimes or light conditions, and surveys during periods outside the required conditions will fail to record them. For species with annual fruit bodies, fruiting might be irregular, so there is a need to revisit sites for comprehensive inventories (e.g. Lindner *et al.*, 2006). Berglund *et al.* (2005) stated that monitoring should be performed on a stand scale and focus on species with durable fruit bodies, for example polypores. This will provide data that can be used both to detect future changes in biodiversity in old-growth spruce forests and to evaluate conservation strategies.

Another complicating factor is the need for mating to occur before fruit bodies can be produced. This can be seen as an example of the Allee effect, resulting in negative population growth at low densities (Courchamp *et al.*, 1999). This may be misinterpreted as substrate specificity if detection is based solely on fruiting. However, the difficulty in fertilizing homokaryons is a challenge for rare basidiomycetes where the low density of airborne spores makes the probabilities for dispersal to new habitats low and the likelihood of two spores reaching the same resource unit even lower. This would be counteracted partly by homokaryotic mycelia defending areas of suitable wood habitat from invasion by other fungi and thereby acting as spore traps over extended periods (Adams *et al.*, 1984; James and Vilgalys, 2001; Edman *et al.*, 2004a, 2004b). However, the dynamics and the genetics of wood colonization has been poorly studied and whether an extended homokaryotic stage is common enough to represent any significant contribution to the basidiomycete life-cycle remains to be proven.

A more direct method for studying decay fungi is to isolate mycelia into pure culture (Table 4). Isolation methods have the potential to be more inclusive than

Table 4 Comparison of different approaches to identification of wood-inhabiting fungi

Method	Strengths	Weaknesses	Opportunities	Threats
Fruiting body inventories	Well-established method; a large number of resource units can be covered in relatively short time; fruiting is a good measure of fitness; low cost	Irregularity in fruiting; need to revisit sites at different seasons etc.; inconspicuous fruiting or non-fruiting mycelia can go undetected; only the surface is sampled; limited possibility to study functional aspects; requires skills in species identification	Large scale comparative studies and surveys are possible; species with reliable fruiting patterns (e.g. perennial polypores) can be used as indicators, also by amateurs	Overemphasizing species groups with conspicuous fruit bodies; misinterpretation of temporal relations
Mycelial identification	Focus on the active mycelia in the resource; relatively low cost	Time consuming; requires high skills in mycelial identification (but can be aided by molecular identification)	Physiological and genetic experiments can be made	Overemphasis on heavily sporulating and fast growing microspecies; missing non-culturable species
Molecular identification	Direct identification of species present in the substrate; allows for precise identification	Time consuming; high cost; requires a molecular lab	Functional studies can be made directly in the resource; comprehensive sampling of the community	Species can be more or less readily amplified (primer bias)

fruit body monitoring although normally they require a higher input of labour. Here the bias is towards species that have a rapid growth and are adapted to growth on artificial media which might not be the same conditions as those for growing in wood (Menkis *et al.*, 2004; Lindahl *et al.*, 2007). For example, heavily sporulating microfungi are typically over represented in isolation work. One important advantage with the isolation approach is that experimental work, for example physiological potential or combative strength (Holmer and Stenlid, 1997; Holmer *et al.*, 1997), can be carried out.

Another approach is to use molecular methods for detection (Table 4). Samples can be taken directly from wood for DNA extraction and subsequent PCR amplification. Typically, ribosomal genes are targeted with the ribosomal internal transcribed spacer region (ITS) giving enough interspecific variation for species identification. In addition, the more conserved large and small subunits give phylogenetic information that are very useful for homing in to the right higher level taxonomic groups. Amplicons can be purified directly from an electrophoresis gel, or cloned into a bacterial library and thereafter sequenced. Identification is carried out by comparing the sequence of the target gene with databases of known sequences. Public databases such as GenBank have a large number of such sequences but wood-inhabiting fungi are not always comprehensively represented, and therefore, complementary information might be needed from local databases. For ectomycorrhizal fungi, a common database UNITE, is published with quality checked information (Koljalg *et al.*, 2005). Amplification from environmental samples will yield a mix of several amplicons. The challenge then is to be able to detect most members of a fungal community from the mixture of several DNA types. One approach is to study RFLP (Johannesson and Stenlid, 1999) or T-RFLPs (Allmér *et al.*, 2006) involving amplification using marked primers followed by restriction enzyme treatment allowing for separation of individual DNA types. Matching the T-RFLPs with predicted databases after treatment with two or three enzymes allow for species-specific detection. To avoid problems with unpredicted discrepancies from theoretical T-RFLP patterns, the reference library for T-RFLPs should be made from actual restriction cuts of the specimen from clone libraries or isolates from the sampling site. An alternative way to separate the primary amplicons is to use a denaturating or temperature gradient gel electrophoresis, DGGE (Vainio and Hantula, 2000) and TGGE (Kulhankova *et al.*, 2006), respectively.

Molecular detection has yielded interesting results: (1) Ascomycota are more common in wood than found from fruit body inventories and from direct isolations (Allmér *et al.*, 2006). (2) Species with prolific conidial stages are less common than what would be deduced from isolation work. (3) A wide range of species are detected that previously were not known from the habitat. For example, species that are normally known to be associated with wood of broad-leaved trees have been found in the decay of conifers (Lygis *et al.*, 2004; Vasiliauskas *et al.*, 2004). (4) New functional roles have been implicated for decay fungi. Interestingly, by using DNA-based identification, on several occasions well-known wood-decay organisms (e.g. *Phlebia centrifuga, Phlebiopsis gigantea*) have been found in mycorrhizal roots and their ability to develop mycorrhizal root tips under axenic conditions has been confirmed (Vasiliauskas *et al.*, 2007).

6. COMMUNITY DEVELOPMENT IN WOOD RESOURCE UNITS

The development of the fungal community can be described at different scales. At the resource unit level, initially, dispersal and ability to utilize fresh wood are important characters shaping the decay community. This has been called primary resource capture (Rayner and Boddy, 1988; Boddy, 2001). As the wood-inhabiting community is building up, the mycelia will start to make contact and microbial interactions (Chapter 7) become increasingly important.

When the initial colonization of the resource is completed and the wood is occupied by primary colonizers, new entries into the unit will be in competition with already established mycelia (Boddy, 2001). This is called secondary resource capture. Mycelia already established in wood have a competitive advantage, not least by having access to a large resource pool (Holmer and Stenlid, 1993; Lindahl *et al.*, 2001). However, new species do enter the closed community, and these are often better at decomposing recalcitrant material, for example lignin, or are good at parasitizing mycelia. Renvall (1995) reported that white-rot fungi that can decompose lignin are more common at later stages of wood decomposition as compared to brown rotters with only marginal ligninolytic abilities.

Species specialized in combative mycelial interactions, leading to takeover of resources, have either a broad spectrum or a narrow range of species that are out-competed. Species with a general high ability to replace others also tend to produce hyphal aggregates, for example cords and rhizomorphs, that can grow through soil and interconnect between resource units (see Chapter 1). This increases the ability to find resources and also allows for import of energy and nutrients into the new resource, thereby increasing the inoculum potential and the chances of out-competing other mycelia. From an evolutionary strategy point of view, the costs of producing extensive mycelial aggregates for foraging can only pay off if the potential gains are high, for example in terms of being able to out-compete other organisms and thereby capture valuable resources. In other words, species with a strong ability to forage might also be expected to be relatively strong in intermycelial encounters with a broad range of target species. Competitive hierarchies can be found among species indicating this (Holmer and Stenlid, 1997). In the everlasting struggle for resources, fungi have various means of defending the resources already captured, including forming outer mycelial boundaries such as thick melanized mycelial rinds called pseudo-sclerotial plates.

Secondary resource capture sometimes also involves species-specific interactions. This typically happens when a species is specifically able to target a common species that frequently appears in the early stages of resource capture, and to invade that species' mycelial domain. This strategy is prompted by targeting species that are successful in primary resource capture and relatively common in the ecosystem. Based on spatiotemporal positioning of fruiting bodies, Niemelä *et al.* (1995) reported on a large number of species pairs where one species was presumably specifically parasitizing the other, and suggested specific interactions at late stages of decay. Holmer *et al.* (1997) confirmed this view and showed that secondary resource capture by mycelia in wood under laboratory conditions was more likely to take place by species close to fruit bodies of primary decay fungi than *vice versa*. Moreover, invasion was more likely to be via the specific primary decayer than by primary decayers from combinations not observed in the field.

With time, the wood resource inevitably disappears. During the latest stages of decomposition, energy will be in short supply and nutrients will be relatively tightly bonded to macromolecules.

The number of basidiomycetes fruiting on wood has been reported to be relatively low at initial decay stages, to be highest at intermediate stages, then decrease again at later stages of decomposition, probably due to a lack of energy resources when the cellulose in the wood is depleted (Table 5; Figure 1). Apparently, the living tree is inimical to colonization by the majority of decay

Table 5 Number of species found in logs of different stages of decomposition in boreal forests

Stage of decay	Early	Intermediate	Late	References
Norway spruce	67	77	60	Lindblad (1998)
Norway spruce	27	68	30	Renvall (1995)
Norway spruce	15	25	10	Bader *et al.* (1995)[a]
Scots pine	11	68	20	Renvall (1995)

[a]Recalculated data.

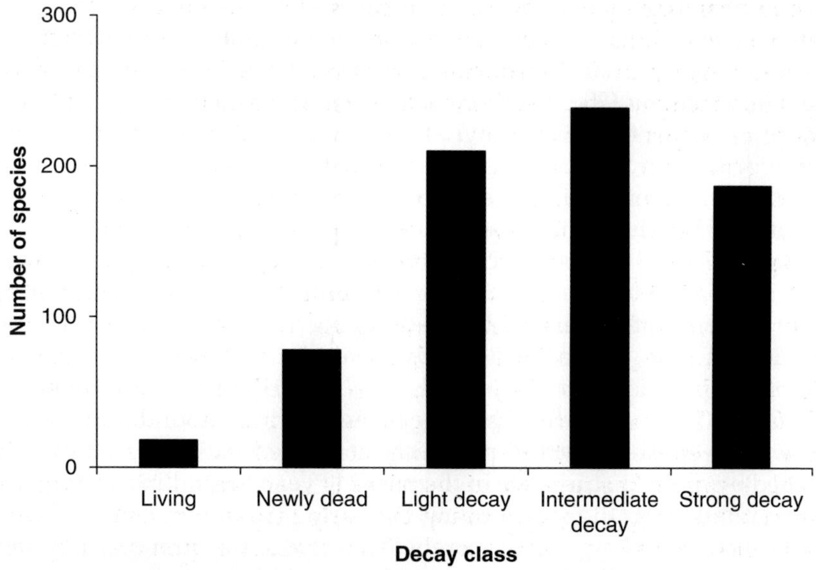

Figure 1 The number of species of wood-decaying basidiomycetes on *Picea abies* occurring at different states of decay. Data are from the Swedish Species Information Centre and are based on 264 species that have *Picea abies* as their principal resource in Sweden.

fungi. The pathogenic decay fungi that are present may contribute to gap formation in the forest (Johannesson and Stenlid, 1999). Once the tree has died it represents an enrichment disturbance in the forest ecosystem and colonization can increase (Figure 1).

A full cycle of decomposition has been reported to take 60–70 years for spruce logs in the southern boreal region (Liu and Hytteborn, 1991) and approximately 150 years in northern areas (Hofgaard, 1993). Mäkinen *et al.* (2006) reported that for spruce and pine stems, it takes 60–80 years and for birch stems 25–40 years for total decomposition in southern Finland. The retention time increases depending on the tree remains standing after death (remaining as snag); snags (white and red pine) remaining for 90 years, fallen trees for 55–60 years (Vanderwel *et al.*, 2006). Storaunet and Rolstad (2002) agreed that decay-rate estimates need to be computed from two regressions, one for the time standing and another for the time fallen. On average decomposition takes 100 years in old growth and 64 years in managed forests in Norway, and an additional 22 years (range 0–91 years) on average to fall. Our own observations indicate that a snag that has been colonized by brown rotters will break faster than those colonized by white-rot fungi, indicating an important role for brown-rotting species, for example *Fomitopsis pinicola*, in shortening the retention time of coarse woody debris.

7. STAND AND LANDSCAPE LEVEL

Looking at the decay community at the forest stand level gives another perspective to wood decay. At this scale, individual logs contribute to a dynamic network of resources. The important question then is whether there is a continuum of resources large enough to support the community or if fluctuations in resource availability will cause local extinctions and recolonization events. Berglund and Jonsson (2003) showed that there is a high degree of nestedness in the occurrence of wood-decay fungi at the stand level, or in other words, small habitats will contain a subset of the species pool of large habitats. Ecological theory tells that the extinction of species is more likely to occur in small habitats than in larger. The persistence of some species is definitely affected by gaps in the continuation of available resource at the stand level (Stokland and Kauserud, 2004). In addition, local dispersal sources (within 3 km) may strongly affect colonization patterns of red-listed wood-decaying fungi (e.g. *Fomitopsis rosea*) in spruce (Edman *et al.*, 2004b, 2004c), and within 1 ha areas, decay fungi seemed not to be dispersal limited (Rolstad *et al.*, 2004). Sverdrup-Thygeson and Lindenmayer (2003) reported that the presence and abundance of *Phellinus nigrolimitatus* were strongly and positively correlated with the area of forest uninterrupted by major disturbance in the past 240 years (equalling 140 years of 'old growth continuity') in the surrounding landscape (80 ha). In contrast, Groven *et al.* (2002) failed to find a positive correlation between the abundance of indicator fungi and continuity of dead wood (number of old logs present) at stand scale, but this could be due to the fact that they did not have good-continuity plots in their study. Taken together, the evidence is clear that anthropogenic influences have reduced the resource availability by 90% (Siitonen, 2001). The disturbance caused by forestry is likely to affect more severely species in later stages of succession (e.g. Stokland and Kauserud, 2004) and those with small populations. Fungal species specialized for large logs have suffered (e.g. Bader *et al.*, 1995; Kruys *et al.*, 1999; Penttilä *et al.*, 2004; Stokland and Kauserud, 2004) since, especially, such logs are removed from managed forests. In a comparative study of species richness in old-growth spruce and managed spruce stands, Penttilä *et al.* (2004) showed that old-growth stands on average contained 80% more polypore species than mature managed stands. Threatened species were confined to old-growth stands and to managed stands where the amount of dead wood exceeded 20 m/ha. Lindblad (1998), Jonsson and Jonsell (1999) and Sippola *et al.* (2001) also found that the species richness and number of red-listed polypore species are much higher in old-growth forests than in managed forests. In another study comparing forest landscapes east and west of the Finnish Russian border in Karelia, strikingly higher numbers of species were found on the eastern side with higher amounts of dead wood and higher continuity between high quality habitats (Siitonen *et al.*, 2001). In addition, Edman *et al.* (2004c) and Penttilä *et al.* (2006) have shown that both the deposition of spores and occurrence of fruit bodies in rare spruce-living polypore species are highly dependent on the forestry and fragmentation history at landscape level (see also Chapter 13).

8. DISTURBANCE REGIMES

The main factors disturbing natural boreal forest landscapes are fire, wind throws and disturbance associated with gaps, insect outbreaks, fungal diseases, periodic drought or excess water (Gromtsev, 2002; Selikhovkin, 2005). Fire, which is considered to be the most important disturbance factor in natural boreal forests, determines the structure and dynamics of boreal forests especially in drier pine-dominated forests and in more continental areas (e.g. Sannikov and Goldammer, 1996; Gromtsev, 2002). On the other hand, gap-phase dynamics and wind throws are of utmost importance in moister (e.g. spruce-dominated forests) forest types (Hofgaard, 1993; Kuuluvainen *et al.*, 1998) as well as in temperate, broad-leaved forests (Runkle, 1985). In addition, insect outbreaks and pathogens can be important forces creating gaps and modifying forest structure especially in natural forests dominated by gap-phase dynamics (Castello *et al.*, 1995; Perry and Amaranthus, 1997).

Natural boreal forests typically show a wide variation in the quality, size, severity and repeatability of disturbances in space and time (Engelmark, 1999; Bergeron *et al.*, 2002; Kuuluvainen, 2002). In contrast, in managed forests the disturbance (harvesting) areas and the harvest rotation are relatively constant (Kuuluvainen, 2002). Natural disturbances typically leave large amounts of biological legacies (both living organisms and organically derived structures like dead trees) from the pre-disturbance period which facilitate the recovery of the post-disturbance communities while biological legacies left by silvicultural disturbances (e.g. clear-cutting) are typically less diverse, less abundant and exhibit lower levels of spatial heterogeneity than those left by natural disturbances (Franklin *et al.*, 2000; Bengtsson *et al.*, 2003). Consequently, managed forests with low levels of biological legacies (i.e. dead wood) show much lower species diversity of wood-decaying fungi than natural forests and very seldom host threatened and demanding fungal species (e.g. Lindblad, 1998; Kruys *et al.*, 1999; Sippola *et al.*, 2001; Penttilä *et al.*, 2004; Junninen *et al.*, 2006). The continuity of dead wood is important for maintaining the species diversity of wood-decaying fungi, both in natural and managed forests (Stokland, 2001). For example, polypores from the old growth can remain for a long time on logs after final cutting of a stand and also on logging waste (Sippola and Renvall, 1999). A high number of species (and findings of threatened species) is often found in young, logged successional stages in pine-dominated forests which can be partly explained by the residual remains from the old, uncut forest (Junninen *et al.*, 2006).

Besides decreasing diversity and heterogeneity of wood-decaying fungi, silvicultural practises may also enhance the impact and spread of fungal pathogens like *Armillaria* spp., *Heterobasidion annosum* and *P. weirii* (Castello *et al.*, 1995). In natural forests both fire disturbance and gap-phase dynamics, creating spatially heterogenous and diverse stand structures, often control the spread and impact of these pathogenic basidiomycetes (e.g. Froelich *et al.*, 1978; Worrall and Harrington, 1988; Dickman and Cook, 1989). In managed forests, on the contrary, increased cutting and fire suppression in addition to homogenization of tree species composition have increased the spread and thus also the problems caused by these necrotrophic pathogens in many areas (e.g. Sherman and Warren, 1988; Byler *et al.*, 1990; Korhonen and Stenlid, 1998).

In natural boreal forests the prevailing disturbance regimes, and especially the frequency and severity of these disturbances, set the stage for the occurrence of fungal species and communities after disturbance. For example, wood-decaying fungi growing on pine and in pine-dominated forests, which have burned more often than spruce-dominated forests (Zackrisson, 1977; Gromtsev, 2002), are likely to be more adapted to fire disturbance than species preferring spruce as a host. Listing species with preferences for burned areas and charred wood (Table 6), gives evidence for this hypothesis, since most of the coniferous-inhabiting species prefer pine as their host. In addition, many of the fire-favoured species are very rare or red-listed in the Nordic countries (e.g. *A. primaeva, Dichomitus squalens, G. carbonarium* and *protractum, Physisporinus rivulosus, Crustoderma dryinum*), which most probably is a consequence of effective fire control and rarity of burned areas with large dead trees.

The immediate, short-term effect of fire on fungal communities is typically destructive (Pugh and Boddy, 1988; Watling, 1988). Intense fire destroys fungal mycelia and decreases the inoculum potential of many fungi by reducing the amount and quality of dead woody material and by creating extreme environmental conditions (Pugh and Boddy, 1988). The few existing studies on the effects of fire on communities of wood-decaying fungi (Penttilä and Kotiranta, 1996; Penttilä, 2004; Junninen *et al.*, 2007) show, that immediately (1 year) after the fire the number of fruiting species is much lower than in the pre-fire communities. However, although fire destroys or inhibits fruiting of resident fungal

Table 6 Boreal wood-decaying fungi that seem to favour burned areas and charred wood (Anthracophilous Species *sensu* Moser, 1949)

Species group	Species name	Main host
Polypores	*Antrodia primaeva*	Coniferous
	A. sinuosa	Coniferous
	A. xantha	Coniferous
	Ceriporiopsis subvermispora	Coniferous (deciduous)
	Dichomitus squalens	Coniferous
	Gloeophyllum carbonarium	Coniferous
	G. protractum	Coniferous
	G. sepiarium	Coniferous
	Physisporinus rivulosus	Coniferous (deciduous)
	Postia placenta	Coniferous
	Pycnoporus cinnabarinus	Deciduous
	Trametes hirsuta	Deciduous
Corticoid fungi	*Crustoderma dryinum*	Coniferous
	Phanerochaete raduloides	Deciduous
	Hyphoderma sp.	Coniferous
Ascomycota	*Daldinia loculata*	Deciduous

Source: The main sources of information were Eriksson (1958), Penttilä and Kotiranta (1996), Johannesson *et al.* (2001), Penttilä (2004) and Junninen *et al.* (unpublished), but also other published and unpublished information dealing with forest fires and habitat preferences of wood-decaying fungi. Nomenclature of polypores follows Ryvarden and Gilbertson (1993, 1994), of corticoid fungi Hansen and Knudsen (1997).

communities, it also provides a large input of new resources for decomposers in the ecosystem, and in doing so, acts as enrichment disturbance (Pugh and Boddy, 1988). Indeed, the amount of dead wood in natural boreal forests is usually at its highest in early successional stages immediately after large-scale disturbances, such as forest fires and storms (Spies *et al.*, 1988; Siitonen, 2001). A long-term study on the effects of fire on wood-decaying fungi in eastern Finland (Penttilä, 2004) clearly shows the enriching effect of fire disturbance on fungal communities in the long run. In this study, a pine-dominated old-growth forest stand with considerable amounts of dead wood was burned in 1989, and for 2 years after the fire the number of polypore species was lower than before the fire. However, 6 years after the fire the burned forest hosted an equal number of species as before the fire, and finally 13 years after the fire the number of species had increased strongly, including a very large number of threatened and near-threatened species in Finland. This increase in species number results partly from the enrichment given by the fire-killed trees but, especially in the case of red-listed species, is also due to the rich legacy of pre-fire dead trees both in the burned stand and in the surrounding old-growth forest.

Examination of fungal life strategies may also be helpful in understanding the species composition and structure of fungal communities after disturbance (Pugh and Boddy, 1988). In general, disturbances are expected to favour ruderal species at the cost of competitive species (Rayner and Boddy, 1988). Also stress-tolerant species (Cooke and Rayner, 1984) may flourish following some types of disturbance (e.g. burned). Evidence from the existing studies (Penttilä and Kotiranta, 1996; Penttilä, 2004; Junninen *et al.*, 2007) mainly support these general patterns, for example many non-competitive pioneer colonizers of fresh wood flourish in burned and in clear-cut areas, while competitive species usually utilizing logs in more advanced stages of decay decrease (fruiting is prohibited) after fire disturbance, at least in the short term. In addition, good examples of stress-tolerant species thriving in disturbed areas are the heat tolerant *Gloeophyllum* species (Loman, 1965) occurring in burned and clear-cut areas (Table 6). However, species life strategies are not always straightforward and there is evidence that the same species can exhibit different life strategies at different times during its life-cycle (Pugh and Boddy, 1988). For example, the post-fire ascomycete *Daldinia loculata* (see Table 6) seems to show such a behaviour. According to the tentative life-cycle presented in Johannesson *et al.* (2001), *D. loculata* lives outside fire areas as a latent mycelia established by sexual ascospores, which disperse from fruit bodies (stromata) emerging in scattered burned forest sites. Besides *D. loculata*, a corticoid species *Phanerochaete raduloides* (see Table 6) may show a similar pattern, since it often produces extensive fruit bodies on dead birch after fire, but is very seldom found outside fire areas (Penttilä and Kotiranta, 1996; Johannesson *et al.*, 2001).

ACKNOWLEDGEMENTS

We are grateful to the Swedish Research Council and the Swedish Research Council for Environment, Agricultural Sciences and Spatial Planning for providing funds for parts of the work. Heikki Kotiranta and Tuomo Niemelä are thanked for their help dealing with the distribution of wood-decaying fungi.

REFERENCES

Adams, J.M. and Woodward, F.I. (1989). Patterns in tree species richness as a test of the Glacial Extinction Hypothesis. *Nature*, **339**, 699–701.

Adams, T.J.H., Williams, E.N.D., Todd, N.K. and Rayner, A.D.M. (1984). A species-specific method of analyzing populations of basidiospores. *Transactions of the British Mycological Society*, **82**, 359–361.

Ahti, T., Hämet-Ahti, L. and Jalas, J. (1968). Vegetation zones and their sections in northwestern Europe. *Annales Botanici Fennici*, **5**, 169–211.

Allmér, J., Vasiliauskas, R., Ihrmark, K., Stenlid, J. and Dahlberg, A. (2006). Wood-inhabiting fungal communities in woody debris of Norway spruce (*Picea abies* (L.) Karst.), as reflected by sporocarps, mycelial isolations and T-RFLP-identification. *FEMS Microbiology Ecology*, **55**, 57–67.

Bader, P., Jansson, S. and Jonsson, B.G. (1995). Wood-inhabiting fungi and substratum decline in selectively logged boreal spruce forests. *Biological Conservation*, **72**, 355–362.

Baier, R., Ettl, R., Hahn, C. and Gottlein, A. (2006). Early development and nutrition of Norway spruce (*Picea abies* (L.) Karst.) seedlings on different seedbeds in the Bavarian limestone Alps—A bioassay. *Annals of Forest Science*, **63**, 339–348.

Bengtsson, J., Angelstam, P., Elmqvist, T., Emanuelsson, U., Folke, C., Ihse, M., Moberg, F. and Nyström, M. (2003). Reserves, resilience and dynamic landscapes. *Ambio*, **32**, 389–396.

Bergeron, Y., Leduc, A., Harvey, B.D. and Gauthier, S. (2002). Natural fire regime: A guide for sustainable management of the Canadian boreal forest. *Silva Fennica*, **36**(1), 81–95.

Berglund, H., Edman, M. and Ericson, L. (2005). Temporal variation of wood-fungi diversity in boreal old-growth forests: Implications for monitoring. *Ecological Applications*, **15**, 970–982.

Berglund, H. and Jonsson, B.G. (2003). Nested plant and fungal communities: The importance of area and habitat quality in maximizing species capture in boreal old-growth forests. *Biological Conservation*, **112**, 319–328.

Boddy, L. (2001). Fungal community ecology and wood decomposition processes: From standing tree to complete decay of coarse woody debris. In: *Ecology of Woody Debris in Boreal Forests* (B.G. Jonsson and N. Kruys, eds.). *Ecological Bulletins* **49**, 43–56.

Boddy, L. and Watkinson, S.C. (1995). Wood decomposition, higher fungi, and their role in nutrient redistribution. *Canadian Journal of Botany*, **73**(Suppl. 1 E–H), S1377–S1383.

Bonan, G.B. and Vancleve, K. (1992). Soil-temperature, nitrogen mineralization, and carbon source sink relationships in boreal forests. *Canadian Journal of Forest Research*, **22**, 629–639.

Brais, S., Pare, D. and Lierman, C. (2006). Tree bole mineralization rates of four species of the Canadian eastern boreal forest: Implications for nutrient dynamics following stand-replacing disturbances. *Canadian Journal of Forest Research*, **36**, 2331–2340.

Brunner, A. and Kimmins, J.P. (2003). Nitrogen fixation in coarse woody debris of *Thuja plicata* and *Tsuga heterophylla* forests on northern Vancouver Island. *Canadian Journal of Forest Research*, **33**, 1670–1682.

Byler, J.W., Marsden, M.A. and Hagle, S.K. (1990). The probability of root disease on the Lolo National Forest, Montana. *Canadian Journal of Forest Research*, **20**, 987–994.

Castello, J., Leopold, D. and Smallidge, P. (1995). Pathogens, patterns, and processes in forest ecosystems. *Bioscience*, **45**(1), 16–24.

Cooke, R.C. and Rayner, A.D.M. (1984). *Ecology of Saprotrophic Fungi*. Longman, London.

Courchamp, F., Clutton-Brock, T. and Grenfell, B. (1999). Inverse density dependence and the Allee effect. *Trends in Ecology and Evolution*, **14**, 405–410.

Dahlberg, A. and Stokland, J.N. (2004). Substrate requirements of wood-inhabiting species. (In Swedish with English summary). Reports 7: 2004 Skogsstyrelsen, Jönkping, Sweden.

Dai, Y.C. (1996). *Polypore flora and diversity in Ghangbai Mts., northeastern China*. Academic dissertation, Publications in Botany from the University of Helsinki, **25**, 1–23.

Dai, Y.C. (2000). A checklist of polypores from Northeast China. *Karstenia*, **40**, 23–29.

Dai, Y.C. and Penttilä, R. (2006). Polypore diversity of Fenglin Nature Reserve, northeastern China. *Annales Botanici Fennici*, **43**, 81–96.

Dickman, A. and Cook, S. (1989). Fire and fungus in a mountain hemlock forest. *Canadian Journal of Botany*, **67**, 2005–2016.

DeLuca, T.H., Zackrisson, O., Nilsson, M.C. and Sellstedt, A. (2002). Quantifying nitrogen-fixation in feather moss carpets of boreal forests. *Nature*, **419**, 917–920.

Edman, M., Gustafsson, M., Stenlid, J. and Ericson, L. (2004a). Abundance and viability of fungal spores along a forestry gradient: Responses to habitat loss and isolation? *Oikos*, **104**, 35–42.

Edman, M., Gustafsson, M., Stenlid, J., Jonsson, B.G. and Ericson, L. (2004b). Spore deposition of wood-decaying fungi: Importance of landscape composition. *Ecography*, **27**, 103–111.

Edman, M., Kruys, N. and Jonsson, B.G. (2004c). Local dispersal sources strongly affect colonisation patterns of wood-decaying fungi on spruce logs. *Ecological Applications*, **14**, 893–901.

Edman, M., Moller, R. and Ericson, L. (2006). Effects of enhanced tree growth rate on the decay capacities of three saprotrophic wood-fungi. *Forest Ecology and Management*, **232**, 12–18.

Engelmark, O. (1999). Boreal forest disturbances. In: *Ecosystems of Disturbed Ground. Ecosystems of the World 16* (L.R. Walker, ed.), pp. 161–186. Elsevier, Amsterdam.

Eriksson, J. (1958). Studies in the heterobasidiomycetes and homobasidiomycetes—Aphyllophorales of Muddus National Park in North Sweden. *Symbolae Botanicae Upsaliensis*, **16**(1), 1–172.

Esseen, P.-A., Ehnström, B., Ericson, L. and Sjöberg, K. (1997). Boreal forests. *Ecological Bulletins*, **46**, 16–47.

Franklin, J.F., Lindenmayer, D.B., MacMahon, J.A., McKee, A., Magnusson, J., Perry, D.A., Waide, R. and Foster, D.R. (2000). Threads of continuity: Ecosystem disturbances, biological legacies and ecosystem recovery. *Conservation Biology in Practice*, **1**, 8–16.

Fridman, J. and Walheim, M. (2000). Amount, structure, and dynamics of dead wood on managed forestland in Sweden. *Forest Ecology and Management*, **131**, 23–36.
Froelich, R.C., Hodges, C.S. Jr. and Sackett, S.S. (1978). Prescribed burning reduces severity of annosus root rot in the south. *Forest Science*, **24**(1), 93–100.
Gärdenfors, U., Hall, R. Hallingbäck, T., Hansson, H.G. and Hedström, L. (2003). *Animals, Fungi and Plants in Sweden.* (In Swedish with English summary). Swedish Species Information Center, Uppsala, Sweden ISBN 92-88506-27-4, pp. 1–89.
Gilbertson, R. (1980). Wood-rotting fungi of North America. *Mycologia*, **72**, 1–50.
Gilbertson, R.L. (1981). North American wood-rotting fungi that cause brown rots. *Mycotaxon*, **XII**, 372–416.
Gilbertson, R.L. and Ryvarden, L. (1986). *North American Polypores 1*. Fungiflora, Oslo.
Gilbertson, R.L. and Ryvarden, L. (1987). *North American Polypores 2*. Fungiflora, Oslo.
Goodale, C.L., Apps, M.J., Birdsey, R.A., Field, C.B., Heath, L.S., Houghton, R.A., Jenkins, J.C., Kohlmaier, G.H., Kurz, W., Liu, S.R., Nabuurs, G.J., Nilsson, S. and Shvidenko, A.Z. (2002). Forest carbon sinks in the Northern Hemisphere. *Ecological Applications*, **12**, 891–899.
Gower, S.T., Krankina, O., Olson, R.J., Apps, M., Linder, S. and Wang, C. (2001). Net primary production and carbon allocation patterns of boreal forest ecosystems. *Ecological Applications*, **11**, 1395–1411.
Gray, A. and Spies, T. (1997). Microsite controls on tree seedling establishment in conifer forest canopy gaps. *Ecology*, **78**(8), 2458–2473.
Gromtsev, A. (2002). Natural disturbance dynamics in the boreal forests of European Russia: A review. *Silva Fennica*, **36**, 41–55.
Groven, R., Rolstad, J., Storaunet, K.O. and Rolstad, E. (2002). Using forest stand reconstructions to assess the role of structural continuity for late-successional species. *Forest Ecology and Management*, **164**, 39–55.
Hallenberg, N. (1991). Speciation and distribution in Corticiaceae (Basidiomycetes). *Plant Systematics and Evolution*, **177**, 93–110.
Hallenberg, N. (1998). Evolutionary processes on species level in wood-inhabiting Basidiomycetes. *Folia Cryptogamica Estonica*, **33**, 35–40.
Hansen, L. and Knudsen, H. (eds.). (1997). *Nordic Macromycetes Vol. 3, Heterobasidioid, Aphyllophporoid and Gastromycetoid Basidiomycetes*. Nordsvamp, Copenhagen.
Harmon, M.E., Ferrell, W.K. and Franklin, J.F. (1990). Effects on carbon storage of conversion of old-growth forests to young forests. *Science*, **247**, 699–702.
Harper, K.A., Bergeron, Y., Drapeau, P., Gauthier, S. and De Grandpre, L. (2006). Changes in spatial pattern of trees and snags during structural development in *Picea mariana* boreal forests. *Journal of Vegetation Science*, **17**, 625–636.
Harvey, A., Larsen, M. and Jurgenson, M. (1976). Distribution of ectomycorrhizae in a mature Douglas-fir/larch forest soil in western Montana. *Forest Science*, **22**, 393–398.
Hofgaard, A. (1993). Structure and regeneration patterns in a virgin *Picea-abies* forest in Northern Sweden. *Journal of Vegetation Science*, **4**, 601–608.
Heijari, J., Nerg, A.M., Kaakinen, S., Vapaavuori, E., Raitio, H., Levula, T., Viitanen, H., Holopainen, J.K. and Kainulainen, P. (2005). Resistance of Scots pine wood to Brown-rot fungi after long-term forest fertilization. *Trees: Structure and Function*, **19**, 728–734.
Holmer, L., Renvall, P. and Stenlid, J. (1997). Selective replacement between species of wood-rotting basidiomycetes, a laboratory study. *Mycological Research*, **101**, 714–720.
Holmer, L. and Stenlid, J. (1993). The importance of inoculum size for the competitive ability of wood decomposing fungi. *FEMS Microbiology Ecology*, **12**, 169–176.
Holmer, L. and Stenlid, J. (1997). Competitive hierarchies of wood decomposing basidiomycetes in artificial systems based on variable inoculum sizes. *Oikos*, **79**, 77–84.
Holub, S.M., Spears, J.D.H. and Lajtha, K. (2001). A reanalysis of nutrient dynamics in coniferous coarse woody debris. *Canadian Journal of Forest Research*, **31**, 1894–1902.
Hörnberg, G., Ohlson, M. and Zackrisson, O. (1997). Influence of bryophytes and microrelief conditions on *Picea abies* seed regeneration patterns in boreal old-growth swamp forests. *Canadian Journal of Forest Research*, **27**, 1015–1023.

Hyvönen, R., Ågren, G.I. and Dalias, P. (2005). Analysing temperature response of decomposition of organic matter. *Global Change Biology*, **11**, 770–778.

James, T.Y. and Vilgalys, R. (2001). Abundance and diversity of *Schizophyllum commune* spore clouds in the Caribbean detected by selective sampling. *Molecular Ecology*, **10**, 471–479.

Johannesson, H. and Stenlid, J. (1999). Molecular identification of wood-inhabiting fungi from a primeval forest in Sweden. *Forest Ecology and Management*, **115**, 203–211.

Johannesson, H., Vasiliauskas, R., Dahlberg, A., Penttilä, R. and Stenlid, J. (2001). Genetic differentiation in Eurasian populations of the postfire ascomycete *Daldinia loculata*. *Molecular Ecology*, **10**, 1665–1677.

Jonsson, B.G. and Jonsell, M. (1999). Exploring potential biodiversity indicators in boreal forests. *Biodiversity and Conservation*, **8**, 1417–1433.

Jonsson, B.G., Kruys, N. and Ranius, T. (2005). Ecology of species living on dead wood—Lessons for dead wood management. *Silva Fennica*, **39**, 289–309.

Junninen, K., Kouki, J. and Renvall, P. (2007). Restoration of natural legacies of fire in European boreal forests: Experimental approach to the effects on wood-decaying fungi. Submitted manuscript, (Unpublished).

Junninen, K., Similä, M., Kouki, J. and Kotiranta, H. (2006). Assemblages of wood-inhabiting fungi along the gradients of succession and naturalness in boreal pine-dominated forests in Fennoscandia. *Ecography*, **29**, 75–83.

Jurgenson, M., Larsen, M. and Harvey, A. (1977). Effects of timber harvesting on soil biology. In: *Forests for People. A Challenge in World Affairs* (eds.), pp. 244–250. Proceedings of Annual Meetings, Society of American Foresters.

Kirby, J.J.H., Stenlid, J. and Holdenrieder, O. (1990). Population structure and responses to disturbance of the basidiomycete *Resinicium bicolor*. *Oecologia*, **85**, 178–184.

Koljalg, U., Larsson, K.H., Abarenkov, K., Nilsson, R.H., Alexander, I.J., Eberhardt, U., Erland, S., Hoiland, K., Kjoller, R., Larsson, E., Pennanen, T., Sen, R., Taylor, A.F.S., Tedersoo, L., Vralstad, T. and Ursing, B.M. (2005). UNITE: A database providing web-based methods for the molecular identification of ectomycorrhizal fungi. *New Phytologist*, **166**, 1063–1068.

Korhonen, K. and Stenlid, J. (1998). Biology of *Heterobasidion annosum* In: *Heterobasidion annosum Biology, Ecology, Impact and Control* (S. Woodward, J. Stenlid, R. Karjalainen and A Hüttermann, eds.), pp. 43–70. CAB International, Wallingford, UK.

Krankina, O.N. and Harmon, M.E. (1995). Dynamics of the dead wood carbon pool in northwestern Russian boreal forests. *Water Air and Soil Pollution*, **82**, 227–238.

Kruys, N., Fries, C., Jonsson, B.G., Lämås, T. and Ståhl, G. (1999). Wood-inhabiting cryptogams on dead Norway spruce (*Picea abies*) trees in managed Swedish boreal forests. *Canadian Journal of Forest Research*, **29**, 178–186.

Kueppers, L.M., Southon, J., Baer, P. and Harte. J. (2004). Dead wood biomass and turnover time, measured by radiocarbon, along a subalpine elevation gradient. *Oecologia*, **141**, 641–651

Kuuluvainen, T. (2002). Natural variability of forests as a reference for restoring and managing biological diversity in boreal Fennoscandia. *Silva Fennica*, **36**, 97–125.

Kuuluvainen, T. and Kalmari, R. (2003). Regeneration microsites of *Picea abies* seedlings in a windthrow area of a boreal old-growth forest in southern Finland. *Annales Botanici Fennici*, **40**, 401–413.

Kuuluvainen, T., Syrjanen, K. and Kalliola, R. (1998). Structure of a pristine *Picea abies* forest in northeastern Europe. *Journal of Vegetation Science*, **9**, 563–574.

Kulhankova, A., Beguiristain, T., Moukoumi, J., Berthelin, J. and Ranger, J. (2006). Spatial and temporal diversity of wood decomposer communities in different forest stands, determined by ITS rDNA targeted TGGE. *Annals of Forest Science*, **63**, 547–556.

Laiho, R. and Prescott, C.E. (2004). Decay and nutrient dynamics of coarse woody debris in northern coniferous forests: A synthesis. *Canadian Journal of Forest Research*, **34**, 763–777.

Larsen, M., Jurgenson, M. and Harvey, A. (1979). N_2 fixation associated with wood decayed by some common fungi in western Montana. *Canadian Journal of Forest Research*, **8**, 341–345.

Latham, R.E. and Ricklefs, R.E. (1993). Continental comparisons of temperate-zone tree species diversity. In: *Species Diversity in Ecological Communities: Historical and Geographical Perspectives* (R.E. Ricklefs and D. Schluter, eds.), pp. 294–314. The University of Chicago Press, Chicago, IL.

Leake, J.R., Donnelly, D.P., Saunders, E.M., Boddy, L. and Read, D.J. (2001). Rates and quantities of carbon flux to ectomycorrhizal mycelium following C-14 pulse labelling of *Pinus sylvestris* seedlings: Effects of litter patches and interaction with a wood-decomposer fungus. *Tree Physiology*, **21**, 71–82.
Lindblad, I. (1998). Wood-inhabiting fungi on fallen logs of Norway spruce: Relations to forest management and substrate quality. *Nordic Journal of Botany*, **18**, 243–255.
Lindahl, B.D. and Finlay, R.D. (2006). Activities of chitinolytic enzymes during primary and secondary colonisation of wood by basidiomycetous fungi. *New Phytologist*, **169**, 389–397.
Lindahl, B.D., Ihrmark, K., Boberg, J., Trumbore, S.E., Högberg, P., Stenlid, J. and Finlay, R.D. (2007). Spatial separation of litter decomposition and mycorrhizal nitrogen uptake in a boreal forest. *New Phytologist*, **173**, 611–620.
Lindahl, B.D., Stenlid, J. and Finlay, R.D. (2001). Effects of resource availability on mycelial interactions and P-32 transfer between a saprotrophic and an ectomycorrhizal fungus in soil microcosms. *FEMS Microbiology Ecology*, **38**, 43–52.
Lindahl, B.D., Stenlid, J., Olsson, S. and Finlay, R.D. (1999). Interactions between ectomycorrhizal fungi and wood decomposing fungi. *New Phytologist*, **144**, 183–193.
Lindner, D.L., Burdsall, H.H. and Stanosz, G.R. (2006). Species diversity of polyporoid and corticioid fungi in northern hardwood forests with differing management histories. *Mycologia*, **98**, 195–217.
Liski, J., Lehtonen, A., Palosuo, T., Peltoniemi, M., Eggers, T., Muukkonen, P. and Makipaa, R. (2006). Carbon accumulation in Finland's forests 1922–2004: An estimate obtained by combination of forest inventory data with modelling of biomass, litter and soil. *Annals of Forest Science*, **63**, 687–697.
Liu, G.H. and Hytteborn, H. (1991). Gap structure, disturbance and regeneration in a primeval *Picea-abies* forest. *Journal of Vegetation Science*, **2**, 391–402.
Loman, A. (1965). The lethal effect of periodic high temperatures on certain lodgepole pine slash decaying basidiomycetes. *Canadian Journal of Botany*, **43**, 334–338.
Lygis, V., Vasiliauskas, R., Stenlid, J. and Vasiliauskas, A. (2004). Silvicultural and pathological evaluation of Scots pine plantations, established to avoid the infections by *Heterobasidion annosum*. *Forest Ecology and Management*, **201**, 275–285.
Mäkinen, H., Hynynen, J., Siitonen, J. and Sievänen, R. (2006). Predicting the decomposition of Scots pine, Norway spruce and birch stems in Finland. *Ecological Applications*, **16**, 1865–1879.
Menkis, A., Allmér, J., Vasiliauskas, R., Lygis, V., Stenlid, J. and Finlay, R. (2004). Ecology and molecular characterization of dark septate fungi from roots, live stems, coarse and fine woody debris. *Mycological Research*, **108**, 965–973.
Montes, F. and Canellas, I. (2006). Modelling coarse woody debris dynamics in even-aged Scots pine forests. *Forest Ecology and Management*, **221**, 220–232.
Moser, M. (1949). Untersuchungen über den Einfluss von Waldbränden auf die Pilzvegetation 1. Summary: Observations on the influence of forest-fire in the vegetation of fungi 1. *Sydowia*, **3**(1), 336–383.
Narukawa, Y., Iida, S., Tanouchi, H., Abe, S. and Yamamoto, S. (2003). State of fallen logs and the occurrence of conifer seedlings and saplings in boreal and subalpine old-growth forests in Japan. *Ecological Research*, **18**, 267–277.
Niemelä, T., Renvall, P. and Penttilä, R. (1995). Interactions of fungi at late stages of wood decomposition. *Annales Botanici Fennici*, **32**, 141–152.
Niemelä, T., Wallenius, T. and Kotiranta, H. (2002). The kelo tree, a vanishing substrate of specified wood-inhabiting fungi. *Polish Botanical Journal*, **47**, 91–101.
Niemelä, T. (2005). *Polypores, lignicolous fungi.* (In Finnish with English summary). Helsinki University Press, Helsinki, Finland.
Nunez, M. and Ryvarden, L. (2000). East Asian Polypores 1. Ganodermataceae and Hymenochaetaceae. *Synopsis Fungorum*, **13**, 1–168.
Nunez, M. and Ryvarden, L. (2001). East Asian Polypores 2. *Synopsis Fungorum*, **14**, 170–522.
Nunez, M. and Stokland, J. (2000). Biogeographical affinities of East Asian polypores. *Karstenia*, **40**, 123–128.
Osono, T. and Takeda, H. (2002). Comparison of litter decomposing ability among diverse fungi in a cool temperate deciduous forest in Japan. *Mycologia*, **94**, 421–427.
Parmasto, E. (1963). On the Fungus-Flora of Kamtchatka. *Issledovanie prirody Dalnego Vostoka*. 1, 221–289.

Penttilä, R. (2004). *The Impacts of Forestry on Polyporous Fungi in Boreal Forests*. Academic dissertation, University of Helsinki, Faculty of Biosciences, Department of Biological and Environmental Sciences.

Penttilä, R. and Kotiranta, H. (1996). Short-term effects of prescribed burning on wood-rotting fungi. *Silva Fennica*, **30**, 399–419.

Penttilä, R., Lindgren, M., Miettinen, O., Rita, H. and Hanski, I. (2006). Consequences of forest fragmentation for polyporous fungi at two spatial scales. *Oikos*, **114**, 225–240.

Penttilä, R., Siitonen, J. and Kuusinen, M. (2004). Polypore diversity in managed and old-growth boreal *Picea abies* forests in southern Finland. *Biological Conservation*, **117**, 271–283.

Perry, D. and Amaranthus, M. (1997). Disturbance, recovery, and stability. In: *Creating a Forestry for the 21st Century* (K.A. Kohm and J.F. Franklin, eds.), pp. 31–56. Island Press, Washington, DC.

Pugh, G.J.F. and Boddy, L. (1988). A view of disturbance and life strategies in fungi. In: *Fungi and Ecological Disturbance* (L. Boddy, R. Watling and A.J.E. Lyon, eds.). *Proceedings of the Royal Society of Edinburgh*, **94B**, 3–11.

Qian, H. and Ricklefs, R.E. (2000). Large-scale processes and the Asian bias in species diversity of temperate plants. *Nature*, **407**, 180–182.

Rayner, A. and Boddy, L. (1988). Fungal decomposition of wood: Its biology and ecology. Wiley, Bath.

Renvall, P. (1995). Community structure and dynamics of wood-rotting Basidiomycetes on decomposing conifer trunks in northern Finland. *Karstenia*, **35**, 1–51.

Rolstad, J., Saetersdal, M., Gjerde, I. and Storaunet, K.O. (2004). Wood-decaying fungi in boreal forest: Are species richness and abundances influenced by small-scale spatiotemporal distribution of dead wood? *Biological Conservation*, **117**, 539–555.

Romero, L.M., Smith, T.J. and Fourqurean, J.W. (2005). Changes in mass and nutrient content of wood during decomposition in a south Florida mangrove forest. *Journal of Ecology*, **93**, 618–631.

Rouvinen, S., Kuuluvainen, T. and Siitonen, J. (2002). Tree mortality in a *Pinus sylvestris* dominated boreal forest landscape in Vienansalo wilderness, eastern Fennoscandia. *Silva Fennica*, **36**, 127–145.

Runkle, J. (1985). Disturbance regimes in temperate forests. In: *The Ecology of Natural Disturbance and Patch Dynamics* (S.T.A. Pickett and P.S. White, eds.), pp. 17–33. Academic Press, Orlando, FL.

Ryvarden, L. (1993a). Distribution of Aphyllophoroid fungi in the taiga region of Fennoscandia. In: *Fungi of Europe: Investigation, Recording and Conservation* (D.N. Pegler, L. Boddy, B. Ing and P.M. Kirk, eds.), pp. 71–76. Royal Botanic Gardens, Kew.

Ryvarden, L. (1993b). *Haploporus odorus* and *Amylocystis lapponica*, two taiga species in Norway. (In Norwegian with an English abstract). *Blyttia*, **51**, 145–159.

Ryvarden, L. and Gilbertson, R.L. (1993). European polypores 1. *Synopsis Fungorum*, **6**, 1–387.

Ryvarden, L. and Gilbertson, R.L. (1994). European polypores 2. *Synopsis Fungorum*, **7**, 388–743.

Sannikov, S. and Goldammer, J. (1996). Fire ecology of pine forests of Northern Eurasia. In: *Fire in Ecosystems of Boreal Eurasia* (J.G. Goldammer and V.V. Furyaev, eds.), pp. 151–167. Kluwer Academic Publishers, Dordrecht.

Selikhovkin, A.V. (2005). Main disturbance factors in north-west Russian forests: Structure and databases. *Scandinavian Journal of Forest Research*, **20**, 27–32.

Sherman, R.J. and Warren, R.K. (1988). Factors in *Pinus ponderosa* and *Calocedrus decurrens* mortality in Yosemite Valley, USA. *Vegetatio*, **77**, 79–85.

Siitonen, J. (2001). Forest management, coarse woody debris and saproxylic organisms: Fennoscandian boreal forests as an example. *Ecological Bulletins*, **49**, 11–41.

Siitonen, J., Martikainen, P., Punttila, P. and Rauh, J. (2000). Coarse woody debris and stand characteristics in mature managed and old-growth boreal mesic forests in southern Finland. *Forest Ecology and Management*, **128**, 211–225.

Siitonen, J., Penttilä, R. and Kotiranta, H. (2001). Coarse woody debris, polyporous fungi and saproxylic insects in an old-growth spruce forest in Vodlozero National Park, Russian Karelia. *Ecological Bulletins*, **49**, 231–242.

Sippola, A.-L., Lehesvirta, T. and Renvall, P. (2001). Effects of selective logging on coarse woody debris and diversity of wood-decaying polypores in eastern Finland. *Ecological Bulletins*, **49**, 243–254.

Sippola, A.-L. and Renvall, P. (1999). Wood-decomposing fungi and seed-tree cutting: A 40-year perspective. *Forest Ecology and Management*, **115**, 183–201.

Spies, T.A., Franklin, J.F. and Thomas, T.B. (1988). Coarse woody debris in Douglas-fir forests of western Oregon and Washington. *Ecology*, **69**(6), 1689–1702.

Stokland, J.N. (2001). The coarse woody debris profile: An archive of recent forest history and an important biodiversity indicator. *Ecological Bulletins*, **49**, 71–83.

Stokland, J.N. and Kauserud, H. (2004). *Phellinus nigrolimitatus*—A wood-decomposing fungus highly influenced by forestry. *Forest Ecology and Management*, **187**, 333–343.

Storaunet, K. and Rolstad, J. (2002). Time since death and fall of Norway spruce logs in old-growth and selectively cut boreal forest. *Canadian Journal of Forest Research*, **32**, 1801–1812.

Sverdrup-Thygeson, A. and Lindenmayer, D.B. (2003). Ecological continuity and assumed indicator fungi in boreal forest: The importance of the landscape matrix. *Forest Ecology and Management*, **174**, 353–363.

Szewczyk, J. and Szwagrzyk, J. (1996). Tree regeneration on rotten wood and on soil in old-growth site. *Vegetatio*, **122**, 37–46.

Takahashi, M., Sakai, Y., Ootomo, R. and Shiozaki, M. (2000). Establishment of tree seedlings and water-soluble nutrients in coarse woody debris in an old-growth *Picea-abies* forest in Hokkaido, northern Japan. *Canadian Journal of Forest Research*, **30**, 1148–1155.

Tanesaka, E., Masuda, H. and Kinugawa, K. (1993). Wood degrading ability of basidiomycetes that are wood decomposers, litter decomposers, or mycorrhizal symbionts. *Mycologia*, **85**, 347–354.

Tedersoo, L., Koljalg, U., Hallenberg, N. and Larsson, K.H. (2003). Fine scale distribution of ectomycorrhizal fungi and roots across substrate layers including coarse woody debris in a mixed forest. *New Phytologist*, **159**, 153–165.

Toljander, Y.K., Lindahl, B.D., Holmer, L. and Högberg, N.O.S. (2006). Environmental fluctuations facilitate species co-existence and increase decomposition in communities of wood decay fungi. *Oecologia*, **148**, 625–631.

Tuhkanen, S. (1980). Climatic parameters and indices in plant geography. *Acta Phytogeographica Suecica*, **67**, 1–105.

Tunlid, A., Jansson, H.B. and Nordbringhertz, B. (1992). Fungal attachment to nematodes. *Mycological Research*, **96**, 401–412.

Urcelay, C. and Robledo, G. (2004). Community structure of polypores (Basidiomycota) in Andean alder wood in Argentina: Functional groups among wood-decay fungi? *Austral Ecology*, **29**, 471–476.

Vainio, E.J. and Hantula, J. (2000). Direct analysis of wood living fungi using denaturing gradient gel electrophoresis of amplified ribosomal DNA. *Mycological Research*, **104**, 927–936.

Vanderwel, M.C., Malcolm, J.R. and Smith, S.M. (2006). An integrated model for snag and downed woody debris decay class transitions. *Forest Ecology and Management*, **234**, 48–59.

Vasiliauskas, R., Lygis, V., Thor, M. and Stenlid, J. (2004). Impact of biological (Rotstop) and chemical (urea) treatments on fungal community structure in freshly cut *Picea abies* stumps. *Biological Control*, **31**, 405–413.

Vasiliauskas, R., Menkis, A., Finlay, R.D. and Stenlid, J. (2007). Wood-decay fungi in fine living roots of conifer seedlings. *New Phytologist*, 16: in press doi: 10.1111/j.1469-8137.2007.02014.

Watling, R. (1988). Larger fungi and some of Earth's major catastrophies. In: *Fungi and Ecological Disturbance* (L. Boddy, R. Watling and A.J.E. Lyon, eds.), pp. 49–59. *Proceedings of the Royal Society of Edinburgh*, **94B**.

Worrall, J.J. and Harrington, T.C. (1988). Etiology of canopy gaps in spruce-fir forests at Crawford Notch, New Hampshire. *Canadian Journal of Forest Research*, **18**, 1463–1469.

Zackrisson, O. (1977). Influence of forest fires on the North Swedish boreal forest. *Oikos*, **29**, 22–32.

Zhang, W., Cui, Z.J. and Li, Y.H. (2006). Review of the timing and extent of glaciers during the last glacial cycle in the bordering mountains of Tibet and in East Asia. *Quaternary International*, **154**, 32–43.

Zimmerman, J.K., Pulliam, W.M., Lodge, D.J., Quinonesofila, V., Fetcher, N., Guzmangrajales, S., Parrotta, J.A., Asbury, C.E., Walker, L.R. and Waide, R.B. (1995). Nitrogen immobilization by decomposing woody debris and the recovery of tropical wet forest from hurricane damage. *Oikos*, **72**, 314–322.

CHAPTER 14

Distribution Patterns of Wood-Decay Basidiomycetes at the Landscape to Global Scale

Jacob Heilmann-Clausen and **Lynne Boddy**

Contents

Abstract

Distribution patterns of fungi and other organisms are influenced by several factors over various scales in time and space. With their microscopic, often wind-dispersed spores, fungi are potentially able to disperse between continents, and many wood-inhabiting fungi with broad host ranges have been thought to have more or less global distribution patterns. With increased insight in fungal taxonomy outside Europe, and the use of molecular methods and mating experiments, it is becoming increasingly clear that many species, previously thought to have a wide distribution, actually circumscribe several biological taxa, each with a much more restricted distribution. Thus, continental drift, glaciations and other long-term geological and geographical factors have more impact on the current distribution patterns of fungi than believed earlier. At the continental scale, climate and host tree distribution patterns are important factors influencing the distribution of wood-inhabiting species, and climate change is likely to affect the distribution patterns of wood-inhabiting fungi considerably in the coming centuries. In the short time, man has had a strong impact on the abundance and distribution of dead wood habitat types, and this has clearly affected

British Mycological Society Symposia Series
Published by Elsevier Ltd.

current distribution and frequency of many species. Most importantly, species strictly associated with large decaying logs have decreased in many parts of Europe, while common species associated with coniferous wood have expanded in many regions, due to widespread planting of coniferous trees, where such species are naturally absent or infrequent.

1. INTRODUCTION

Several factors influence fungal distribution patterns at geographical scales from landscapes to continents, and these are reviewed here. Some relate to the conditions for fungal establishment and growth as influenced by climate and host tree distribution, and others relate to dispersal dynamics at various scales in time and space. Over geological time continental drift and glaciations have influenced distribution patterns of species and species groups, and have allowed allopatric speciation in populations that become geographically isolated. In the shorter time perspective man has caused very significant changes to the distribution, abundance and composition of dead-wood habitats at various geographical scales, and consequently to communities of wood-decay fungi.

2. CONTINENTS, GLACIATIONS, BIOREGIONS: WHAT DO WE KNOW ABOUT THEIR RELEVANCE FOR DISTRIBUTION PATTERNS OF WOOD-DECAY FUNGI?

While the biogeography of plants and animals has been a research topic for more than a century, biogeography patterns of fungi are only beginning to emerge (Knudsen and Ryman, 2000; Hibbet, 2001). Considering the potential of fungi for long distance dispersal it would be expected that major prehistoric processes, for example continental drift and major glaciations, would have had relatively low impact on the present distribution of fungi compared to terrestrial plant and animal taxa, which generally have much lower potential for long range dispersal. Accordingly, many fungal species have traditionally been believed to have very wide if not global distribution patterns (e.g. Hallenberg, 1995). Modern molecular studies increasingly indicate that this is not the case to the extent earlier believed.

Among wood-inhabiting fungi, species previously thought to have very wide distribution in fact comprise separate taxa, each with a much more restricted distribution (e.g. Nilsson *et al.*, 2003; Fischer and Binder, 2004; Zervakis *et al.*, 2004). For instance, *Hyphoderma setigurum* has been reported from all tree bearing continents but, in fact, comprises a species complex with at least nine putatively identified taxa (Nilsson *et al.*, 2003). Four of those investigated were restricted to northern Europe, one to Greenland, two to North America and two to East Asia. One lineage, on the other hand, was found in Greenland, North America, Caucasus and Eastern Asia. Even at the intraspecific level genetic as well as morphological and mating differentiation has been reported between regions (e.g. Petersen and Bermudes, 1992; James *et al.*, 2001), although there are also examples of very low genetic differentiation in wood-inhabiting fungi over thousands of kilometres (e.g. Johannesson *et al.*, 2001).

So far the true biographical distribution has been determined for very few wood-inhabiting species and it is currently difficult to evaluate how common various distribution patterns are. It is, however, evident that biogeographical patterns reported for plant and animal taxa are also relevant for fungi, such that current distribution patterns reflect major prehistoric events, for example continental drift, formation of land bridges etc. (e.g. Hibbet, 2001). In particular the distribution of major forest types seem to have a strong influence on species distribution patterns, with large and consistent differences in species composition occurring among forests dominated by, for example *Pinaceae, Betulaceae, Fagaceae* or *Polycarps* over large geographical scales (e.g. Christensen, 2006). Many species in the boreal zone thus have a circumglobal distribution, occurring in Europe, North America and possibly throughout Siberia (Hallenberg, 1995), probably reflecting the fact that the taiga zone has been coherent for many millennia. Many species occurring in the temperate deciduous forest zone in Europe are also reported in similar forest types in North America and East Asia but in most cases compatibility has not been tested between continents. The same applies to many species with a southern or Mediterranean distribution widely reported from subtropical and tropical regions, for example in East Africa and southern North America (Ryvarden and Gilbertson, 1992, 1993). On the other hand, there are also a large number of described species which are known from only very few records in geographically small areas, though they may have a wider distribution yet to be resolved.

3. CLIMATE AS A FACTOR AFFECTING DISTRIBUTION OF WOOD-DECAY FUNGI

The influence of macroclimatic conditions on current distribution patterns of wood-decay fungi has not been investigated in detail. Thus, it is not known whether gradients of temperature and precipitation are as important to wood-decay fungi as they are to vegetation (Mathiessen and Økland, 2007). In boreal Scandinavia there seems to be a higher ratio of polypores to corticoid Basidiomycota in continental sites compared to more Atlantic forests (Høiland and Bendiksen, 1996), which could reflect the polypore sporocarp type being better adapted than the corticoid type to dry conditions. Equally, differences in physiological ecology may be significant. In a comparison of fungal communities on decaying beech (*Fagus sylvatica*) wood in six European countries there was a close similarity between Sweden and Slovenia, even though these two study areas were the most remote geographically (Heilmann-Clausen, 2005a). The studied regions of Slovenia and Sweden were both characterized by high precipitation ($>$1,000 mm/year) and low winter temperatures (mean of coldest month less than -1 °C) which is probably the main explanation for the recorded similarities.

Many polypores causing heart-rot in living deciduous trees have a southern distribution in Europe, for example *Ganoderma pfeifferi, Inonotus cuticularis* and

Spongipellis delectans (Ryvarden and Gilbertson, 1992, 1993). This could partly reflect the distribution of host trees, but it is noteworthy that these species are often confined to places with a warm microclimate, for example south facing cliffs or grazing forests, in climatically marginal regions, even though host trees show a much wider amplitude in the same landscapes (Martikainen *et al.*, 2000; Heilmann-Clausen, 2005a). Similarly, Ódor *et al.* (2006), reported that certain threatened polypores and agarics causing heart-rot in beech were more common than expected in Hungarian forest reserves than in Slovenian reserves, probably because of the more continental climate (lower precipitation, larger difference between summer and winter temperatures) in the former, which was assumed to be beneficial for stress-tolerant infection strategies of the species concerned.

Effects of longer term climate changes are now becoming manifest. Most temperate wood-inhabiting fungi produce sporocarps primarily in late summer and autumn, which must reflect mycelial activity, at least to some extent (Chapter 5). Interestingly, over the last 30 years the fruiting season of many species in the U.K. has become extended by earlier first fruiting and later last fruiting, with spring fruiting in some cases, associated with global climate change (Gange *et al.*, 2007; Chapter 5). Similarly there have been reports of wood-decay fungi spreading northwards in Europe, for example *Hyphodontia flavipora* and *Trametes cervina*, perhaps as a response to a warmer climate (Keizer and Becker, 2003; Heilmann-Clausen *et al.*, 2006).

4. IMPACT OF MAN ON CURRENT DISTRIBUTION OF WOOD-DECAY FUNGI IN EUROPE

4.1 Changes in Habitat Availability

Potential habitats for wood-inhabiting fungi have decreased considerably almost everywhere in Europe during the last 2–5 millennia, due to forest clearance and cutting operations in extant forests. The extent of decline in forest cover differs markedly between countries and regions: current forest cover in the deciduous forest zone in Europe varies between 8% of the land area in Ireland to about 50% in Austria, Bosnia-Herzegovina and Greece (Parviainen, 1999). The amount of dead wood in long-established European beech forest reserves has been reported to average between 131 and 220 m^3/ha, in lowland and highland areas respectively, while managed stands in the same regions typically contain less than 10 m^3/ha (Christensen *et al.*, 2005). Thus, beech forest reserves typically contain at least 10 times the amount of woody resource present in similar managed stands in temperate Europe. If these values are generalised to be representative of other forest types in temperate Europe, habitats for wood-decay fungi can then be estimated to have suffered a decrease of 90–99.5% at the landscape scale. Within countries the extent of decrease is much more variable, with urbane and highly cultivated regions showing an even higher decline, while remote mountain forest regions generally have suffered a more modest change.

Apart from the overall decline in available resources, wood-inhabiting organisms are also faced with fragmentation of suitable habitat patches. This is evident both at the local scale, where management in most forests has increased distances between individual wood units (especially for large diameter wood) and at the regional scale, where remaining forests typically form more or less isolated patches in a matrix of farmland and urban areas. At the European scale the most extensive deciduous forest areas, including the majority of coarse woody debris (cwd) rich forest reserves, are restricted to mountainous parts of Eastern, Central and South Europe (Christensen *et al.*, 2005).

In addition to the decrease and fragmentation of cwd habitats there has been a shift in general forest composition. In most parts of Europe, forestry has favoured conifers to the detriment of deciduous tree species, and at the same time forest management has focussed almost exclusively on timber production. Forestry activities deliberately remove certain wood types; especially logs and larger branches, while twigs and cut stumps are mostly left for natural decay. Managed forests thereby present a smaller, but also different, selection of microhabitats for saproxylic organisms compared to unmanaged forest (Figure 1). It is noteworthy that many, now vanishing, traditional forest practices had a very different impact on dead-wood habitats compared to modern forestry. Most notably, forest grazing (wood pasture) allows or even favours the presence of old trees, due to their high mast or leaf production, and along with pollarding, results in high densities of old trees with heart-rot or dead parts, even compared to long unmanaged old grown forest reserves (Figure 2).

4.2 Local Scale Effects

Changes in local habitat composition is most detrimental to species associated with habitats facing a strong decline, while species associated with less declining or even increasing habitats are less negatively or even positively affected. This implies that species depending on large diameter dead wood (standing or lying) are likely to be very sparsely represented in managed forests, while species preferring small diameter dead wood (e.g. attached or fallen branches) and cut dead wood are likely to be frequent (Figure 1). Cut surfaces are interesting as they do not occur under natural conditions, though splintered wood surfaces in fallen trunks and branches are comparable to some degree. As described in Chapter 11, cut surfaces are relatively open resources, inviting establishment of ruderal primary or competitive secondary invading decay fungi to the detriment of latent decay fungi and heart-rot agents.

Several studies have compared the species composition among dead-wood habitats along the gradient from natural to managed forests, but mostly in boreal coniferous forests (e.g. Bader *et al.*, 1995; Sippola and Renvall, 1999; Pentillä *et al.*, 2004; Junninen *et al.*, 2006). Logging waste and cut stumps in managed forest attracted species that were infrequent in natural forests (Sippola and Renvall, 1999; Pentillä *et al.* 2004). These included species confined to early stages of wood decay, for example *Stereum sanguinolentum*, but also a number of species preferring cut stumps, for example *Porpomyces mucidum*, *Postia fragilis*, *P. stiptica*,

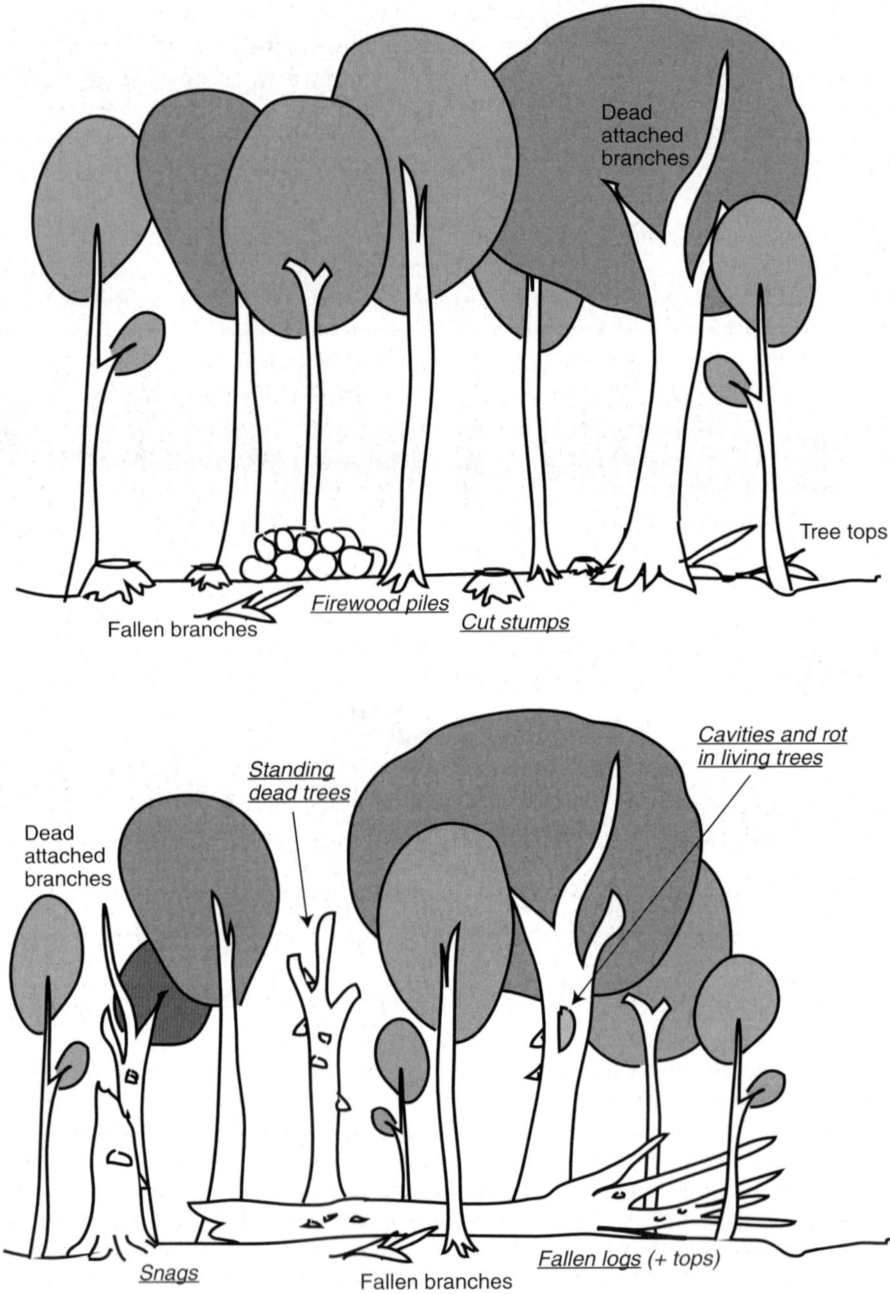

Figure 1 Location and type of dead wood in managed (top) and natural forests (lower). Underlined texts indicate wood types largely restricted to either situation, while non-underlined texts indicate wood types that occur in both managed and unmanaged stands. (*Source*: Modified from Heilmann-Clausen (2005b).) See also Figure 2.

Figure 2 Dead wood habitats differ markedly among sites depending on management history. (1) Natural forests with minimal signs of any management contain a variety of dead wood habitats, including fallen logs, branches and twigs as well as attached dead wood, mostly subject to shady and cool conditions. (a) Beech (*Fagus sylvatica*) forest at Fontainebleau, France. (b) Mixed deciduous forest in Suserup Skov, Denmark. (2) Forests managed for timber production have little dead wood, mainly occurring as fallen or cut branches and twigs, cut stumps and attached dead wood. (c) Managed beech forest near Silkeborg, Denmark (© Jacob Heilmann-Clausen). (3) Large, well spaced, often pollarded veteran trees, are often prominent along roadsides and rivers and in wood pasture. Such trees typically have attached dead components and contain heart rots supporting stress-tolerant fungi adapted to decay in living, often sun-exposed wood. Due to a pronounced competition for light and space similar veteran trees are mostly infrequent in natural forest. (d) Beech pollards from Bertizarana in Navarra, northern Spain. (e) Beech pollard from Epping, U.K.
Source: (a) © Morten Christensen; (b, c and d) © Jacob Heilmann-Clausen; (e) © Martyn Ainsworth. **(See Colour Section)**

Phlebiopsis gigantea and the pathogenic *Heterobasidion parvisporum*. The list of species more or less confined to natural forests was much longer including a number of proposed old-growth indicators, for example *Amylocystis lapponica*, *Fomitopsis rosea* and *Phellinus ferrugineofuscus* all confined to decaying logs.

In deciduous forests there is also evidence that cut stumps and tree tops left after cutting develop different communities of decay fungi than in wood made available more naturally. In a study of Danish beech forests cut stumps hosted sporocarps of the pyrenomycetes *Kretzschmaria deusta* and *Xylaria hypoxylon*, and the agaric *Kuehneromyces mutabilis* significantly more often than other types of dead wood (Heilmann-Clausen and Aude, 2006). Tree tops left after cutting similarly hosted a number of pyrenomycetes, for example *Eutypella quaternata* and *Hypoxylon* spp., but also some aphyllophorales, for example *Skeletocutis nivea* and *Hyphodontia paradoxa*, which were significantly less frequent on decaying logs. Logs situated in natural forests hosted sporocarps of a number of species which were practically absent from cut tree tops and stumps, such as polypores causing heart-rot, for example *Fomes fomentarius*, *Ganoderma lipsiensis* and *Ischnoderma resinosum*, but also cord-formers, for example *Lycoperdon pyriforme*, *L. perlatum* and *Ramaria stricta*, and late stage agarics, for example *Mycena haematopus*, *Galerina marginata* and *Psathyrella piluliformis*.

In conclusion, communities of wood-decay fungi in managed forests have different composition to those in more natural forests. Stress tolerant heart-rot fungi and other species involved with decay of fallen logs are much less frequent in managed forest, whereas ruderal/competitive species infecting through cut surfaces are abundant.

4.3 Fragmentation Effects

Within a metapopulation context the decline and fragmentation of dead-wood habitats is likely to cause a decline in species-diversity of wood-inhabiting fungi. Based on general species–area relations, Sittonen (2001) estimated that the observed decrease in available cwd in Fennoscandia is likely to cause a regional species loss of at least 22–32% in the long run. A comparable estimate for potential species loss was given by Berglund and Jonsson (2001) based on extrapolations from field observations of sporocarps in a natural forest wetland mosaic in northern Sweden. In contrast to Fennoscandia, where the loss of dead-wood habitats is fairly recent, the decrease in cwd has a longer history in the landscapes of Central and Western Europe. Even if the Scandinavian estimates of possible extinctions are overestimates, by failing to account for local survival of populations in protected forest reserves, it still seems likely that many wood-inhabiting species have suffered extinction in the deciduous forest zones of Europe, or at least have a much more restricted distribution than under natural conditions.

Fragmentation is likely to be most detrimental to species allocating limited energy to long distance dispersal, that is competitive and stress-tolerant species (see Chapter 11), though effective local dispersal may counteract this tendency in small areas with plentiful appropriate habitats. There are, unfortunately no available data to test whether allocation to spore dispersal differs between wood-decay fungi having different ecological strategies. It has been suggested that poor dispersal ability is a trait of species adapted to persistent and predictable habitat types with a random distribution pattern in the primeval landscape, while species depending on naturally patchy or temporally discontinuous habitats are assumed to have good dispersal abilities (e.g. Nilsson and Baranowski, 1997). Therefore it can be speculated that fungi depending on old living trees, presenting a very common and persistent habitat in primeval forests, are more likely to have poor ability for dispersal, compared with species living on decaying dead wood on the forest floor (a very common, but less persistent habitat), while species depending on highly patchy habitats, for example burned wood, are predicted to be superior dispersers. There is some evidence from boreal forests that such differences in dispersal potential occur. Thus, species confined to pine tend to be less sensitive to habitat fragmentation than species growing on spruce (Stokland, 2001; Pentillä *et al.*, 2006), which could reflect the fact that pine forests are more prone to natural forest fires than spruce-forests. Hence old-growth pine stands tend to occur more clumped in time and space and be more prone to sudden destruction, compared to the ecologically more stable spruce-dominated stands.

4.4 Combined Effects of Changes in Habitat Composition and Fragmentation

Several studies from boreal Fennoscandia (Siitonen *et al.*, 2001; Sverdrup-Thygeson and Lindenmayer, 2003; Stokland and Kauserud, 2004; Pentillä *et al.*, 2006) have documented differences in species diversity (based on sporocarp surveys) and composition of communities of wood-decay fungi depending on the fragmentation and management history at the landscape scale. In all cases certain rare or endangered species were found to be significantly less frequent in long fragmented landscapes compared to more recently, or less fragmented landscapes, even in the presence of suitable habitats. For instance, the red-listed polypore *Phellinus nigrolimitatus* was five times less frequent on suitable conifer logs in managed forests compared to similar logs in natural forests (Stokland and Kauserud, 2004).

In the European deciduous forest zone actual studies on fragmentation effects are few, but there is some evidence from beech forests, at various geographical scales, that changes in species composition similar to those in boreal forest has occurred (Holec, 2003; Òdor *et al.*, 2006). In a study of old-growth forest remnants in Denmark, the incidence of red-listed species (as evidenced by sporocarps) declined along a geographical gradient, related to differences in gross forest history and climate (Heilmann-Clausen and Christensen, 2005). However, in contrast, overall species density (i.e. the average number of species per fallen tree) did not differ or was even higher in less natural sites. Similarly, along a gradient from SE to NW Europe the relative fraction of threatened species similarly declined from largely coherent forest landscapes in Hungary and Slovenia to highly fragmented landscapes in Belgium and The Netherlands (Òdor *et al.*, 2006).

The selective decline of specialized fungi in relation to fragmentation of old-growth forests is likely to be strengthened synergistically by the relative increase in other dead-wood resources, and their associated fungi, in managed forests, which make up the matrix in most forested landscapes. Thus, the overall and asymmetric scarcity of large logs at the landscape scale, compared with, for example stumps and branches, implies that the relative fraction of spores from species depending on large logs will decrease relative to spores of generalists or stump or branch specialists (*cf.* Edman *et al.*, 2004). This implies that, even if a

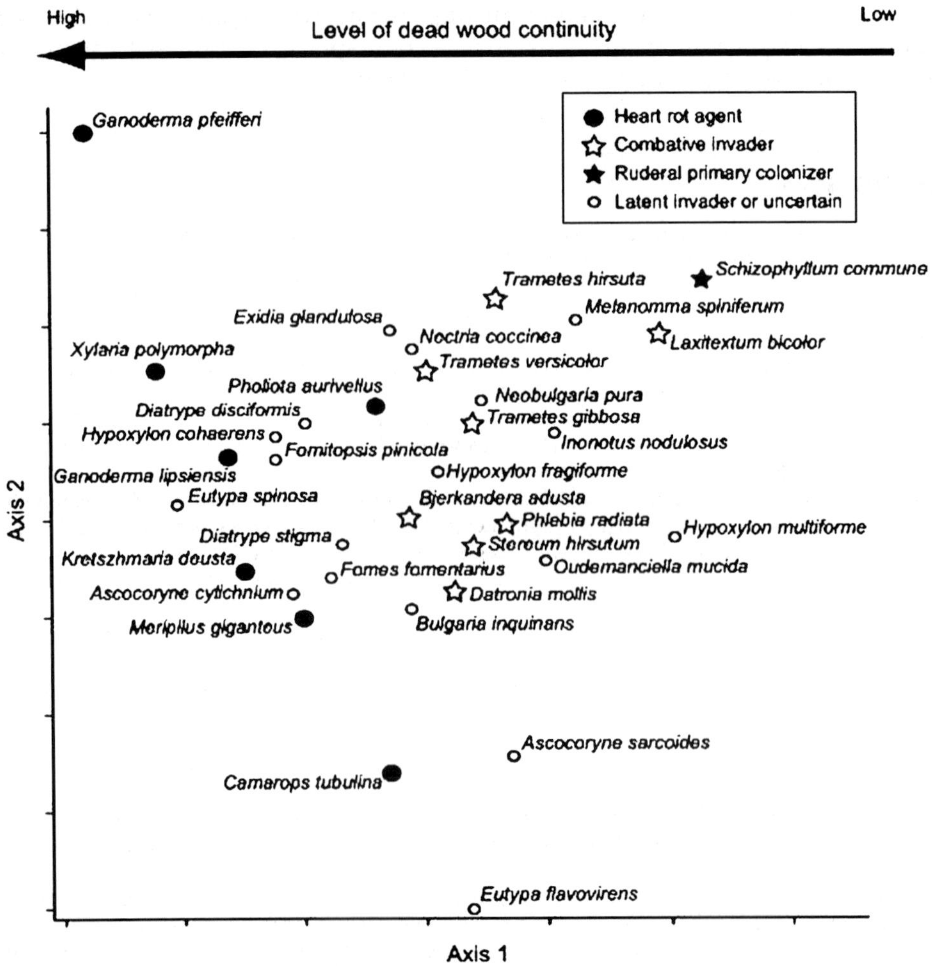

Figure 3 Optima (dead wood continuity) for heart-rot fungi, early combative invaders, ruderal primary colonizers and latent invaders (see Table 1 of Chapter 12) on beech logs along a dead-wood continuity gradient in Denmark. Species optima are based on a DCA-ordination based on frequency data for each species in 14 localities across Denmark. The preference for sites with long dead wood continuity is distinct for heart-rot agents, while many early combative invaders were most frequent in sites with lower continuity in the presence of dead wood. (*Source*: Modified from Heilmann-Clausen, 2004).

suitable resource for species with special requirements is created, the chance of establishment is decreased because non-selective species are likely to arrive first or with such a majority of propagules, that the selective species is out-competed (Gourbiere and Gourbiere, 2002). Along a gradient of old to recent forest reserves in Denmark the species composition of early decay agents did differ according to this hypothesis (Heilmann-Clausen, 2004). Thus, heart-rot agents were much more frequent on beech logs in old forest reserves, compared to recent reserves while early secondary invaders showed the opposite pattern (Figure 3). This has important implications for conservation of rare wood-inhabiting species, because restoration of dead-wood habitats in highly fragmented (in time and space) landscapes may have much less impact on rare specialists than enlargement of reserves in less fragmented areas (see Chapter 18).

REFERENCES

Bader, P., Jansson, S. and Jonsson, B.G. (1995). Wood-inhabiting fungi and substratum decline in selectively logged boreal spruce forests. *Biological Conservation*, **72**, 355–362.

Berglund, H. and Jonsson, B.G. (2001). Predictability of plant and fungal species richness of old-growth boreal forest islands. *Journal of Vegetation Science*, **12**, 857–866.

Christensen, M. (2006). Diversity of ecto-mycorrhiza fungi in Nepal: Relation to forest types and management. In: *8th International Mycological Congress, Cairns (Australia)*, August 21–25 (M. Wieland, and C. Pearce, eds.), pp. 41–47.

Christensen, M., Hahn, K., Mountford, E., Ódor, P., Standovár, T., Rozenbergar, D., Diaci, J., Wijdeven, S., Meyer, P., Winter, S. and Vrska, T. (2005). Dead wood in European beech forest reserves. *Forest Ecology and Management*, **210**, 267–282.

Edman, M., Kruys, N. and Jonsson, B.G. (2004). Local dispersal sources strongly affect colonisation patterns of wood-decaying fungi on experimental logs. *Ecological Applications*, **14**, 893–901.

Fischer, M. and Binder, M. (2004). Species recognition, geographic distribution and host-pathogen relationships: A case study in a group of lignicolous basidiomycetes, *Phellinus s.l. Mycologia*, **96**, 799–811.

Gange, A.C., Gange, E.G., Sparks, T.H. and Boddy, L. (2007). Rapid and recent changes in fungal fruiting patterns. *Science*, **316**, 71.

Gourbiere, S. and Gourbiere, F. (2002). Competition between unit-restricted fungi: A metapopulation model. *Journal of Theoretical Biology*, **207**, 351–368.

Hallenberg, N. (1995). Dispersal abilities and distributional patterns in Aphyllophorales, with emphasis on corticioid fungi. *Acta Universitatis Upsaliensis, Symbolae Botanicae Upsaliensis*, **30**(3), 95–100.

Heilmann-Clausen, J. (2004). Wood-inhabiting fungi in Danish deciduous forests: Diversity, habitat preferences and conservation, PhD Thesis, Danish Research Academy/Danish Forest and Landscape Research Institute.

Heilmann-Clausen, J. (2005a). Diversity of saproxylic fungi on decaying beech wood in protected forests in the county of Halland. *Information från Länsstyrelsen Halland, Meddelande*, **7**.

Heilmann-Clausen, J. (2005b). Naturnær skovdrift og biodiversitet. *Dansk Skovbrugs Tidsskrift*, **90**, 78–113.

Heilmann-Clausen, J. and Aude, E. (2006). Forvaltning af dødt ved i naturnære bøgebevoksninger—til gavn for biodiversiteten? *HabitatVision, Rapport*, **06-01**.

Heilmann-Clausen, J. and Christensen, M. (2005). Wood-inhabiting fungi in Danish beech-forests: Conflicting diversity patterns and their implications in a conservation perspective. *Biological Conservation*, **122**, 633–642.

Heilmann-Clausen, J., Læssøe, T. and Vesterholt, J. (2006). Nye eller sjældent rapporterede danske barksvampe (Corticiaceae s.l., Thelephoraceae). *Svampe*, **53**, 21–32.

Höiland, K. and Bendiksen, E. (1996). Biodiversity of wood-inhabiting fungi in a boreal coniferous forest in Sör-Tröndelag County, Central Norway. *Nordic Journal of Botany*, **16**, 643-659.

Holec, J. (2003). Auf natürliche, vom Menschen nur minimal beeinflusste Vegetation beschränkte Großpilze. *Fritschiana*, **42**, 25–27.

James, T.Y., Moncalvo, J.-M., Li, S. and Vilgalys, R. (2001). Polymorphism at the ribosomal DNA spacers and its relation to breeding structure of the widespread mushroom *Schizophyllum commune*. *Genetics*, **157**, 149–161.

Johannesson, H., Vasiliauskas, R., Dahlberg, A., Penttila, R. and Stenlid, J. (2001). Genetic differentiation in Eurasian populations of the postfire ascomycete *Daldinia loculata*. *Molecular Ecology*, **10**, 1665–1677.

Junninen, K., Simila, M., Kouki, J. and Kotiranta, H. (2006). Assemblages of wood-inhabiting fungi along the gradients of succession and naturalness in boreal pine-dominated forests in Fennoscandia. *Ecography*, **29**, 75–83.

Keizer, P.J. and Becker, A. (2003). *Trametes cervina* (Schwein.) Bres. in Nederland. *Coolia*, **46**, 6–8.

Knudsen, H. and Ryman, S.G. (2000). Vegetation zones and mycogeography of the area. In: *Nordic Macromycetes, Vol. 1. Ascomycetes* (L. Hansen and H. Knudsen, eds.), pp. 21–33. Nordsvamp, Copenhagen.

Martikainen, P., Penttilä, R., Kotiranta, H. and Miettinen, O. (2000). New records of *Funalia trogii*, *Perenniporia tenuis* and *Polyporus pseudobetulinus* from Finland, with notes on their habitat requirements and conservation implications. *Karstenia*, **40**, 79–92.

Mathiessen, G. and Økland, R.H. (2007). Relative importance of host tree species and environmental gradients for epiphytic species composition, exemplified by *pyrenomycetes* s. lat. (*Ascomycota*) on *Salix* in central north Scandinavia. *Ecography*, in press.

Nilsson, R.H., Hallenberg, N., Nordén, B., Maekawa, N. and Wu, S.H. (2003). Phylogeography of *Hyphoderma setigerum* (Basidiomycota) in the Northern Hemisphere. *Mycological Research*, **107**, 645–652.

Nilsson, S.G. and Baranowski, R. (1997). Habitat predictability and the occurrence of wood beetles in old-growth beech forests. *Ecography*, **20**, 491–498.

Ódor, P., Heilmann-Clausen, J., Christensen, M., Aude, E., van Dort, K.W., Piltaver, A., Siller, I., Veerkamp, M.T., Walleyn, R., Standovár, T., van Hees, A.F.M., Kosec, J., Matočec, N., Kraigher, H. and Grebenc, T. (2006). Diversity of dead wood inhabiting fungal and bryophyte assemblages in semi-natural beech forests in Europe. *Biological Conservation*, **131**, 58–71.

Parviainen, J. (1999). Strict forest reserves in Europe: Efforts to enhance biodiversity and strengthen research related to natural forests in Europe. In: *Research in Forest Reserves and Natural Forests in European Countries — Country Reports for the COST Action E4: Forest Reserves Research Network, EFI Proceedings 16* (J. Parviainen, D. Little, M. Doyle, A. O'Sullivan, M. Kettunen and M. Korhonen, eds.), pp. 7–33.

Penttila, R., Lindgren, M., Miettinen, O., Rita, H. and Hanski, I. (2006). Consequences of forest fragmentation for polyporous fungi at two spatial scales. *Oikos*, **114**, 225–240.

Penttila, R., Siitonen, J. and Kuusinen, M. (2004). Polypore diversity in managed and old-growth boreal *Picea abies* forests in southern Finland. *Biological Conservation*, **117**, 271–283.

Petersen, R.H. and Bermudes, D. (1992). Intercontinental compatibility in *Panellus stypticus* with a note on bioluminescence. *Persoonia*, **14**, 457–463.

Ryvarden, L. and Gilbertson, R.L. (1992). *European Polypores. I.* Fungiflora, Oslo.

Ryvarden, L. and Gilbertson, R.L. (1993). *European Polypores. II.* Fungiflora, Oslo.

Siitonen, J. (2001). Forest management, coarse woody debris and saproxylic organisms: Fennoscandian boreal forests as an example. *Ecological Bulletins*, **49**, 11–41.

Siitonen, J., Penttilä, R. and Kotiranta, H. (2001). Coarse woody debris, polyporous fungi and saproxylic insects in an old-growth spruce forest in Vodlozero National Park, Russian Karelia. *Ecological Bulletins*, **49**, 231–242.

Sippola, A.-L. and Renvall, P. (1999). Wood-decomposing fungi and seed-tree cutting: A 40 year perspective. *Forest Ecology and Management*, **115**, 183–201.

Stokland, J.N. (2001). The coarse woody debris profile: An archive of recent forest history and an important biodiversity indicator. *Ecological Bulletins*, **49**, 71–83.

Stokland, J. and Kauserud, H. (2004). *Phellinus nigrolimitatus*: A wood-decomposing fungus highly influenced by forestry. *Forest Ecology and Management*, **187**, 333—343.

Sverdrup-Thygeson, A. and Lindenmayer, D.B. (2003). Ecological continuity and assumed indicator fungi in boreal forest: The importance of the landscape matrix. *Forest Ecology and Management*, **174**, 353–363.

CHAPTER **15**

Saprotrophic Basidiomycetes in Grasslands: Distribution and Function

Gareth W. Griffith and **Kevin Roderick**

Contents

Abstract

Natural and semi-natural grasslands dominate many terrestrial ecosystems, with succession prevented by herbivore grazing, low rainfall and fire. Inputs to grassland soils are typically low in lignin, often comminuted and in the form of dung with below-ground inputs from roots being important. The several hundred basidiomycete species which are preferentially found in grassland can be placed into four functional groupings: litter decomposers, dung fungi, terricolous species and root endophytes. However, detection of these in the absence of basidiocarps has hampered their study, an exception being the fairy ring-forming species. It is clear that basidiomycetes contribute to lignocellulose degradation in grassland soil and litter, though it is likely that ascomycetes play a relatively greater part in this process than in woodland systems. Changes in agricultural management have led to the loss of many semi-natural grasslands in Europe and there are concerns about losses of several grassland taxa, such as *Hygrocybe* and *Entoloma* spp.

British Mycological Society Symposia Series
Published by Elsevier Ltd.

1. INTRODUCTION

Fungi are key agents of nutrient cycling and thus of central importance to any understanding of carbon sequestration and nutrient cycling processes in all terrestrial ecosystems. However, mycologists have historically tended to have a sylvan bias and most fungal ecologists (as evidenced by several chapters in this book) have focused on woodland systems, resulting in wide knowledge of wood-decay fungi (Rayner and Boddy, 1988) and ectomycorrhizal taxa (Smith and Read, 1997). Similarly most fungal forays are held in woodland habitats where a diverse array of resources and host plants contribute to much higher levels of fungal diversity than are found in other habitats such as grasslands.

Of the 3,600 macrofungi found in the Netherlands, 80% are prevalent in woodlands (Arnolds and de Vries, 1989). Of the 20% of taxa generally found in non-wooded habitats, 10% (ca. 360 species) showed a preference for grasslands. Grassland basidiomycetes have received greater attention in recent decades, initially in Scandinavia (Rald, 1985; Arnolds, 1992a), and more recently in the UK (Rotheroe *et al.*, 1996) and other parts of Europe (Adamcik and Kautmanova, 2005). This increase in attention was spurred by the precipitous loss of semi-natural grassland habitats ('traditional' lowland haymeadows) due to modern mechanized agriculture, mainly through ploughing of permanent pastures and application of synthetic fertilizers. Thus, the study of the ecology of grassland basidiomycetes has largely been driven by conservation concerns (Chapter 17), although it is clear that elucidation of the role played by basidiomycetes and other fungi in grassland nutrient cycling is important for understanding the dynamics of carbon sequestration in the context of global climate change.

2. WHAT IS GRASSLAND?

The most intensively studied grasslands are those of Northern Europe and North America, and this review will focus mainly on these habitats. However, it is apposite to provide an overview of the global diversity of grassland systems and how they differ from the other main habitat types. Globally, grassland habitats cover ca. 20% of the terrestrial land area (Swift *et al.*, 1979; Parton *et al.*, 1995), occurring mainly where low or seasonal rainfall (250–1,500 mm $year^{-1}$) has prevented the establishment of woodland, due to the actions of grazing mammals, drought and fire (Ford *et al.*, 2004). Distinctive grassland ecosystems occur at a range of latitudes, for example, in the tropics (e.g. East African savanna, *Los Llanos* in Colombia/Venezuela) and in temperate climes (steppes, prairies and pampas). At higher altitudes, montane grasslands occur (e.g. in Andean Páramo and alpine meadows), often merging into tundra and heathland. It is very likely that human activity, through livestock farming, fire-setting and logging for fuel, has extended these grassland areas at the expense of woodlands. Such anthropogenic plagioclimax grasslands have in the past two centuries increased in distribution, due to the migration of Europeans and their agricultural practices, for instance, in New Zealand and North America.

The semi-natural grasslands which dominate many parts of Northern Europe (e.g. covering >50% of the UK land area (DEFRA, 2005)) are generally believed to be the result of millennia of anthropogenic deforestation. Palynological evidence indicates that most of Northern Europe was under continuous forest cover until ca. 4,000 BP and that there has been progressive deforestation. However, it has been suggested (Vera, 2000; Bakker *et al.*, 2004) that in pre-human (quaternary) times, Northern Europe comprised significant areas of grassland (large 'forest glades'), with larger grazing mammals playing a key role in the maintenance of habitat heterogeneity. These ideas remain controversial (Kirby, 2003; Mitchell, 2005), not least because very little grass pollen is detected in cores dating from quaternary times, with the main area of disagreement relating to the extent of these pre-historic grasslands (wood pasture with grassy glades, e.g. the New Forest in England or much larger open areas). Long-held views of the distinction between grasslands and woodlands may require reappraisal, with cycling of plant cover over century timescales (grassland → scrub → woodland → parkland → grassland) being a potential successional scenario. From a soil perspective, such thinking is intriguing since it raises the possibility that grasslands and woodlands are less different than is usually perceived. Mycologically this is not a huge surprise since several macrofungal taxa, which are predominantly found in grasslands in Europe (e.g. *Hygrocybe* spp.), are typically associated with woodland habitats in most other parts of the world (Cantrell and Lodge, 2000; Griffith *et al.*, 2004).

For understanding the ecology of decomposer basidiomycetes in grassland systems, it is important to consider how grasslands differ from other ecosystems, notably woodland. First, soil respiration in grasslands (by decomposers and plant roots in approximately equal measure) tends to be ~20% higher than in comparable woodlands (Raich and Tufekcioglu, 2000), mainly because the temperature of grassland soils fluctuates more widely and is higher in summer, due to greater insolation (Morecroft *et al.*, 1998). In Kansas woodland summer soil temperatures are 5 °C lower than in adjacent grassland, with soil carbon flux 38% lower as a consequence (Smith and Johnson, 2004). Seasonal droughts cause fluctuations in soil moisture and lead to root penetration to depths of several metres (Baker *et al.*, 2007). Following death *in situ* this leads to significant accumulation of SOM at depth (Reijs *et al.*, 2003).

The main theatre of fungal activity in grasslands is at or beneath the soil surface. Low and fluctuating moisture can limit microbial processes, with surface litter often being particularly inhospitable. For basidiomycetes, dry periods (especially in prairie-type grasslands) constrain fruiting, potentially disguising the existence/abundance of macrofungi in these habitats. It has been suggested that grassland species exhibit adaptations to reduce transpiration from basidiocarps (e.g. the slimy caps of *Hygrocybe* spp.; Friedrich, 1940). Organic matter inputs into grassland soils differ in several fundamental ways from woodland, influencing the prevalence of different types of decomposer organisms:

(1) Litter inputs into grassland soils are of smaller unit size with a greater surface area for microbial attack and with much lower amounts of secondarily thickened resource units (branches twigs, etc.).

(2) Investment in secondary metabolite production, including lignins, is also lower in grasses than in other plants, so the fungitoxic extractives formed in woody tissues are absent.
(3) Mammalian herbivores consume 43–73% of above-ground net primary production (NPP) in grasslands (compared to <10% in woodlands; Swift *et al.*, 1979) and consequently a large proportion of plant litter (ca. 50% of ingested C) enters the soil system in highly comminuted and partially digested form as dung. Although herbivore activity increases NPP (Stark and Grellmann, 2002), mineralization of vegetation in the digestive tracts of grazers and in dung reduces microbial biomass in grassland soil by up to 30% (Sankaran and Augustine, 2004). Variations in grazing intensity also influence surface litter accumulation, with litter accumulation leading to increased occurrence of fire, and thereby reduced N retention (Holdo *et al.*, 2007).
(4) A high proportion of plant biomass in grasslands (60–70%; Swift *et al.*, 1979) is below ground, especially under higher grazing pressure (Augustine and Frank, 2001).
(5) Regular defoliation by grazers leads to a high turnover of root tissues (a process still not well understood), so a greater proportion of plant biomass enters the soil system from roots (Turner *et al.*, 1993).

Grasslands in areas of high human population are among the most disturbed habitats, being susceptible to destruction by ploughing and also abandonment (removal of grazing). Such transitions are usually linked to political/social or economic changes, for instance, the redistribution of land following the French revolution (Dutoit *et al.*, 2004) leading to ploughing up of grasslands, or conversely the abandonment of arable farming (Highland Clearances in Scotland and the Great Depression in the US). It is likely that similar cycles have occurred in earlier periods of history, but even recent shifts can be difficult to discern (e.g. ridge and furrow evidence of historic ploughing), although with the exception of some upland and wooded areas it is quite likely that most North European grasslands have been cultivated at some point in the past. Shifts in populations of higher plants on grasslands in response to such changes have been well studied, but comparable investigations of higher fungi have been much more limited. There are, however, some historical descriptions of fungi which provide some useful clues, for instance, the association of some basidiomycetes with old pastures (Davies, 1813).

Natural grassland systems are maintained by grazing, and removal of herbivores usually leads to gradual afforestation. Amenity grasslands such as lawns and road verges are maintained by mowing. The nutrient cycles in such grasslands are dependent on management strategy. Removal of clippings removes nutrients from the system, a process which broadly mimics grazing (with N often added as fertilizer). Where clippings are returned, there is a thick litter layer and changes in plant diversity ensue due to nutrient enrichment. Supplementary feeding of stock in grasslands also represents a comparable form of nutrient addition, only partially offset by grazing activity.

3. FUNCTIONAL GROUPS OF GRASSLAND FUNGI

There has been a tendency in fungal ecology to assign species of known function to particular groupings and to use the term saprotrophic as a 'dustbin' group for the remainder. With the exception of the rust and smut fungi (which have no contact with dead organic matter), all basidiomycetes have some saprotrophic ability and for many involved in mutualistic associations with plants, their ability to release nutrients from organic matter (Read and Perez-Moreno, 2003) is a crucial part of the mutualism. Furthermore, the situation is confused by the occurrence of species which inhabit recently dead plant tissues having first colonized the living host (latent endophytism). Examples of such establishment strategies in woodland systems include *Oudemansiella mucida* on beech (Rayner and Boddy, 1988) and some members of the genus *Crinipellis* (Griffith and Hedger, 1994). For these examples there is circumstantial evidence that biotrophic infection by basidiospores occurs even though the dominant phase of the life cycle is saprotrophic.

The niche occupied by basidiomycetes is usually ascribed to the resources upon which they fruit but fruiting on dead tissues does not exclude some biotrophic/endophytic capability. The degree to which such a life strategy is necrotrophic is also difficult to establish. For instance, some fairy ring-forming basidiomycetes (see below) are occasionally termed 'weakly pathogenic'. Many asymptomatic endophytic fungi are known, though basidiomycetes have tended to be overlooked due to their slow growth on agar media. The potential diversity of basidiomycetes associated with grass leaves was highlighted in bamboo by Zhang *et al.* (1997) and a similar situation was also found in cocoa leaves (Arnold *et al.*, 2003). Thus, many predominantly saprotrophic basidiomycetes may have life cycles that are more complex than previously suspected. Hibbett *et al.* (2000) have estimated that ca. 50% of saprotrophic homobasidiomycetes (including many agarics) may have evolved from ectomycorrhizal ancestors. As such, several species may belong to more than one of the groups defined below.

3.1 Litter Decomposers

Primary above-ground inputs into grasslands (depending on the grazing regime) are in the form of plant litter, often forming a 'thatch' layer on the soil surface. Culture-based studies of grassland litter have tended to focus on ascomycetes (Hudson, 1968) but some basidiomycetes, usually forming small basidiocarps, are also abundant (e.g. *Mycena* spp. on grass litter, *Galerina* spp. on mosses), with others such as *Crinipellis stipitaria*, a possible latent invader, associated with more xerophytic grass tussocks but never soil (Warcup, 1951a; Parker-Rhodes, 1952). There is significant fungal translocation of N from soil to surface litter (Frey *et al.*, 2000), and there are likely to be fungi which colonize and decompose litter but only fruit on soil. Microcosm studies using grass litter have demonstrated the effectiveness of *Mycena* spp. in lignin decomposition but also that decay rates are reduced when species compete (Deacon *et al.*, 2006). In temperate grasslands, litter is rapidly incorporated into soil, largely through earthworm activity.

However, in African savanna termites, notably *Macrotermes michaelseni*, consume a high proportion of grass litter (Dangerfield and Schuurman, 2000). These eusocial insects cultivate lignolytic basidiomycete mutualists belonging to the genus *Termitomyces* in conspicuous nests, providing the fungus combs with partially digested faecal material and consuming the resulting hyphae (Chapter 9).

3.2 Dung Decomposers

In grazed grasslands, dung from herbivorous mammals is a major input to the soil and is initially decomposed by distinctive communities of fungi and invertebrates. Dung fungi play a key role in the catabolism of the lignocellulose and the microbial polymers (from intestinal bacteria, protozoa and fungi), although leaching/dispersal by rainfall and invertebrate activity leads to the rapid incorporation of dung into soil (Dickinson and Craig, 1990). Relative to plant litter or soil organic matter, dung is a high quality resource (C:N ratio ranging from 20 to 40 depending on host and diet; Richardson, 2001; Reijs *et al.*, 2003). Enhanced resource quality, partial digestion of plant polymers (with the exception of lignin) by gut microbes and increased access to microbial exoenzymes (due to comminution) lead to rapid decomposition (Nagy and Harrower, 1980).

Basidiomycetes and other fungi adapted to growth on dung tend to have pigmented spores, permitting them to withstand ingestion and digestion by herbivores (*enterophilic*), so they are already present in the faeces on excretion (Harper and Webster, 1964; Webster, 1970). It was originally thought that the fruiting of dung fungi exhibited a succession, but with regard to biomass and activity, it is more likely that the various groups of dung fungi all develop in parallel but achieving critical biomass for fruiting at different times (Webster, 1970), and culminating, for example, in the formation of basidiocarps (mostly *Coprinus* spp.) after 10–50 days (Richardson, 2001). Significant decomposition of lignin occurs in dung (Waksman *et al.*, 1939), and the activity of basidiomycetes is correlated with this process (Wicklow *et al.*, 1980b). Dung also comprises a proportion of debris from intestinal microbes, and some grassland basidiomycetes can decompose bacterial cell wall polymers effectively (Fermor, 1988). Dung from different herbivore species exhibits different patterns of fungal colonization (Ebersohn and Eicker, 1997). However, it is unclear whether this is due to variations in the unit resource size, differences in fungal inoculum present or differences in the resource quality of the dung (Wicklow *et al.*, 1980a).

Dung invertebrates generally inhibit fungal activity (Lussenhop and Wicklow, 1985), through nutrient competition, grazing by larvae on hyphae and physical disruption of the resource (McGranaghan *et al.*, 1999). However, invertebrates are susceptible to freezing in winter, possibly explaining the increased abundance of fruit bodies in winter (Richardson, 2001). Application of anthelminthics, some with selective antifungal activity (Edgington *et al.*, 1971), also inhibits invertebrates (Hutton and Giller, 2003; Warren and Paul, 2006). There can be considerable competition between microbial colonizers (Harper and Webster, 1964; Safar and Cooke, 1988), dung microcosms inoculated with combinations of fungi showing slower decomposition than when singly inoculated (Wicklow and

Yocom, 1981). Several dung fungi, notably *Coprinus* spp. (Ikediugwu and Webster, 1970), are able to disrupt the hyphae of competing species and there are several examples of production of inhibitory metabolites.

Coprinus spp. fruit abundantly in laboratory microcosms (Webster, 1970), whereas in nature there is a greater diversity of basidiomycetes, for example, species of *Conocybe, Panaeolus, Psathyrella, Psilocybe* and *Stropharia*. The highly fluctuating moisture conditions of the grassland environment (possibly providing triggers for primordium formation; Chapter 5) and interaction with underlying soil, absent from microcosms, may explain this difference. Wicklow and Moore (1974) did not find any significant colonization by soil microbes, suggesting that competition from the enterophilic dung fungi prevented subsequent colonization by soil fungi. However, several species, termed subcoprophilous, are more often associated with dunged fields rather than dung itself (Lisiewska, 1992) and these species generally have melanized spores (e.g. *Panaeolina foenisecii, Psilocybe semilanceata*), potentially able to tolerate gut passage.

3.3 Terricolous/Lignicolous Decomposers and Fairy Rings

For most terricolous (i.e. fruiting on soil) basidiomycetes, ecological information is largely reliant on spatiotemporal analysis of fruiting, though some studies on mycelia, notably those of Warcup (see below), have provided valuable insights. Vertical stratification of grassland soils is usually less than in woodland due to invertebrate activity and it is not known whether mycorrhizal fungi dominate the deeper soil horizons, as is the case in woodland (Lindahl *et al.*, 2007).

The most obvious manifestations of basidiomycete activity in grasslands are fairy rings, which are more visible in close-cropped and homogeneous vegetation than in other habitats where they also occur (Dowson *et al.*, 1989; Chapter 5). Radial expansion rates of fairy rings range from 8 cm year^{-1} for *Marasmius oreades* (Smith, 1980) to over 100 cm year^{-1} for *Lepista sordida* (Terashima *et al.*, 2004). Maximal ring diameters of 100–300 m (Shantz and Piemeisel, 1917; Kreisel and Ritter, 1985) have been reported with estimated ages of up to 200–700 years (Shantz and Piemeisel, 1917; Burnett and Evans, 1966; Kreisel and Ritter, 1985). Fairy rings are classified according to whether vegetation is killed at the ring margin (type 1), grows more vigorously (type 2) or is unaffected (type 3) (Shantz and Piemeisel, 1917). More than 50 species of grassland basidiomycetes have been reported to form type 1 or 2 fairy rings, mostly belonging to the genera *Marasmius, Lepista, Agaricus, Clitocybe, Lycoperdon* and *Calvatia* (Couch, 1995), with others, such as *Hygrocybe* and *Panaeolus* spp., forming type 3 rings occasionally (Figure 1). Several studies have demonstrated the genetic integrity of fairy rings, by investigation of mating type factor distribution (Burnett and Evans, 1966), molecular markers (Abesha *et al.*, 2003) or mycelial pairings to determine somatic compatibility (K. Roderick, unpublished). There has been speculation, but no experimentation as to whether unmated mycelia (homokaryotic primary mycelia) can form rings without fruiting (Parker-Rhodes, 1955).

Fairy rings or arcs are formed by the annular growth of a mycelial system with apparent dieback of mycelium internal to the growth front. It has often been

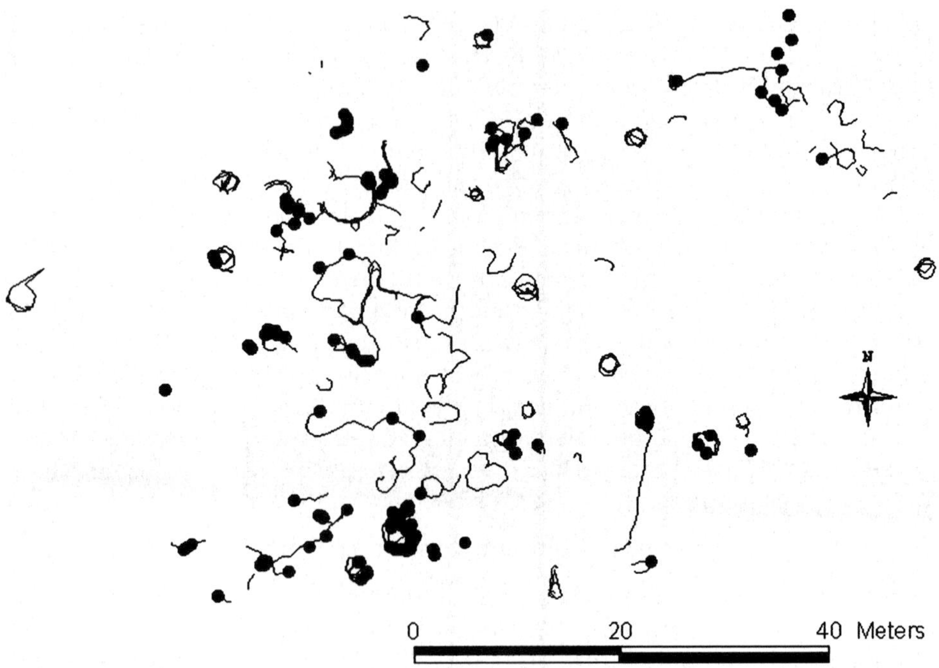

Figure 1 Differential GPS mapping of *Agaricus campestris* fairy rings on the UW Aberystwyth campus (SN595818), showing altered *Lolium/Festuca* vegetation (lines) and basidiocarps (dots), during the summer of 2004. Note that many rings did not produce basidiocarps and the heterogeneous distribution of the rings.

noted that such growth is simply an emergent property of localized nutrient depletion/toxin accumulation behind the growth front. Soil organic matter and nitrogen are depleted within the rings of several species (Lawes *et al.*, 1883; Edwards, 1984, 1988; Kaiser, 1998). It is likely that nutrient redistribution via dung would increase nutrient levels within larger rings but there would still be an annular region of depleted nutrient internal to the mycelial front. The study of Dowson *et al.* (1989) is the only study to our knowledge to have explored the reasons for continued outward expansion of a ring, demonstrating that polarity of growth of *Lepista nebularis* was maintained after translocation of ring fragments, though why this should be the case remains unclear (Chapter 1). Their observation that mycelia disappeared when their orientation was reversed to face mycelia from their original ring is consistent with observations that when adjacent rings intersect the underlying mycelium degenerates (Parker-Rhodes, 1955).

Excavation of soil at the margin of type 1 and 2 rings reveals dense mycelial growth visible to the naked eye, whereas for species forming type 3 rings (e.g. *Hygrocybe* spp.) even microscopic observation of soil beneath the basidiocarps does not reveal an abundance of clamped mycelia (Warcup, 1959; G.W. Griffith and G.L. Easton, unpublished). The areas of mycelial abundance in type 1 and 2 rings coincide with areas of more luxuriant or killed/'scorched' vegetation,

Figure 2 Type 3 fairy ring of *Hygrocybe pratensis* (Ystumtuen, Aberystwyth; SN731799). (Inset) An ISSR fingerprint gel showing the genetic identity of basidiocarps from the same ring and difference compared with basidiocarps from an adjacent ring (10 m away). One sample (*, arrow) contained additional bands due to the presence of an endophyte (*Paecilomyces marquandii*).

sometimes in concentric rings (Edwards, 1984; Terashima *et al.*, 2004). Appearance of vegetation symptoms is highly seasonal and linked to soil moisture conditions, with rings (both basidiocarps and vegetation effects; Figure 2) often visible only in certain years (Shantz and Piemeisel, 1917), and possibly linked to growth or reproductive phases of the fungal life cycle (Fisher, 1977). This has led to some confusion regarding the classification (type 1 or 2) of species (Halisky and Peterson, 1970). Soil respiration and nutrient content are elevated beneath zones of luxuriant vegetation associated with *Agaricus arvensis* (Edwards, 1984), suggesting that enhanced decomposition of SOM is responsible for increased plant nutrient availability. In the same rings, symptoms of K deficiency were observed in associated grasses, suggesting that fungal tissues concentrated nutrients (with basidiocarps containing 6% N, 3% K and 1% P by dry weight), at the expense of adjacent plants. Edwards (1984, 1988) estimated that basidiocarps contained ca. 25% of all the K present in areas of dense fungal growth. The ability of grassland basidiomycetes to concentrate K and related elements in fruit bodies has subsequently received attention in the context of radiocaesium (^{137}Cs) accumulation following the Chernobyl reactor explosion (Dighton *et al.*, 1991; Anderson *et al.*, 1997).

The high mycelial density in annular areas of rings causes changes in the hydrological properties of the soil (Warcup, 1959; Terashima and Fujiie, 2005). Increased soil hydrophobicity is linked to hyphal secretions, possibly hydrophobins, which coat soil particles. On managed turfgrasses the resulting 'dry patch' symptoms can be alleviated by use of surfactants and fungicides (York and Canaway, 2000). Under suitable climatic conditions (rings are usually most visible in dry summers), these localized changes in soil hydrology can alter growth of vegetation, potentially masking the beneficial effects of elevated soil nutrients described above. However, in rings of some type 1 species, secretion of toxins (such as cyanide) has been implicated (Blenis *et al.*, 2004), while other species (e.g. *M. oreades*, *Vascellum curtisii* and *Bovista dermoxantha*) have a necrotrophic ability following colonization of healthy root and leaf tissues (Filer, 1965; Terashima *et al.*, 2004). Several type 1 species exhibit host specificity with regard to symptom production, with Terashima and Fujiie (2005) reporting a ring of *L. sordida* causing type 2 symptoms on *Zoysia japonica*, but disappearing on reaching an area vegetated by *Lolium perenne*.

3.4 Root Endophytes/Pathogens

In addition to the facultative necrotrophic abilities of fairy ring fungi, other agarics, e.g. *P. semilanceata*, are able to colonize healthy cortical tissues of grasses but without clear evidence of any deleterious symptoms in the host (Keay and Brown, 1990). Similarly colonization of grass roots has been observed under field or microcosm conditions by species such as *Melanoleuca grammopodia* and *Conocybe dunensis* (McKay, 1968). In both there was some evidence of host specificity, with *P. semilanceata* exhibiting a preference for *Agrostis tenuis* and *Poa annua* over *L. perenne*, and infection rates by basidiomycete (clamped) hyphae being much higher for *Ammophila arenaria* than other sand dune grasses. *Thanatephorus cucumeris* (anamorph *Rhizoctonia solani*) is commonly isolated from grassland and arable soil (Garrett, 1951; Warcup and Talbot, 1962) and is a capable cellulolytic saprotroph. It is also an economically important necrotrophic pathogen in grassland, causing various diseases (e.g. 'brown patch', root rot and aerial blight) in turfgrasses and other grassland plants (Couch, 1995). However, *T. cucumeris* and related species in the Ceratobasidiaceae are detected in healthy roots (Jumpponen and Johnson, 2005), and are also able to form mycorrhizal symbioses with orchids, and *Carex* spp. (Haselwandter and Read, 1982; Roberts, 1999).

Presence of basidiomycetes is occasionally revealed by culture-based examination of healthy roots from grasslands but at low frequency (Warcup, 1959; Wilberforce *et al.*, 2003). However, use of fungal-specific PCR primers has recently shown a great diversity of basidiomycetes in healthy root tissues. Wilberforce (2003) found that basidiomycetes comprised 15% of clones from an oligotrophic temperate grassland in the UK, while Jumpponen and Johnson (2005) found ca. 30% of clones in a library derived from tallgrass prairie roots to be basidiomycete in origin. However, like many aspects of root biology, decomposition of these organs is poorly understood and further work is required to elucidate the function of many of these endophytes. The recent discovery, by Harrington and

Mitchell (2002), of ectomycorrhiza-like structures formed by *Cortinarius cinnamomeus* on the roots of *Carex flacca* and *C. pilulifera* in calcareous grassland, consistent with earlier observation of the association of *Tricholoma melaleucum* with *Carex glauca* (Wilkins and Patrick, 1939), illustrates that mycorrhizal associations involving agarics and non-woody hosts may be more common in temperate habitats than previously thought. Distinctive assemblages of ectomycorrhizal fungi do occur with shrubs in grasslands (e.g. *Helianthemum nummularium*), but association of agarics with non-woody hosts is usually restricted to Arctic-alpine habitats (Gardes and Dahlberg, 1996). Thus, the assignment of mycorrhizal status can be problematic especially in the absence of evidence of distinctive morphological structures.

4. DETECTION OF GRASSLAND FUNGI

The question of what role is played by particular species or groups in relation to ecosystem function is fundamental to microbial ecology. The technological and conceptual challenge required by any attempt to answer this has led to an obsession with methods. For unit-restricted taxa (see Chapter 1) such as many dung fungi, it would appear to be a relatively simple question, though current data are based almost exclusively on basidiocarp presence. However, most fungal activity in grasslands takes place in the soil, the physicochemical complexity and small scale heterogeneity of which make it difficult to map the location of hyphae (Feeney *et al.*, 2006). For terricolous basidiomycetes in particular, this presents a challenge, since their distribution can be addressed at a range of spatial scales from soil crumb to field level (from a few micrometres to many metres). For most species (excepting fairy ring-forming fungi) such detailed spatial information is largely absent, and without this information it is difficult to elucidate what resources are being decomposed by particular species.

Standard dilution plating seldom recovers basidiomycete colonies, mainly because they are slow-growing but also because their hyphae are tightly associated with soil particles (Warcup, 1951b; Thorn *et al.*, 1996). However, Warcup (1959) was able to isolate several taxa from pasture soil and roots by plating soil crumbs or micromanipulating individual hyphae. Among the diverse basidiomycetes isolated by these methods were several resupinate taxa, including *Peniophora* and *Athelia* spp. (Warcup and Talbot, 1962), which only rarely fruit (on the underside of soil clods or in worm tunnels; Eriksson, 1949). Direct counts of fungal hyphae by microscopy have been informative with regard to fungal standing crop, showing the increase in fungal biomass in a grassland chronosequence following arable cultivation (van der Wal *et al.*, 2006). Quantification of basidiomycete mycelium, identifiable to some degree by the presence of clamp connections, has been achieved in woodland systems (Frankland, 1982; Robinson *et al.*, 2005) but not to our knowledge in grasslands. Current biochemical approaches (e.g. ergosterol, phospholipid fatty acids (PLFA), etc.), while informative about overall fungal activity/biomass, are hitherto unable to dissect out the basidiomycete component. We refer the reader to the excellent review on the merits of these approaches by Robinson *et al.* (2005). The activity of saprotrophic basidiomycetes has also been investigated by study of lignolytic enzymes from grassland soils (Gramss, 1997).

Although lacking in specific biomarkers or reliable isolation methods, the study of basidiomycetes is distinctly advantaged by the fact that many species form macroscopic fruit bodies. Indeed, with the exception of a limited number of well-studied species, inferences about the ecology of basidiomycetes are largely derived from the spatiotemporal distribution of these reproductive structures. However, fruiting patterns of grassland fungi present if anything a greater challenge than those of woodland taxa since environmental conditions in grasslands are generally less conducive to basidiocarp formation and persistence, especially the often low and fluctuating levels of atmospheric humidity. In drier grasslands especially, fruiting data are very sparse (e.g. North American mycologists seldom conduct grassland forays; Leon Shernoff, personal communication), but recent data from molecular studies suggest that many of the species present fruit only very rarely (Lynch and Thorn, 2006).

Most data of basidiocarp occurrence are collected informally and non-quantitatively in fungus forays and thus are not easily interpretable in any ecological context. Gilbert's (1875) study of basidiomycetes in response to various agricultural treatments at Park Grass Rothamsted is probably the first systematic survey of grassland fungi, finding that rings of *M. oreades* were most abundant on plots treated with lime superphosphate (either alone or in combination with sodium and magnesium sulphates) and mostly absent from plots treated with N (ammonium or manured) or K. A broadly similar pattern was found for *Hygrocybe* spp., which were present in greatest diversity on untreated plots. Wilkins and Patrick (1939, 1940) were the first to apply a more quantitative approach, recording basidiocarp numbers in fixed quadrats (ca. 700 m^2) visited repeatedly over 2 years. When assessing basidiomycete diversity in different habitat types, they found ca. 20% of the 620 species encountered were present in grassland compared to ca. 60% in deciduous woodland but only 38 spp. exclusive to grassland (e.g. *Hygrocybe, Lycoperdon* and *Panaeolus* spp.) and fewer species being found on clay soils compared to chalk or sand. The most common species at the 20 grassland sites was *H. virginea*, present on all soil types at '80–100% constancy'. After 70 years of agricultural intensification, it would be interesting to examine whether these fruiting patterns have changed at these sites. Arnolds (1989) found that diversity of grassland fungi was much greater in fields where there had been no addition of synthetic fertilizer, a finding confirmed by more recent surveying of permanent quadrats at a range of replicated grassland field experiments (Griffith *et al.*, 2002, 2004). This is consistent with a decrease in the ratio of fungal:bacterial biomass (based on PLFA profiles) following fertilization (Bardgett *et al.*, 1999).

The vagaries of basidiocarp production have been noted many times and several studies have illustrated discrepancies between patterns of fruiting and mycelial abundance below ground (Horton and Bruns, 2001). Even basidiocarp surveys repeated over several years may provide an incomplete picture of below-ground diversity (see Chapter 5; Parker-Rhodes, 1951), although information can be gathered for large areas in a very time- and cost-efficient manner. The potential pitfalls of basidiocarp surveys of grasslands are lucidly described by Arnolds (1992b) and Watling (1995), including consideration of differential longevity of basidiocarps, fruiting periodicity, annual fluctuations and succession.

DNA-based approaches have transformed our understanding of microbial ecology, for instance, with regard to ectomycorrhizal fungi in woodlands (Horton and Bruns, 2001; Lindahl *et al.*, 2007; Chapter 10). The most useful data currently available are from sequencing of clone libraries based on PCR amplification with fungal-specific primers. These provide a useful snapshot of the species present, often revealing the presence of unexpected taxa (compared to basidiocarp data). Use of taxon-specific primers has revealed that basidiomycetes are two- to three-fold less abundant (relative to total fungal abundance) in prairie grassland soil than woodland (Fierer *et al.*, 2005; O'Brien *et al.*, 2005). The most detailed study to date (Lynch and Thorn, 2006) identified almost 300 basidiomycete species in adjacent pasture and arable plots, with up to 9 species in some 10 g soil samples. These comprised 45 species of clavarioid fungi (20% of the total), as well as other taxa (e.g. *Hygrocybe* and *Entoloma* spp.) typically observed in oligotrophic grassland in Europe. Thus, the diversity revealed by genetic analysis greatly exceeded both the limited range of basidiocarps found at the site (http://lter.kbs.msu.edu/) and the 51 morphospecies isolated on selective media (Thorn *et al.*, 1996). A similar disparity between molecular data, culture-based approaches and basidiocarp surveys was also observed in Welsh grasslands (Hunt *et al.*, 2004).

Cloning and sequencing is costly when scaled up and more rapid fingerprinting approaches, such as terminal restriction fragment length polymorphism (T-RFLP) or fungal automated ribosomal intergenic spacer analysis (FARISA), can robustly reveal treatment effects, for example, along grassland fertilization gradients (Brodie *et al.*, 2003; Kennedy *et al.*, 2006). More powerful still is a dual approach allowing peaks in T-RFLP profiles to be identified from sequence data. However, the possibility of bias (due to primer specificity or differential efficiency of DNA extraction) can skew data (Anderson *et al.*, 2003; Avis *et al.*, 2006). A potential problem with genetic approaches relates to effective sampling, given the often very heterogeneous distribution of grassland basidiomycetes (Figure 2). One hectare of grassland contains ca. 1,000 t of topsoil (crudely assuming 10 cm soil depth and bulk density of 1 g cm^{-3}) and it is very difficult to devise an effective sampling strategy to ensure representative coverage (when DNA extraction methods are limited to 1–10 g soil) without a very large budget. Technological advances, possibly soil fungus microarray chips (Sessitsch *et al.*, 2006) or metagenomics, will increase efficiency of genetic approaches but basidiocarp surveys will remain a valuable complement of grassland research.

5. CONTRIBUTION OF SAPROTROPHIC BASIDIOMYCETES TO NUTRIENT CYCLING AND SOIL STRUCTURE

The main input into grassland decomposition systems is lignocellulose. As described by Baldrian (Chapter 2), saprotrophic basidiomycetes are able to secrete batteries of extracellular enzymes but our knowledge of lignocellulose decay in soil and the organisms involved is less detailed than for larger woody resources. While there have been detailed studies of decomposition in woodland systems (Frankland, 1982; Steffen *et al.*, 2000, 2002), the only comparable studies in grasslands have focused on fairy rings (see above).

It is the decomposition of lignin that is generally accepted to be the rate-limiting stage in carbon and nutrient cycling in terrestrial ecosystems. In addition to containing less lignin, the composition of grass lignin contains 10–20% phenolic units, a higher proportion than in wood (Lapierre *et al.*, 1989). This may allow easier catabolism by laccase and manganese peroxidase that directly degrade only phenolic units (Camarero et al., 1994). Grass lignins are also more extensively cross-linked with polysaccharides cell wall polymers (via *p*-coumaryl subunits to hemicelluloses) than are wood lignins (Iiyama *et al.*, 1990; Lam *et al.*, 1992). These factors make grass lignins more readily degradable (Lapierre *et al.*, 1989). As is the case for woodland litter/soil, most lignolytic basidiomycetes in grasslands belong to the Agaricales, though as noted above Aphyllophorales are also present. There is evidence that the role of ascomycete fungi in lignin degradation may be relatively more important (Kluczek-Turpeinen *et al.*, 2003; Deacon *et al.*, 2006), with several soil-inhabiting species having been shown to be able to mineralize grass lignin more rapidly than wood lignin (Rodriguez *et al.*, 1996).

Lignins in soil are a major source material for the formation of humic compounds. There is a correlation between the lignin content of organic inputs and the amount of humus formed (Hammel, 1997; Heal *et al.*, 1997) but it is difficult to assess the degree to which plant lignins are transformed through humification. The enzymes involved in ligninolysis can also mediate formation and degradation of humic compounds (Gramss *et al.*, 1999; Scheel *et al.*, 1999; Steffen *et al.*, 2002). These phenoloxidases can mediate covalent binding of aromatic compounds and it is suggested that humic compounds are the partially oxidized products of phenoloxidase activity in soil (quinones condensed with peptides, amino sugars and aromatics; Gramss *et al.*, 1999). While the energetic benefits of degrading complex aromatic polymers are considered to be marginal, humic compounds (unlike lignin) contain N (much soil N is present in this form), so for basidiomycetes in oligotrophic grassland such sources may be important. However, the mobilization of the recalcitrant organic N pool in soil is a poorly understood process (O'Connor, 1983).

Through mucilage secretion and mycelial entanglement of soil particles, fungi are considered to be important in the formation of water-stable aggregates by binding microaggregates (50–250 μm) into macroaggregates (>250 μm) (Tisdall and Oades, 1982). The role of the glycoprotein glomalin, secreted by AM fungi, in this process is well established (Rillig and Mummey, 2006) but basidiomycetes including *R. solani* also contribute to aggregate stabilization (Tisdall *et al.*, 1997). An unidentified grassland basidiomycete, closely related to *Peniophora*, has also been shown to secrete large quantities of a polysaccharide with significant soil-binding properties (Caesar-TonThat and Cochran, 2000; Caesar-TonThat *et al.*, 2001). Antibodies raised against cell walls of this fungus reacted strongly with larger (>2 mm) soil aggregates from dry grassland soils and to a lesser extent in adjacent arable soils. A less desirable effect of basidiomycetes on soil texture is due to the water repellent properties of their hyphae (White *et al.*, 2000), thought to be associated with the secretion of hydrophobin proteins (Rillig and Mummey, 2006).

6. EFFECTS OF GRASSLAND MANAGEMENT AND CLIMATE CHANGE

In intensive modern farms, grassland areas are ploughed and reseeded (usually with *L. perenne* in Europe) on a 5–10 year cycle, and their soils in consequence bear more similarity to arable fields than permanent grasslands. Additionally, the past 50 years have seen the widespread use of synthetic fertilizers to improve grassland productivity. Thus, disturbance and eutrophication have led to the demise of most macrofungal fruiting in these habitats, although it has yet to be demonstrated that the mycelia are also absent. Losses of fungal diversity generally mirror declines in plant and invertebrate diversity, and in the case of these better studied groups changes in grassland management can also lead to loss of diversity (Rook and Tallowin, 2003). Shifts from haymaking to silage production or from cattle and sheep to sheep only grazing have also altered patterns of abundance of higher plants and insects. For soil dwelling fungi such changes might be anticipated to have a lesser effect, although changes in patterns of root death and photosynthate translocation will affect the nutrition of soil microbes (Turner *et al.*, 1993). Conversely, microclimatic conditions for basidiocarp formation are altered by sward height variation, and macrofungal fruiting in rank grassland is much reduced compared to adjacent grazed areas (Griffith *et al.*, 2006). However, as appears to be the case in many prairie grasslands where vegetation is much longer than the 3–15 cm sward height typical of European pastures, the health of the underlying mycelium may be little affected by aboveground vegetation height. Mown grasslands, especially historic lawns, represent important refugia for grassland fungi. While these habitats are often spared fertilizer application, the failure to remove clippings can cause eutrophication and loss of diversity, especially in areas of higher nitrogen deposition (i.e. most of Europe).

Fungi are seldom considered in issues of land use but there is a growing body of evidence that sites with diverse fungal communities do not necessarily host diverse plant communities. This is consistent with the idea that soil nutrient conditions are far more important than sward management. While many sites with diverse grassland fungal communities receive some legal protection (SSSI, etc.), fungal diversity is seldom mentioned in the notification statements (Chapter 8). Since site visits by nature conservation staff generally occur in the summer, there is little information about macrofungal diversity. Recent UK legislation (EIA (Agriculture) Regulations, 2001) controls change of use of agricultural land (e.g. ploughing of pasture), but since biodiversity assessments are generally conducted in the summer, low plant diversity can lead to destruction of valuable fungal sites.

With prospective changes in agricultural support, the re-establishment of semi-natural habitats is gaining attention. Dispersal of fungi is not perceived to be a significant factor limiting recolonization but reductions in soil nutrient status, coupled with a latent period between colony establishment and fruiting, can lead to delays in reappearance. Our work at various restoration sites, consistent with other studies (Lange, 1991), suggests that fruiting of the more common member of the more prized grassland taxa (*Hygrocybe, Entoloma* spp., etc.) may occur within a decade of cessation of nutrient addition. We note, however, that some of the most diverse sites for grassland fungi were subject to significant disturbance in recent centuries (e.g. post-industrial sites such as iron works, canal/reservoir embankments).

Since grasslands contain 12% of the world's SOM (33 kg m^{-2} in temperate grasslands; Conant *et al.*, 2001), factors that affect the activity of saprotrophic basidiomycetes in grasslands can impact on atmospheric CO_2 levels and consequent climate change (Freibauer *et al.*, 2004). Global warming and changing rainfall patterns combined with changes in agricultural subsidies are likely to lead to changes in climax vegetation types (Raich and Tufekcioglu, 2000), with scrub invasion and afforestation of grasslands generally resulting in increased soil C pools (Smith and Johnson, 2004). However, there are examples where the opposite has occurred. Planting of exotic pines in Andean *paramo* grasslands has caused loss of SOM, apparently due to the saprotrophic activity (soil C mineralization) of the usually ectomycorrhizal symbiont, *Suillus luteus* (Chapela *et al.*, 2001). There is already evidence of changes in phenology of basidiocarp production in UK grasslands since the 1970s with grassland species showing contrasting patterns to woodland saprotrophs (Gange *et al.*, 2007; Chapter 5).

Many parts of the world now experience high levels of aerial deposition of 'fixed' N (from intensive agriculture and vehicle emissions), a consequence of anthropogenic fixation of nitrogen (Haber–Bosch process), which has increased 10-fold since pre-industrial times (Fowler *et al.*, 2004), and now exceeds natural fixation by bacteria (Galloway *et al.*, 1995). Even modest nitrogen deposition (5–10 kg N ha^{-1} $year^{-1}$) reduces diversity of ectomycorrhizal agarics in boreal forests (Lilleskov *et al.*, 2002), probably due to alteration of soil nitrogen cycles (especially mobilization of organic nitrogen), which are very likely also to affect saprotrophic species. Although critical N loads for grasslands are higher than for woodlands, loss of plant diversity in UK grasslands (receiving 6–50 kg N ha^{-1} $year^{-1}$) is correlated with nitrogen deposition (Stevens *et al.*, 2004; Chapter 17). Projected N deposition in 2050 for the world's 34 biodiversity hotspots suggests that half of these, including grassland systems such as the Brazilian *cerrado*, will be subjected to >15 kg N ha^{-1} $year^{-1}$ (Phoenix *et al.*, 2006).

7. CONCLUSION

Almost 60 years have elapsed since Chesters (1949) postulated that the basidiomycetes were "the missing link in soil mycology". Our colleagues focusing on woodland ecosystems have made great advances in elucidating the role of these fungi in plant nutrition and decomposition processes. While the specialized catabolic functions performed by lignolytic basidiomycetes are relatively less important and partly mediated by ascomycete fungi, several lines of evidence suggest that grassland basidiomycetes may play a more important role in plant nutrition than previously suspected. With respect to fungal conservation, grasslands outside Europe merit more detailed study, given the unexpectedly high diversity revealed by molecular investigations. There is some urgency to this last point. As evidenced in Europe by the past 50 years of agricultural intensification, future uncertainty with respect to climate change and agricultural practices places remaining semi-natural grasslands at high risk of destruction.

ACKNOWLEDGEMENTS

We are grateful to the BMS for award of a summer studentship to KR and to the Grasslands' Trust, especially Charlie Evans, for ongoing support. We also thank Gary Easton, Lizzie Wilberforce, Andrew Detheridge and many colleagues at IGER for useful discussions.

REFERENCES

Abesha, E., Caetano-Anolles, G. and Hoiland, K. (2003). Population genetics and spatial structure of the fairy ring fungus *Marasmius oreades* in a Norwegian sand dune ecosystem. *Mycologia*, **95**(6), 1021–1031.

Adamcik, S. and Kautmanova, I. (2005). *Hygrocybe* species as indicators of conservation value of grasslands in Slovakia. *Cathatelasma*, **6**, 25–34.

Anderson, I.C., Campbell, C.D. and Prosser, J.I. (2003). Potential bias of fungal 18S rDNA and internal transcribed spacer polymerase chain reaction primers for estimating fungal biodiversity in soil. *Environmental Microbiology*, **5**, 36–47.

Anderson, P., Davidson, C.M., Littlejohn, D., Ure, A.M., Shand, C.A. and Cheshire, M.V. (1997). The translocation of caesium and silver by fungi in some Scottish soils. *Communications in Soil Science and Plant Analysis*, **28**, 635–650.

Arnold, A.E., Mejia, L.C., Kyllo, D., Rojas, E.I., Maynard, Z., Robbins, N. and Herre, E.A. (2003). Fungal endophytes limit pathogen damage in a tropical tree. *Proceedings of the National Academy of Sciences of the United States of America*, **100**, 15649–15654.

Arnolds, E. (1989). The influence of increased fertilization on the macrofungi of a sheep meadow. *Opera Botanica*, **100**, 7–21.

Arnolds, E. (1992a). Macrofungal communities outside forests. In: *Fungi in Vegetation Science* (W. Winterhoff, ed.), pp. 113–149. Kluwer Academic Publishers, Dordrecht, Netherlands.

Arnolds, E. (1992b). The analysis and classification of fungal communities with special reference to macrofungi. In: *Fungi in Vegetation Science* (W. Winterhoff, ed.), pp. 7–47. Kluwer Academic Publishers, Dordrecht, Netherlands.

Arnolds, E. and de Vries, B. (1989). Oecologische statistiek van de Nederlandse macrofungi. *Coolia*, **32**, 72–86.

Augustine, D.J. and Frank, D.A. (2001). Effects of migratory grazers on spatial heterogeneity of soil nitrogen properties in a grassland ecosystem. *Ecology*, **82**, 3149–3162.

Avis, P.G., Dickie, I.A. and Mueller, G.M. (2006). A 'dirty' business: Testing the limitations of terminal restriction fragment length polymorphism (T-RFLP) analysis of soil fungi. *Molecular Ecology*, **15**, 873–882.

Baker, J.M., Ochsner, T.E., Venterea, R.T. and Griffis, T.J. (2007). Tillage and soil carbon sequestration—What do we really know? *Agriculture, Ecosystems and Environment*, **118**, 1–5.

Bakker, E.S., Olff, H., Vandenberghe, C., De Maeyer, K., Smit, R., Gleichman, J.M. and Vera, F.W.M. (2004). Ecological anachronisms in the recruitment of temperate light-demanding tree species in wooded pastures. *Journal of Applied Ecology*, **41**, 571–582.

Bardgett, R.D., Lovell, R.D., Hobbs, P.J. and Jarvis, S.C. (1999). Seasonal changes in soil microbial communities along a fertility gradient of temperate grasslands. *Soil Biology & Biochemistry,* **31**, 1021–1030.

Blenis, P.V., Chow, P.S., Duncan, I. and Knowles, N.R. (2004). Cyanide levels near fairy rings affect the growth of grasses and fungi. *Canadian Journal of Botany,* **82**, 1324–1329.

Brodie, E., Edwards, S. and Clipson, N. (2003). Soil fungal community structure in a temperate upland grassland soil. *FEMS Microbiology Ecology,* **45**, 105–114.

Burnett, J.H. and Evans, E.J. (1966). Genetical homogeneity and the stability of the mating-type factors of 'fairy rings' of *Marasmius oreades. Nature, London,* **210**(June), 1368–1369.

Caesar-TonThat, T.C. and Cochran, V.L. (2000). Soil aggregate stabilization by a saprophytic lignin— Decomposing basidiomycete fungus — I. Microbiological aspects. *Biology and Fertility of Soils,* **32**, 374–380.

Caesar-TonThat, T.C., Shelver, W.L., Thorn, R.G. and Cochran, V.L. (2001). Generation of antibodies for soil aggregating basidiomycete detection as an early indicator of trends in soil quality. *Applied Soil Ecology,* **18**, 99–116.

Camarero, S., Galletti, G.C. and Martínez, A.T. (1994). Preferential degradation of phenolic lignin units by two white rot fungi. *Applied and Environmental Microbiology,* **60**, 4509–4516.

Cantrell, S.A. and Lodge, D.J. (2000). Hygrophoraceae of the Greater Antilles: *Hygrocybe* subgenus *Hygrocybe. Mycological Research,* **104**, 873–878.

Chapela, I.H., Osher, L.J., Horton, T.R. and Henn, M.R. (2001). Ectomycorrhizal fungi introduced with exotic pine plantations induce soil carbon depletion. *Soil Biology & Biochemistry,* **33**, 1733–1740.

Chesters, C.G.C. (1949). Concerning fungi inhabiting soil. *Transactions of the British Mycological Society,* **32**, 197–216.

Conant, R.T., Paustian, K. and Elliott, E.T. (2001). Grassland management and conversion into grassland: Effects on soil carbon. *Ecological Applications,* **11**, 343–355.

Couch, H.B. (1995). *Diseases of Turfgrasses.* Krieger Publishing Company, Malabar, FL.

Dangerfield, J.M. and Schuurman, G. (2000). Foraging by fungus-growing termites (Isoptera: Termitidae, Macrotermitinae) in the Okavango Delta, Botswana. *Journal of Tropical Ecology,* **16**, 717–731.

Davies, H. (1813). *Welsh Botanology; Part of the First. A Systematic Catalogue of the Native Plants of the Isle of Anglesey in Latin, English and Welsh (with the Habitats of the Rarer Species and a Few Observations).* W. Marchant, London.

Deacon, L.J., Pryce-Miller, E.J., Frankland, J.C., Bainbridge, B.W., Moore, P.D. and Robinson, C.H. (2006). Diversity and function of decomposer fungi from a grassland soil. *Soil Biology & Biochemistry,* **38**, 7–20.

DEFRA. (2005). *Environmental Protection, e-Digest Statistics. Land and Land Use. Key Facts about: Land Use and Land Cover.* Land by Agricultural and other Uses: 1998 and 2005, http://www.defra.gov.uk/environment/statistics/land/kf/ldkf08.htm.

Dickinson, C.H. and Craig, G. (1990). Effects of water on the decomposition and release of nutrients from cow pats. *New Phytologist,* **115**, 139–147.

Dighton, J., Clint, G.M. and Poskitt, J. (1991). Uptake and accumulation of Cs-137 by upland grassland soil fungi — A potential pool of Cs immobilization. *Mycological Research,* **95**, 1052–1056.

Dowson, C.G., Rayner, A.D.M. and Boddy, L. (1989). Spatial dynamics and interactions of the woodland fairy ring fungus, *Clitocybe nebularis. New Phytologist,* **111**, 699–705.

Dutoit, T., Buisson, E., Roche, P. and Alard, D. (2004). Land use history and botanical changes in the calcareous hillsides of Upper-Normandy (north-western France): New implications for their conservation management. *Biological Conservation,* **115**, 1–19.

Ebersohn, C. and Eicker, A. (1997). Determination of the coprophilous fungal fruit body successional phases and the delimitation of species association classes on dung substrates of African game animals. *Botanical Bulletin of Academia Sinica,* **38**, 183–190.

Edgington, L.V., Khew, K.L. and Barron, G.L. (1971). Fungitoxic spectrum of benzimidazole compounds. *Phytopathology,* **61**, 42–44.

Edwards, P.J. (1984). The growth of fairy rings of *Agaricus arvensis* and their effect upon grassland vegetation and soil. *Journal of Ecology,* **72**, 505–513.

Edwards, P.J. (1988). Effects of the fairy ring fungus *Agaricus arvensis* on nutrient availability in grassland. *New Phytologist*, **110**, 377–381.

Eriksson, J. (1949). Notes on *Corticium muscicola* Bres. and its taxonomical position. Svensk Botanisk Tidskrift, **43**, 310–315.

Feeney, D.S., Crawford, J.W., Daniell, T., Hallett, P.D., Nunan, N., Ritz, K., Rivers, M. and Young, I.M. (2006). Three-dimensional microorganization of the soil–root–microbe system. *Microbial Ecology*, **52**(1), 151–158.

Fermor, T.R. (1988). Significance of micro-organisms in the composting process for cultivation of edible fungi. In: *Treatment of Lignocellulosics with White Rot Fungi* (F. Zadrazil and P. Reiniger, eds.), pp. 21–30. Elsevier Applied Science Publishers Ltd., Essex, UK.

Fierer, N., Jackson, J.A., Vilgalys, R. and Jackson, R.B. (2005). Assessment of soil microbial community structure by use of taxon-specific quantitative PCR assays. *Applied and Environmental Microbiology*, **71**, 4117–4120.

Filer, T.H. (1965). Parasitic aspects of a fairy ring fungus *Marasmius oreades. Phytopathology*, **55**, 1132–1134.

Fisher, R.F. (1977). Nitrogen and phosphorus mobilization by fairy ring fungus, *Marasmius oreades* (Bolt.) Fr. *Soil Biology & Biochemistry*, **9**, 239–241.

Ford, P.L., Potter, D.U., Pendleton, R., Pendleton, B., Robbie, W.A. and Gottfried, G.J. (2004). Southwestern grassland ecology. In: *Assessment of Grassland Ecosystem Conditions in the Southwestern United States. Volume 1. General Technical Report RMRS-GTR-135-vol. 1* (D.M. Finch, ed.), pp. 18–48. U.S. Department of Agriculture, Forest Service, Rocky Mountain Research Station, Fort Collins, CO.

Fowler, D., O'Donoghue, M., Muller, J.B.A., Smith, R.I., Dragosits, U., Skiba, U., Sutton, M.A. and Brimblecombe, P. (2004). A chronology of nitrogen deposition in the UK between 1900 and 2000. *Water, Air and Soil Pollution: Focus*, **4**, 9–23.

Frankland, J.C. (1982). Biomass and nutrient cycling by decomposer basidiomycetes. In: *Decomposer Basidiomycetes: Their Biology and Ecology* (J.C. Frankland, J.N. Hedger and M.J. Swift, eds.), pp. 241–261. Cambridge University Press, Cambridge.

Freibauer, A., Rounsevell, M.D.A., Smith, P. and Verhagen, J. (2004). Carbon sequestration in agricultural soils in Europe. *Geoderma*, **122**, 1–23.

Frey, S.D., Elliott, E.T., Paustian, K. and Peterson, G.A. (2000). Fungal translocation as a mechanism for soil nitrogen inputs to surface residue decomposition in a no-tillage agroecosystem. *Soil Biology & Biochemistry*, **32**, 689–698.

Friedrich, K. (1940). Untersuchungen zur Okologie der hoheren Pilze. *Pflanzenforschung*, **22**, 1–52. Jena. Gustav Fischer.

Galloway, J.N., Schlesinger, W.H., Levy, H., Michaels, A. and Schnoor, J.L. (1995). Nitrogen fixation—Anthropogenic enhancement–environmental response. *Global Biogeochemical Cycles*, **9**, 235–252.

Gange, A.C., Gange, E.G., Sparks, T.H. and Boddy, L. (2007). Rapid and recent changes in fungal fruiting patterns. *Science* **316**, 71, in press.

Gardes, M. and Dahlberg, A. (1996). Mycorrhizal diversity in arctic and alpine tundra: An open question. *New Phytologist*, **133**, 147–157.

Garrett, S.D. (1951). Ecological groupings of soil fungi: A survey of substrate relationships. *New Phytologist*, **50**, 149–166.

Gilbert, J.H. (1875). The occurrence of 'fairy rings'. *Linnean Society's Journal – Botany*, **15**, 17–24.

Gramss, G. (1997). Activity of oxidative enzymes in fungal mycelia from grassland and forest soils. *Journal of Basic Microbiology*, **37**, 407–423.

Gramss, G., Ziegenhagen, D. and Sorge, S. (1999). Degradation of soil humic extract by wood- and soil-associated fungi, bacteria, and commercial enzymes. *Microbial Ecology*, **37**, 140–151.

Griffith, G.W., Aron, C., Evans, D.A., Evans, S., Graham, A., Holden, L. and Mitchel, D. (2006). *Mycological Survey of Selected Semi-Natural Grasslands in Wales*. Final Report August 2006, Contract FC-73-01-403, 55 pp.

Griffith, G.W., Bratton, J.L. and Easton, G.L. (2004). Charismatic megafungi: The conservation of waxcap grasslands. *British Wildlife*, **15**, 31–43.

Griffith, G.W., Easton, G.L. and Jones, A.W. (2002). Ecology and diversity of waxcap (*Hygrocybe* spp.) fungi. *Botanical Journal of Scotland*, **54**, 7–22.

Griffith, G.W. and Hedger, J.N. (1994). Spatial distribution of mycelia of the liana (L-) biotype of the agaric *Crinipellis perniciosa* (Stahel) Singer in tropical forest. *New Phytologist*, **127**, 243–259.
Halisky, P.M. and Peterson, J.L. (1970). Basidiomycetes associated with fairy rings in turf. *Bulletin of the Torrey Botanical Club*, **97**, 225–227.
Hammel, K.E. (1997). Fungal degradation of lignin. In: *Driven by Nature—Plant Litter Quality and Decomposition* (G. Cadisch and K.E. Giller, eds.). CAB International, Wallingford, UK, 33–45.
Harper, J.E. and Webster, J. (1964). An experimental analysis of coprophilous fungus succession. *British Mycological Society Transactions*, **47**, 511–530.
Harrington, T.J. and Mitchell, D.T. (2002). Colonization of root systems of *Carex flacca* and *C. pilulifera* by *Cortinarius (Dermocybe) cinnamomeus*. *Mycological Research*, **106**, 452–459.
Haselwandter, K. and Read, D.J. (1982). The significance of a root–fungus association in two *Carex* species of high alpine plant communities. *Oecologia*, **53**, 352–354.
Heal, O.W., Anderson, J.M. and Swift, M.J. (1997). Plant litter quality and decomposition: An historical review. In: *Driven by Nature—Plant Litter Quality and Decomposition* (G. Cadisch and K.E. Giller, eds.), pp. 3–30. CAB International, Wallingford, UK.
Hibbett, D.S., Gilbert, L.B. and Donoghue, M.J. (2000). Evolutionary instability of ectomycorrhizal symbioses in basidiomycetes. *Nature*, **407**, 506–508.
Holdo, R.M., Holt, R.D. and Coughen, M.B. (2007). Plant productivity and soil nitrogen as a function of grazing, migration and fire in an African savanna. *Journal of Ecology*, **95**, 115–128.
Horton, T.R. and Bruns, T.D. (2001). The molecular revolution in ectomycorrhizal ecology: Peeking into the black-box. *Molecular Ecology*, **10**, 1855–1871.
Hudson, H.J. (1968). The ecology of plant remains above the soil. *New Phytologist*, **67**, 837–874.
Hunt, J., Boddy, L., Randerson, P.F. and Rogers, H.J. (2004). An evaluation of 18S rDNA approaches for the study of fungal diversity in grassland soils. *Microbial Ecology*, **47**, 385–395.
Hutton, S.A. and Giller, P.S. (2003). The effects of the intensification of agriculture on northern temperate dung beetle communities. *Journal of Applied Ecology*, **40**, 994–1007.
Iiyama, K., Lam, T.B.T. and Stone, B.A. (1990). Phenolic acid bridges between polysaccharides and lignin in wheat internodes. *Phytochemistry*, **29**, 733–737.
Ikediugwu, F.E.O. and Webster, J. (1970). Antagonism between *Coprinus heptemerus* and other coprophilous fungi. *Transactions of the British Mycological Society*, **54**, 181–187.
Jumpponen, A. and Johnson, L.C. (2005). Can rDNA analyses of diverse fungal communities in soil and roots detect effects of environmental manipulations? A case study from tallgrass prairie. *Mycologia*, **97**, 1177–1194.
Kaiser, P. (1998). Relations of *Leucopaxillus giganteus*, basidiomycete of fairy rings, with soil microflora and grassland plants. *Cryptogamie Mycologie*, **19**, 45–61.
Keay, S.M. and Brown, A.E. (1990). Colonization by *Psilocybe semilanceata* of roots of grassland flora. *Mycological Research*, **94**, 49–56.
Kennedy, N., Brodie, E., Connolly, J. and Clipson, N. (2006). Seasonal influences on fungal community structure in unimproved and improved upland grassland soils. *Canadian Journal of Microbiology*, **52**, 689–694.
Kirby, K.J. (2003). What might a British forest-landscape driven by large herbivores look like? *English Nature Research Reports Number 530*.
Kluczek-Turpeinen, B., Tuomela, M., Hatakka, A. and Hofrichter, M. (2003). Lignin degradation in a compost environment by the deuteromycete *Paecilomyces inflatus*. *Applied Microbiology and Biotechnology*, **61**, 374–379.
Kreisel, H. and Ritter, G. (1985). Okologie der Grosspilze. In: *Handbuch Pilzfr* (M. Hennig and H. Kresiel, eds.), pp. 9–47. Gustav Fischer, Jena.
Lam, T.B.T., Iiyama, T.K. and Stone, B.A. (1992). Cinnamic acid bridges between cell wall polymers in wheat and *Phalaris* inter-nodes. *Phytochemistry*, **31**, 1179–1183.
Lange, M. (1991). Fleshy fungi of grass fields. 3. Reaction to different fertilizers and to age of grass turf—Periodicity of fruiting. *Nordic Journal of Botany*, **11**, 359–368.
Lapierre, C., Jouin, D. and Monties, B. (1989). On the molecular origin of the alkali solubility of Gramineae lignins. *Phytochemistry*, **28**, 1401–1403.

Lawes, J.B., Gilbert, J.H. and Warington, R. (1883). Contributions to the chemistry of "fairy rings". *Journal of the Chemical Society (London)*, **43**, 208–223.
Lilleskov, E.A., Fahey, T.J., Horton, T.R. and Lovett, G.M. (2002). Belowground ectomycorrhizal fungal community change over a nitrogen deposition gradient in Alaska. *Ecology*, **83**, 104–115.
Lindahl, B.D., Ihrmark, K., Boberg, J., Trumbore, S.E., Högberg, P., Stenlid, J. and Finlay, R.D. (2007). Spatial separation of litter decomposition and mycorrhizal nitrogen uptake in a boreal forest. *New Phytologist*, **173**, 611–620.
Lisiewska, M. (1992). Macrofungal on special substrates. In: *Fungi in Vegetation Science* (W. Winterhoff, ed.), pp. 151–182. Kluwer Academic Publishers, Dordrecht, Netherlands.
Lussenhop, J. and Wicklow, D.T. (1985). Interaction of competing fungi with fly larvae. *Microbial Ecology*, **11**, 175–182.
Lynch, M.D.J. and Thorn, R.G. (2006). Diversity of basidiomycetes in Michigan agricultural soils. *Applied and Environmental Microbiology*, **72**, 7050–7056.
McGranaghan, P., Davies, J.C., Griffith, G.W., Davies, D.R. and Theodorou, M.K. (1999). The survival of anaerobic fungi in cattle faeces. *FEMS Microbiology Ecology*, **29**, 293–300.
McKay, R.I. (1968). The association of higher fungi and sand-dune grasses, PhD Thesis, Department of Botany, University of St. Andrews, St. Andrews.
Mitchell, F.J.G. (2005). How open were European primeval forests? Hypothesis testing using palaeo-ecological data. *Journal of Ecology*, **93**, 168–177.
Morecroft, M.D., Taylor, M.E. and Oliver, H.R. (1998). Air and soil microclimates of deciduous woodland compared to an open site. *Agricultural and Forest Meteorology*, **90**, 141–156.
Nagy, L.A. and Harrower, K.M. (1980). Coprous and non-coprous decomposition. *Transactions of the British Mycological Society*, **74**, 639–641.
O'Brien, H.E., Parrent, J.L., Jackson, J.A., Moncalvo, J.M. and Vilgalys, R. (2005). Fungal community analysis by large-scale sequencing of environmental samples. *Applied and Environmental Microbiology*, **71**, 5544–5550.
O'Connor, K.F. (1983). Nitrogen balances in natural grasslands and extensively managed grassland systems. *New Zealand Journal of Ecology*, **6**, 1–18.
Parker-Rhodes, A.F. (1951). The basidiomycetes of Skokholm Island. VII. Some floristic and ecological calculations. *New Phytologist*, **50**, 227–243.
Parker-Rhodes, A.F. (1952). The basidiomycetes of Skokholm Island. VIII. Taxonomic distributions. *New Phytologist*, **51**, 216–228.
Parker-Rhodes, A.F. (1955). Fairy ring kinetics. *Transactions of the British Mycological Society*, **38**, 59–72.
Parton, W.J., Scurlock, J.M.O., Ojima, D.S., Schimel, D.S. and Hall, D.O. (1995). Impact of climate change on grassland production and soil carbon worldwide. *Global Change Biology*, **1**, 13–22.
Phoenix, G.K., Hicks, W.K., Cinderby, S., Kuylenstierna, J.C.I., Stock, W.D., Dentener, F.J., Giller, K.E., Austin, A.T., Lefroy, R.D.B., Gimeno, B.S., Ashmore, M.R. and Ineson, P. (2006). Atmospheric nitrogen deposition in world biodiversity hotspots: The need for a greater global perspective in assessing N deposition impacts. *Global Change Biology*, **12**, 470–476.
Raich, J.W. and Tufekcioglu, A. (2000). Vegetation and soil respiration: Correlations and controls. *Biogeochemistry*, **48**, 71–90.
Rald, E. (1985). Vokshatte som indikatorarter for mykologisk vaerdifulde overdrevslokaliteter. *Svampe*, **11**, 1–9.
Rayner, A.D.M. and Boddy, L. (1988). *Fungal Decomposition of Wood: Its Biology and Ecology*. Wiley, New York, NY.
Read, D.J. and Perez-Moreno, J. (2003). Mycorrhizas and nutrient cycling in ecosystems—A journey towards relevance? *New Phytologist*, **157**, 475–492.
Reijs, J., Meijer, W.H., Bakker, E.J. and Lantinga, E.A. (2003). Explorative research into quality of slurry manure from dairy farms with different feeding strategies. *Njas-Wageningen Journal of Life Sciences*, **51**, 67–89.
Richardson, M.J. (2001). Diversity and occurrence of coprophilous fungi. *Mycological Research*, **105**, 387–402.
Rillig, M.C. and Mummey, D.L. (2006). Mycorrhizas and soil structure. *New Phytologist*, **171**, 41–53.
Roberts, P. (1999). *Rhizoctonia-Forming Fungi: A Taxonomic Guide*. Kew, Royal Botanic Gardens, London.

Robinson, C.H., Pryce Miller, E.J. and Deacon, L. (2005). Biodiversity of saprotrophic fungi in relation to their function: Do fungi obey the rules? In: *Biological Diversity and Function of Soils* (R.D. Bardgett, M.B. Usher and D.W. Hopkins, eds.), pp. 189–215. Cambridge University Press, Cambridge.

Rodriguez, A., Perestelo, F., Carnicero, A., Regalado, V., Perez, R., DelaFuente, G. and Falcon, M.A. (1996). Degradation of natural lignins and lignocellulosic substrates by soil-inhabiting fungi imperfecti. *FEMS Microbiology Ecology,* **21**, 213–219.

Rook, A.J. and Tallowin, J.R.B. (2003). Grazing and pasture management for biodiversity benefit. *Animal Research,* **52**, 181–189.

Rotheroe, M., Newton, A., Evans, S. and Feehan, J. (1996). Waxcap-grassland survey. *The Mycologist,* **10**, 23–25.

Safar, H.M. and Cooke, R.C. (1988). Interactions between bacteria and coprophilous Ascomycotina and a *Coprinus* species on agar and in copromes. *Transactions of the British Mycological Society,* **91**, 73–80.

Sankaran, M. and Augustine, D.J. (2004). Large herbivores suppress decomposer abundance in a semiarid grazing ecosystem. *Ecology,* **85**, 1052–1061.

Scheel, T., Holker, U., Ludwig, S. and Hofer, M. (1999). Evidence for and expression of a laccase gene in three basidiomycetes degrading humic acids. *Applied Microbiology and Biotechnology,* **52**, 66–69.

Sessitsch, A., Hackl, E., Wenzl, P., Kilian, A., Kostic, T., Stralis-Pavese, N., Sandjong, B.T. and Bodrossy, L. (2006). Diagnostic microbial microarrays in soil ecology. *New Phytologist,* **171**, 719–736.

Shantz, H.L. and Piemeisel, R.L. (1917). Fungus fairy rings in eastern Colorado and their effect on vegetation. *Journal of Agricultural Research,* **11**, 191–245.

Smith, J.D. (1980). Is biologic control of *Marasmius oreades* fairy rings possible? *Plant Disease,* **64**, 348–355.

Smith, D.L. and Johnson, L. (2004). Vegetation-mediated changes in microclimate reduce soil respiration as woodlands expand into grasslands. *Ecology,* **85**, 3348–3361.

Smith, S.E. and Read, D.J. (1997). *Mycorrhizal Symbiosis*. Academic Press, London, UK.

Stark, S. and Grellmann, D. (2002). Soil microbial responses to herbivory in an arctic tundra heath at two levels of nutrient availability. *Ecology,* **83**, 2736–2744.

Steffen, K.T., Hatakka, A. and Hofrichter, M. (2002). Degradation of humic acids by the litter-decomposing basidiomycete *Collybia dryophila. Applied and Environmental Microbiology,* **68**, 3442–3448.

Steffen, K.T., Hofrichter, M. and Hatakka, A. (2000). Mineralisation of C^{14}-labelled synthetic lignin and ligninolytic enzyme activities of litter-decomposing basidiomycetous fungi. *Applied Microbiology and Biotechnology,* **54**, 819–825.

Stevens, C.J., Dise, N.B., Mountford, O. and Gowing, D.J. (2004). Impact of nitrogen deposition on the species richness of grasslands. *Science,* **303**, 1876–1879.

Swift, M.J., Heal, O.W. and Anderson, J.M. (1979). *Decomposition in Terrestrial Ecosystems*. Blackwell Scientific Publications, Oxford.

Terashima, Y. and Fujiie, A. (2005). Development of a fairy ring caused by *Lepista sordida* on zoysigrass over and eight year period. *International Turfgrass Society Research Journal,* **10**, 251–257.

Terashima, Y., Fukiharu, T. and Fujiie, A. (2004). Morphology and comparative ecology of the fairy ring fungi, *Vascellum curtisii* and *Bovista dermoxantha*, on turf of bentgrass, bluegrass, and Zoysiagrass. *Mycoscience,* **45**, 251–260.

Thorn, R.G., Reddy, C.A., Harris, D. and Paul, E.A. (1996). Isolation of saprophytic basidiomycetes from soil. *Applied and Environmental Microbiology,* **62**, 4288–4292.

Tisdall, J.M. and Oades, J.M. (1982). Organic-matter and water-stable aggregates in soils. *Journal of Soil Science,* **33**, 141–163.

Tisdall, J.M., Smith, S.E. and Rengasamy, P. (1997). Aggregation of soil by fungal hyphae. *Australian Journal of Soil Research,* **35**, 55–60.

Turner, C.L., Seastedt, T.R. and Dyer, M.I. (1993). Maximization of aboveground grassland production — The role of defoliation frequency, intensity, and history. *Ecological Applications,* **3**, 175–186.

van der Wal, A., van Veen, J.A., Smant, W., Boschker, H.T.S., Bloem, J., Kardol, P., van der Putten, W.H. and de Boer, W. (2006). Fungal biomass development in a chronosequence of land abandonment. *Soil Biology & Biochemistry,* **38**, 51–60.

Vera, F.W.M. (2000). *Grazing Ecology and Forest History.* CABI Publishing, New York, NY.

Waksman, S.A., Cordon, T.C. and Hulpoi, N. (1939). Influence of temperature upon the microbiological population and decomposition processes in composts of stable manure. *Soil Science*, **47**, 83–113.

Warcup, J.H. (1951a). Studies on the growth of basidiomycetes in soil. *Annals of Botany*, **15**, 305–318.

Warcup, J.H. (1951b). The ecology of soil fungi. *Transactions of the British Mycological Society*, **34**, 376–399.

Warcup, J.H. (1959). Studies on the growth of basidiomycetes in soil. *Annals of Botany*, **15**, 305–317.

Warcup, J.H. and Talbot, P.H.B. (1962). Ecology and identity of mycelia isolated from soil. *Transactions of the British Mycological Society*, **45**, 495–518.

Warren, J. and Paul, A. (2006). A comparison of invertebrate populations and their relationship with rates of faecal decomposition in organic and conventionally managed pastures. *Biological Agriculture and Horticulture*, **24**, 61–71.

Watling, R. (1995). Assessment of fungal diversity—Macromycetes, the problems. *Canadian Journal of Botany*, **73**(Suppl. 1), S1–S24.

Webster, J. (1970). Coprophilous fungi. *Transactions of the British Mycological Society*, **54**, 161–180.

White, N.A., Hallett, P.D., Feeney, D., Palfreyman, J.W. and Ritz, K. (2000). Changes to water repellence of soil caused by the growth of white-rot fungi: Studies using a novel microcosm system. *FEMS Microbiology Letters*, **184**, 73–77.

Wicklow, D.T., Angel, S.K. and Lussenhop, J. (1980a). Fungal community expression in lagomorph versus ruminant feces. *Mycologia*, **72**, 1015–1021.

Wicklow, D.T., Detroy, R.W. and Adams, S. (1980b). Differential modification of the lignin and cellulose components in wheat straw by fungal colonists of ruminant dung—Ecological implications. *Mycologia*, **72**, 1065–1076.

Wicklow, D.T. and Moore, V. (1974). Effect of incubation temperature on coprophilous fungal succession. *Transactions of the British Mycological Society*, **62**, 411–415.

Wicklow, D.T. and Yocom, D.H. (1981). Fungal species numbers and decomposition of rabbit faeces. *Transactions of the British Mycological Society*, **76**, 29–32.

Wilberforce, E.M. (2003). The ecology of root endophytic fungi in semi-natural grasslands, PhD Thesis, University of Wales, Wales.

Wilberforce, E.M., Boddy, L., Griffiths, R. and Griffith, G.W. (2003). Agricultural management affects communities of culturable root—Endophytic fungi in temperate grasslands. *Soil Biology & Biochemistry*, **35**, 1143–1154.

Wilkins, W.H. and Patrick, S.H.M. (1939). The ecology of the larger fungi. III. Constancy and frequency of grassland species with special reference to soil types. *Annals of Applied Biology*, **26**, 25–46.

Wilkins, W.H. and Patrick, S.H.M. (1940). The ecology of the larger fungi. IV. The seasonal frequency of grassland fungi with special reference to the influence of environmental factors and soil types. *Annals of Applied Biology*, **27**, 17–34.

York, C.A. and Canaway, P.M. (2000). Water repellent soils as they occur on UK golf greens. *Journal of Hydrology*, **231**, 126–133.

Zhang, W., Wendel, J.F.W. and Clark, L.G. (1997). Bamboozled again! Inadvertent isolation of fungal rDNA sequences from bamboos (Poaceae: Bambusoideae). *Molecular Phylogenetics and Evolution*, **8**, 205–217.

CHAPTER 16

Ecology of Marine and Freshwater Basidiomycetes

E. B. Gareth Jones and **Rattaket Choeyklin**

Contents

Abstract

Marine and freshwater basidiomycetes are few in number compared to their terrestrial counterparts and colonize a wide range of substrata: sea-grasses, feathers, wood associated with sand, free floating in the sea, but most occur on mangrove wood or timbers submerged in the sea (boats, piling, sea defences), and leaves and twigs in streams and rivers. They are an ecological group and taxonomically diverse (Agariomycotina, Uredinomycotina and Ustilaginomycotina). Most are able to utilize simple carbohydrates, while filamentous species can decompose cellulose, hemicellulose and lignin. Aquatic basidiomycetes are well adapted to their habitats, with reduced basidiomata. Marine species are known only as teleomorphs with basidiospores generally released passively. Freshwater basidiomycetes are primarily known by their anamorphs on decaying leaves, with conidia that are much branched, while their teleomorphs occur on land and on woody substrata.

1. INTRODUCTION

Aquatic basidiomycetes are an ecological group and taxonomically very diverse (Table 1). Numerically they are not numerous and this is dictated by environmental conditions. Although a variety of substrata are available for colonization

British Mycological Society Symposia Series
Published by Elsevier Ltd.

Table 1 Summary of the characteristics of marine and freshwater basidiomycetes

Classification	Teleomorph	Anamorph	Clamp	Substratum	Cellulose utilization	Lignin activity	Reference
Marine							
Agaricomycotina Agaricomycetes Agaricales *Nia* clade	*Calathella mangrovei*	—	+	Mangrove wood	0	0	Jones and Agerer (1992)
Agaricomycotina Agaricomycetes Agaricales *Nia* clade	*Halocyphina villosa*	—	+	Mangrove wood	+	+	Kohlmeyer and Kohlmeyer (1965)
Agaricomycotina Agaricomycetes Agaricales *Nia* clade	*Nia vibrissa*	—	+	Many timber species mangrove wood	+	+	Moore and Meyers (1959)
Agaricomycotina Agaricomycetes Agaricales *Nia* clade	*N. globospora*	—	+	*Spartina maritima* culm baits	0	0	Barata *et al.* (1997)
Agaricomycotina Agaricomycetes Agaricales *Nia* clade	*N. epidermoidea*	—	+	Feathers in sand	0	0	Rossello *et al.* (1993)
Agaricomycotina Agaricomycetes Agaricales Physalacriaceae	*Mycaureola dilseae*	—	+	Red alga *Dilsea carnosa*	0	0	Porter and Farnham (1986)

Agaricomycotina Agaricomycetes Agaricales Physalacriaceae	*Physalacria maipoensis*	—	+	Mangrove substrata	0	0	Inderbitzin and Desjardin (1999)
Agaricomycotina Agaricomycetes Atheliaceae *Digitatispora* clade	*Digitatispora lignicola*	—	+	Wood	0	0	Jones (1986)
Agaricomycotina Agaricomycetes Atheliaceae *Digitatispora* clade	*D. marina*	—	+	Wood	+	+	Doguet (1962)
Agaricomycotina Agaricomycetes Russulales Peniophoraceae	*Haloaleurodiscus mangrovei*	—	+	Mangrove wood	0	0	Maekawa *et al.* (2005)
Agaricomycotina Agaricomycetes Polyporales Hyphodermataceae	*Bulbillomyces sp.*	*Aegerita sp.*	—	Mangrove wood, *Acanthus ilicifolius* stems	0	0	Sadaba *et al.* (1995), Jones (unpublished data)
Agaricomycotina Tremellomycetes Cystofilobasidiales	*Cystofilobasidium bisporidii*	—	+	Seawater	0	0	Fell *et al.* (1973, 2001)
Agaricomycotina Tremellomycetes Cystofilobasidiales	*C. capitatum*	—	—	Seawater	0	0	Fell *et al.* (1973, 2001)
Agaricomycotina Tremellomycetes Cystofilobasidiales	*C. infirmominiatum*	—	+ few	Seawater	0	0	Fell *et al.* (1973, 2001)

(continued)

Table 1 Continued

Classification	Teleomorph	Anamorph	Clamp	Substratum	Cellulose utilization	Lignin activity	Reference
Agaricomycotina Tremellomycetes Cystofilobasidiales	*Rhodosporidium diobvatum*	—	?	Seawater, mangrove detritus	0	0	Fell *et al.* (1973, 2001)
Agaricomycotina Tremellomycetes Cystofilobasidiales	*Rh. paludigenum*	—	?	Mangrove and *Juncus roemerianus*, marshes	0	0	Fell *et al.* (1973, 2001)
Agaricomycotina Tremellomycetes Cystofilobasidiales	*Rh. sphaerocarpum*	—	?	Seawater	0	0	Fell *et al.* (1973, 2001)
Ustilaginomycotina Urtilaginomycetes Melanotaeniaceae	*Melanotaenium ruppiae*	—	—	Marine angiosperm *Ruppia maritima*	0	0	Feldmann (1959)
Pucciniomycotina Agaricostilbomycetes Agaricostilbales Agaricostilbaceae	—	*Sterigmatomyces halophilus*	—	Seawater	0	0	Fell *et al.* (2001), Kurtzman and Fell (2006)
Puccinimycotina Microbotryomycetes Sporidiobolales? Microbotryales?	*Sakaguchia dacryoidea*	—	+	Seawater	0	0	Fell *et al.* (1973, 2001), Kurtzman and Fell (2006)
Freshwater							
Agaricomycotina Agaricomycetes Agaricales *Nia* clade	*Peyronelina glomerulata*	+	?	Wood	0	0	Yamaguchi *et al.* (2006)

Agaricomycotina Agaricomycetes Agaricales Cyotellaceae	*Limnoperdon incarnatum*	—	+	Hardwood twigs, marsh water	0	0	Escobar *et al.* (1976)
Agaricomycotina Agaricomycetes Agaricales Eugarics clade Physalacriaceae clade	*Gloiocephala aquatica*	—	+	Submerged culms	—	—	Desjardin *et al.* (1995)
Agaricomycotina Agaricomycetes Polyporales Hyphodermataceae	*Bulbillomyces farinosus*	*Aegerita candida*	+	Wood	0	0	
	Subulicystidium longisporum	*A. tortuosa*	+	Wood	0	0	
Agaricomycotina Agaricomycetes Cantharellales Sistotremataceae	*Sistotrema hamatum*	*Ingoldiella hamata*	+	Wood	0	0	Shaw (1972), Nawawi and Webster (1982)
Agaricomycetes Polyporales Hyphodermataceae	—	*I. fibulata*	+	Foam	0	0	Nawawi (1973, 1985)
Agaricomycetes Polyporales Hyphodermataceae	—	*I. nutans*	+		0	0	Bandoni and Marvanová (1989)
Agaricomycotina Tremellomycetes Tremellaels Sirobasidiaceae	*Xenolchne flagellifera*	—	+	Logs	0	0	Rogers (1947)

(continued)

Table 1 Continued

Classification	Teleomorph	Anamorph	Clamp	Substratum	Cellulose utilization	Lignin activity	Reference
Agaricomycotina Tremellomycetes Tremellales Sirobasidiaceae? Filobasidiales Filobasidiaceae	—	*Cryptococcus aquaticus*	—	Foam	0	0	Marvanová and Bärlocher (1998), Fell *et al*. (2001)
Agaricomycotina Tremellomycetes Cystofilobasidiales		*Rhodotorula ferulica*	?	Freshwater	0	0	Kurtzman and Fell (1998)
Basidiomycota Incertae sedis	—	*Stauriella aquatica*	+	Submerged test blocks of *Dipterocarpus alatus*	0	0	Sivichai and Jones (2004)
Basidiomycota Incertae sedis	—	*Dendrosporomyces prolifer*	Dolipore septum	Freshwater foam	0	0	Nawawi *et al*. (1977b)
Basidiomycota Incertae sedis	—	*D. splendens*	Dolipore septum	Decaying leaves, twigs, foam	0	0	Nawawi and Webster (1982)
Basidiomycota Incertae sedis	—	*Tricladiomyces geniculatus*	Dolipore septum	On leaves submerged in a stream	0	0	Nawawi and Kuthubutheen (1988)
Basidiomycota Incertae sedis	—	*T. malaysianus*	Dolipore septum	Submerged decaying leaves	0	0	Nawawi *et al*. (1977a), Nawawi (1985)
Basidiomycota Incertae sedis	*Fibulomyces crucelliger*	*Taeniospora descalsii*	+		0	0	Marvanová and Stalpers (1987)
Basidiomycota Incertae sedis	—	*T. gracilis var. enecta*	+	Foam	0	0	Marvanová (1997), Nawawi *et al*. (1977a)
Basidiomycota Incertae sedis	—	*T. gracilis var. gracilis*	+	Water, foam	0	0	Nawawi *et al*. (1977a)

Basidiomycota Incertae sedis	—	*T. nasifera*	+	Foam	0	0	Marvanová and Bärlocher (1988)
Basidiomycota Incertae sedis	—	*Fibulotaeniella canadensis*	+	Foam	0	0	Marvanová and Bärlocher (1988)
Basidiomycota Incertae sedis	—	*Anguillomyces acadiensis*	+?	Foam	0	0	Marvanová and Bärhlocher (2000)
Basidiomycota Incertae sedis	—	*Nodulospora inconstans*	+	Foam,	0	0	Marvanová and Bärlocher (2000)
Pucciniomyctoina Microbotryomycetes Sporidiobolales Sporidiobolaceae	*Rogersiomyces okefenokeensis*	—	+	Decaying leaves	0	0	Crane and Schoknecht (1978)
Pucciniomycotina Pucciniomycetes Platygloeales Platygloeaceae Incertae sedis	*Camptobasidium hydrophilum*	*Crucella subtilis*	+	Decaying leaves in a stream	0	0	Marvanová and Suberkropp (1990)
Pucciniomycotina Classiculomycetes Classiculales Classiculaceae	*Classicula fluitans*	*Naiadella fluitans*	+	*Scirpus microcarpus* leaf litter, foam	0	0	Marvanová and Bandoni (1987)
Basidiomycota Incertae sedis	—	*Jaculispora submersa*	—	Leaf litter	0	0	Hudson (1961), Bauer *et al.* (2003)
Basidiomycota Platygloeaceae? Incertae sedis	*Helicogloea angustispora*	*Infundibura adhaerens*	?	Foam	0	0	Kirschner (2004)
Basidiomycota Incertae sedis	—	*Reniforma strues*	?	Wastewater	0	0	Pore and Sorenson (1990), Kurtzman and Fell (2006)

Note: +, Positive character; –, negative; 0, not determined.

in aquatic habitats, wood and leaves have received the greatest attention. Anamorphic fungi and ascomycetes appear to be better able to withstand the saturated conditions prevailing in such habitats, far more so than basidiomycetes (Jones, 1972; Shearer, 1993). Microfungi cause soft-rot decay of wood in such environments with hyphae penetrating the S2 layer of the wood cell walls (Mouzouras *et al.*, 1987) with penetration from cell to cell and the formation of T-branches leading to typical diamond shaped cavities (Mouzouras, 1986). This feature may confer on them a competitive advantage over basidiomycetes who degrade the wood from the cell lumen inwards (Mouzouras *et al.*, 1987). Thus enzymes may be leached into the surrounding water.

Materials that fall into aquatic systems are colonized by a variety of organisms. In the marine environment bacteria and actinomycetes, are the initial colonizers but do not penetrate deeply into wood, this being accomplished by microfungi (ascomycetes and anamorphic fungi). Basidiomycetes generally appear later in the succession (Byrne and Jones, 1974; Alias and Jones, 2000a). Little is known of the sequence of wood colonization by basidiomycetes in freshwater as few studies have examined this aspect in any detail (Sivichai *et al.*, 2000, 2002; Kane *et al.*, 2002).

2. AQUATIC BASIDIOMYCETES AND THEIR TAXONOMIC RELATIONSHIP

Aquatic Basidiomycota are a minority group when compared with the ascomycetes, anamorphic fungi and Chromista (Table 2). Taxonomically they are unrelated, belonging to such diverse groups as the Agaricomycotina, Uredinomyctoina and Ustilaginomycotina (Table 1). Filamentous Agaricomycotina are represented by the orders Agaricales, Atheliales, Cantharellales, Polyporales, Russulales and Tremellales. Basidiomycete yeasts are found in four lineages in the Uredinomycotina (Agaricostilbomycetes, Microbotryomycetes, Naohidea clade, Urediniomycetes), and the Ustilaginomycotina (Scorzetti *et al.*, 2002; Kurtzman and Fell, 2006).

Although only 11 marine filamentous basidiomycetes are documented, others occur on terrestrial mangrove timbers: 26 species (Chalermpongse, 1991; Schmidt and Shearer, 2003). At the time of writing we had collected 12 basidiomycetes on the intertidal bases of the palm *Nypa fruticans*, for example *Grammothele fuligo*

Table 2 Numbers of freshwater and marine fungi

Taxonomic group	Freshwater	Marine
Basidiomycota	29	20+
Ascomycota	650	439
Anamorphic fungi	660	72
Total	1,337	530

Source: After Hyde *et al*. (2000) and Tsui and Hyde (2003).

Figure 1 *Grammothele fuligo* growing on the decayed frond base, in ground contact, of the mangrove palm *Nypa fruitcans*.

(Figure 1). Mangrove trees of *Xylocarpus granatum* at Khanom National Park, south Thailand, were infected with butt rot with *Phellinus* species the possible causative agents (Figure 2). Agarics have also been found on mangrove soils, especially *Coprinus, Cortinarius* and *Mycena* species; they are short lived, fruit during the intertidal period and shed their spores before the tide returns (Jones, personal observation). These basidiomycetes have received little attention to date and further investigation for their adaptation to semi-aquatic habitats is warranted.

Figure 2 Butt rot of the mangrove tree *Xylocarpus granatum* at Hat Khanom-Mu Ko Thale Tai National Park, Thailand caused by *Phellinus* species.

3. OCCURRENCE AND DISTRIBUTION OF AQUATIC BASIDIOMYCETES

Freshwater basidiomycetes have been isolated from foam samples (Jones and Sloof, 1966) and from senescent decaying leaf litter (Nawawi *et al.*, 1977a, 1977b; Marvanová and Stalpers, 1987; Nawawi and Kuthubutheen, 1988), while *Stauriella aquatica* was recovered from test blocks of *Dipterocarpus alatus* submerged for 9 months in a stream in Khao Yai National Park, Thailand (Sivichai and Jones, 2004). In a study of fungal colonization of senescent palms submerged in a peat swamp at Narathiwat, Thailand, only *Ingoldiella hamata* was recovered, although other basidiomycetes did occur on fronds in ground contact (Pinruan, 2004; Pinnoi *et al.*, 2006). An aero-aquatic hyphomycete, *Peyronelina glomerulata* (Fisher *et al.*, 1976; Kane *et al.*, 2002), sporulated on submerged wood after one week's incubation in the laboratory and has been shown to have a basidiomycete teleomorph (Yamaguchi *et al.*, 2006). Most freshwater basidiomycetes occur as the anamorph stage, with prolific conidial production on leaf material, ensuring wide dispersal and their entrapment to suitable substrata (Read *et al.*, 1992; Jones, 1994). Freshwater basidiomycetes however, are few in number compared to the Ascomycota and anamorphic taxa (Bärlocher, 1992; Tsui and Hyde, 2003).

Marine species have been described from wood (*Digitatispora lignicola*, *D. marina*, *Nia vibrissa*) including mangrove (*Calathella mangrovei*, *Haloaleurodiscus marina*), *Spartina maritima* culm baits (*N. globospora*), feathers in contact with sand (*N. epidermoidea*), and as a parasite of the marine angiosperm grass *Ruppia maritima* (*Melanoetenium ruppiae*) (see Table 1 for references). *Mycaureola dilseae* and *Melan. ruppiae* are unique amongst marine fungi in being host-specific on *Dilsea carnosa* and *R. maritima*, respectively (Stanley, 1992; Binder *et al.*, 2006).

Biogeographically marine basidiomycetes are either: (1) restricted to temperate cold water zones, for example *D. lignicola, D. marina, M. dilseae* (e.g. Henningsson, 1974; Booth, 1983; Koch and Peterson, 1996); (2) restricted to tropical mangrove locations, for example *C. mangrovei, Haloc. villosa, Haloc. marina* (e.g. Hyde, 1986, 1988; Maekawa *et al.*, 2005; Jones and Puglisi, 2006; Jones *et al.*, 2006); (3) cosmopolitan, for example *N. vibrissa* (Hyde, 1986; Jones and Vrijmoed, 2003; Jones *et al.*, 2006); (4) known only from their original locations (Rossello *et al.*, 1993; Barata *et al.*, 1997).

Of the 11 marine basidiomycetes, most information is on *C. mangrovei, H. villosa* and *N. vibrissa* with Schmidt and Shearer (2003) reporting them from 10, 31 and 20 countries, respectively, and from the Indian, Pacific and Atlantic Oceans, with the exception of *C. mangrovei* which has not been reported from the latter. *Halocyphina villosa* has been reported from a wide range of mangrove timbers: 15 in Brunei (Hyde, 1990a), 4 in Singapore but more frequently on *Avicennia alba* (Leong *et al.*, 1991), 9 on the west coast of India with an average frequency of occurrence of 13.7% (Mari and Sridhar, 2002), 3 on the south west coast of India (Mari and Sridhar, 2003), 3 and more frequently on *Nypa. fruticans* in the Philippines (Besitulo *et al.*, 2002). In Malaysia, *Haloc. villosa* and *C. mangrovei* had a frequency of occurrence of 11.6 (590 collections) and 1.2% (60 collections) respectively (Alias *et al.*, 2007). *Halocyphina villosa* occurs widely within different geographical locations, for example at six of eight sites in Brunei, and commonly at sites in the Andaman and Nicobar Islands (Hyde, 1988; Chinnaraj, 1993).

Few studies have followed the pattern of basidiomycete colonization of submerged wood. In temperate locations *N. vibrissa* was recorded on submerged beech test blocks in the UK at Port Erin (24, 36 weeks submergence), Newton Ferrers (16 weeks), *D. marina* occurred on beech test blocks at Newton Ferrers (32, 40 weeks) while neither were reported on similar test blocks in Langstone Harbour (Byrne and Jones, 1974). On submerged test blocks of four mangrove species, *Haloc. villosa* occurred frequently on *A. alba* and *A. lanata*, but was infrequent on *B. cylindrica*, and *Rh. apiculata* (Tan *et al.*, 1989; Leong *et al.*, 1991). *H. villosa* was present at all stages in the succession on *Bruguiera. cylindrica*, but occurred only on the intermediate and later stages of the succession on *Rh. apiculata* (Leong *et al.*, 1991). However, in other studies basidiomycetes have not featured in the colonization of submerged wood (Abdel-Wahab and El-Sharouny, 2002). *Physalacria maipoensis* may be an early colonizer of senescent *Acanthus illicifolius* in mangrove habitats (Jones, personal observations). *N. vibrissa* occurred frequently on timbers of the Tudor ship, The Mary Rose, in the UK, especially when sprayed with chilled water (Jones and Jones, 1993). The occurrence of *D. lignicola, D. marina,* and *M. dilseae* may be dictated by low water temperature, that is below 15°C (Jones, 1985, 1986; Stanley, 1992). This may also govern enzyme activity (see below).

Different species have been shown to be vertically distributed on prop roots of *Rhizophora apiculata* at Morib, Malaysia, *Haloc. villosa* was the only marine basidiomycete found and occurred in the upper (1.8–2.2 m) and middle (0.8–1.8 m) zones (Alias, 1996; Alias and Jones, 2000b); while it occurred at all levels in the intertidal zone at Kampong Kapok, Brunei and Ranong mangrove, Thailand (Hyde, 1989, 1990a, 1990b; Hyde *et al.*, 1993). Similarly, *Aegerita* sp. was observed at three levels on the herbaceous mangrove *Avicennia. ilicifolius*: basal (directly above soil level), middle (0.5–1.0 m above soil level) and apical (1.5–2.0 m), with frequencies of occurrence of 27.5, 15.0 and 5.0% respectively (Sadaba *et al.*, 1995).

Alias (1996) followed the colonization of submerged test blocks of *Avicennia. marina* and *Bruguiera parviflora* at two mangroves in Malaysia and at three different levels (Tables 3 and 4). *C. mangrovei* and *Haloc. villosa* were the only basidiomycetes recorded. At Kuala Selangor *C. mangrovei* was an early and intermediate colonizer and was absent in the latter exposure period, while *Haloc. villosa* was a late colonizer (Table 3). *C mangrovei* was more frequent on *Bruguiera* wood while *Haloc. villosa* showed a preference for *Avicennia* wood. Both occurred at all levels within the mangrove. At Morib mangrove, *C. mangrovei* was not collected and *Haloc. villosa* was an intermediate to late colonizer (Table 4). Examination of drift and attached mangrove showed that *Haloc. villosa* was common at the three test sites, while *C. mangrovei* occurred only at Kuala Selangor where the water had a lower salinity and a muddy shore (Table 5).

Sarma and Vittal (2001) recorded the mangrove fungi on different substrata of *R. apiculata* (seedlings, wood, prop roots) and Avicennia spp. (roots, pneumatophores, wood) in India. *H. villosa* was the only basidiomycete observed and was present on all the substrata, infrequently on *R. apiculata* but frequent on *Avicennia spp*. However, Pasannarai and Sridhar (2001) reported *C. mangrovei, H. villosa* and *N. vibrissa* at three, five and four sites, and one, two and two sites in a subsequent study (Pasannarai and Sridhar, 2003) respectively, on the west coast of India. *Haloc. villosa* was the only species regarded as frequent with a frequency of occurrence of 5.3%. They also observed that all required 6–18 months to sporulate on incubated intertidal wood in the laboratory. *Haloc. villosa* was listed as very frequent (>10%) in 9 studies out of 41 and frequent in 9 (5–10%) (Sarma and Hyde, 2001).

Freshwater basidiomycetes are primarily known by their anamorphs, which are elaborate and range from sigmoid, tetra-radiate to elaborately branched conidia (Webster, 1992; Marvanová, 1997).

Basidiomycete yeasts have been reported from freshwater lakes, streams, brackish waters to fully saline seawater, sewage contaminated waters and wastewater (Ahearn and Meyers, 1976; Cooke, 1976; Fell, 1976; Pore and Sorenson, 1990). Many marine basidiomycetes have a unicellular budding haploid state alternating with a dikaryotic hyphal state (dimorphic with a yeast state). There are few studies of freshwater basidiomycete yeasts, although extensive data are available on their marine counterparts, obtained from cruises to the Indian, Indo-Pacific and Pacific Oceans, and to the Black and North Seas (Fell, 1976). Table 1 lists selected examples of freshwater and marine basidiomycetous yeasts. That such habitats are rich in yeasts is supported by the unpublished data of Statzell-Tallman and Fell (cited in Kurtzman and Fell, 2006) who reported 23 genera and 120 species (ascomycetes and basidiomycetes) from subtropical Everglade habitats, of which 54% were new taxa to science.

The number of yeasts documented has steadily increased—341 species in 1970, 500 in 1984 and 700 in 1998 (Kurtzman and Fell, 1998), however few of these

Table 3 Frequency of occurrence of marine basidiomycetes on *Avicennia marina* and *Bruguiera parviflora* test blocks exposed at Kuala Selangor mangrove, Malaysia

	A. marina stand		*B. gymnorrhiza stand*		*Rhizophora apiculata stand*		
	A. marina	*B. parviflora*	*A. marina*	*B. parviflora*	*A. marina*	*B. parviflora*	No. of samples
6–18 weeks							18
Calathella mangrovei	16	16	16	22	22	22	
Halocyphina villosa	–	16	–	–	–	–	
26–54 weeks							36
C. mangrovei	11	44	6	19	–	3	
H. villosa	–	–	–	–	15	–	
72–90 weeks							27
C. mangrovei	4	–	–	–	–	–	
H. villosa	15	4	22	7	–	30	

Source: After Alias (1996).
Note: Dash indicates no fruit bodies detected.

Table 4 Frequency of occurrence of marine basidiomycetes fruiting on *Avicennia marina* and *Bruguiera parviflora* test blocks exposed at morib mangrove, Malaysia

	Zone		Zone		Zone		
	A. marina	*B. parviflora*	*A. marina*	*B. parviflora*	*A. marina*	*B. parviflora*	No. of samples
6–8 weeks							18
Calathella mangrovei	–	–	–	–	–	–	
Halocyphina villosa	–	–	–	–	–	17	
26–54 weeks							36
C. mangrovei	–	–	–	–	–	–	
H. villosa	–	–	3	8	25	64	
72–90 weeks							27
C. mangrovei	–	–	–	–	–	–	
H. villosa	4	37	22	7	22	19	

Source: After Alias (1996).
Note: Dash indicates no fruit bodies detected.

Table 5 Frequency of occurrence of *Calathella mangrovei* and *Halocyphina villosa* in general collections at three mangroves in Malaysia

Fungus	Kuala Selangor	Port Dickson	Morib
H. villosa	66 (top species)	1 (ranked 56)	44 (ranked 2)
C. mangrovei	9 (ranked 15)	–	–

Source: After Alias (1996).

are aquatic. Marine yeast densities range from 5 to 10 cells per litre (Fell, 1976), and 118–1,228 L^{-1} (Van Uden and Castello-Branco, 1963) to 3,000 L^{-1} (Meyers *et al.*, 1967). Many yeasts have been reported from shell fish, food brines and blooms of marine algae, but these remain outside the scope of this chapter. Most are free floating, saprobes of seaweeds and dead marine animals, while others may be animal parasites. *Leucospordium* spp., *Rhodosporidium* spp., *Candida austromarina* (Ascomycota), *C. natalensis* and *Sympodiomyces parvus* (Ascomycota) are undoubtedly autochthonous marine species as they are recovered in relatively high cell concentrations from open ocean samples (Fell, 1976; Lachance and Starmer, 1998). They are able to survive for long periods in seawater and can utilize a wide range of carbon compounds (Lachance and Starmer, 1998). However, few yeasts secrete cellulases or chitinases, therefore their contribution to the breakdown of complex organic material is limited (Lachance and Starmer, 1998). While considerable progress has been made in the isolation, identification and phylogenetic studies of aquatic yeasts, there is a singular lack of any detailed investigation of their ecology (Lachance and Starmer, 1998). In fact few microbiological books cover them in any detail, as indeed other aquatic fungi.

4. LIMITATIONS OF AN AQUATIC HABITAT FOR BASIDIOMYCETES

Typically large putrescent fruit bodies of aquatic basidiomycetes are 'impractical' due to constant wave action in the sea and turbulent running water in rivers and streams. There would also be problems for basidiospore development, and their forcible release under these conditions (Jones, 2000). Teleomorphs of many freshwater basidiomycetes are primarily terrestrial and thus not subject to constant submergence and exposure to water. Marine species have 'overcome' this problem and this is discussed in the next section. However, typical agarics can survive in mangrove habitats subject to infrequent inundation, for example a high spring tide (E.B.G. Jones, personal observation).

5. ADAPTATION TO AN AQUATIC HABITAT

All 11 filamentous marine basidiomycetes have a reduced fruit body and this has hampered their assignment to higher taxonomic groups (Hibbett and Binder, 2001; Binder et al., 2006). Like marine ascomycetes and anamorphic fungi, they are secondary invaders of the marine environment (Spatafora *et al.*, 1998; Binder *et al.*, 2006). Two independent lineages amongst the true agarics have been demonstrated: the *Nia* clade (related to the cyphelloid agarics) and the Physalariaceae clade (highly reduced stipitate-pileate agarics) (Binder *et al.*, 2006). Other lineages include the *Digitatispora* clade (Atheliales?) and *Haloal. mangrovei* (Russulales) (Hibbett and Donoghue, 1998; Binder *et al.*, 2001; Hibbett and Binder, 2001; Maekawa *et al.*, 2005). It remains to be determined if *Haloal. mangrovei* is truly marine as it was reported from intertidal mangrove wood. A fifth lineage includes *Melanotaenium ruppia* (Ustilaginomycotina, Urocystales, Melanotaeniaceae).

Of the 11 filamentous marine basidiomycetes, 6 have appendaged basidiospores that aid in their dispersal and entrapment (Jones, 1994), while in *C. mangrovei* and *H. villosa* they are not appendaged but are also discharged passively (Kohlmeyer and Kohlmeyer, 1979; Nakagiri and Ito, 1991). Unlike their freshwater counterparts, marine basidiomycetes do not produce anamorphs (Webster, 1992).

In *Haloc. villosa* the basidiome is cyphelloid and the upper opening covered by three layers of hyphal tissue, which subsequently separate or uncoil as the basidiome develops (Nakagiri and Ito, 1991). These hyphal layers protect the hymenium from exposure to water and allow basidiospores to develop. At maturity, basidiospores are pushed through the opening in the hyphal layer and accumulate at the tip of the basidiome and are dispersed by the incoming tide. It is not known whether they are also released under water.

Hibbett and Binder (2001) speculated as to the possible evolution of the marine filamentous basidiomycetes and considered *N. vibrissa* as the most derived member, with a shift from terrestrial to periodically immersed to fully submerged substrata, loss of basidiospory, evolution of appendaged spores and an enclosed fruit body. They hypothesized that the mangrove-inhabiting *Haloc. villosa* may be morphologically and ecologically intermediate between *N. vibrissa* and terrestrial cyphelloid forms such as *Cyphellopsis anomala*.

The freshwater basidiomycete *Limnoperdon incarnatum* has sessile non-ballistosporic basidiospores and cyphelloid basidiomes, but lacks the hyphal layers found in *Haloc. villosa* (Nakagiri and Ito, 1991). In basidiomes that develop on the surface of water, the peridial wall is hydrophobic and the hymenium is not exposed to water. Dispersal is then by flotation, as in aero-aquatic hyphomycetes and the basidiomycetes *P. glomerulata* and *N. vibrissa*. In the latter species the peridial wall breaks open and the spores are released passively (Fazzani and Jones, 1977), but in *L. incarnatum* spore release has not been reported (Nakagiri and Ito, 1991).

6. UTILIZATION OF SUBSTRATA

Substrate utilization by aquatic basidiomycetes depends on the substrata and environment from which they were isolated. Probably the least studied group in this respect is the marine basidiomycetous yeasts. Most have been isolated from water columns (Fell *et al.*, 1973, 2001) or from water scums or the internodes of

Table 6 Mean percentage loss in dry weight of five timbers exposed to marine basidiomycetes for 24 weeks at 10 and 22°C

Fungus species	Timber species	10°C	22°C
Digitatispora marina	*Ochroma lagopus*	14.33	4.94
	Fagus sylvatica	10.07	2.68
	Pinus sylvestris	−0.33	0.57
Halocyphina villosa	*O. lagopus*	−0.58	22.98
	F. sylvatica	3.43	7.98
	P. sylvestris	−0.59	−0.09
	Avicennia officinalis	NA	5.78
	Xylocarpus granatum	NA	0.51
Nia vibrissa	*O. lagopus*	13.75	27.91
	F. sylvatica	4.33	5.55
	P. sylvestris	−1.26	0.08
Monodictys pelagica	*O. lagopus*	28.30	40.87
Marine soft-rot ascomycete	*F. sylvatica*	11.03	20.76
	P. sylvestris	0.39	0.68

Source: After Mouzouras (1986, 1989).
Note: data for gain in weight we adjusted with +/− loss in dry weight of controls; NA not tested.

Equisetum fluviatile (Jones and Sloof, 1966; Webster and Davey, 1975). These in general were able to utilize simple sugars (glucose, maltose, sucrose) and also cellobiose, D-xylose and D-mannitol (Fell *et al.*, 1973). There is no evidence that these marine basidiomycetous yeasts can utilize more complex polymers, for example lignocellulose.

Most studies on lignocellulose decomposition by marine basidiomycetes have been on *D. marina, Haloc. villosa* and *N. vibrissa* (Mouzouras, 1986, 1989; Mouzouras *et al.*, 1987). *D. marina, Haloc. villosa* and *N. vibrissa* were grown on test blocks of *Fagus sylvatica, Pinus sylvestris* and *Ochroma lagopus* and at two temperatures (10, 22°C). All caused weight loss of wood (Table 6) although this was much lower than in their terrestrial counterparts, for example more than 75% after 90 days incubation on birch wood, *Ganoderma applanatum, Lenzites betulina, Trametes hirsuta* (Käärik, 1974). Temperature affected their ability to cause decay, with *Haloc. villosa* causing a greater weight loss at the higher temperature (22°C) and conversely *D. marina* at the lower temperature 10°C (Mouzouras, 1986). The nature of the substratum also affected the weight loss caused by *Haloc. villosa*, for example 5.7% loss of *A. alba* and none in *X. granatum*, both mangrove timbers (Mouzouras, 1989), while on *O. lagopus* (a soft timber with a low lignin content) losses of 30% have been noted (Mouzouras, 1986).

The ability of these basidiomycetes to cause decay of wood is dependent on the enzymes they possess and the salinity of the media they are grown on (Table 7). *N. vibrissa* produced cellulase, peroxidase and laccase on seawater and deionized water, but *D. marina* produced these enzymes only on seawater media (Rohrmann and Molitoris, 1992). Growth and enzyme production of *Haloc. villosa*

Table 7 Comparison of wood-degrading enzymes of marine and selected terrestrial basidiomycetes

	Deionized water				Seawater			
Fungus species	Cellulase	Laccase	Tyrosinase	Peroxidase	Cellulase	Laccase	Tyrosinase	Peroxidase
Digitatispora marina	w	—	—	—	1	+	—	+
Bjerkandera fumosa	w	3	w	1	1	2	—	2
Phellinus igniarius	w	2	2	3	3	2	1	3

Source: After Rohrmann and Molitoris (1992).
Note: Activity: +, presence of enzyme; –, no enzyme detected; w, weak; 1, clear reaction; 2, strong reaction; 3, very strong reaction.

was almost identical in natural and artificial seawater media (Rohrmann *et al.*, 1992). Different strains of *N. vibrissa* produce amylase, caseinase, cellulase, gelatinase, laminarinase, lipase, nitrate reductase, peroxidase and xylanase, but none produced tyrosinase (Schimpfhauser and Molitoris, 1991).

Freshwater basidiomycetes have been isolated from decaying leaves, foam, woody debris, submerged test blocks and are able to breakdown major plant cell wall polysaccharides: pectins, cellulose and hemicellulose (Chamier, 1985). Their ability to degrade lignin remains to be established, although Marvanová and Bandoni (1987) listed *Naiadella fluitans* and *Taeniospora* spp. as laccase positive in an alpha-naphthol test (Stalpers, 1978).

7. CONCLUDING REMARKS

Aquatic basidiomycetes have been recorded infrequently compared with ascomycetes and anamorphic taxa (Hyde *et al.*, 2000), and of these only *C. mangrovei, D. marina, Haloc. villosa* and *N. vibrissa* have been studied in detail with respect to their ecological distribution and role in aquatic habitats (Hyde *et al.*, 1998). All marine basidiomycetes sporulating under submerged conditions have reduced fruit bodies that are adaptations to the marine milieu. None to date have been reported with an anamorphic stage, unlike their freshwater counterparts. The teleomorphs of freshwater basidiomycetes are generally terrestrial with active discharge of their small, ovoid or round, hyaline basidiospores, while their anamorphs are variously branched. Future studies should focus on: (1) the occurrence of agarics in mangrove soils and whether they can survive saline water; (2) on the physiology and biochemistry of aquatic basidiomycetes; (3) greater attention to the ecology of basidiomycetous yeasts and on (4) a better understanding of the ecology of freshwater species.

ACKNOWLEDGEMENTS

We are grateful to a number of colleagues for their assistance with the ever changing taxonomy of fungi, and drawing attention to relevant literature: Manfred Binder, Jack Fell, Ludmilla Marvanová and Peter Molitoris; Jack Fell for his willing assistance and readiness to answer questions on the ecology of aquatic yeasts; to Aisyah Alias for permission to reproduce data on the ecology of mangrove basidiomycetes from her Ph. D. thesis; to Kaoru Yamaguchi, Yousuke Degawa and Akira Nakagiri for permission to refer to their data on *Peyronelina glomerulata*; to Tsutomu Hattori for the tentative identification of *Phellinus* species on *Xylocarpus granatum*; to Morakot Tanticharoen and Kanyawim Kirtikara (BIOTEC) for continued support; and for the financial support of BRT research grants: R_245002 and R_248002 to enable the study of Thai marine fungi. RC thanks TIGIST for the award of a research scholarship and Somtawin Jaritkhuan for her continued support.

REFERENCES

Abdel-Wahab, M.A. and El-Sharouny, H.M. (2002). Ecology of subtropical mangrove fungi with emphasis on *Kandelia candel* mycota. *Fungal Diversity Research Series*, **7**, 247–265.

Ahearn, D.G. and Meyers, S.P. (1976). Fungal degradation of oil in the marine environment. In: *Recent Advances in Aquatic Mycology* (E.B.G. Jones, ed.), pp. 125–133. Elek Science, London.
Alias, S.A. (1996). Ecological and taxonomic studies of lignicolous marine fungi in Malaysian mangrove, PhD Thesis, University Portsmouth, 391 pp.
Alias, S.A. and Jones, E.B.G. (2000a). Colonization of mangrove wood by marine fungi at Kuala Selangor mangrove stand. *Fungal Diversity*, **5**, 9–21.
Alias, S.A. and Jones, E.B.G. (2000b). Vertical distribution of marine fungi on *Rhizophora apiculata* at Morib mangrove, Selangor, Malaysia. *Mycoscience*, **41**, 431–436.
Alias, A.S., Zainuddin, N. and Jones, E.B.G. (2007). Biodiversity of marine fungi from mangroves in Malaysia. *Botanica Marina*, in press.
Bandoni, R.J. and Marvanová, L. (1989). On a new species of *Ingoldiella*. *Mycologia*, **81**, 42–46.
Barata, M., Basilio, M.C. and Baptista-Ferrerra, J.L. (1997). *Nia globospora*, a new marine gasteromycete on baits of *Spartina maritima* in Portugal. *Mycological Research*, **101**, 687–690.
Bärlocher, F. (1992). *The Ecology of Aquatic Hyphomycetes* (ed.), Springer-Verlag, Berlin.
Bauer, R., Begerow, D., Oberwinkler, F. and Marvanová, L. (2003). *Classicula*: The teleomorph of *Naiadella fluitans*. *Mycologia*, **95**, 756–764.
Besitulo, A.D., Sarma, V.V. and Hyde, K.D. (2002). Mangrove fungi from Siargao Island, Philippines. In: *Fungi in Marine Environments* (K.D. Hyde, ed.), pp. 267–283. Fungal Diversity Research Series, Hong Kong SAR.
Binder, M., Hibbett, D.S. and Molitoris, H.P. (2001). Phylogenetic relationships of the marine gasteromycete *Nia vibrissa*. *Mycologia*, **93**, 679–688.
Binder, M., Hibbet, D.S., Wang, Z. and Farnham, W.F. (2006). Evolutionary relationships of *Mycaureola dilsea* (Agaricales), a basidiomycete pathogen of a sub tidal Rhodophyte. *American Journal of Botany*, **93**, 547–556.
Booth, T. (1983). Lignicolous marine fungi from São Paulo, Brazil. *Canadian Journal of Botany*, **61**, 488–506.
Byrne, P.J. and Jones, E.B.G. (1974). Lignicolous marine fungi. *Veröffentlichungen des Instituts für Meeresforschung, Bremerhaven*, **5**(Suppl.), 301–320.
Chalermpongse, A. (1991). Fungal diseases in mangrove ecosystem. In: *Proceedings of the 5th Silviculture Seminar in Thailand*, pp. 307–338.
Chamier, A.C. (1985). Cell-wall degrading enzymes of aquatic Hyphomycetes: A review. *Botanical Journal of the Linnean Society*, **91**, 67–81.
Chinnaraj, S. (1993). Higher marine fungi from mangroves of Andaman and Nicobar Islands. *Sydowia*, **45**, 109–115.
Cooke, W.B. (1976). Fungi in sewage. In: *Recent Advances in Aquatic Mycology* (E.B.G. Jones, ed.), pp. 389–434. Elek Science, London.
Crane, J.L. and Schoknecht, J.D. (1978). *Rogersiomyces*: A new genus in the Filobasidiaceae (Homobasidiomycetes) from an aquatic habitat. *American Journal of Botany*, **65**, 902–906.
Desjardin, D.E., Martinez-Peck, L. and Raichenberg, M. (1995). An unusual psychrophilic aquatic agaric from Argentina. *Mycologia*, **87**, 547–550.
Doguet, G. (1962). *Digitatispora marina* n. g., n. sp. Basidiomycète marin. *Compte rendu Hebdomadaire des Séances Acadmémie des Sciences, serie D*, **254**, 4336–2338.
Escobar, G.A., McCabe, D.E. and Harpel, C.W. (1976). *Limnoperdon*, a floating gasteromycete isolated from marshes. *Mycologia*, **68**, 874–880.
Fazzani, K. and Jones, E.B.G. (1977). Spore release and dispersal in marine and brackish water fungi. *Material and Organismen*, **12**, 235–248.
Feldmann, G. (1959). Une ustilaginale marine, parasite du *Ruppia maritima* L. *Revue Général Botanique*, **66**, 35–38.
Fell, J.W. (1976). Yeasts in oceanic regions. In: *Recent Advances in Aquatic Mycology* (E.B.G. Jones, ed.), pp. 93–124. Elek Science, London.
Fell, J.W., Boekhout, T., Fonseca, A. and Sampaio, J.P. (2001). Basidiomycetous yeasts. In: *Mycota, Systematics and Evolution* (D.J. McLaughlin, E.G. McLaughlin and P.A. Lemke, eds.), Part VII, pp. 3–35. Springer-Verlag, Berlin.

Fell, J.W., Hunter, I.L. and Tallman, A.S. (1973). Marine basidiomycetous yeasts (*Rhodosporidium* spp. n.) with tetrapolar and multiple allelic bipolar mating systems. *Canadian Journal of Microbiology*, **19**, 643–657.

Fisher, P.J., Webster, J. and Kane, D.F. (1976). *Peyronelina glomerulata* from submerged substrata in Britain. *Transactions of the British Mycological Society*, **67**, 351–354.

Henningsson, M. (1974). Aquatic lignicolous fungi in the Baltic and along the West Coast of Sweden. *Svensk Botanisk Tidskrift*, **68**, 401–425.

Hibbett, D.S. and Binder, M. (2001). Evolution of marine mushrooms. *Biological Bulletin*, **201**, 319–322.

Hibbett, D.S. and Donoghue, M.J. (1998). Integrating phylogenetic analysis and classification in fungi. *Mycologia*, **90**, 347–356.

Hudson, J.H. (1961). *Heliscus submersus* sp. nov. an aquatic hyphomycete from Jamaica. *Transactions of the British Mycological Society*, **44**, 91–94.

Hyde, K.D. (1986). Frequency of occurrence of lignicolous marine fungi in the tropics. In: *The Biology of Marine Fungi* (S.T. Moss, ed.), pp. 311–322. Cambridge University Press, UK.

Hyde, K.D. (1988). Studies on the tropical marine fungi of Brunei. *Botanical Journal of the Linnean Society*, **98**, 135–151.

Hyde, K.D. (1989). Vertical zonation of intertidal mangrove fungi. In: *Recent Advances in Microbial Ecology* (T. Hattori, Y. Ishida, Y. Maruyama, R.Y. Morita and A. Uchida, eds.), pp. 3032–3036. Japan Scientific Societies Press, Tokyo.

Hyde, K.D. (1990a). A comparison of the intertidal mycota of five mangrove tree species. *Asian Marine Biology*, **7**, 93–107.

Hyde, K.D. (1990b). A study of the vertical zonation of intertidal fungi on *Rhizophora apiculata* at Kampong Kapok mangrove, Brunei. *Aquatic Botany*, **36**, 255–262.

Hyde, K.D., Chalermpongse, A. and Boonthavikoon, T. (1993). The distribution of intertidal fungi on *Rhizophora apiculata*. In: *Marine Biology of the South China Sea* (B. Morton, ed.), pp. 643–651. Hong Kong University Press, Hong Kong.

Hyde, K.D., Jones, E.B.G., Leano, L., Pointing, S.B., Poonyth, A.D. and Vrijmoed, L.L.P. (1998). Role of fungi in marine ecosystems. *Biodiversity and Conservation*, **7**, 1147–1161.

Hyde, K.D., Sarma, V.V. and Jones, E.B.G. (2000). Morphology and taxonomy of higher marine fungi. In: *Marine Mycology: A Practical Approach 1* (K.D. Hyde and S.B. Pointing, eds.), pp. 172–204. Fungal Diversity Research Series, Fungal Diversity Press, Hong Kong.

Inderbitzin, P. and Desjardin, D.E. (1999). A new halotolerant species of *Physalacria* from Hong Kong. *Mycologia*, **91**, 666–668.

Jones, A.M. and Jones, E.B.G. (1993). Observations on the marine gasteromycete *Nia vibrissa*. *Mycological Research*, **97**, 1–6.

Jones, E.B.G. (1972). The decay of timber in aquatic environments. *British Wood Preserving Association*, Annual Convention, 1972, pp. 1–18.

Jones, E.B.G. (1985). Wood-inhabiting marine fungi from San Juan Island with special reference to ascospore appendages. *Botanical Journal of the Linnean Society*, **91**, 219–231.

Jones, E.B.G. (1986). *Digitatispora lignicola* sp. nov. A new marine lignicolous Basidiomycotina. *Mycotaxon*, **27**, 155–159.

Jones, E.B.G. (1994). Fungal adhesion. Presidential address 1992. *Mycological Research*, **98**, 961–981.

Jones, E.B.G. (2000). Marine fungi: Some factors influencing biodiversity. *Fungal Diversity*, **4**, 53–73.

Jones, E.B.G. and Agerer, R. (1992). *Calathella mangrovei* sp. nov. and observations on the mangrove fungus *Halocyphina villosa*. *Botanica Marina*, **35**, 83–91.

Jones, E.B.G., Pilantanapak, A., Chatmala, I., Sakayaroj, J., Phongpaichit. S. and Choeyklin, R. (2006). Thai marine fungal diversity. *Songklanakarin Journal of Science and Technology*, **28**, 687–708.

Jones, E.B.G. and Puglisi, M.P. (2006). Marine fungi from Florida. *Florida Scientist*, **69**, 157–164.

Jones, E.B.G. and Sloof, W.C. (1966). *Candida aquatica* sp. nov. isolated from water scums. *Antonie van Leeuwenhoek*, **32**, 223–228.

Jones, E.B.G. and Vrijmoed, L.L.P. (2003). Biodiversity of marine fungi. In: *Perspectives on Marine Environment Change in Hong Kong and Southern China Hong Kong Coastal Waters* (B. Morton, ed.), pp. 75–93. Hong Kong University Press, Hong Kong.

Käärik, A.A. (1974). Decomposition of wood. In: *Biology of Plant Litter Decomposition* (C.H. Dickinson and G.J.F. Pugh, eds.), pp. 129–174. Academic Press, New York.
Kane, D.F.R., Tam, W.Y. and Jones, E.B.G. (2002). Fungi colonizing and sporulating on submerged wood in the River Severn, UK. *Fungal Diversity*, **10**, 45–55.
Kirschner, R. (2004). Sporodochial anamorphs of species of *Helicogloea*. In: *Frontiers of Basidiomycota Mycology* (R. Agerer, M. Piepenbring and P. Blanz, eds.), pp. 165–178. IHW-Verlag, Eching.
Koch, J. and Peterson, K.R.L. (1996). A checklist of higher marine fungi on wood from Danish coasts. *Mycotaxon*, **60**, 529–540.
Kohlmeyer, J. and Kohlmeyer, E. (1965). New marine fungi from mangroves and trees along eroding shorelines. *Nova Hedwigia*, **9**, 89–104.
Kohlmeyer, J. and Kohlmeyer, E. (1979). *Marine Mycology: The Higher Fungi*, p. 690. Academic Press, New York.
Kurtzman, C.P. and Fell, F.W. (eds.). (1998). *The Yeasts, A Taxonomic Study*, 4th edn., p. 890. Elsevier, Amsterdam.
Kurtzman, C.P. and Fell, F.W. (2006). Yeast systematics and phylogeny—Implications of molecular identification methods for studies in ecology. In: *Yeast Handbook* (C. Rosa and G. Peter, eds.), *Vol. 1, Biodiversity and Ecophysiology of Yeasts*, pp. 11–30, Springer-Verlag, NJ, USA.
Lachance, M.-A. and Starmer, W.T. (1998). Aquatic habitats, ecology and yeasts. In: *The Yeasts, A Taxonomic Study* (C.P. Kurtzman and F.W. Fell, eds.), 4th edn., pp. 21–30. Elsevier, Amsterdam.
Leong, W.F., Tan, T.K. and Jones, E.B.G. (1991). Fungal colonization of submerged *Bruguiera cylindrica* and *Rhizophora apiculata* wood. *Botanica Marina*, **34**, 69–76.
Maekawa, N., Suhara, H., Kinjo, K., Kondo, R. and Hashi. Y. (2005). *Haloaleuridiscus mangvrovei* gen. sp. nov. (Basidiomycota) from mangrove forests in Japan. *Mycological Research*, **109**, 825–832.
Mari, G.L. and Sridhar, K.R. (2002). Richness and diversity of filamentous fungi on woody litter of mangroves along the west coast of India. *Current Science*, **83**, 1573–1580.
Mari, G.L. and Sridhar, K.R. (2003). Diversity of filamentous fungi on woody litter of five mangrove plant species from the southwest coast of India. *Fungal Diversity*, **14**, 109–126.
Marvanová, L. (1997). Freshwater hyphomycetes: A survey with remarks on tropical taxa. In: *Tropical Mycology* (K.K. Janardhanan, C. Rajendran, K. Natarajan and D.H. Hawksworth, eds.), pp. 169–226. Science Publishers Inc., USA.
Marvanová, L. and Bandoni, R.J. (1987). *Naiadella fluitans* gen. et sp. nov.: A conidial basidiomycete. *Mycologia*, **79**, 578–586.
Marvanová, L. and Bärlchocher, F. (1988). Hyphomycetes from Canadian streams. 1. Basidiomycetous anamorphs. *Mycotaxon*, **32**, 339–351.
Marvanová, L. and Bärlchocher, F. (1998). New species of *Filosporella, Pachycladina* and *Pleuropedium* from Canadian streams. *Mycological Research*, **102**, 750–754.
Marvanová, L. and Bärlchocher, F. (2000). Hyphomycetes from Canadian streams. V. Two new conidial Basidiomycetes. *Mycotaxon*, **75**, 409–423.
Marvanová, L. and Stalpers, F.A. (1987). The genus *Taeniospora* and its teleomorph. *Transactions of the British Mycological Society*, **89**, 489–498.
Marvanová, L. and Suberkropp, K. (1990). *Camptobasidium hydrophilum* and its teleomorph, *Crucella subtilis*, a new heterobasidiomycete from streams. *Mycologia*, **82**, 208–217.
Meyers, S.P., Ahearn, D.G., Gunkel, W. and Roth, F.J. (1967) Yeasts from the North Sea. *Marine Biology*, **1**, 118–123.
Moore, R.T. and Meyers, S.P. (1959). Thalassiomycetes 1. Principles of delimitation of the marine mycota with the description of a new aquatically adapted deuteromycete genus. *Mycologia*, **51**, 871–876.
Mouzouras, R. (1986). Patterns of timber decay caused by marine fungi. In: *The Biology of Marine Fungi* (S.T. Moss, ed.), pp. 341–353. Cambridge University Press, UK.
Mouzouras, R. (1989). Decay of mangrove wood by marine fungi. *Botanica Marina*, **32**, 65–69.
Mouzouras, R., Jones, E.B.G., Venkatasamy, R. and Moss, S.T. (1987). Decay of wood by microorganisms in marine environments. *Record Annual Convention, British Wood Preserving Association*, Cambridge, UK, pp. 27–45.
Nakagiri, A. and Ito, T. (1991). Basidiocarp development of the cyphelloid gasteroid aquatic basidiomycetes *Halocyphina villosa* and *Limnoperdon incarnatum*. *Canadian Journal of Botany*, **69**, 2320–2327.

Nawawi, A. (1973). Two clamp-bearing fungi from Malaysia. *Transactions of the British Mycological Society*, **61**, 521–528.
Nawawi, A. (1985). Basidiomycetes with branched, water-borne conidia. *Botanical Journal of the Linnean Society*, **91**, 51–60.
Nawawi, A. and Kuthubutheen, A.J. (1988). *Tricladiomyces geniculatus* sp. nov., a conidial basidiomycete. *Transactions of the British Mycological Society*, **90**, 670–673.
Nawawi, A. and Webster, J. (1982). *Sistotrema hamatum* sp. nov., the teleomorph of *Ingoldiella hamata*. *Transactions of the British Mycological Society*, **78**, 287–291.
Nawawi, A., Webster, J. and Davey, R.A. (1977a). *Leptosporomyces glazinnii*, the basidial state of a clamped branched conidium in freshwater. *Transactions of the British Mycological Society*, **68**, 32–36.
Nawawi, A., Webster, J. and Davey, R.A. (1977b). *Dendrosporomyces prolifera* gen et. sp. nov., a basidiomycete with branched conidia basidial state of a clamped branched conidium in freshwater. *Transactions of the British Mycological Society*, **68**, 59–63.
Pasannarai, K. and Sridhar, K.R. (2001). Diversity and abundance of higher marine fungi on woody substrates along the west coast of India. *Current Science*, **81**, 304–311.
Pasannarai, K. and Sridhar, K.R. (2003). Abundance and diversity of marine fungi on intertidal woody litter of the west coast of India on prolonged incubation. *Fungal Diversity*, **14**, 127–141.
Pinnoi, A., Lumyong, S., Hyde, K.D. and Jones, E.B.G. (2006). Biodiversity of fungi on the palm *Eleiodoxa conferta* in Sirindhorn peat swamp forest, Narathiwat, Thailand. *Fungal Diversity*, **22**, 205–218.
Pinruan, U. (2004). Palm fungi, their ecophysiology and biochemical diversity, MSc Thesis, University of Chiang Mai, Thailand.
Pore, R.S. and Sorenson, W.G. (1990). *Reniforma strues*, a new yeast from wastewater. *Mycologia*, **82**, 549–553.
Porter, D. and Farnham, W.F. (1986). *Mycaureola edulis*, a marine basidiomycete parasite of the red alga, *Dilsea carnosa*. *Transactions of the British Mycological Society*, **87**, 575–582.
Read, S.J., Moss, S.T. and Jones, E.B.G. (1992). Attachment and germination of conidia. In: *The Ecology of Aquatic Hyphomycetes* (F. Bärlocher, ed.), pp. 135–151. Ecological Studies 94, Springer-Verlag, Berlin.
Rogers, D.P. (1947). A new gymnocarpus Heterobasidiomcyete with gasteromycetous basidia. *Mycologia*, **39**, 556–564.
Rohrmann, S., Lorenz, R. and Molitoris, H.P. (1992). Use of natural and artificial seawater for investigation of growth, fruit body production, and enzymes activities in marine fungi. *Canadian Journal of Botany*, **70**, 2106–2110.
Rohrmann, S. and Molitoris, H.P. (1992). Screening for wood-degrading enzymes in marine fungi. *Canadian Journal of Botany*, **70**, 2116–2123.
Rossello, M.A., Descals, E. and Cabrer, R. (1993). *Nia epidermoidea*, a new marine gasteromycete. *Mycological Research*, **97**, 68–70.
Sadaba, R.B., Vrijmoed, L.L.P., Jones, E.B.G. and Hodgkiss, I.J. (1995). Observations on vertical distribution of fungi associated with standing senescent *Acanthus ilicifolius* stems at Mai Po mangrove, Hong Kong. *Hydrobiologia*, **295**, 119–126.
Sarma, V.V. and Hyde, K.D. (2001). A review on frequently occurring fungi in mangroves. *Fungal Diversity*, **8**, 1–34.
Sarma, V.V. and Vittal, B.P.R. (2001). Biodiversity of mangrove fungi on different substrata of *Rhizophora apiculata* and *Avicennia* spp. from Godavari and Krishna deltas, East coast of India. *Fungal Diversity*, **5**, 23–41.
Schimpfhauser, G. and Molitoris, H.P. (1991). Enzyme activities of monokaryotic and dikaryotic strains of the marine Basidiomycete *Nia vibrissa*. *Kieler Meeresforschung Sonderh*, **8**, 361–368.
Schmidt, J.P. and Shearer C.A. (2003). A checklist of mangrove-associated fungi, their geographical distribution and known host plants. *Mycotaxon*, **80**, 423–477.
Scorzetti, G., Fell, J.W., Fonseca, A. and Statzell-Tallman, A. (2002). Systematics of basidiomycetous yeasts: A comparison of large subunit D1/D2 and internal transcribed spacer rDNA regions. *FEMS Yeast Research*, **2**, 495–517.
Shaw, D.E. (1972). *Ingoldiella hamata* gen. et sp. nov., a fungus with clamp-connections from a stream in North Queensland. *Transactions of the British Mycological Society*, **59**, 255–259.

Shearer, C.A. (1993). The freshwater Ascomycetes. *Nova Hedwigia*, **56**, 1–33.
Sivichai, S. and Jones, E.B.G. (2004). *Stauriella* gen. nov., proposed for a new lignicolous basidiomycetous anamorph from freshwater in Thailand. *Sydowia*, **56**, 131–136.
Sivichai, S., Jones, E.B.G. and Hywel-Jones, N.L. (2000). Fungal colonisation of wood in a freshwater stream at Khao Yai National Park, Thailand. In: *Aquatic Mycology across the Millennium* (K.D. Hyde, W.H. Ho and S.B. Pointing, eds.). *Fungal Diversity*, **5**, 71–88.
Sivichai, S., Jones, E.B.G. and Hywel-Jones, N.L. (2002). Lignicolous freshwater higher fungi with reference to their teleomorph and anamorph stages. In: *Tropical Mycology* (R. Watling, J.C. Frankland, S. Isaac, A.M. Ainsworth and C. H. Robinson, eds.), *Vol. 2. Micromycetes*, pp. 42–49. CABI Publishing, London, UK.
Spatafora, J.F., Volkmann-Kohlmeyer, B. and Kohlmeyer, J. (1998). Independent terrestrial origins of the Halosphaeriales (marine Ascomycota). *American Journal of Botany*, **85**, 1569–1580.
Stalpers, J.A. (1978). Identification of wood-inhabiting Aphyllophorales in pure culture. *Studies in Mycology*, **16**, 248.
Stanley, S.J. (1992). Observations on the seasonal occurrence of marine endophytic and parasitic fungi. *Canadian Journal of Botany*, **70**, 2089–2096.
Tan, T.K., Leong, W.F. and Jones, E.B.G. (1989). Succession of fungi on wood of *Avicennia alba* and *A. lanata* in Singapore. *Canadian Journal of Botany*, **67**, 2686–2691.
Tsui, C.K.M. and Hyde, K.D. (2003). Freshwater Mycology. *Fungal Diversity Research Series*, **10**, 1–350.
Van Uden, N. and Castello-Branco, R. (1963). Distribution and population densities of yeast species in Pacific water, air, animals and kelp off Southern California. *Limnology and Oceanography*, **8**, 323–329.
Webster, J. (1992). Anamorph–teleomorph relationships. In: *The Ecology of Aquatic Hyphomycetes* (F. Bärlocher, ed.), pp. 99–117. Ecological Studies 94, Springer-Verlag, Berlin.
Webster, J. and Davey, R.A. (1975). Sedimentation rates and trapping efficiency of cells of *Candida aquatica*. *Transactions of the British Mycological Society*, **83**, 43–52.
Yamaguchi, K., Degawa, Y. and Nakagiri, A. (2006). Teleomorph of *Peyronelina glomerulata*, an aero-aquatic fungus. Poster, 50th Anniversary meeting of the Japanese Mycological Society, Chiba, Japan.

CHAPTER 17

Conservation: Selection Criteria and Approaches

Jacob Heilmann-Clausen and **Jan Vesterholt**

Contents

Abstract

Conservation of fungi faces the challenge of high species diversity, limited knowledge and a general lack of public awareness. In practical conservation the high species diversity makes it necessary to focus on a limited number of indicator species. Indicator species schemes are burdened by shortcomings: some are experience based and flimsy in their definition of indicator goals, whereas others are scientific in their approach, but with disputable results or an irrelevant indicator goal. The IUCN criteria for red-listing organisms are not specifically designed for fungi, and red-listing fungi, that is calculating the risk of their extinction, is complicated by a limited knowledge on population sizes, lifespan and spatial dynamics in fungi. In this chapter both approaches are discussed from an overall perspective, and with respect to two groups of saprotrophic basidiomycetes which are decreasing in Europe—grassland and wood-inhabiting fungi.

1. INTRODUCTION

Compared with vascular plants and many groups of animals, the conservation of fungi has received limited interest. Very few nature reserves have been declared in order to conserve fungal biodiversity (see Anonymous, 2004), and the number

British Mycological Society Symposia Series
Published by Elsevier Ltd.

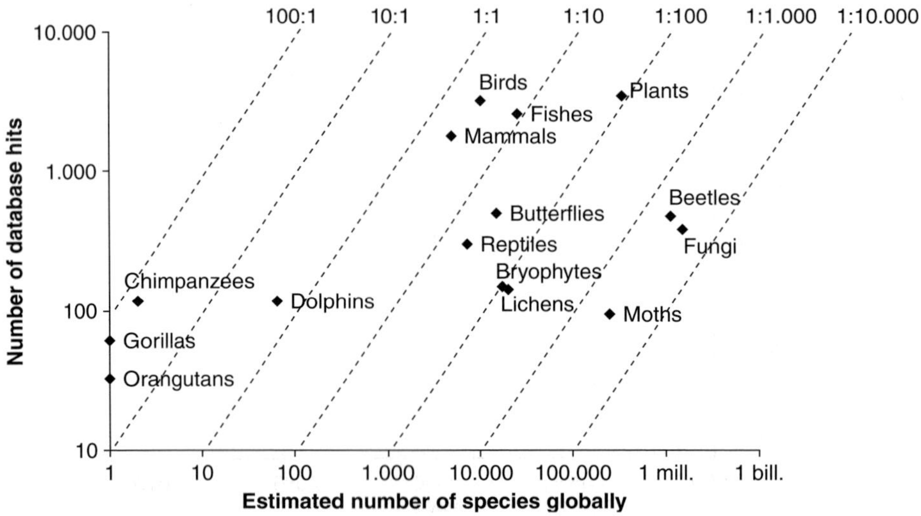

Figure 1 The relation between estimated global species numbers and hits returned by the article database offered by Web of Science® covering most peer-revived scientific journals using the search string "conservation and X not gene", where X denotes the name of a particular organism group included in the figure. Not all hits returned by the search string represented articles actually dealing with nature conservation, focussing on the particular organism group. Irrelevant hits were not sorted out in any group, and, similarly, no efforts were made to trace conservation relevant articles not picked up by the search string. The dashed lines Indicate different ratios between estimated species numbers and database hits. The search was carried out on 22nd September 2006 (Updated from Heilmann-Clausen, 2003).

of scientific publications dealing with conservation of fungi is comparatively low but increasing (Figures 1 and 2). The European Council for the Conservation of Fungi (ECCF) was founded in 1985. In 2001, the council suggested a list of 33 macromycetes to be included in the annexes of the Bern Convention as worthy of protection at the European level (Dahlberg and Croneborg, 2003). If the suggestion had been accepted by the European Union Habitat Committee all 33 species would have been included in EU nature monitoring programmes under Natura 2000, but the suggestion was dismissed and no fungi are currently protected at the European scale. As a slight concession, the EU Commission has published a report giving detailed information on all 33 species (Dahlberg and Croneborg, 2006), claiming that they "certainly deserve the attention of conservation agencies". At the national scale, official initiatives to protect specific fungi have been launched in some countries (e.g. Antonín and Bieberová, 1995; Naturvårdsverket, 2006; UK Biodiversity Action Plan, 2006), but in many countries fungi are still completely neglected in official nature conservation programmes. A major obstacle for fungal conservation may be that fungi are not considered attractive and worthy of protection by the general public in the western world. Unlike furry animals, butterflies and orchids they do not awake an immediate protective response. Rather, people tend to perceive fungi as representative of darkness, decay and death.

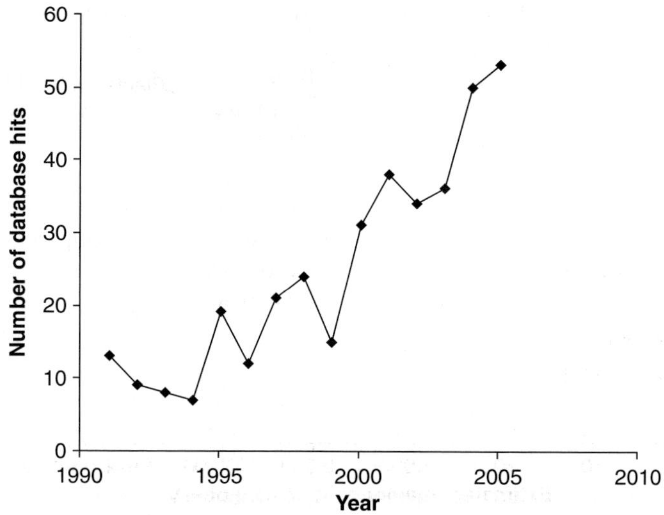

Figure 2 The number of hits returned by the article database offered by Web of Science® using the search string "fungi and conservation not gene" from 1991 to 2005. Not all hits returned by the search string represented articles actually dealing with nature conservation, focussing on the particular organism group. Irrelevant hits were not sorted out in any group, and, similarly, no efforts were made to trace conservation relevant articles not picked up by the search string. The search was carried out on 22nd September 2006.

Fungal communities are more difficult to investigate for assessment of conservation than vascular plants and some animal communities. With macrofungi, sporocarps indicate presence of a species, but they do not indicate which species are present that are not fruiting. Sporocarp production is typically highly variable in intensity and timing between years, and it may take decades to record all species fruiting at a particular site (Watling, 1995; Straatsma *et al.*, 2001; Chapter 5). In addition, species with competitive or stress-tolerant strategies (see Chapter 11) may put considerably less energy into sporocarp production, compared with more ruderal species, and hence are likely to be systematically under-represented at sporocarp level. Considerable differences in species composition between sporocarps and mycorrhizal roots have been reported (Gardes and Bruns, 1996; Peter *et al.*, 2001b), and also studies of decomposer fungi in dead wood have shown that sporocarp-based inventories may not adequately reflect the diversity and community structure at the mycelial level (Allmér *et al.*, 2005). Knowledge of population dynamics in fungi is poor: very simple data relating to average lifespan and size of mycelia, population sizes and dispersal dynamics are missing or very scarce for the vast majority of species, making estimation of population trends and extinction risks difficult. Finally, many groups of macrofungi are still poorly resolved taxonomically. Even in well-investigated regions like Europe the taxonomy of important groups of decomposer basidiomycetes, e.g. *Pluteus* and *Psathyrella,* is far from resolved, and in many other regions, e.g. the tropics, a basic understanding of species diversity is only starting to emerge.

Techniques for sampling fungi at the mycelial level are steadily improving, especially with the development of efficient PCR techniques allowing direct identification of fungi in soil and wood samples (Vainio and Hantula, 2000; Pennanen *et al.*, 2001; Allmér *et al.*, 2005). Molecular techniques are increasingly applied to increase insight into population sizes and dynamics in fungi, to explore the dynamics of mycelia development, activity and sporocarp formation and to increase phylogenetic resolution in poorly known species groups. In relation to management and monitoring they are less relevant at present, because they are labour intensive and unsuited for detection of species occurring with very low frequencies (Allmér *et al.*, 2005). In addition, sampling is more or less destructive, which may be a problem in relation to endangered species. In contrast, sporocarp surveys are generally non-destructive and allow investigation of many hectares of habitat in a single day, depending on how conspicuous are the sporocarps of the species in question. Sporocarp-based and molecular techniques should thus be seen as complementary rather than mutually exclusive in conservation mycology.

Here we review the most important approaches to use of fungi as focal species in conservation. Problems connected with evaluation of extinction risk for fungi by applying the criteria of the International Union for the Conservation of Nature and Natural Resources (IUCN) are discussed; as is the use of fungi as indicator species of valuable habitats. Decomposer, grassland fungi and saproxylic (wood-inhabiting) fungi are used to illustrate practical challenges and possibilities in conservation mycology.

2. FUNGI AS FOCAL SPECIES IN CONSERVATION

Conservation of biodiversity is a complex task, and it is broadly accepted that means to reduce complexity are necessary if we want to make conservation biology practicable. This is equally relevant in a mega diverse group such as fungi, where global species numbers are estimated to be close to 1.5 million (Hawksworth, 2001). In most cases it will be unrealistic to make complete or even semi-complete species inventories to guide fungal conservation priorities. A way forward is, therefore, to focus on particular species or species groups, commonly termed *focal* species (Caro and O'Doherty, 1999). Focal species definitions and terminology are confusing and widely different classification systems have been suggested (Heilmann-Clausen, 2003). Often the terms indicator species, surrogate species, umbrella species and focal species are used interchangeably which does not improve clarity. Below, we review two different approaches to reduce complexity in fungal conservation, by focussing on a smaller fraction of the total species diversity: (i) red-listing and (ii) the use of indicators of habitat quality. Both approaches have been widely applied in Europe.

2.1 Applying the IUCN Red-Listing Criteria to Fungi

The aim of red-lists is to "provide an explicit, objective ... classification of species according to their extinction risk" (IUCN, 2001). Earlier, red-lists were compiled

Table 1 Threat categories following IUCN criteria

Extinct (EX)
Extinct in the wild (EW)
Regionally extinct (RE)
Critically endangered (CR)
Endangered (EN)
Vulnerable (VU)
Near threatened (NT)
Least concern (LC)
Data deficient (DD)
Not applicable (NA)
Not evaluated (NE)

following a range of different national and regional approaches, but in general, recent national or regional red-lists of fungi follow the guidelines of IUCN (2003). However, national and regional red-lists do show varying interpretations, especially in evaluation of rare species (with limited data). Red-lists for fungi following current IUCN criteria have been published in several countries, e.g. Norway (Brandrud *et al.*, 2006), Sweden (Gärdenfors, 2005) and Finland (Kytövuori *et al.*, 2005). Red-list evaluation of major groups of Danish macrofungi is ongoing, with 1,500 species having been evaluated by 2005 (DMU, 2006), and the discussion following is based mainly on this work.

Basically, evaluating an organism, typically a species, for red-listing involves determining the risk that it may disappear from the region in question. The species is then referred to a threat category (Table 1). Each species is evaluated by four sets of different criteria (Box 1): (A) population decrease; (B) decrease in geographical range; (C) small population size and decline; and (D) very small or restricted population size. Alternatively (criterion E), available data may be used for a so-called population viability analysis that uses an algorithm to estimate risk of extinction. However, to our knowledge the E criterion has not yet been applied to fungi.

The IUCN red-list criteria are not specifically designed for fungi, and applying them to fungi (Box 1) involves a number of challenges, such as defining: (i) what an (mature) individual is; (ii) how population sizes are estimated? (iii) what is the generation time in different fungi; and (iv) how population trends are estimated? These challenges are discussed below.

2.1.1 What is an Individual?

In all macrofungi an individual is to be understood as a mycelium. In the field it is easy to observe fruit bodies, but not mycelia; hence, it is often difficult to decide how many mycelia there are in a given area. One piece of wood may host several individuals belonging to the same species (e.g. Kauserud and Schumacher, 2002, 2003), but in other cases a mycelium may become very large and inhabit several hectares (Smith *et al.*, 1992). Old mycelia may be fragmented and split up into

Box 1. IUCN Red-List Criteria Applied to Fungi.

Criterion A: Population decrease

The criterion concerning population decrease basically deals with the percentage decrease in population size over a period of three generations (within a range of 10–100 years).

The population decrease may either be an observed decrease in recent years or an expected decrease in the coming years, i.e. caused by an expected loss of habitat quality. The threat category obtained by using criterion A depends on the rate of decrease in the species and, to some extent, on whether the process is understood and reversible. However, it is independent of the population size. Species must have decreased or expected to decrease by at least 30% within three generations (or 10 years). Criterion A is accordingly most relevant for common fungi with a long generation time where the quality or quantity of habitats has decreased significantly or is expected to decrease significantly in the near future. Examples may be where increasing nitrogen deposition is threatening vulnerable habitats, or where changes in land use result in a dramatic change in habitat quality and quantity. If a species is decreasing and already rare, it will normally qualify as more threatened when evaluated under criterion C.

Criterion B: Decrease in geographical range

Species with a small area of occurrence may qualify as threatened under criterion B if at least two of the following are met: (a) the distribution is strongly fragmented; (b) a decrease takes place in distribution/habitat quality/population size; and (c) there is a fluctuation in distribution/population size. The actual threat category obtained by using criterion B depends on the area of occurrence and on how many localities the species is known from.

Criterion B was designed to identify populations that are severely fragmented, undergoing a form of continuing decline and/or are subject to extreme fluctuations (IUCN, 2003). Fragmentation of populations may well constitute a problem for fungi, but fungal spores may cross long distances, and the problem is likely to be smaller than for organisms with a more limited dispersal potential.

For some fungi it has, however, been shown that area of distribution has decreased. The best example is fungi restricted to nutrient-poor soil, which have disappeared from areas with a nitrogen deposit exceeding the critical load for the species in question. This has been documented for several mycorrhizal species, including species of *Hydnellum*, *Phellodon*, *Bankera*, *Sarcodon*, *Tricholoma* (Arnolds, 1989, 1991; Otto, 1992; Vesterholt *et al.*, 2000), but is also likely to be relevant for saprotrophic basidiomycetes that occur on naturally nutrient-poor soils. For such species criterion B is relevant if the regional geographic range of the species in question is restricted.

Criterion C: Small population size and decline

This criterion is justified by the fact that small populations are more sensitive to decline than larger populations. Like criterion A, criterion C covers species which are declining and species that are expected to decline in the near future. Species with less than 10,000 mature individuals qualify for one of the threatened categories, provided there is an estimated continuing decrease in the number of mature individuals of at least 10% over a period of one to three generations (max. 100 years). Species may also qualify if a smaller or slower decline is taking place, if the species is also restricted to very few, very small or highly fluctuating subpopulations.

Criterion C can be applied to strongly or moderately decreasing fungi which are already rare. This decrease may be inferred, either because there has been an actual decrease in population size, or because an observed or expected loss of habitat quality is assumed to have a negative influence on the number of mature individuals. A general loss of habitat quality is observable for many fungi. One example is grassland species that are threatened by eutrophication and management changes in many parts of Europe, which also qualify for a threat category under criterion B, if the regional population size is small.

Criterion D: Very small or restricted population size

Criterion D covers species with very small populations of mature individuals, because species with small populations have a risk of extinction as result of unforeseen incidents and inbreeding. There is no requirement for a species to be declining. Species with less than 1,000 mature individuals qualify for one of the threat categories, the actual threat category depending on the number of mature individuals. Also species with larger population sizes, known from only up to five localities, or with an area of occurrence below 20 km^2, may also be red-listed. Criterion D is relevant for all types of rare fungi, but it is most important for species that do not qualify using other criteria, i.e. very rare species that are not known to be decreasing. Species with less than 1,000 mature individuals qualify to a threat category irrespective of the size of the area considered.

Criteria A–D are quantitative measures, and the different criteria may point to different threat categories (Table 1). The actual status of the species is then determined by the most severe of the threat categories (see IUCN, 2001, 2003).

several ramets. Somatic incompatibility and molecular methods may help to delimit and quantify individual mycelia (e.g. Boddy and Rayner, 1983; Kirby *et al.*, 1990; Kauserud and Schumacher, 2002, 2003), but for monitoring, the number of mycelia occurring on a locality still has to be estimated based on counting groups of sporocarps. In Sweden, each mycelium has been assumed to consist of 10 ramets (counting as individuals) in ground-living and 2 ramets in saproxylic species (Gärdenfors, 2005), but this approach has not been followed in Denmark, partly because we have assumed that ramets of a mycelium with restricted spatial distribution (<10 m^2) are all liable to be wiped out if a negative factor affects a habitat.

With some of the IUCN criteria, the number of *mature* individuals is relevant. It can be assumed that a fungus is mature when it starts producing fruit bodies, and that it remains mature as long as it is capable of doing so. In general fungi that can be observed in the field, must be considered as mature.

2.1.2 Estimation of Population Size

Databases often show how many localities a species is known from, but give no clues on the number of localities which host the species. To estimate the actual number of individuals in a region, it is necessary to have an idea of the magnitude of yet unknown localities of each species in question and of the number of mycelia in each locality. A possible way of estimating this is to multiply the known number of localities by a factor representing the likely proportion of undiscovered localities for the species (locality factor) and by a factor representing the likely number of mycelia per locality (mycelium factor). When a species is very conspicuous, fruits every year, has very exact ecological demands and occurs in a well-investigated biotope, these factors may be very low, and the estimated total number may be close to the actual number of mycelia. An example of such a species could be the conspicuous orange polypore *Hapalopilus croceus*, which is restricted to very old oaks. Inconspicuous species that are not well investigated may, on the other hand, be assumed to have large locality and mycelium factors giving a high estimated total number of mycelia, even when the known number of localities is small.

2.1.3 Generation Time

Another problem relates to estimating the generation time for a fungus, which has to be known in criteria A–C. In the IUCN context a generation time equals the average time from a fungus reaching maturity and producing spores, to the next generation deriving from those spores reaching maturity and starting to produce spores. How old are the fungi we see? It seems very obvious that certain fungi, growing on ephemeral resources, e.g. dung or herbaceous stems, do not reach a great age, but their spores may have been dormant for a long time before germinating, and many of such species probably have a generation time of 1 year. On the other hand, mycelia of species associated with very stable habitats, e.g. mycorrhizal species growing with old trees, saprotrophic cord and rhizomorphic formers (Chapter 1) and probably also some of the saprotrophic grassland fungi (Chapter 14), may persist for tens or even hundreds of years. For species associated with stable habitats it seems reasonable to use 10 years as a proxy of average generation time, but in some cases a considerably longer time may be appropriate.

With most of the IUCN criteria the percentage decrease in population is to be evaluated over a period of three generations; however, this interval must be within the range of 10–100 years. For the Swedish red-list (Gärdenfors, 2005) it has been decided that three generations equals 50 years for mycorrhizal and litter-inhabiting fungi, 20 years for wood-inhabiting fungi, and 10 years for other species.

2.1.4 Estimation of Population Trends

As fungi are rarely included in monitoring programmes, there is often a shortage of data that can document an actual decrease in population size. A sudden decline may also be difficult to detect, because the production of fruit bodies is strongly influenced by climatic factors (Chapter 5). A species fruiting in summer may not be observed for several years if the summers are hot and dry, and then, when favourable conditions again occur, it may fruit in abundance. Luckily there are few examples, if any, of common fungi which have been documented to decrease very rapidly during the recent years. In The Netherlands substantial work has been carried out to investigate changes in the frequency of macromycetes, based on foray reports, mapping projects and repeated analyses of permanent plots, and a strong decrease, mainly for ectomycorrhizal taxa, has been documented (e.g. Arnolds, 1991; Nauta and Vellinga, 1993). In most other countries similar data are lacking and gradual changes in population sizes of particular species have to be estimated on a more uncertain basis. In many cases however, negative population trends can be assumed with reference to overall changes in habitat quality, land use or forestry practices, which may strengthen the understanding of former, present and even future population changes.

Short-term fluctuations in population levels are also evaluated under some of the IUCN criteria. Fluctuations in fruit body production, caused by climatic factors, make it almost impossible to detect if fluctuation at the population level takes place. Strongly fluctuating population sizes may be an obvious threat for other groups of organisms, but are probably less relevant for fungi.

2.1.5 Adjusting Categories to the Regional Level

The IUCN criteria are designed to evaluate the risk of extinction at a global level. To adjust to the regional level requires further considerations as the regional status depends on the exchange taking place with populations in other regions. The threat category must be adjusted to a lower category if populations outside the region support the regional population. Mycelia do not migrate between regions, but their spores do, and if the species occur in an adjacent region, it is likely that there will be an influx of viable spores to the region. Such spore influx is likely to enhance the regional survival of small populations, especially for species occurring at the margin of their distribution area or close to an area where the species is more common. In the same way the category may be adjusted to a higher level if the regional population fitness is believed to depend on spore exchange with populations in adjacent regions, when these extra-marginal populations are known to be declining.

2.2 Fungi as Indicators of Habitat Quality

Saprotrophic fungi have been tested and applied as indicators of valuable habitats following two different approaches. In the first attempts to use fungi as indicators (e.g. Rald, 1985) the suggested indicator species were simply stated to indicate "nature value" or, circularly, sites valuable for the protection of the indicator species themselves. Others have mentioned ecological continuity or overall conservation value as indication goals (e.g. Karström, 1992; Bredesen *et al.*, 1997). Most of the proposed indicator schemes have been suggested by experienced field mycologists, but few have been tested scientifically to determine if the indicator species really indicate what they are assumed to indicate. For this reason the relevance of some indicator schemes has been questioned. Most importantly, the use of selected polypores to indicate ecological continuity in boreal forests has been challenged (Nordén and Appelqvist, 2001; Rolstad *et al.*, 2002), due to lack of evidence, both theoretically and in practice. Partly based on this critique, more sophisticated approaches to test and use fungi as indicators of overall species richness have been developed. Some of those approaches have focussed solely on fungi, whereas others have investigated the potential of fungi as indicators of diversity in other organism groups—and *vice versa*.

Balmford *et al.* (2000) tested the usefulness of the so-called higher taxon approach in which the diversity at higher taxonomical levels (e.g. genera, families) is assumed to reflect diversity at lower taxonomic levels (typically species). Based on inventory data from 19 sites in England, a very high correlation between genus and species level richness was found, while richness at family and order level predicted species richness slightly less well. When data were used to select hypothetical reserve networks, richness at both the family and genus level was useful for indicating richness at the species level. Balmford *et al.* (2000) thus suggested inventories of genera rather than species to identify fungal diversity hotspots in a region, e.g. for inclusion in reserve networks. A related idea is the nested subset approach, which is based on the fact that species often form nested patterns across landscape habitat patches (Wright *et al.*, 1998; Jonsson and Jonsell, 1999). Nested patterns occur if locally rare species tend to be confined to the most species rich localities/patches, while species poor localities/patches, in contrast, only host locally common species. In the presence of nested patterns, subsets of species with intermediate or low numbers of occurrences accordingly may indicate overall species richness (Jonsson and Jonsell, 1999). The approach has been tested in relation to saproxylic fungi in boreal forests, but with inconsistent results. Jonsson and Jonsell (1999) failed to identify nested patterns among polypores in 10 selectively logged spruce forests, but nested patterns were identified in a similar study conducted in old growth forest patches in a natural wetland matrix (Berglund and Jonsson, 2003). Also Sætersdal *et al.* (2005) identified nested patterns among boreal wood-decay fungi, but patterns were highly inconsistent among three inventoried forest regions in Norway. Thus, indicator species, identified by nested subset analysis in one region, were typically non-indicative of species-rich stands in other regions.

Approaches focusing on the relation between fungal species diversity and diversity in other organism groups, typically analyse patterns of: (i) species richness coincidence and/or (ii) complementarities between groups. The first concept is based on the assumption that richness in one taxonomical or functional group may reflect richness in other groups, so that a well-known and easily surveyed group, e.g. birds or higher plants, can be used as a surrogate of diversity in less easily surveyed groups, e.g. fungi. Similarly, the complementarity approach assumes that complementarity or species turnover in one taxonomic group across a number of sites reflects complementarity or turnover in other groups. The typical aim is to facilitate the design of reserve networks protective also for less known species groups.

Most attempts to investigate species richness coincidence and complementarity patterns involving fungi have focussed on saproxylic species (Jonsson and Jonsell, 1999; Virolainen *et al.*, 2000; Berglund and Jonsson, 2001; Sætersdal *et al.*, 2004). In all cases the species richness coincidence among fungi and other inventoried groups of organisms was absent or low, but even complementarity patterns were vague. It thus seems that inventories based on other organism groups are poorly suited to detect the most species-rich sites for saproxylic fungi—and *vice versa*. This reflects different habitat requirements among groups. The diversity of saproxylic fungi is influenced mainly by the amounts and diversity of dead-wood habitats present, while diversity in other groups is influenced by other factors, e.g. soil productivity, forest climate or local stand continuity (e.g. Ohlson *et al.*, 1997; Berglund and Jonsson, 2001). For other fungal groups Chiarucci *et al.* (2005) found that vascular plants were poorly suited to capture macrofungal species richness in hypothetical reserve networks in Italy. On the other hand, Schmit *et al.* (2005) found that tree species richness was a promising surrogate for fungal species richness at a large scale in a meta-analysis of 25 datasets from temperate and boreal forests from three different continents.

A general problem for all approaches focussing on species richness is that the most species-rich sites are not necessarily the most valuable for conservation (e.g. Gjerde *et al.*, 2004; Heilmann-Clausen and Christensen, 2005). Threatened fungal communities are not necessarily species rich, and human disturbances threatening rare species may often increase rather than decrease overall diversity. If, for instance, half of an ancient beech forest hosting rare saproxylic polypores is cleared and planted with spruce, this is likely to introduce a large number of new species associated with this new host, while only a few very uncommon species associated with the ancient beech forest will disappear in the short term. Similarly eutrophication is likely to increase species richness but certainly not "naturalness" in naturally nutrient-poor habitat types, e.g. sphagnum bogs. There is accordingly a need for novel approaches bridging rigorous scientific testing with the early intensions of indicators as species pointing to threatened and conservation demanding fungal habitats.

One way forward could be to focus directly on red-listed species. Monitoring of all red-listed species is not practicable, because of the high number of species involved, e.g. 632 in Sweden alone (Gärdenfors, 2005), and their rarity and spatial patchiness (Straatsma and Krisai-Greilhuber, 2003). It would be very useful to identify easily identifiable and not too rare indicators of sites with high numbers of red-listed species, e.g. by applying nested subset approaches with red-listed rather than total species richness in focus. Another possibility is to investigate further the relations between habitat qualities and the occurrence of rare or red-listed species, e.g. relations between habitat continuity at various geographical scales and the presence of certain fungal communities. Such analyses may or may not show that easily monitorable habitat parameters are more efficient predictors of the fungal conservation value than a suite of indicator species.

In summary, the use of fungal indicator species is rarely scientifically proved, and the most important aspect of fungal indicator species may be that they are attractive to interested fungal amateurs. Thus, they may be very powerful in increasing the awareness and knowledge of valuable fungal habitats, which would be difficult to achieve if the same group of people were encouraged to investigate habitat parameters, such as soil productivity, abundance of veteran trees or local habitat continuity through interviews with land owners or costly historical studies. This has, for instance, been the case with *Hygrocybe* species and other saprotrophs in grasslands, as discussed below.

3. CONSERVATION OF SAPROTROPHIC FUNGI IN EUROPE

Saprotrophic basidiomycetes are found in most terrestrial habitat types with forests being particularly species rich. Natural processes constantly affect populations, and species naturally go extinct from time to time, even in the absence of human intervention, while others split into several genetic distinct lineages resulting in the formation of new biological species.

The human use of European landscapes through recent millennia has dramatically influenced these processes. Most natural habitats have decreased in quality or quantity due to clearing of woods, agriculture, urbanisation and industrialisation. Sometimes humans have also created new habitat types or have increased the abundance of naturally rare habitats to the benefit of certain groups of fungi. For instance, the expansion of animal husbandry through the last 5,000 years strongly increased the extent of grasslands to the benefit of dung and grassland fungi. Similarly, the current use of woodchips in horticulture has lead to a strong increase for a number of saproxylic fungi, including invasive alien species (Shaw *et al.*, 2004). Interactions between human and natural processes affecting fungal habitats are complex, and it is not feasible to distinguish sharply between natural versus artificial habitats. Rather these form a continuum from "virgin" habitat types (e.g. virgin forests) through semi-natural habitats (e.g. most grasslands and wooded meadows) to highly artificial habitats (e.g. crop fields, wood-chip covered garden strips). In this context also the time factor is important. Habitat types with a long local or regional history are more likely to host a distinct suite of locally adapted habitat specialists compared with more recent, typically human introduced habitats. Currently the most important forces affecting decomposer basidiomycetes in Europe are:

1. Intensified silviculture (removing of dead wood and veteran trees, conversion from native to exotic tree species, drainage of swamp forests, diminished forest grazing).
2. Intensified agriculture ("improving" grasslands by fertiliser additions and conversion to arable crops).
3. Decreasing animal husbandry in low fertile areas leading to forest re-growth.
4. Increased nitrogen deposition.
5. Climate change.

These factors affect fungal communities directly or indirectly in different ways. For the two latter factors probable effects on decomposer basidiomycetes in the field are uncertain, though there is evidence of changes in fruiting patterns as a result of climate change (Chapter 5). Nitrogen deposition is known to affect ectomycorrhizal species considerably (e.g. Peter *et al.*, 2001a; Avis *et al.*, 2003), and fertilisation experiments have even shown effects on decomposer communities

(Rühling and Tyler, 1991). Below we review conservation approaches for two important groups of basidiomycetes, grassland and saproxylic saprotrophs which include many species believed to be decreasing in Europe.

3.1 Grassland Species

Most grasslands can be considered semi-natural because they depend on management to prevent them developing into scrub or forest. The history of grasslands in Europe, before the first animal husbandry has been much discussed (e.g. Svenning, 2002), but there is little doubt that the habitat types reached a maximum in the pre-industrialised farming era, from the 15th through to the beginning of the 20th century. Since then semi-natural grasslands has decreased dramatically in many countries (e.g. Bruun and Ejrnæs, 1998).

Without the use of artificial fertilisers most dry grasslands are nutrient poor because nutrients (phosphorus and nitrogen) are removed by the grazing animals. Grasslands may have a continuity of hundreds or maybe even thousands of years. Characteristic fungi of dry grasslands with long continuity are not least wax-caps (*Hygrocybe*, including *Camarophyllus*) and *Entoloma* species, especially of subgenus *Leptonia*, each group with ~50 species occurring in grasslands. In addition, a number of club fungi (*Clavaria*, *Clavulinopsis*, *Ramariopsis*) and Geoglossaceae (*Geoglussum*, *Microglossum*, *Trichoglossum*) occur in grasslands. Many other groups may be represented in dry grasslands, like puff balls and species of *Galerina*, *Agaricus* and *Macroleptiota*, but such species generally have broader ecological amplitudes or are associated with more nutrient-rich grassland types, and are not considered as indicators of valuable grassland sites with a long continuity.

In most European countries, the wax-caps are well investigated, and wax-cap species are rarely recorded as new to a country. The checklists of *Entoloma* species, however, are still incomplete in most countries. Most *Hygrocybe* species fruit in autumn, and continue until early winter. Some *Hygrocybe* species fruit from summer, but only when it rains. *Entoloma* species most commonly fruit in summer, and therefore their fruiting is less constant.

3.1.1 Conservation Status and Threats

In most European countries the quality of many dry grassland localities has decreased as a result of more intensive land use, including use of fertilisers (Arnolds, 1988; Nitare, 1988; Nauta and Vellinga, 1993; Boertmann, 1995; Evans, 2003; Chapter 15). In many areas of Europe grasslands are now more or less restricted to remote areas, steep slopes, etc. where a more intensive land use is impossible or non-profitable. Therefore, many grassland species, including fungi, occur in the red-lists of various countries. In some countries, the changes in grassland management may be considerable despite conservation efforts to protect the habitat type. In 2005 a survey of 188 valuable grassland localities with recent records of red-listed or rare organisms was carried out in Vejle County, Denmark (Vesterholt and Levesen, 2006). Only 81 localities (43%) were grazed, and 83 localities (44%) were in more or less urgent need of resuming traditional management. Grazing of dry and wet grasslands has been subsidised by Danish government/European community means, but after a change of policy by the Danish government it has only been possible to apply for subsidies in Natura 2000 areas. In the survey in Vejle County only 48 areas (26%) qualified as such. Such national changes in policy may result in a decline of dry grasslands, and a serious threat to the associated species.

3.1.2 Methods of Identifying Valuable Grassland Sites

In several countries grassland fungi have been proposed as indicators of valuable grassland localities. Rald (1985) suggested that the number of *Hygrocybe* species (including *Camarophyllus*) found in a locality was a good indicator of the mycological importance of the locality: localities with 17 species or more were of national importance, and sites with fewer species were of local, regional or no importance (Table 2). Rald and Boertmann (1989) found that 12 Danish localities qualified as being of national importance. Ten years later the number had increased to 35, due to extensive field work (Vesterholt *et al.*, 1999), while 12 Danish grassland localities with 22 *Hygrocybe* species or more were considered as being of international importance following Rald (1985).

The exact habitat requirements of most grassland fungi are still poorly understood (Chapter 14), but it is broadly accepted that some grassland species are better indicators of valuable grassland localities than others. Such good indicators are typically rare and restricted to localities with high species diversity. Some of the grassland species seem to occur only in old grasslands with a long continuity, e.g. *Entoloma anatinum, E. longistriatum, E. mougeotii, Hygrocybe intermedia, H. citrinovirens, H. ovina, H. aurantiosplendens* and *H. ingrata* (McHugh *et al.*, 2001; Vesterholt, 2002; Newton *et al.*, 2003). On the other hand, some rare species, e.g. *Entoloma formosum, E. hispidulum, E. xanthochroum, Hygrocybe subpapillata, H. glutinipes, H. vitellina* and *H. spadicea* are also found in grasslands with a shorter continuity. Based on such experiences, Jordal and Gaarder (1993) proposed a classification system, where an individual indicator value was assigned to each grassland species, from 0 to 8 points, so that the value of a locality was calculated by summing the value assigned to each of the species found in the locality. The system includes several groups of fungi, and it is therefore more broadly applicable than the systems based only on *Hygrocybe* species, which is important since grasslands are sometimes richer in *Entoloma* than in *Hygrocybe*

Table 2 Erik Ralds system for classification of dry grassland localities, amended by Vesterholt *et al.* (1999)

Of limited importance	1–3 *Hygrocybe* species
Of local importance	4–8 *Hygrocybe* species
Of regional importance	9–16 *Hygrocybe* species
Of national importance	17–21 *Hygrocybe* species
Of international importance	≥22 *Hygrocybe* species

species (Newton *et al.*, 2003). A similar system has been suggested by McHugh *et al.* (2001) for Irish grassland fungi where each species is given an indicator value on a scale from 1 to 4 points.

Grassland fungi do not fruit in abundance every year, and some species fruit after intervals of several years. This is especially the case with many *Entoloma* species that fruit mainly in the summer. Thus, it is necessary to visit a locality several times, and preferably at least once in a very good season, to assess the real value of a locality. Accordingly, Rald's (1985) method of classifying a site after a single visit is not recommended.

3.2 Saproxylic Fungi

Saproxylic fungi constitute a highly diverse group, comprising thousands of species in Europe (Chapter 11). Well-known, species-rich groups include polypores and corticoid basidiomycetes, but also agarics and pyrenomycetes and many others. Unlike grassland fungi, saproxylic fungi are associated with naturally relatively short-lived resources that are spatially restricted. Therefore, individual mycelia are usually short-lived (<100 years, often much less), good dispersal ability being paramount for survival in the long run. Some, e.g. cord- and rhizomorph-forming basidiomycetes, may however become very old and large (e.g. Ferguson *et al.*, 2003; Chapter 1). Under natural conditions dead-wood habitats are common. Thus, dispersal over relatively short distances (1–1,000 m) is more important than over longer distances. Nevertheless, spores of several saproxylic species have been shown to be able to germinate after travelling hundreds or even thousands of kilometres from their source (Vilgalys and Sun, 1994; Hallenberg and Küffer, 2001).

Little is known of population densities of wood decay fungi. The few studies conducted so far have demonstrated that the number of mycelia in unit-restricted fungi is often higher than expected from the distribution of sporocarps. Thus, single pieces of dead wood are commonly inhabited by several individuals of the same species (Boddy and Rayner, 1983; Hendry *et al.*, 1998; Kauserud and Schumacher, 2002, 2003). Also, most living trees host saproxylic fungi resting latently, waiting for appropriate conditions for active growth (Chapter 11). Population density is therefore underestimated by sporocarp surveys. On the other hand, dead-wood communities are often very species rich pointing to a high level of niche differentiation (Heilmann-Clausen *et al.*, 2005).

3.2.1 Conservation Status and Threats

The abundance and quality of dead-wood habitats has changed considerably through historical time in temperate Europe, and similar changes have occurred, or are happening in most other parts of the world. These changes can be summarised for temperate Europe (Chapter 11):

1. Decline in the amount of dead wood. The actual extent and timing of decline is highly variable among forest regions in Europe, ranging between 90 and 99.5% of that in primeval forests.

2. Fragmentation of remaining dead-wood habitats. This is evident both at the local scale, where management in most forests has increased distances between individual wood units (especially for large-diameter wood fractions), and at the regional scale, where remaining forests typically form more or less isolated patches in a matrix of farmland and urban areas. In Europe the majority of forest reserves rich in dead wood are restricted to mountainous and other cultivation-hostile areas.
3. Change in dead-wood composition, i.e. an increase in amount of coniferous wood and modified or cut wood, a modest decrease in small-diameter dead wood and a strong decrease in large-diameter wood.

The effects for communities of saproxylic fungi have been dramatic (Chapter 11): the decline in amounts of dead wood has resulted in a reduction in population size of most saproxylic fungi, most dramatically for species with a natural low population density. The change in availability of certain types of dead wood (e.g. large decomposing logs, dead wood in veteran trees) has resulted in a decline in species adapted to the conditions within these, while species with broad ecological amplitudes or adaptations to less decreasing or even increasing habitats (cut wood, small diameter wood, coniferous wood) has declined less or even increased. These trends are reflected in red-lists from several European countries, e.g. Sweden, where saproxylic fungi are over-represented among species showing a distinct preference for large-diameter decaying trunks or veteran trees (Dahlberg and Stokland, 2004).

New forest reserves have been declared in many European countries, as a response to the decline of dead-wood habitats and their associated biodiversity (Parviainen, 1999), and initiatives have been taken to increase dead-wood quantity in managed forests (e.g. Skov- og Naturstyrelsen, 2005). For some species these initiatives might come too late or be too limited to counteract further species extinctions, due to a lack of congruence between population and habitat dynamics, resulting in a so-called *extinction debt* (e.g. Jonsson *et al.*, 2005). One example could be the polypore *Hapalopilus croceus* which is currently known from less than 150 sites in Europe, all with very low local population sizes (Dahlberg and Croneborg, 2006). The species is associated with very old living or decaying oaks, and even though dead wood is increasing in many landscapes, the specific habitat is highly fragmented.

3.2.2 Identifying Valuable Sites

The first attempts to identify important areas for the conservation of saproxylic fungi in Europe were launched in boreal Scandinavia in the early 1990s (Høiland and Bendiksen, 1991; Karström, 1992). The approach was based on lists of indicator species, including fungi, assumed to indicate not only valuable habitats for the organisms themselves, but forest with high local continuity and conservation value in general. The concept has since been widely applied in Scandinavia (Kotiranta and Niemelä, 1996; Bredesen *et al.*, 1997; Nitare, 2000), and more recently in several countries across Europe (Parmasto and Parmasto, 1997; Heilmann-Clausen and Christensen, 2000; Nordstedt *et al.*, 2001; Holec, 2003; Łuszczyński, 2003; Ainsworth, 2004; Walleyn and Veerkamp, 2005).

In Scandinavia the use of saproxylic fungi as indicators of ecological continuity has been heavily criticised (Rolstad *et al.*, 2002; Nordén and Appelqvist, 2001), due to lack of statistical support from field studies and the obvious potential of fungi for long distance dispersal. Nordén and Appelquist (2001) thus noted that saproxylic fungi in general are likely to depend on a rich supply of dead-wood habitats, rather than the presence of long local forest continuity. Following this argument, and backed by actual results from field studies, several authors (e.g. Similä *et al.*, 2004; Sætersdal *et al.*, 2004, 2005) have argued that structural indicators (e.g. amount and quality of dead wood) rather than indicator species are relevant in identification of valuable sites for conservation of saproxylic organisms, including fungi.

Other studies, mainly over larger geographical scales (tens to hundreds of kilometres), have however shown that the frequency of threatened saproxylic fungi cannot be predicted solely based on easily monitored dead-wood parameters, but that factors related to forest history at the landscape scale have a marked effect on species composition (Sverdrup-Thygeson and Lindenmayer, 2003; Stokland and Kauserud, 2004; Heilmann-Clausen and Christensen, 2005; Ódor *et al.*, 2006; Pentillä *et al.*, 2006). Christensen *et al.* (2005) suggested saproxylic fungi to be particularly suited as indicators of dead wood and veteran tree continuity at the landscape scale (25–10,000 ha), and Strange *et al.* (2004) used indicator species to model the efficacy of fungal conservation in reserve networks in time and space. Both applications may prove very useful in a conservation context: the first, because dead wood continuity in space and time is difficult or even impossible to investigate at the appropriate scale; the second, because localised conservation efforts are likely to prove unsuccessful for species showing population dynamics at the landscape scale. Scientific testing of indicator schemes based on saproxylic fungi is much needed to determine how relevant they are for monitoring and management of threatened saproxylic fungi and other groups of organisms associated with old growth forest habitats. In Figure 3, an example of practical application of saproxylic fungi as indicators of valuable habitats in Denmark is illustrated.

3.3 Recommendations for the Future

Conservation mycology is still in its infancy, and has not yet received the same attention as animal and plant conservation. Above we have discussed different challenges and approaches relevant for fungal conservation and reviewed the current situation for saprotrophic basidiomycetes in grassland and on dead wood. We believe that indicator species and red-lists are useful means to put fungi on the overall conservation agenda and, in practice, to identify valuable sites for conservation. However, focussed initiatives to increase both the scientific and more broad understanding of conservation mycology are required. Relevant issues include:

(1) Lobbying to stimulate public and political awareness of fungi as organisms worthy of protection. Positive examples, such as successful

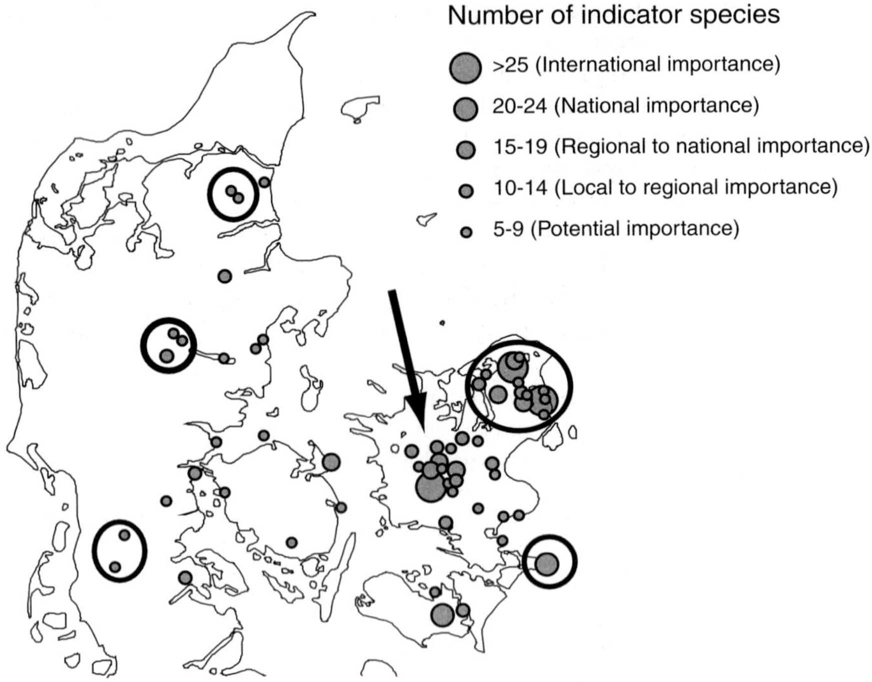

Figure 3 Map showing localities with five or more indicator species of valuable beech forest habitats (Heilmann-Clausen and Christensen, 2000) in Denmark. Encircled areas contain one to several larger forest reserve, >50 ha, containing protected, old grown beech stands. The map shows that valuable sites occur clumped at the landscape scale and that one important region containing several valuable localities (arrowed) currently is insufficiently protected (adapted from Heilmann-Clausen, 2005).

protection schemes for fungi (e.g. Naturvårdsverket, 2006) and protection of important fungus localities, deserve promotion to a broader public as demonstration projects. Inclusion of fungi in international conventions, e.g. in the EU Natura 2000 program, must be a goal.

(2) Elucidation of the extent to which existing conservation schemes such as nature reserves, National Parks and the Natura 2000 program protect threatened fungi.

(3) Increased focus on identifying fungi threatened at the international scale is much needed. Work is underway to create a European red-list, coordinated by ECCF (2006).

(4) Many research issues are relevant but the following are among the most urgent questions: (i) How should valuable sites be identified for inclusion in reserve networks—are indicators (species or structures) useful, or is it more effective to focus directly on selected priority species (e.g. internationally threatened species)? (ii) As pointed out by Jonsson *et al.* (2005), wood-inhabiting fungi are structured as metapopulations. This implies that local population sizes are unstable and variable over time even

though the total population size is stable. How does this affect populations of threatened species in fragmented landscapes, where dead-wood habitats are highly agglomerated spatially in protected forest reserves, and what are the consequences for design of reserve networks? (iii) To what degree do fungi associated with more stable habitats (e.g. grassland fungi) function as metapopulations and what is the level of genetic exchange among sites in a fragmented landscape? Do we risk a widespread local extinction of threatened species in small, isolated protected sites, even if local habitat qualities are protected? (iv) How important are resting non-sporulating mycelia for the resilience of local populations of threatened fungi, and how long is the typical lifespan of mycelia of various species?

REFERENCES

Ainsworth, A.M. (2004). Developing Tools for Assessing Fungal Interest in Habitats. 1: Beech Woodland Saprotrophs. *English Nature, Report No. 597*. English Nature, Peterborough, UK.

Allmér, J., Vasiliauskas, R., Ihrmark, K., Stenlid, J. and Dahlberg, A. (2005). Wood-inhabiting fungal communities in woody debris of Norway Spruce (*Picea abies* (L.) Karst.), as reflected by sporocarps, mycelial isolations and T-RFLP identification. *FEMS Microbiology Ecology*, **55**, 57–67.

Anonymous. (2004). Réserve mycologique de la Chanéaz [The Mycological Reserve la Chanéaz]. www.pilzreservat.ch (1st Feb. 2007).

Antonín, V. and Bieberová, Z. (1995). *Chráněné houby ČR* [Protected Fungi of the Czech Republic]. Agentura OP, Brno.

Arnolds, E. (1988). The changing macromycete flora in the Netherlands. *Transactions of the British Mycological Society*, **90**, 391–406.

Arnolds, E. (1989). Former and present distribution of stipitate hydnaceous fungi (Basidiomycetes) in the Netherlands. *Nova Hedwegia*, **48**, 107–142.

Arnolds, E. (1991). Decline of ectomycorrhizal fungi in Europe. *Agriculture, Ecosystems and Environment*, **35**, 209–244.

Avis, P.G., McLaughlin, D.J., Dentinger, B.C. and Reich, P.B. (2003). Long-term increase in nitrogen supply alters above- and below-ground ectomycorrhizal communities and increases the dominance of *Russula* spp. in a temperate oak savanna. *New Phytologist*, **160**, 239–253.

Balmford, A., Lyon, A.J.E. and Lang, R.M. (2000). Testing the higher-taxon approach to conservation planning in a megadiverse group: The macrofungi. *Biological Conservation*, **93**, 209–217.

Berglund, H. and Jonsson B.G. (2001). Predictability of plant and fungal species richness of old-growth boreal forest islands. *Journal of Vegetation Science*, **12**, 857–866.

Berglund, H. and Jonsson B.G. (2003). Nested plant and fungal communities: The importance of area and habitat quality in maximizing species capture in boreal old-growth forests. *Biological Conservation*, **112**, 319–328.

Boddy, L. and Rayner, A.D.M. (1983). Ecological roles of basidiomycetes forming decay communities in attached oak branches. *New Phytologist*, **93**, 77–88.

Boertmann, D. (1995). *The Genus Hygrocybe. Fungi of Northern Europe*, Vol. 1. Svampetryk, Greve.

Brandrud, T.E., Bendiksen, E., Hofton, T.H., Høiland, K. and Jordal, J.B. (2006). Sopp [Fungi]. In: *Norsk Rødliste 2006* [Norwegian Red List] (J.A. Kålås, Å Viken and T. Bakken, eds.), pp. 103–128. Artsdatabanken, Norway.

Bredesen, B., Haugan, R., Aanderaa, R., Lindblad, I., Økland, B. and Røsok, Ø. (1997). Vedlevende sopp som indikatorarter på kontinuitet i østnorske granskoger [Wood-inhabiting fungi as indicators of continuity in spruce forests in eastern Norway]. *Blyttia*, **54**, 131–140.

Bruun, H.H. and Ejrnæs, R. (1998). *Overdrev—En Beskyttet Naturtype* [*Dry Grasslands—A Protected Habitat Type*]. Skov- og Naturstyrelsen, København.

Caro, T.M. and O'Doherty, G. (1999). On the use of surrogate species in conservation biology. *Conservation Biology,* **13**, 805–814.
Chiarucci, A., D'Auria, F., De Dominicis, V., Lagana, A., Perini, C. and Salerni, E. (2005). Using vascular plants as a surrogate taxon to maximize fungal species richness in reserve design. *Conservation Biology,* **19**, 1644–1652.
Christensen, M., Heilmann-Clausen, J., Walleyn, R. and Adamcik, S. (2005). Wood-inhabiting fungi as indicators of nature value in European beech forests. *EFI Proceedings,* **51**, 218–226.
Dahlberg, A. and Croneborg, H. (eds.). (2003). *33 Threatened Fungi in Europe. Complementary and Revised Information on Candidates for Listing in Appendix I of the Bern Convention.* Artsdatabanken, Uppsala.
Dahlberg, A. and Croneborg, H. (eds.). (2006). The 33 Threatened Fungi in Europe. *Nature and Environment No. 136,* Council of Europe Publishing, Strasbourg.
Dahlberg, A. and Stokland, J.N. (2004). Vedlevande arters krav på substrat—sammanställning och analys av 3.600 arter [Habitat requirements of wood-inhabiting species—Synthesis and analysis of 3600 species]. *Skogsstyrelsen, Rapport,* **7**.
DMU. (2006). *The Danish Red Data Book.* http://redlist.dmu.dk (1st Feb. 2007).
ECCF (2006). Resolutions from the 7th Meeting in Cordoba, 9–12 November 2005, in Newsletter 14 - February 2006, www.wsl.ch/eccf/newsletter14.pdf.
Evans, S. (2003). Waxcap-Grasslands—An assessment of English sites. *English Nature Report 555.* English Nature, Peterborough, UK.
Ferguson, B.A., Dreisbach, T.A., Parks, C.G., Filip, G.M. and Schmitt, C.L. (2003). Coarse-scale population structure of pathogenic *Armillaria* species in a mixed-conifer forest in the Blue Mountains of northeast Oregon. *Canadian Journal of Forest Research,* **33**, 612–623.
Gardes, M. and Bruns T.D. (1996). Community structure of ectomycorrhizal fungi in a *Pinus muricata* forest: Above- and below-ground views. *Canadian Journal of Botany,* **74**, 1572–1583.
Gärdenfors, U. (ed.). (2005). *Rødlistade Arter i Sverige 2005. The 2005 Red List of Swedish Species.* Art-Databanken, Uppsala.
Gjerde, I., Satersdal, M., Rolstad, J., Blom, H.H. and Storaunet, K.O. (2004). Fine-scale diversity and rarity hotspots in northern forests. *Conservation Biology,* **18**, 1032–1042.
Hallenberg, N. and Küffer, N. (2001). Long-distance spore dispersal in wood-inhabiting Basidiomycetes. *Nordic Journal of Botany,* **21**, 431–436.
Hawksworth, D.L. (2001). The magnitude of fungal diversity: The 1.5 million species estimate revisited. *Mycological Research,* **105**, 1422–1432.
Heilmann-Clausen, J. (2003). Wood-inhabiting fungi in Danish deciduous forests—Diversity, habitat preferences and conservation, PhD Thesis, Danish Research Academy/Danish Forest and Landscape Research Institute.
Heilmann-Clausen, J. (2005). Svampe, dødt ved og naturbeskyttelse [Fungi, dead wood and nature conservation]. In: *Svampe i verden [Fungi in the World]* (D. Hammelev, R. Macholm, J.H. Petersen, T. Læssøe and M. Holm, eds.), pp. 56–63. Foreningen til Svampekundskabens Fremme, Hornbæk.
Heilmann-Clausen, J., Aude, E. and Christensen, M. (2005). Cryptogam communities on decaying deciduous wood—Does tree species diversity matter? *Biodiversity and Conservation,* **14**, 2061–2078.
Heilmann-Clausen, J. and Christensen, M. (2000). Svampe på bøgestammer—indikatorer for værdifulde løvskovslokaliteter [Fungi on beech logs—Indicators of habitat quality]. *Svampe,* **42**, 35–47.
Heilmann-Clausen, J. and Christensen, M. (2005). Wood-inhabiting fungi in Danish beech-forests—Conflicting diversity patterns and their implications in a conservation perspective. *Biological Conservation,* **122**, 633–642.
Hendry, S.J., Lonsdale, D. and Boddy, L. (1998). Strip-cankering of beech (*Fagus sylvatica*): Pathology and distribution of symptomatic trees. *New Phytologist,* **140**, 549–565.
Høiland, K. and Bendiksen, E. (1991). Problems concerning lignicolous fungi in boreal forests in Norway. In: *Conservation of Fungi in Europe, Proceedings of the second meeting of the European Council for the Conservation of Fungi, Vilm, 13–18th Sept. 1991*(Arnolds, E. and Kreisel, H. eds.). Ernst-Moritz-Arndt-Universität, Greifswald.
Holec, J. (2003). Auf natürliche, vom Menschen nur minimal beeinflusste Vegetation beschränkte Grosspilze. Eine Modellstudie aus dem Böhmerwald [On macrofungi restricted to forests with minimal human influence. A case study from Böhmerwald]. *Fritschiana,* **42**, 25–27.

IUCN. (2001). *IUCN Red List Categories and Criteria: Version 3.1.* IUCN Species Survival Commission, Gland and Cambridge.

IUCN. (2003). *Guidelines for Application of IUCN Red List Criteria at Regional Levels: Version 3.0.* IUCN Species Survival Commission, Gland and Cambridge.

Jonsson, B.G. and Jonsell, M. (1999). Exploring potential biodiversity indicators in boreal forests. *Biodiversity and Conservation*, **8**, 1417–1433.

Jonsson, B.G., Kruys, N. and Ranius, T. (2005). Ecology of species living on dead wood—Lessons for dead wood management. *Silva Fennica*, **39**, 289–309.

Jordal, J.B. and Gaarder, G. (1993). Soppfloraen i en del naturbeitemarker og naturenger i Møre og Romsdal og Trøndelag [Fungi in some unimproved grasslands in Møre and Romsdal and Trøndelag]. *Fylkesmannen i Møre og Romsdal, Miljøvernavd. rapport* **9-1993**.

Karström, M.P. (1992). Steget före—en presentation [The project one step ahead—A presentation]. *Svensk Botanisk Tidskrift*, **86**, 103–114.

Kauserud, H. and Schumacher, T. (2002). Population structure of the endangered wood decay fungus *Phellinus nigrolimitatus* (*Basidiomycota*). *Canadian Journal of Botany*, **80**, 597–606.

Kauserud, H. and Schumacher, T. (2003). Regional and local population structure of the pioneer wood-decay fungus *Trichaptum abietinum*. *Mycologia*, **95**, 416–425.

Kirby, J.J.H., Stenlid, J. and Holdenrieder, O. (1990). Population structure and responses to disturbance of the basidiomycete *Resinicium bicolor*. *Oekologia*, **85**, 178–184.

Kotiranta, H. and Niemelä, T. (1996). *Uhanalaiset käävät Suomessa [The threatened Aphyllophorales in Finland].* Finnish Environmental Institute, Helsinki.

Kytövuori, I., Nummela-Salo, U., Ohenoja, E., Salo, P. and Vauras, J. (2005). Helttasienten ja tattien ekologiataulukko [Ecology table of agarics boletes in Finland]. In: *Suomen helttasienten ja tattien ekologia, levinneisyys ja uhanalaisuus* (P. Salo, T. Niemelä, U. Nummela-Salo and E. Ohenoja, eds.), pp. 225–426. Soumen ympäristökeskus, Helsinki.

Łuszcyński, J. (2003). Relict fungi of primeval forests in the Świętokrzyskie mountains (Central Poland). *Botanica Lithuanica*, **9**, 243–251.

McHugh, R., Mitchel, D., Wright, M. and Anderson, R. (2001). The fungi of Irish grasslands and their value for nature conservation—Biology and environment. *Proceedings of the Royal Irish Academy*, **101B**, 225–242.

Naturvårdsverket. (2006). *Åtgärdsprogram för Hotade Arter [Action Plans for Threatened Species].* http://www.naturvardsverket.se/index.php3?main=/dokument/natur/atgardpr/atgardpr.htm (1st Feb. 2007).

Nauta, M. and Vellinga, E.C. (1993). Distribution and decline of macrofungi in The Netherlands. In: *Fungi of Europe: Investigation, Recording and Conservation* (D.N. Pegler, L. Boddy, B. Ing and P.M. Kirk, eds.), pp. 21–46. Royal Botanical Gardens, Kew.

Newton, A.C., Davy, L.M., Holden, E., Silverside, A, Watling, R. and Ward, S.D. (2003). Status, distribution and definition of mycologically important grasslands in Scotland. *Biological Conservation*, **111**, 11–23.

Nitare, J. (1988). Jordtungor, en svampgrupp i tilbakagång i naturlige fodermarker [Earthtongues, a group of fungi in decline in natural grasslands]. *Svensk Botanisk Tidskrift*, **82**, 341–368.

Nitare, J. (2000). *Signalarter. Indikatorer på skyddsvärd skog—flora över kryptogamer [Indicator species for assessing the nature conservation value of woodland sites—A Flora of selected cryptogams].* Skogsstyrelsen, Jönköbing.

Nordén, B. and Appelqvist, T. (2001). Conceptual problems of ecological continuity and its bioindicators. *Biodiversity and Conservation*, **10**, 779–791.

Norstedt, G., Bader, P. and Ericson, L. (2001). Polypores as indicators of conservation value in Corsican pine forests. *Biological Conservation*, **99**, 347–354.

Ódor, P., Heilmann-Clausen, J., Christensen, M., Aude, E., van Dort, K.W., Piltaver, A., Siller, I., Veerkamp, M.T., Walleyn, R., Standovár, T., van Hees, A.F.M., Kosec, J., Matočec, N., Kraigher, H. and Grebenc, T. (2006). Diversity of dead wood inhabiting fungal and bryophyte assemblages in semi-natural beech forests in Europe. *Biological Conservation*, **131**, 58–71.

Ohlson, M., Söderström, L., Hörnberg, G., Zakcrisson, O. and Hermansson, J. (1997). Habitat qualities versus long-term continuity as determinants of biodiversity in boreal old-growth swamp forests. *Biological Conservation*, **81**, 221–231.

Otto, P. (1992). Verbreitung und Rückgang der terrestrischen Stachelpilze Ostdeutschlands [Distribution and decline of terrestrial tooth fungi in East Germany]. *Gleditschia*, **20**, 153–202.
Parmasto, E. and Parmasto, I. (1997). Lignicolous Aphyllophorales of old and primeval forests in Estonia. 1. The forests of northern Central Estonia with a preliminary list of indicator species. *Folia Cryptogamia Estonica*, **31**, 38–45.
Parviainen, J. (1999). Strict forest reserves in Europe—Efforts to enhance biodiversity and strengthen research related to natural forests in Europe. *EFI Proceedings*, **16**, 7–33.
Pennanen, T., Paavolainen, L. and Hantula, J. (2001). Rapid PCR-based method for the direct analysis of fungal communities in complex environmental samples. *Soil Biology and Biochemistry*, **33**, 697–699.
Penttilä, R., Lindgren, M., Miettinen, O., Rita, H. and Hanski, I. (2006). Consequences of forest fragmentation for polyporous fungi at two spatial scales. *Oikos*, **114**, 225–240.
Peter, M., Ayer, F. and Egli, S. (2001a). Nitrogen addition in a Norway spruce stand altered macromycete sporocarp production and below-ground ectomycorrhizal species composition. *New Phytologist*, **149**, 311–325.
Peter, M., Ayer, F., Egli, S. and Honegger, R. (2001b). Above- and below-ground community structure of ectomycorrhizal fungi in three Norway spruce stands in Switzerland. *Canadian Journal of Botany*, **79**, 1628–1642.
Rald, E. (1985). Vokshatte som indikatorarter for mykologisk værdifulde overdrevslokaliteter [Waxcaps as indicator species of mycologically important grassland localities]. *Svampe*, **11**, 1–9.
Rald, E. and Boertmann, D. (1989). Overdrevssvampe—en truet flora [Grassland fungi—A threatened flora]. *Urt*, **13**, 39–45.
Rolstad, J., Gjerde, I., Gundersen, V.S. and Sætersdal, M. (2002). Use of indicator species to assess forest continuity: A critique. *Conservation Biology*, **16**, 253–257.
Rühling, Å. and Tyler, G. (1991). Effects of simulated nitrogen deposition to the forest floor on the macrofungal flora of a beech forest. *Ambio*, **20**, 261–263.
Sætersdal, M., Gjerde, I. and Blom, H.H. (2005). Indicator species and the problem of spatial inconsistency in nestedness patterns. *Biological Conservation*, **122**, 305–316.
Sætersdal, M., Gjerde, I., Blom, H.H., Ihlen, P.G., Myrseth, E.W., Pommeresche, R., Skartveit, J., Solhøy, T. and Aas, O. (2004). Vascular plants as a surrogate species group in complementary site selection for bryophytes, macrolichens, spiders, carabids, staphylinids, snails, and wood living polypore fungi in a northern forest. *Biological Conservation*, **115**, 21–31.
Schmit, J.P., Mueller, G.M., Leacock, P.R., Mata, J.L., Wu, Q. and Huang, Y. (2005). Assessment of tree species richness as a surrogate for macrofungal species richness. *Biological Conservation*, **121**, 99–110.
Senn-Irlet, B. (1997). Provisorische Rote Liste der gefährdeten Höheren Pilze der Schweiz [A preliminary list of threatened fungi in Switzerland]. *Mycologia Helvetica*, **9**, 81–110.
Shaw, P.J.A., Butlin, J. and Kibby, G. (2004). Fungi of ornamental woodchips in Surrey. *Mycologist*, **18**, 12–15.
Similä, M., Kouki, J., Mönkkönen, M, Sippola, A.-L. and Huhta, E. (2004). Co-variation and indicators of species diversity: Can richness of forest-dwelling species be predicted in northern boreal forests? *Ecological Indicators*, **6**, 686–700.
Skov- og Naturstyrelsen (2005). *Handlingsplan for Naturnær Skovdrift i Statsskoven [Actionplan for Near-Natural Forestry in the State Forests]*. Skov- og Naturstyrelsen, Miljøministeriet, København.
Smith, M.L., Bruhn, J.N. and Anderson, J.B. (1992). The fungus *Armillaria bulbosa* is among the largest and oldest living organisms. *Nature*, **356**, 428–431.
Stokland, J and Kauserud, H. (2004). *Phellinus nigrolimitatus*—A wood-decomposing fungus highly influenced by forestry. *Forest Ecology and Management*, **187**, 333–343.
Straatsma, G., Ayer, F. and Egli, S. (2001). Species richness, abundance and phenology of fungal fruit bodies over 21 years in a Swiss forest plot. *Mycological Research*, **105**, 515–523.
Straatsma, G. and Krisai-Greilhuber, I. (2003). Assemblage structure, species richness, abundance, and distribution of fungal fruit bodies in a seven year plot-based survey near Vienna. *Mycological Research*, **107**, 632–640.
Strange, N., Christensen, M. and Heilmann-Clausen, J. (2004). Policy implication in selection of forest reserves—Combining economic and biological information. *Scandinavian Journal of Forestry Research*, **19**(Suppl. 4), 138–149.

Svenning, J.C. (2002). A review of natural vegetation openness in north-western Europe. *Biological Conservation*, **104**, 133–148.

Sverdrup-Thygeson, A. and Lindenmayer, D.B. (2003). Ecological continuity and assumed indicator fungi in boreal forest: The importance of the landscape matrix. *Forest Ecology and Management*, **174**, 353–363.

UK Biodiversity Action Plan. (2006). *Species Action Plans*. http://www.ukbap.org.uk/species.aspx (1st Feb. 2007).

Vainio, E.J. and Hantula, J. (2000). Direct analysis of wood-inhabiting fungi using denaturing gradient gel electrophoresis of amplified ribosomal DNA. *Mycological Research*, **104**, 927–936.

Vesterholt, J. (2002). Contribution to the knowledge of species of *Entoloma* subgenus *Leptonia* growing in dry grassland. *Fungi non Delineati Pars*, **21**.

Vesterholt, J., Asman, W.A.H. and Christensen, M. (2000). Kvælstofnedfald og tilbagegang for svampe på mager bund [Nitrogen deposition and decline for fungi on poor and sandy soils]. *Svampe*, **42**, 53–60.

Vesterholt, J., Boertmann, D. and Tranberg, H. (1999). 1998—et usædvanlig godt år for overdrev-ssvampe [1998—An exceptionally good year for grassland fungi]. *Svampe*, **40**, 36–44.

Vesterholt, J. and Levesen, B. (2006). Alvorlig tilbagegang for enge og overdrev [Serious decline for Danish grasslands]. *Svampe*, **53**, 33–36.

Vilgalys, R. and Sun, B.L. (1994). Assessment of species distributions in Pleurotus based on trapping of airborne basidiospores. *Mycologia*, **86**, 270–274.

Virolainen, K.M., Ahlroth, P., Hyvärinen, E., Korkeamäki, Mattila, J., Päivinen, J., Rintala, T., Suomi, T. and Suhonen, J. (2000). Hot spots, indicator taxa, complementarity and optimal networks of taiga. *Proceedings of the Royal Society of London, Series **B***, **267**, 1143–1147.

Walleyn, R. and Veerkamp, M. (2005). Houtzwammen op beuk. Kensoorten voor soortenrijke bossen in België en Nederland [Characteristic species of important sites for beech-inhabiting fungi in Belgium and The Netherlands]. *Natuur. Focus*, **4**, 82–88.

Watling, R. (1995). Assessment of fungal diversity—Macromycetes, the problems. *Canadian Journal of Botany*, **73**(Suppl. 1 A–D), S15–S24.

Wright, D.H., Patterson, B.D., Mikkelson, G.M., Cutler, A. and Atmar, W. (1998). A comparative analysis on nested subset patterns of species composition. *Oecologia*, **113**, 1–20.

Colour Section

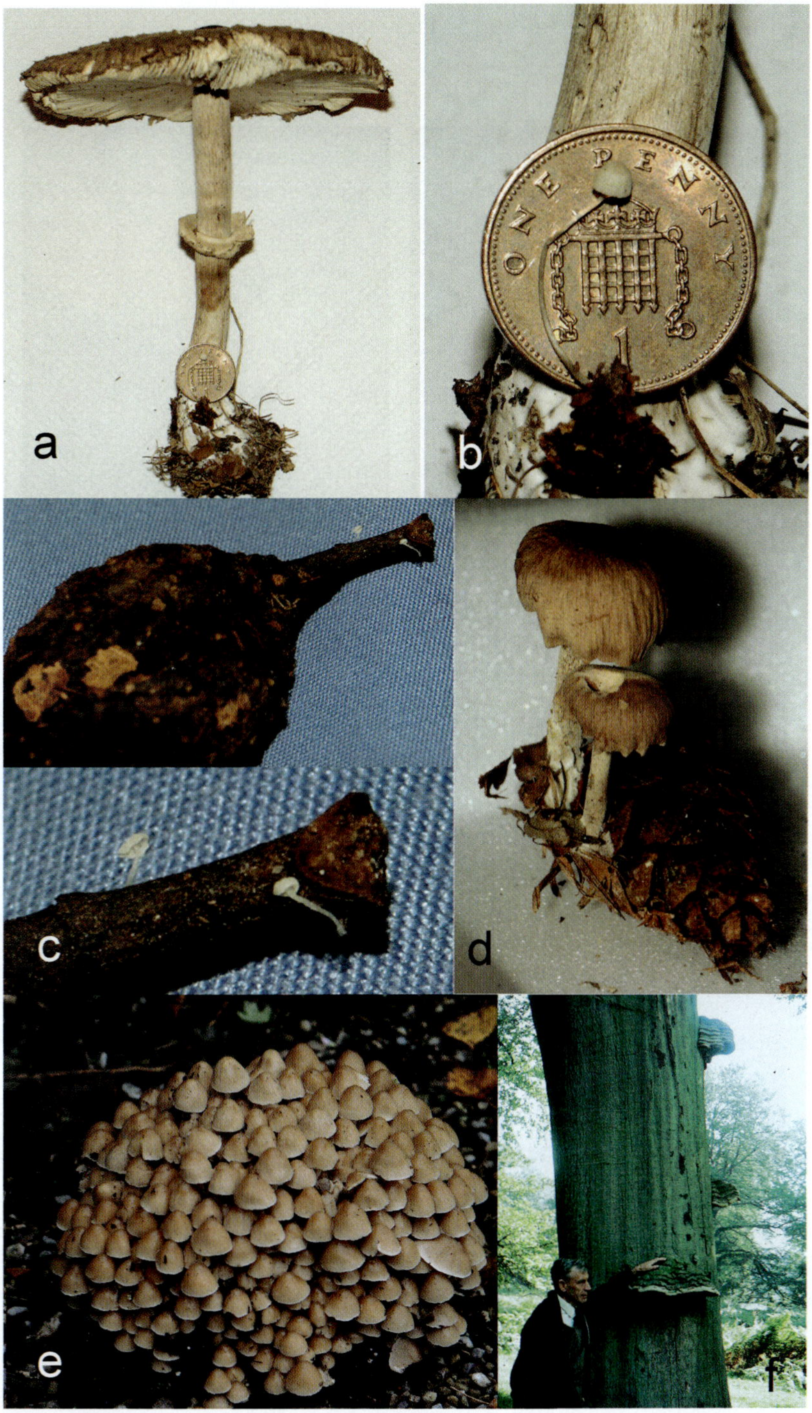

Please refer to Fig. 2 p. 89 in text for figure legend.

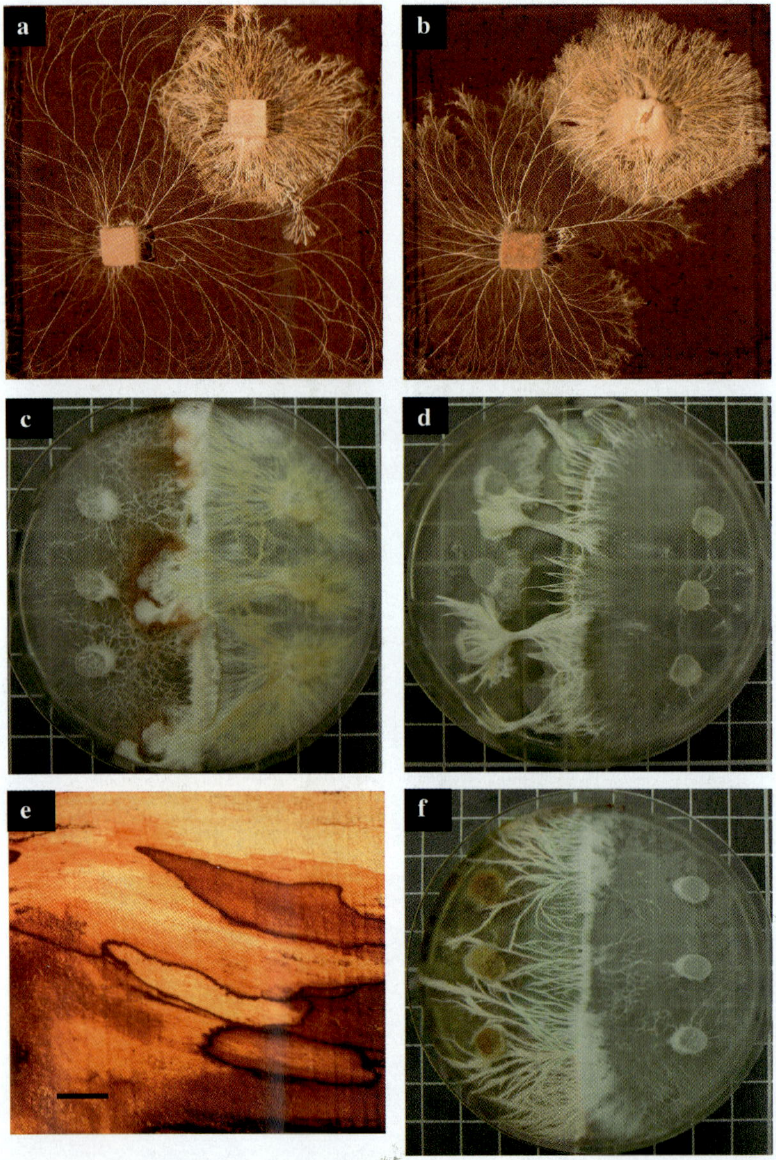

Please refer to Fig. 1 p. 127 in text for figure legend.

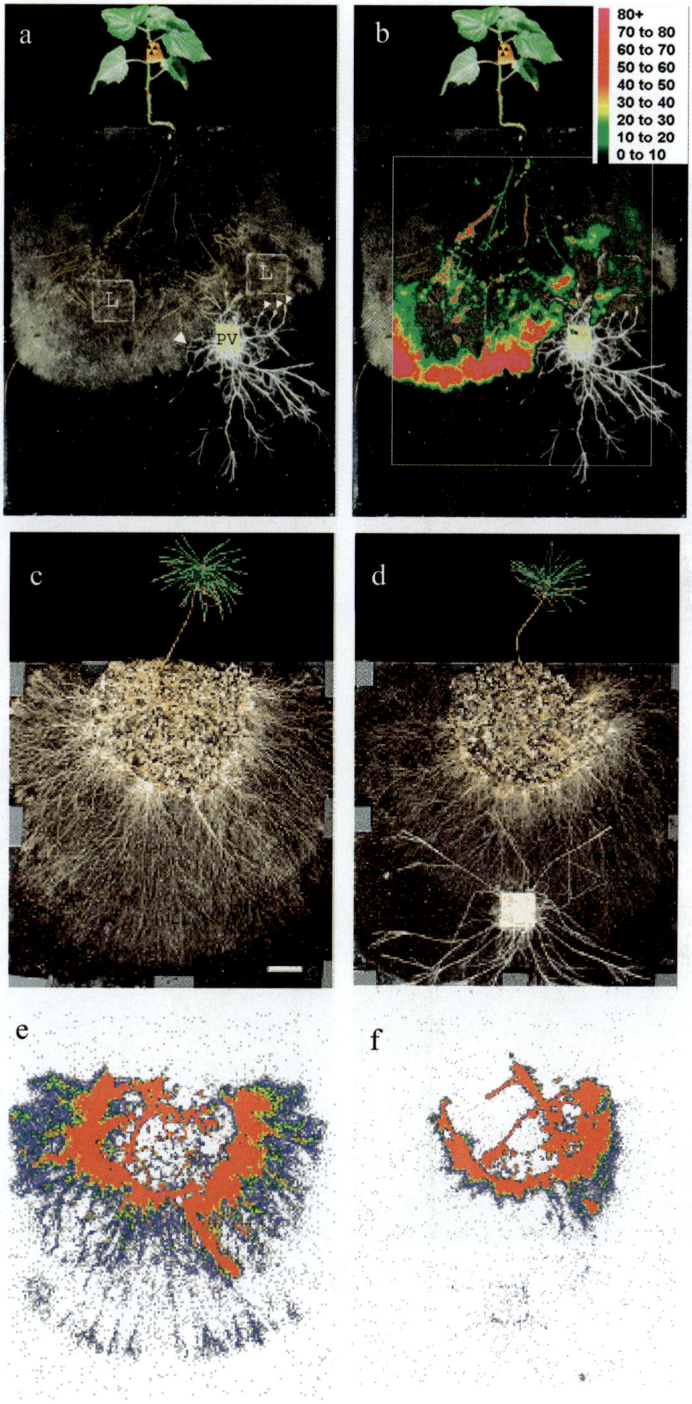

Please refer to Fig. 3 p. 132 in text for figure legend.

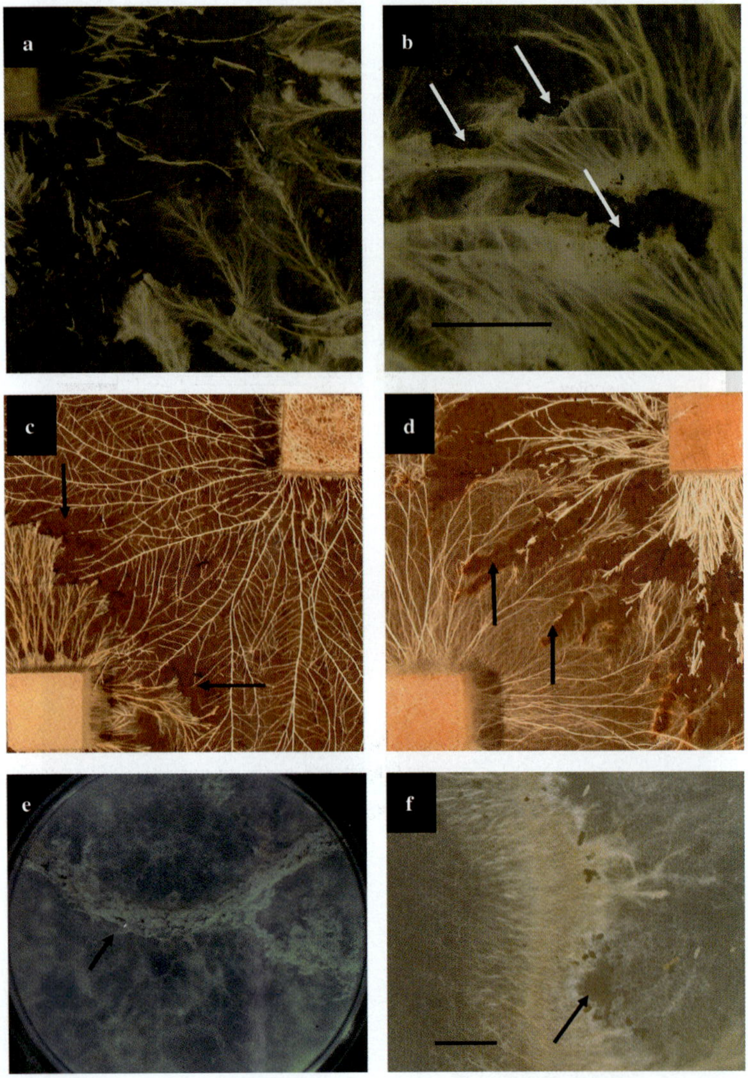

Please refer to Fig. 2 p. 168 in text for figure legend.

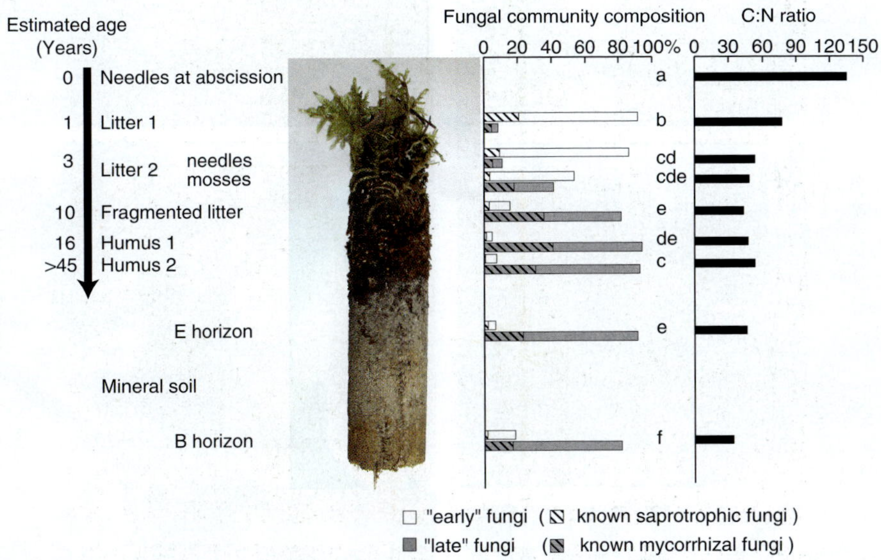

Please refer to Fig. 2 p. 188 in text for figure legend.

Please refer to Fig. 2 p. 269 in text for figure legend.

Species Index

Ascomycota

Zygomycota

Bacteria

Plantae

Animalae

Nematoda

Subject Index